Environmental UV Photobiology

Edited by

Antony R. Young
St. John's Institute of Dermatology
London, United Kingdom

Johan Moan
The Norwegian Radium Hospital
Oslo, Norway

Lars Olof Björn
Lund University
Lund, Sweden

Wilhelm Nultsch
Philipps University
Marburg, Germany

PLENUM PRESS • NEW YORK AND LONDON

The quotation from Robert A. Heinlein's *Time Enough for Love* appearing on page 89 of this
volume is reproduced by permission of Virginia Heinlein, c/o Ralph M. Vicinanza Ltd.
© 1973 Robert A. Heinlein

ISBN 0-306-44443-7

© 1993 Plenum Press, New York
A Division of Plenum Publishing Corporation
233 Spring Street, New York, N.Y. 10013

Printed in the United States of America

Contributors

Lars Olof Björn • Plant Physiology, Lund University, S-220 07 Lund, Sweden

Janet F. Bornman • Plant Physiology, Lund University, S-220 07 Lund, Sweden

Arne Dahlback • Norwegian Institute for Air Research, 2001 Lillestrøm, Oslo, Norway

Edward C. De Fabo • Laboratory of Photoimmunology and Photobiology, Department of Dermatology, The George Washington University Medical Center, Washington, DC 20037

Frank R. de Gruijl • Institute of Dermatology, University Hospital, 3508 GA Utrecht, The Netherlands

Adèle Green • Epidemiology Department, Queensland Institute of Medical Research, Brisbane, 4029 Australia

E. Walter Helbling • Marine Research Division, Scripps Institution of Oceanography, University of California at San Diego, La Jolla, California 92093-0202

Osmund Holm-Hansen • Marine Research Division, Scripps Institution of Oceanography, University of California at San Diego, La Jolla, California 92093-0202

Weine A. P. Josefsson • Research and Development Division, Swedish Meteorological and Hydrological Institute, S-601 76 Norrköping, Sweden

Deneb Karentz • Department of Biology, University of San Francisco, San Francisco, California 94117

Tiina Karu • Laser Technology Center of the Russian Academy of Sciences, 142092 Moscow Region, Troitzk, Russia

Dan Lubin • California Space Institute, Scripps Institution of Oceanography, University of California at San Diego, La Jolla, California 92093-0202

v

Sasha Madronich • Atmospheric Chemistry Division, National Center for Atmospheric Research, Boulder, Colorado 80307-3000

David L. Mitchell • Department of Carcinogenesis, University of Texas System Cancer Center, Science Park/Research Division, Smithville, Texas 78957

Johan Moan • Department of Biophysics, Institute for Cancer Research, The Norwegian Radium Hospital, Montebello, N-0310 Oslo, Norway

Frances P. Noonan • Laboratory of Photoimmunology and Photobiology, Department of Dermatology, The George Washington University Medical Center, Washington, DC 20037

Alan H. Teramura • Department of Botany, University of Maryland, College Park, Maryland 20742

Jan C. Van der Leun • Institute of Dermatology, University Hospital, 3508 GA Utrecht, The Netherlands

Ann R. Webb • Department of Meteorology, University of Reading, Reading RG6 2AU, England

Martin A. Weinstock • Dermatoepidemiology Unit, Departments of Medicine, VA Medical Center, Roger Williams Medical Center, and Brown University, Providence, Rhode Island 02908

Gail Williams • Master of Public Health Program, University of Queensland, Brisbane, 4029, Australia

Seymour Zigman • Department of Ophthalmology and Biochemistry, University of Rochester School of Medicine and Dentistry, Rochester, New York 14642

Preface

Under the unifying theme of environmental photobiology, the 14 chapters of this book gather together many important aspects of our current state of knowledge of the diverse effects of ultraviolet radiation on terrestrial and aquatic life. This knowledge and its future development is vital if we, as photobiologists, are to address the increasing concern about the possible short- and long-term effects of ozone depletion. In order to understand, let alone predict, these effects, it is essential to have quantitative data on terrestrial sunlight and its modification by ozone depletion, as well as the often complex photobiological consequences of exposure to ultraviolet radiation. As might be expected, much of the emphasis of this book is on UV-B effects, but UV-A effects are also covered.

Environmental UV Photobiology is sponsored by the European Society for Photobiology, and on behalf of the Society and my co-editors I gratefully acknowledge the contributions of all authors.

Antony R. Young

London

Contents

4. Influence of Ozone Depletion on the Incidence of Skin Cancer: Quantitative Prediction

Frank R. de Gruijl and Jan C. Van der Leun

5. UV-Induced Immunosuppression: Relationships between Changes in Solar UV Spectra and Immunologic Responses

Frances P. Noonan and Edward C. De Fabo

6. Ocular Damage by Environmental Radiant Energy and Its Prevention

Seymour Zigman

7. Vitamin D Synthesis under Changing UV Spectra

Ann R. Webb

8. Can Cellular Responses to Continuous-Wave and Pulsed UV Radiation Differ?

Tiina Karu

9. **Ultraviolet Radiation and Skin Cancer: Epidemiological Data
    from Australia**

Adèle Green and Gail Williams

10. Ultraviolet Radiation and Skin Cancer: Epidemiological Data from Scandinavia

Johan Moan and Arne Dahlback

11. Ultraviolet Radiation and Skin Cancer: Epidemiological Data from the United States and Canada

Martin A. Weinstock

14. Effects of Ultraviolet-B Radiation on Terrestrial Plants

Janet F. Bornman and Alan H. Teramura

Environmental
UV Photobiology

The Atmosphere and UV-B Radiation at Ground Level

Sasha Madronich

1. Overview

Ultraviolet (UV) radiation emanating from the sun travels unaltered until it enters the earth's atmosphere. Here, absorption and scattering by various gases and particles modify the radiation profoundly, so that by the time it reaches the terrestrial and oceanic biospheres, the wavelengths which are most harmful to organisms have been largely filtered out. Human activities are now changing the composition of the atmosphere, raising serious concerns about how this will affect the wavelength distribution and quantity of ground-level UV radiation.

The objective of this chapter is to give the reader familiarity with the basic concepts related to quantifying environmental UV radiation. Section 2 discusses the UV output of the sun and the geometric factors which relate the earth's orbit and rotation to the sun's illumination. Section 3 describes some aspects of the earth's atmosphere, with emphasis on those atmospheric constituents which affect UV transmission. Section 4 presents fundamental concepts of absorption and scattering of atmospheric radiation and some techniques for estimating their influence on UV radiation reaching the biosphere.

Sasha Madronich • Atmospheric Chemistry Division, National Center for Atmospheric Research, Boulder, Colorado 80307-3000.

Environmental UV Photobiology, edited by Antony R. Young *et al.* Plenum Press, New York, 1993.

Biologically weighted radiation and its sensitivity to atmospheric variability are discussed in Section 5. Section 6 summarizes recent trends in UV radiation resulting from atmospheric changes.

2. Extraterrestrial Solar Radiation and Earth–Sun Geometry

2.1. Solar Output

The sun is a yellow main sequence star (type G2 V) composed primarily of hydrogen and helium. Nuclear reactions in its interior generate energy which is propagated outward by convection and radiative transfer. Although the entire sun is gaseous, the *photosphere* is the part most clearly visible from the earth. There, electromagnetic energy is radiated with a continuous spectral distribution corresponding approximately to a thermal source of about 5,800 K. Atomic absorption and emission occurs in the photosphere and the outer layers (the chromosphere and the corona), and produces the characteristic *Fraunhofer lines* in the visible and ultraviolet wavelengths.

The total amount of solar electromagnetic energy reaching the earth can be estimated from the law of thermal emission, the area of the sun, and the average earth–sun distance, as

$$C = \sigma_{SB} T^4 R_s^2 / R_o^2 \tag{1}$$

where σ_{SB} is the Stefan–Boltzmann constant (5.670×10^{-8} W m^{-2} K^{-4}), T is the temperature of the solar photosphere (ca. 5,800 K), R_s is the solar radius (6.60×10^8 m at the solar equator), R_o is the average earth–sun distance (1 astronomical unit, au, or 1.496×10^{11} m), and C is the solar constant estimated from the above expression as 1,390 W m^{-2}. Direct measurements of C from the Nimbus-7 satellite between 1978 and 1991 give a value of about 1,372 $\pm$ 2 W m^{-2} (Hoyt *et al.*, 1992).

The *ultraviolet* (UV) region spans the 10- to 400-nm wavelength range and accounts for less than 9% of the total solar energy output. This wavelength range can be broadly divided into extreme UV (EUV, 10–120 nm), far UV (120–200 nm), vacuum UV (<240 nm), middle UV (MUV), or UV-C (200–280 nm), UV-B (280–315 nm), and UV-A (315–400 nm).† Wavelengths shorter than about 280 nm are absorbed almost completely by the earth's

---

† Alternative definitions of UV-B use the ranges 280–320 nm or 290–320 nm. The exact definition is usually not a problem when the terms UV-B and UV-A are used qualitatively and detailed spectral analysis is performed.

atmosphere (see Section 4) and are therefore unimportant for biological processes at the surface. The UV radiation relevant to environmental biology is therefore restricted to the combined UV-B and UV-A ranges, 280–400 nm. Figure 1 shows the spectral distribution of solar energy, F_∞, for the average earth–sun distance of 1 astronomical unit. These are *extraterrestrial* values, rather than *top of the atmosphere* values, since they do not include reflections from the earth's atmosphere and the surface. F_∞ may be converted to equivalent photon units Q_∞ (e.g., quanta m^{-2} s^{-1} nm^{-1}) through the relation

$$Q_\infty = F_\infty \lambda / hc \tag{2}$$

where λ is the wavelength, h is Planck's constant (6.626×10^{-34} J s), and c the speed of light (2.998×10^8 m s^{-1}). The data shown in Fig. 1 have been obtained by instruments from various satellite, balloon, and rocket platforms. At the present time, uncertainties in the measurements vary between $\pm 10\%$ at 280 nm and $\pm 3\%$ or better at 400 nm (World Meteorological Organization [*WMO*], 1985).

The solar electromagnetic output exhibits some temporal variability associated with the 27-day *apparent solar rotation,* the 11-year *cycle of sunspot activity,* and occasional *solar flares.* This variability affects mostly the UV-C and shorter wavelengths, with essentially negligible perturbations to the extraterrestrial UV-B and UV-A. However, the variability at the shorter wavelengths can change the rates of photochemical production of ozone in the stratosphere, and can therefore have an indirect but significant influence on the transmission of UV-B radiation from the top of the atmosphere to the surface.

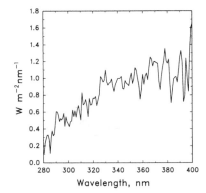

Figure 1. Extraterrestrial solar spectrum for the UV-B and UV-A wavelengths, at a distance of 1 au from the sun. Values from WMO (1985), modified to higher resolution by Madronich (1992a).

2.2. Variation in the Earth–Sun Distance

The earth's orbit is slightly elliptical with the sun at one of the foci, so the average separation varies with a yearly cycle. This variation can be computed in terms of the day number d_n (0 for 1 January, 364 for 31 December):

$$(R_o/R_n)^2 = a_0 + a_1 \cos\theta_n\, a_2 \sin\theta_n + a_3 \cos2\theta_n + a_4 \sin2\theta_n \qquad (3)$$

where R_o is the average earth–sun distance, R_n is the earth–sun distance on day d_n, the coefficients a_0–a_4 are given in Table 1, and

$$\theta_n \equiv 2\pi d_n/365 \qquad \text{radians} \qquad (4)$$

R_n varies by about 3.4% from minimum (perihelion, on about 3 January) to maximum (aphelion, on about July 5). The variation in R_n^2, and therefore in the intensity of extraterrestrial radiation, is about 6.9%, and is significant especially when considering seasonal differences in UV intensities between Southern and Northern Hemispheres.

2.3. The Solar Zenith Angle

The illumination of the earth varies with time of the day, season, and geographic location (latitude and longitude). All of these variations may be ascribed to changes in a single parameter, the *solar zenith angle* θ_o, which is the angle between the *local vertical* direction and the direction of the center

Table 1. Coefficients for the Earth–Sun Distance (a_i), Solar Declination (b_i), and the Equation of Time (c_i)[a]

i	a_i	b_i	c_i
0	1.000110	0.006918	0.000075
1	0.034221	−0.399912	0.001868
2	0.001280	0.070257	−0.032077
3	0.000719	−0.006758	−0.014615
4	0.000077	0.000907	−0.040849
5		−0.002697	
6		0.001480	

[a] Values from Spencer (1971). Alternate formulations may be found in manuals and textbooks on spherical astronomy, e.g., Duffett-Smith (1988), Smart (1979).

of the solar disk.‡ The complementary angle, between the horizon and the sun, is the *solar elevation angle*. The solar zenith angle is fundamental in computations of radiation reaching the surface, and must be calculated for any specific situation.

Suppose it is of interest to find θ_o at a location having latitude Φ, longitude Ψ, on day d_n of the year (see definition above), and a specific local clock time. The general expression for θ_o is

$$\cos\theta_o = \sin\delta \, \sin\Phi + \cos\delta \, \cos\Phi \, \cos t_h \tag{5}$$

where δ is the *solar declination,* and t_h is the *local hour angle*. The declination is the angle between the sun's direction and the earth's equatorial plane, and varies between $+23.45°$ (21 June) and $-23.45°$ (21 December), crossing $0°$ at the spring and fall equinoxes. A reasonably accurate expression for δ in radians is

$$\begin{aligned}\delta = b_0 &+ b_1 \cos\theta_n + b_2 \sin\theta_n + b_3 \cos2\theta_n \\ &+ b_4 \sin2\theta_n + b_5 \cos3\theta_n + b_6 \sin3\theta_n\end{aligned} \tag{6}$$

where θ_n was defined earlier, and the coefficients b_0–b_6 are given in Table 1. The local hour angle is the angle between the observer's meridian and the meridian of the sun. It differs from the local clock angle for several reasons, salient among which is that Ψ is not necessarily coincident with the longitude defining the local time zone. Thus, the clock time should first be converted to Greenwich mean time (GMT) using the appropriate time zone difference. The local hour angle (in radians) is then

$$t_h = \pi(\text{GMT}/12 - 1 + \Psi/180) + EQT \tag{7}$$

The last term is the *equation of time,* and accounts for the nonuniformity of the apparent angular speed of the sun in the sky. Values of EQT may be evaluated with good accuracy from the expression

$$EQT = c_0 + c_1 \cos\theta_n + c_2 \sin\theta_n + c_3 \cos2\theta_n + c_4 \sin2\theta_n \tag{8}$$

with the coefficients c_0–C_4 given in Table 1. As an illustration, consider the calculation of θ_o for Boulder, Colorado (latitude $= 40°0'$ N, longitude $= 105°15'$ W), on 1 July at 11:30 A.M. Mountain standard time (MST, equal

---

‡ Because the earth is not perfectly spherical, the angles are actually defined relative to a virtual celestial sphere concentric with the earth.

to GMT − 7). Note that latitudes north of the equator are positive, and longitudes west of the Greenwich meridian are negative. Application of the above equations gives EQT = −0.016 radians and δ = 0.404 radians, GMT = 18.5 hours, t_h = −0.151 radians, and finally θ_o = 0.321 radians or 18.4°.

Figure 2 gives the solar zenith angle for a few locations and times of the year. These values may be applied to additional cases by taking advantage of seasonal and geographical symmetries, e.g., 45°N on 1 June is similar to 11 July at the same latitude, and to 1 December and 10 January at 45°S.

3. The Atmosphere

3.1 Composition and Structure

The composition of the atmosphere near the earth's surface is shown in Table 2. The total number of air molecules in a unit volume, also called the *number density n*, is related to the pressure P and temperature T of an air parcel through the ideal gas law

$$n = P/kT \tag{9}$$

where k is the Boltzmann constant (1.381×10^{-23} J K^{-1}). At standard temperature and pressure (273.15 K, 1,013 hPa), $n = 2.69 \times 10^{25}$ molecules m^{-3} (*Loschmidt number*).

With increasing altitude z, both pressure and number density decrease rapidly due to the hydrostatic balance between gravity and thermal energy. This nearly exponential decrease is described approximately by the equation

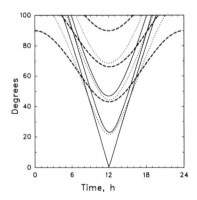

Figure 2. Diurnal variation of the solar zenith angle for different latitudes and times of the year. Solid lines are at the tropic circle (23.45° latitude), dotted lines at midlatitudes (45°), and dashed lines for the polar circle (66.55°). Winter solstice, equinox, and summer solstice have progressively smaller values of the solar zenith angle.

Table 2. Composition of the Atmosphere[a]

Molecule	Fraction[b]	Variability[c]
Nitrogen, N_2	78.084 ± 0.004%[d]	0
Oxygen, O_2	20.946 ± 0.06%[d]	0
Water vapor, H_2O	0–2%	±
Carbon dioxide, CO_2	353 ppm	+
Argon, Ar	9340 ± 10 ppb	0
Neon, Ne	18.18 ± 0.04 ppb	0
Helium, He	5.239 ± 0.002 ppb	0
Krypton, Kr	1.14 ± 0.01 ppb	0
Xenon, Xe	0.086 ± 0.001 ppb	0
Hydrogen, H_2	515 ppb	+
Nitrous oxide, N_2O	310 ppb	+
Carbon monoxide, CO	50–100 ppb	+
Methane, CH_4	1.72 ppb	+
Ozone, O_3	10 ppb–10 ppm[e]	±
Chlorofluorocarbons, CFCs[f]	0.85 ppb	+

[a] Values from Brasseur and Solomon (1986) and IPCC (1990).
[b] 1 ppm = 10^{-6} and 1 ppb = 10^{-9} molar mixing ratio.
[c] Low variability is denoted by 0, high variability by ±, and increasing trend by +.
[d] Values for dry air.
[e] Low values are for troposphere, high values for stratosphere.
[f] Sum of CFC-11, -12, -13, -113, -114, and -115.

$$P(z) \sim P(0)\exp(-z/H) \qquad (10)$$

where H, the *scale height*, is about 8 km in the lower atmosphere. This equation is only approximate, principally because temperature is also a function of altitude. Figure 3 shows the vertical dependence (profiles) of pressure, temperature, and ozone.

The temperature profile defines the vertical structure of the atmosphere and the associated nomenclature. The surface of the earth absorbs sunlight and heats the air in the *planetary boundary layer* (PBL), the lowest 0.5–2 km of the atmosphere. The warmed air rises, and the reduction in pressure (expansion) causes a reduction of the temperature. For typical conditions, this cooling rate is about -6.5 K km^{-1} (*environmental lapse rate*), which is somewhat lower than the theoretical *dry adiabatic lapse rate* of -9.8 K km^{-1} because of temperature changes from condensation and evaporation of water vapor, among other factors. Overall, however, the temperature drops systematically from the surface to about 10 km, as shown in Fig. 3. This region of falling temperatures is called the *troposphere*. Near about 10 km, the average

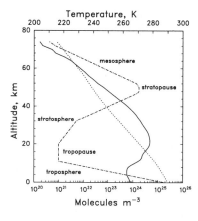

Figure 3. Vertical structure of the atmosphere. Solid line is ozone number density multiplied 10^6. Dotted line is air number density. Dashed line is temperature (top scale). Values from U.S. Standard Atmosphere (1976).

temperature becomes nearly constant for a few kilometers (the *tropopause*), and then increases to about 50 km (the *stratosphere*). The heating is due to the absorption of sunlight (especially UV) by stratospheric ozone. The *stratopause* marks the top of the stratosphere and another downturn in the temperature (the *mesosphere*), followed by several other upper-atmospheric layers. Figure 3 is the midlatitude U.S. Standard Atmosphere (1976). The details of the vertical structure vary with season and latitude. For example, the height of the tropopause is greatest in the tropics (about 14–18 km) and lowest in the polar regions (10–12 km). Small-scale variations in the tropopause also occur, due, e.g., to waves in the atmosphere.

The total number of air molecules in a *vertical column* 1 m² wide, extending from sea level to the top of the atmosphere, is 2.15×10^{29} molecules m⁻². About half of this column amount is contained in the lowest 5 km, and about 90% is in the troposphere. The stratosphere contains most of the remaining air column, with less than 0.1% in the mesosphere and above. In contrast, about 90% of the ozone is in the stratosphere, with about 10% in the troposphere. The *total ozone column* amount for the U.S. Standard Atmosphere shown in Fig. 3 is 9.35×10^{22} molecules m⁻². Very frequently, the ozone column is expressed as the height it would occupy if compressed to 1,013 hPa of pure gaseous O_3 at 0°C. This height, when expressed in millicentimeters, is the Dobson unit (DU). The conversion is 1 DU = 2.69×10^{20} molecules m⁻².

The latitudinal and seasonal variation of the total ozone column is shown in Fig. 4. Because stratospheric ozone accounts for the greater share of the total column, the figure reflects mostly the distribution of the stratospheric amounts. Stratospheric east–west (zonal) winds distribute the ozone rapidly around the globe, so that for the purposes of calculating average UV levels at

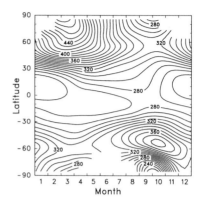

Figure 4. Latitudinal and seasonal variation of total ozone, in Dobson units. Values are averages for the years 1979–1989, from TOMS version 6 (data courtesy R. Stolarski).

the surface, the longitudinal variation of stratospheric ozone is usually unnecessary. However, tropospheric ozone tends to be highest in more polluted, industrialized regions such as the continental United States and Europe, and is transported relatively slowly and ineffectively to global scales.

3.2. Ozone Chemistry

Stratospheric ozone is formed by the action of short-wavelength solar UV radiation ($\lambda < 242$ nm) on oxygen molecules. These photons have sufficient energy to break the oxygen–oxygen bond, as shown in reaction 1 of Table 3. The ground-state oxygen atoms can combine with unbroken O_2 molecules to produce ozone (reaction 2). The production of ozone peaks in the stratosphere. At higher altitudes, the amount of oxygen diminishes. At lower altitudes (i.e., in the troposphere), no radiation at $\lambda < 242$ nm is available, as it has been absorbed by the overhead stratospheric O_2 and O_3.

The amount of ozone in the stratosphere is set by a dynamic balance between this two-step production sequence and a number of *catalytic cycles.* The most important of these, which are shown in Table 3, may be classified as cycles of odd oxygen ($O_x = O + O_3$), odd hydrogen ($HO_x = OH + HO_2$), odd nitrogen ($NO_x = NO + NO_2$), and chlorine ($Cl_x = Cl + ClO$). The key point to notice in these catalytic cycles is that the *active species* NO_x, HO_x, and Cl_x are not lost, so that the ozone-destroying steps can be repeated a large number of times. Other reactions, not shown in Table 3, convert these species into less active *reservoir species* such as H_2O, HNO_3, HCl, and $ClONO_2$. These reservoir species may be gradually transported to the troposphere where they are removed by, e.g., rainout. Alternatively, they can be reconverted to active species by photodissociation, by gas-phase chemical reactions, or by reactions occurring on the surface of stratospheric cloud and aerosol particles.

Table 3. Simplified Atmospheric Chemistry of Ozone

Stratosphere		
Production	$O_2 + h\nu\ (\lambda < 240\ nm) \rightarrow 2\ O(^3P)$	R1
	$O(^3P) + O_2 \rightarrow O_3$	R2
O_x cycle	$O_3 + h\nu\ (\lambda < 800\ nm) \rightarrow O(^3P) + O_2$	R3
	$O(^3P) + O_3 \rightarrow 2\ O_2$	R4
HO_x cycle	$O_3 + OH \rightarrow O_2 + HO_2$	R5
	$O(^3P) + HO_2 \rightarrow O_2 + OH$	R6
	$O_3 + HO_2 \rightarrow 2\ O_2 + OH$	R7
NO_x cycle	$O_3 + NO \rightarrow O_2 + NO_2$	R8
	$O(^3P) + NO_2 \rightarrow O_2 + NO$	R9
Cl_x cycle	$O_3 + Cl \rightarrow O_2 + ClO$	R10
	$O(^3P) + ClO \rightarrow O_2 + Cl$	R11
Troposphere		
NO_x stationary state	$NO_2 + h\nu\ (\lambda < 400\ nm) \rightarrow NO + O(^3P)$	R12
	$O(^3P) + O_2 \rightarrow O_3$	R2
	$O_3 + NO \rightarrow O_2 + NO_2$	R8
Ozone destruction	$O_3 + h\nu\ (\lambda < 330\ nm) \rightarrow O(^1D) + O_2$	R13
	$O(^1D) + H_2O \rightarrow 2\ OH$	R14
	$O_3 + OH \rightarrow O_2 + HO_2$	R5
	$O_3 + HO_2 \rightarrow 2\ O_2 + OH$	R7
Net ozone production	$OH + CO \rightarrow H + CO_2$	R15
	$H + O_2 \rightarrow HO_2$	R16
	$HO_2 + NO \rightarrow NO_2 + OH$	R17

The effects of particles are further discussed in Section 6 in connection with the polar ozone depletion problem.

Tropospheric ozone contributes significantly to the total ozone column, and is of considerable importance as the main source of hydroxyl (OH) radicals, which control the atmospheric lifetimes of many gases. One source of tropospheric O_3 is the downward transport of stratospheric O_3 by atmospheric motions such as those associated with tropical convection and midlatitude fronts. The other main source is the photodissociation of natural and anthropogenic nitrogen dioxide (NO_2) by UV-B and UV-A light, liberating the oxygen atom needed for O_3 formation, as shown in the lower part of Table 3. The photodissociation of NO_2 (reaction 12) is followed by both reaction 2, which makes O_3, and reaction 8, which destroys O_3, with no net O_3 change. Thus, the true sources of tropospheric O_3 are reactions which convert NO to NO_2 without destroying O_3 directly. Many such reactions exist, and they usually involve the *photooxidation* of chemicals such as carbon monoxide (reactions, 15–17), methane, and other hydrocarbons. The main losses for tropospheric O_3 include photodissociation, reaction with OH and HO_2, and loss by contact with soil and other surfaces.

3.3. Other Atmospheric Constituents

A few other atmospheric gases, especially SO_2 and NO_2, have a potential for attenuating UV radiation, but are present in only trace amounts even in moderately polluted air. However, they could be important during high-pollution episodes such as are occasionally encountered in major urban areas. For example, levels as large as 2 ppm (parts per million) for SO_2 and 0.5 ppm for NO_2 have been observed in heavily polluted urban areas (Finlayson-Pitts and Pitts, 1986), but these high values are usually confined to the lowest 0.5–2.0 km of the atmosphere (the PBL).

Particles suspended in the atmosphere (*aerosols*) may include mineral dust, sea-salt aerosols, carbonaceous particles (esp. soot), ammonium sulfate, and diluted sulfuric acid droplets. Their origin can be natural or related to human activity such as fossil fuel combustion and biomass burning. Their occurrence and size distributions depend on specific sources as well as on many meteorological factors such as winds and relative humidity. The number of particles is usually greatest in the PBL and decreases rapidly with increasing altitude. High-altitude aerosols (e.g., the stratospheric Junge sulfate layer) are normally too thin to affect the transmission of UV radiation, except in the direct plume of volcanic eruptions.

Clouds in the lower and middle troposphere are usually composed of liquid water, while in the upper troposphere and stratosphere they are often ice crystals. They can accumulate impurities from the surrounding air and from contamination by the solid or liquid nucleus around which the water is condensed. The morphology of individual clouds and of cloud fields is highly irregular and constantly changing. Classifications of distinct cloud types have been made, and some are given in Table 4. Observers from the ground (e.g., from airports) often report the cloud cover as the fraction of the sky which is covered by cloud, in tenths, so that $0/10^{ths}$ is a completely clear sky and $10/10^{ths}$ is completely overcast, or equivalently in octas (8^{ths}). Unfortunately, these observations cannot distinguish all the factors which affect the transmission of radiation, including single versus stacked clouds, size distributions of cloud droplets, the geometry of cloud–sun reflections, and the like. Very few observations exist which provide good quantification of these various effects, and then only for some special situations, such as uniform marine stratus cover.

**Table 4. Visible Transmission of Different
Cloud Types**[a]

Cloud type	Transmission (%)[b]
Ci, cirrus	84
Cs, cirrostratus	78
Ac, altocumulus	50
As, altostratus	41
Sc, stratocumulus	34
St, stratus	25
Ns, nimbostratus	19
Fog	17

[a] From Haurwitz (1948).
[b] Transmissions are for a solar zenith angle of $60°$.

3.4. Surface Albedo

The surface of the earth represents the lower boundary of the atmosphere. Reflections from different types of soils, vegetation, and water (liquid or frozen) can have a significant effect on environmental UV radiation. Two idealized types of reflection are distinguished: *specular*, in which the angle of the reflected light is the same as the angle of incident light, and *Lambertian*, in which the reflected light is *isotropic* (equally distributed in all directions) independent of the angle of incidence. In practice, most natural surfaces fall somewhere between these two extremes. The *surface albedo* is the ratio of reflected energy to incident energy. The albedo generally depends on the type of surface, the wavelength, and to some extent the angle of incidence. Although many measurements of the visible albedo are available, comparatively few data are available for UV wavelengths. Some typical values of the UV albedo for different surfaces are given in Table 5.

The surface albedo should be distinguished from the *planetary albedo*, which is the fraction of the incident extraterrestrial irradiance reflected back to space by the entire surface–atmosphere system, and includes the effects of surface reflections, atmospheric molecular scattering, and reflections from aerosols and clouds.

3.5. Atmospheric Variability

The geographic and temporal variability of atmospheric gases, aerosols, and clouds is a major problem when attempting to create an average global climatology of UV radiation. Of the factors discussed above, only the air

Table 5. Ultraviolet Albedo of Various Surfaces

Surface	Albedo (%)[a]	References[b]
Liquid water	5–10	1–3
Clean dry snow	30–100	1–4
Dirty wet snow	20–95	1, 2
Ice	7–75	1, 2
Sudan grass	2	1
Maize	2	1
Conifer trees	4–8	1
Grass (unspecified)	1	4
Alfalfa	2–4	5
Rice	2–6	5
Sorghum	2–5	5
Pasture	2–6	2
Grassland	1–3	2
Green farmland	4	3
Brown farmland	4	3
Pine forest	1–2	3
Dry Yolo loam	5–8	5
Wet Yolo loam	5–8	5
Sacramento silt and clay	8–11	5
Black lava	1–3	6, 7
Limestone	8–12	2
Desert sand	4	7
Gypsum sand	16–30	7
White cement	17	4
Blacktop asphalt	4–11	2, 5
Plywood	7	4
Black cloth	2	4

[a] Depending on solar zenith angle and surface roughness.
[b] 1, Kondratyev (1969); 2, Blumthaler and Ambach (1988); 3, Doda and
Green (1980); 4, Dickerson et al. (1982); 5, Coulson and Reynolds (1971);
6, Shetter et al. (1992); 7, Doda and Green (1981).

density is known with excellent accuracy at all locations and times of the year. Measurements from satellites are adding to the data bases for total ozone, clouds, and ground albedo. The data for tropospheric ozone and aerosols are much sparser and tend to be available only in selected geographic regions. All of these constituents have diurnal and seasonal cycles, as well as episodic, nearly random, fluctuations that are extremely difficult to quantify. While the rest of this chapter addresses the calculation of UV radiation for average conditions, it should always be remembered that at any specific time and location, significant deviations from these idealized estimates may occur due to fluctuations in atmospheric composition.

4. Light in the Atmosphere

4.1. Absorption and Scattering

The atmospheric *attenuation* of UV radiation occurs by a combination of *absorption* and *scattering*. When a UV photon is absorbed, the energy is usually lost to chemical bonds or heat, and is removed completely from the radiation field. Scattering, by contrast, changes the direction of propagation of the photon, so that, for the purposes of most atmospheric UV calculations, no energy is lost and the wavelength of the photon is unchanged (elastic scattering).

Consider monochromatic parallel radiation incident at angle θ onto a layer of vertical thickness z(m) which contains an absorbing gas uniformly distributed with number density n (molecules m^{-3}). The relationship between the incident radiation I_o and the transmitted radiation I is given by the *Beer–Lambert law*,

$$I/I_o = e^{-\tau(\lambda,\, z)/\cos\theta} \tag{11}$$

where

$$\tau(\lambda,\, z) = \sigma(\lambda)nz \tag{12}$$

is the *vertical optical depth* (dimensionless) for absorption, and $\sigma(\lambda)$ is the *absorption cross section.* The cross section is generally a function of the specific absorbing gas (e.g., ozone) and of wavelength, and can also be a function of temperature and in a few cases pressure. Cross sections measured in laboratory studies are available in the scientific literature for many compounds over wide wavelength ranges, and are usually reported in units of cm^2. Figure 5 shows the absorption cross sections for several molecules of atmospheric interest.

In cases where n varies with altitude, the optical depth may be rewritten as

$$\tau(\lambda,\, z) = \sigma(\lambda)N \tag{13}$$

where the *vertical column density* (N, molecules m^{-2}) is defined as

$$N \equiv \int ndz \tag{14}$$

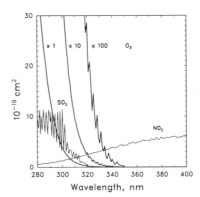

Figure 5. Absorption cross sections for ozone at 273 K (from Molina and Molina, 1986), nitrogen dioxide at 273 K (from Davidson *et al.,* 1988), and for sulfur dioxide at 295 K (from Warneck *et al.,* 1964; McGee and Burris, 1987).

As an illustration, consider the absorption by ozone at two different wavelengths, $\lambda_1 = 290$ nm ($\sigma_1 = 1.40 \times 10^{-18}$ cm^2) and $\lambda_2 = 310$ nm ($\sigma_2 = 9.30 \times 10^{-20}$ cm^2), for incidence at $\theta_o = 60°$ and a vertical ozone column of 300 DU. The fraction transmitted at 310 nm is 0.22, while at 290 nm it is only $\sim 10^{-10}$. The strongly nonlinear dependence on the cross section is due to the exponential nature of the Beer–Lambert expression.

The attenuation of a single light beam by scattering may also be described by a Beer–Lambert-type expression, with the appropriate *scattering cross section* and the number densities of particles and gases which induce scattering. However, in contrast to absorption, where photons are lost, scattering out of the incident beam produces a field of radiation which propagates simultaneously in many different directions. Also, photons may be scattered several times as they traverse the atmosphere (*multiple scattering*).

All gases and particles scatter radiation but differ quantitatively in the overall probability and in the angular distribution of the resulting radiation. For gases (*Rayleigh scattering*), the probability of scattering is extremely small, and is important only for N_2 and O_2 in the lower atmosphere, where air densities are greatest. The Rayleigh scattering cross section for air, σ_r, is given to good accuracy with the expression given by Frölich and Shaw (1980),

$$\sigma_r = 3.90 \times 10^{-28}/\Lambda^x \qquad \text{cm}^2 \qquad (15)$$

where Λ is the wavelength in μm ($=\lambda_{nm}/1000$), and

$$x = 3.916 + 0.074\Lambda + 0.050/\Lambda \qquad (16)$$

Note that the vertical optical depth for Rayleigh scattering changes by about one order of magnitude over the visible solar spectrum (400–700 nm), which accounts for the blue appearance of clear skies.

Scattering by aerosols and cloud particles (*Mie scattering*) is more complex, and less is known about particle numbers and size distributions. Thus, the scattering is usually expressed by estimates of the optical depth, either total (i.e., from surface to top of atmosphere), or as a function of height (e.g., 1-km layers). For aerosols in the PBL, the only information which is commonly available is the horizontal *visible range,* defined as the distance over which a 2% contrast can be detected between an object and the background.

An important difference between Rayleigh scattering (by air) and Mie scattering (by larger particles) is in the angular redistribution of light. The *scattering phase function P* (θ, ϕ, θ', ϕ') is defined as the probability that a photon incoming from the direction defined by the angles θ' and ϕ' will be scattered into angles θ and ϕ. Figure 6 shows several different phase functions. The Rayleigh scattering phase function for unpolarized light is

$$P(\Theta) = \frac{3}{4}(1 + \cos^2\Theta) \qquad (17)$$

where $\Theta = \theta - \theta'$, and is symmetric, with equal probabilities of scattering into the forward and backward hemispheres. Larger particles have complex phase functions, typically with a strong forward-scattering probability and several

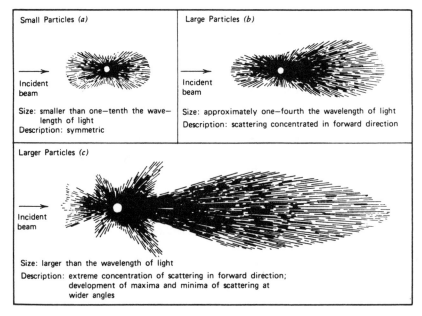

Figure 6. Phase functions for different-size particles. From McCartney (1976).

weaker side lobes. A simple measure of the directionality of phase functions is the *asymmetry factor g,* defined as

$$g = \frac{1}{2} \int_{1}^{+1} P(\Theta)\cos\Theta \; d(\cos\Theta) \tag{18}$$

which ranges from -1 (complete back scattering) to $+1$ (complete forward scattering), and is zero for Rayleigh scattering. Typical values for aerosols range between 0.6 and 0.8, while for the larger cloud droplets values are between 0.8 and 0.9. The dimensionless *size parameter* α of a particle having radius r is defined as

$$\alpha = 2\pi r/\lambda \tag{19}$$

and is one of the key parameters determining the shape of the phase function (and other optical properties). For Rayleigh scattering, $\alpha << 1$, while for Mie scattering, $\alpha \sim 1$ or greater.

All molecules and particles can both scatter and absorb radiation. For molecules, it is usually true that one process far exceeds the other, which can therefore be neglected. For aerosol particles and cloud droplets, it is quite common that both processes occur with measurable probability. The parameter which describes the relative magnitude of scattering and absorption is the *single scattering albedo,* defined as

$$\omega_o = \sigma_s/(\sigma_s + \sigma_a) \tag{20}$$

where σ_s and σ_a are, respectively, the cross sections for scattering and absorption. Extreme values are $\omega_o = 0$ for pure absorption and $\omega_o = 1$ for pure scattering. Note that ω_o is usually a strong function of wavelength. Typical values at UV wavelengths are 0.8 and 0.999 for aerosols and clouds, depending on the level of chemical impurities present in the particles. Mixtures of different gases (e.g., ozone and air) can also result in a range of effective values of ω_o for a given atmospheric layer.

4.2. The Geometry of Radiation

An observer at or near the earth's surface sees several components of radiation: the *direct solar beam,* the *diffuse sky radiation* due to atmospheric scattering, and possibly *reflections* from the underlying surface. The diffuse radiation is particularly important at UV wavelengths where Rayleigh scattering is strong, and at all wavelengths if clouds and aerosols are present. The total radiation field therefore has a complex dependence on the various directions of incidence.

From the point of view of the observer, each sky direction can be defined by the two usual spherical coordinates, θ and ϕ, respectively the zenith and azimuth angles. A small angular field of view, or *solid angle,* centered around this direction is defined by the area element on a sphere of unity radius, $d(\cos\theta)\,d\phi = \sin\theta\,d\theta\,d\phi$. The total field of view is

$$\int_0^{2\pi} \int_{-1}^{1} d(\cos\Theta)d\phi = 4\pi \text{ steradians (sr)} \tag{21}$$

It is often convenient to consider separately the *upper hemisphere (down-welling radiation)* and the *lower hemisphere (up-welling radiation),* each with 2π sr total field of view.

The angular dependence of the radiation is then expressed on the angular coordinates by the *radiance* (sometimes called the *intensity*) $I(\theta, \phi)$, which is the energy arriving from a solid angle element (e.g., a narrow cone) centered around the angular direction (θ, ϕ), per unit area, unit time, and unit solid angle (e.g., W m^{-2} sr^{-1}). If the radiance is known, the total energy impingent from all directions can be calculated for any specific "target." Note that the energy intercepted by the target depends on the target shape and its orientation relative to the incoming radiation. Two idealized targets are of special interest to the biological effects. If the target is a spherical surface, the energy intercepted per unit cross-sectional area is the *scalar irradiance* (sometimes called *fluence rate, actinic flux, beam irradiance, flux density, speed of light* $\times$ *energy density,* or *mean intensity* $\times$ 4π)

$$F_s = \int_0^{2\pi} \int_{-1}^{1} I(\theta, \phi)d(\cos\theta)d\phi$$

$$= F_o + \int_0^{2\pi} \int_0^{1} I_{\downarrow}(\theta, \phi)d(\cos\theta)d\phi + \int_0^{2\pi} \int_{-1}^{0} I_{\uparrow}(\theta, \phi)d(\cos\theta)d\phi \tag{22}$$

where the last terms show separate contributions from the direct solar beam incident at solar zenith angle θ_o (scalar irradiance F_o), from the down-welling diffuse radiance $I_{\downarrow}(\theta, \phi)$, and from the up-welling diffuse radiance $I_{\uparrow}(\theta, \phi)$. If the target is a two-sided horizontal surface, the total energy received (per unit area) is the *vector irradiance* (sometimes called *energy flux,* or simply *irradiance* or *flux*),

$$F_v = \int_0^{2\pi} \int_{-1}^{1} I(\theta,\, \phi)\cos\theta d(\cos\theta)d\phi$$

$$= F_o\cos\theta_o + \int_0^{2\pi} \int_0^{1} I_\downarrow(\theta,\, \phi)\cos\theta d(\cos\theta)d\phi$$

$$+ \int_0^{2\pi} \int_{-1}^{0} I_\uparrow(\theta,\, \phi)\cos\theta d(\cos\theta)d\phi \qquad (23)$$

Although both scalar and vector irradiance have units of W m^{-2} (and are occasionally confused), the vector irradiance has cosine factors to account for the variation of the projected area of the target for different angles of incidence of direct and diffuse light. The distinction between vector and scalar irradiance is of some significance for studies of the irradiation of biological systems. The scalar irradiance may often be more representative of environmental UV exposure for randomly oriented organisms, but the vector irradiance is the quantity most often measured (e.g., with flat plate radiometers). For this reason, the discussion which follows will concern the vector irradiance unless stated otherwise.

The radiance and irradiance are understood to be integrated over wavelengths. They can also be defined as functions of wavelength, in which case their names are preceded by the modifier *spectral*. Corresponding units are W m^{-2} sr^{-1} nm^{-1} for the spectral radiance and W m^{-2} nm^{-1} for the spectral irradiance.

4.3. Radiative Transfer

At any given wavelength, the (spectral) irradiance of the direct solar beam reaching the surface is given simply by the Beer–Lambert law. Calculation of the diffuse irradiance is, however, much more complex. One reason for this is that scattering directions are defined by different phase functions of atmospheric gases and particles, as was shown in Fig. 6. Multiple scattering events further complicate the directional redistribution. Also, the number densities of scatterers and absorbers (and even their relative amounts) generally vary with altitude, so that the atmosphere is not vertically uniform.

The importance of scattered radiation is shown in Fig. 7. At 300 nm, between 40% and 100% of the irradiance at the ground is diffuse, while most of the visible irradiance is still contained in the direct solar beam, at least for high sun. The results shown in Fig. 7 were obtained for clear sky conditions; if aerosols or clouds are present, or if the scalar irradiance is of interest, the relative contribution of diffuse light is even greater at all wavelengths.

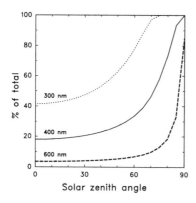

Figure 7. Contributions of diffuse radiation to the spectral irradiance at the surface, for different wavelengths. Calculation for cloud- and aerosol-free conditions with surface albedo of 0.05.

To fully describe the vertical and angular dependence of the diffuse radiance, it is necessary to solve the integrodifferential *radiative transfer equation,*

$$\cos\theta \frac{dI(\tau, \theta, \phi)}{d\tau} = -I(\tau, \theta, \phi) + \frac{\omega_0}{4\pi} F_{\infty} e^{-\tau/\cos\theta_0} P(\theta, \phi; \theta_0, \phi_0)$$

$$+ \frac{\omega_0}{4\pi} \int_0^{2\pi} \int_{-1}^{+1} I(\tau, \theta', \phi') P(\theta, \phi; \theta', \phi') d\cos\theta' d\phi' \quad (24)$$

where θ and ϕ are the usual angular coordinates (θ_0 and ϕ_0 for the direct solar beam); τ is the vertical coordinate measured in optical depth units; F_{∞} is the extraterrestrial irradiance; $P(\theta, \phi, \theta', \phi')$ and ω_0 are defined as before. The first term on the right-hand side is the attenuation of the radiance due to absorption and scattering, while the second and third terms give the increase in the diffuse radiance due to scattering of, respectively, the direct solar beam, and the radiance at other locations and directions. The sum of these terms must balance the vertical change in radiance, given on the left-hand side. The solution of this equation is generally difficult, and analytic solutions exist for only a few simple cases. Many numerical methods have been developed, ranging from accurate but computer-intensive, to approximate but more efficient (see e.g., Chandrasekhar, 1960; Hansen and Travis, 1974; Goody and Yung, 1989).

An important class of solutions is based on the *two-stream approximation* (see, e.g., Meador and Weaver, 1980, and Toon *et al.,* 1989). In these methods, solutions are only for the up-welling and down-welling diffuse irradiances, $F\uparrow$ and $F\downarrow$, not for the fully directional radiance. Specific assumptions, which vary among different two-stream methods, are made about the average direction of propagation of the diffuse radiation, and about the amount of radiation scattered in the forward and backward directions. For example, if the diffuse light is regarded as isotropic (a rough but often reasonable approximation), its average angle of propagation through an atmospheric layer is 60° (measured from the vertical). The direct solar irradiance is computed as usual, by the Beer–Lambert law.

Two-stream methods are simple to implement and can be run quite rapidly even on personal computers. Their accuracy is sufficient for most biological applications, although some limitations should be kept in mind. Figure 8 shows the difference in the spectral irradiance calculated with a popular two-stream method, the delta-Eddington approximation (Joseph *et al.*, 1976), and a more accurate discrete ordinates method (Stamnes *et al.*, 1988). The agreement is excellent for high sun at all wavelengths, but rather poor at low sun for wavelengths where ozone absorption is very strong. In practice, this low-sun, short-wavelength limit corresponds to very small absolute levels of radiation, with transmission less than 10^{-6}, and thus is usually not a problem unless detailed spectral measurements are of interest. The main drawbacks of the more accurate methods are their complexity of formulation and the large computational requirements for realistic atmospheres; the use of two-stream models offers a practical alternative sufficient for most studies.

At the other extreme, *simple parameterizations* have been developed to represent the diffuse radiation under many conditions. One such model, developed by Green and co-workers (Green *et al.*, 1980), has found broad application in environmental photobiology studies. Such models, which are based on fitting the results of a more accurate radiative transfer model with simple analytical functions, can be quite accurate for the conditions under which they are derived. The general disadvantage of these formulations is that they cannot be applied to conditions other than those used in the derivation of the fitting coefficients (e.g., aerosols, tropospheric ozone, or cloud cover). The advent of faster computational machines and improved two-stream models has decreased the relative advantages of the parametric models.

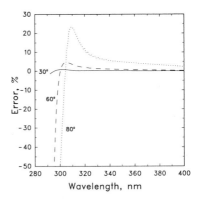

Figure 8. Comparison of sea-level irradiance calculated with the approximate two-stream delta-Eddington radiative scheme and an eight-stream discrete ordinates method. Results are for cloudless skies, no aerosols, surface albedo of 0.05, ozone column of 348 DU, U.S. Standard Atmosphere profiles. Values are shown for different solar zenith angles, as indicated.

4.4. Spectral Irradiance at the Surface

The final result of radiative transfer calculations is the spectral irradiance reaching the surface, $F(\lambda)$, which can then be summarized schematically as the product of the extraterrestrial spectral irradiance and a wavelength-dependent effective transmission function T,

$$F(\lambda) = F_\infty(\lambda)T(\lambda, \theta_{\mathrm{o}}, \tau_{\mathrm{O_3}}, \tau_{\mathrm{r}}, \tau_{\mathrm{a}}, \tau_{\mathrm{c}}, A) \qquad (25)$$

The transmission is a function of solar zenith angle θ_{o}, the ozone optical depth $\tau_{\mathrm{O_3}}$, the Rayleigh scattering optical depth τ_{r}, the optical depth of aerosols and clouds, τ_{a} and τ_{c}, respectively, and reflections from a surface of albedo A. In addition, the vertical distribution of the atmospheric absorbers and scatterers, while not shown explicitly in this expression, can have some effect on the radiation.

The spectral transmission is shown in Fig. 9. The wavelength dependence is due primarily to ozone absorption over 280–330 nm, and to Rayleigh scattering at all wavelengths. The short-wavelength cutoff is most sensitive to changes in the ozone column and in the solar zenith angle, that is, effectively to changes in the *slant ozone column.* Cloud cover, aerosols, and surface albedo have a strong effect on the absolute value of the transmission, but do not significantly alter the wavelength distribution. Perhaps most importantly, Fig. 9 illustrates the range of variability which may be encountered in a realistic environment. This variability is particularly great in the UV-B region, where ozone absorption dominates, and where many biological processes are most sensitive.

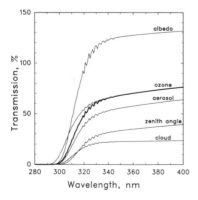

Figure 9. Spectral transmission of the atmosphere. Thick line gives results of a delta-Eddington radiative transfer model for clear sky (no aerosol or clouds), ozone column = 300 DU, solar zenith angle = 30°, and surface albedo = 0.05. Effect of changing these parameters individually is shown by thin solid lines. New values are cloud optical depth = 32, solar zenith angle = 60°, aerosol visible range = 10 km, ozone column = 150 DU, and surface albedo = 0.60.

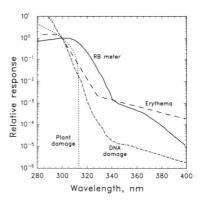

Figure 10. Action spectra for DNA damage (Setlow, 1974), generalized plant damage (Caldwell *et al.*, 1986), erythema induction (McKinlay and Diffey, 1987), and the Robertson–Berger UV radiometer (Urbach *et al.*, 1974). All spectra are normalized to unity at 300 nm and show relative responses per energy unit.

5. Biologically Active UV Radiation

5.1. Dose Rates and Doses

The sensitivity of organisms to UV radiation is in general a function of wavelength. This wavelength dependence must be known accurately if an estimate of the biological responses to changes in atmospheric composition is desired. For example, ozone changes affect mostly short-wavelength UV-B radiation, while cloud cover changes affect both UV-B and UV-A; therefore, the impact of these changes on the rates of biological and chemical processes may be quite different.

The most common representation of the wavelength dependence of biological effects is through *monochromatic action spectra*,† obtained in laboratory studies by exposing a biological target to various isolated wavelengths of radiation and comparing the responses (e.g., damage). A few action spectra are shown in Fig. 10. Action spectra for many other processes of biological interest are also available (United Nations Environmental Programme [*UNEP*], 1991).

The product of a biological action spectrum $B(\lambda)$ and the surface spectral irradiance $F(\lambda)$ is the *spectral dose rate,* illustrated in Fig. 11 for the DNA damage spectrum at three different solar zenith angles. Note the peak response between 300 and 310 nm (depending on the solar zenith angle), and the rapid decrease at shorter wavelengths (due to fewer incident photons) and at longer

---

† In principle, *polychromatic* action spectra may be more desirable, because the simultaneous irradiation by several wavelengths may have a different net effect than that of several separate monochromatic irradiations, due to possible synergisms or antagonisms between complex chemical and biological processes. In practice, however, such polychromatic action spectra are exceedingly difficult to measure.

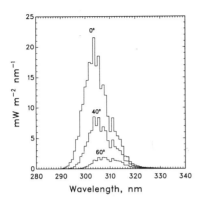

Figure 11. Spectral dose rates weighted for DNA damage at three different solar zenith angles, for clear skies and an ozone column of 300 DU.

wavelengths (due to decreased DNA response). The figure also illustrates a problem which is frequently encountered in computations of biologically weighted UV: the calculations must be performed using a fine wavelength grid (Fig. 11 was calculated using ~1-nm steps), because the spectral dose rate may be a sharply peaked function. The radiative transfer calculation must be carried out at each grid point, resulting in a substantial computational burden.

Integration of the spectral dose rate over wavelength (i.e., the area under the curves of Fig. 11) yields the *dose rate:*

$$\text{Dose rate} = \int B(\lambda)F(\lambda)d\lambda \tag{26}$$

The dose rate is an instantaneous measure of the *biologically weighted UV irradiance,* with units of W m^{-2}.‡ At any given location, the dose rate changes with time of the day, reaching a maximum at solar noon. This is illustrated in Fig. 12 for three different latitudes.

Integration of the dose rate over a full day gives the *daily dose,* and over a full year the *yearly dose,* in units of J m^{-2},

$$\text{Dose} = \int \int B(\lambda)F(\lambda)d\lambda dt \tag{27}$$

‡ $B(\lambda)$ is dimensionless and only measures the *relative* effectiveness of different wavelengths, so it is necessary to specify its normalization point in order to compare different calculations of spectral dose rates, dose rates, and doses. In this chapter, all action spectra are normalized to unity at 300 nm.

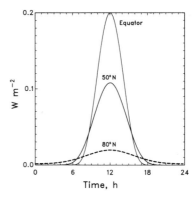

Figure 12. Diurnal dependence of DNA-damaging dose rate on 21 June at three different latitudes.

where t is time. The seasonal and latitudinal dependence of daily DNA-damaging dose rates is shown in Fig. 13. The maximum values occur in the tropics when the sun is directly overhead (23.5°S on 21 December, equator on both equinoxes, and 23.5°N on 21 June). The doses decrease toward higher latitudes but are still non-negligible at the poles during the summer. *Yearly doses* are shown in Fig. 14 for several action spectra. Note that the fall-off toward the poles is strongest for DNA and plant damage, while values for erythema induction at the poles are still about 10% of the equatorial values, due to the larger contributions of UV-A. A slight asymmetry between the Southern and Northern Hemispheres is apparent in Fig. 14. It is due to differences in average ozone over the years 1979–1989 (see Fig. 6) and the fact that the earth–sun separation is smallest during the Southern Hemisphere summer.

5.2. Sensitivity to Clouds, Aerosols, Surface Reflections, and Elevation

The results presented in Figs. 11–14 apply to clear skies, that is, without aerosols or clouds, and for a Lambertian surface albedo of 5%. Brighter surfaces

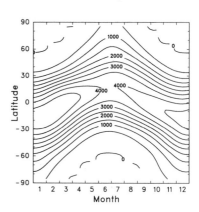

Figure 13. Seasonal and latitudinal dependence of the daily dose (J m^{-2}day^{-1}) for DNA damage, calculated for clear skies, using ozone column averages over 1979–1989. From Madronich (1992b).

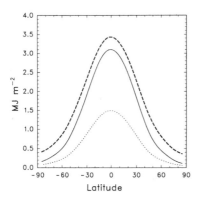

Figure 14. Yearly doses for DNA damage (dotted line), plant damage (solid line), and erythema induction (dashed line).

always increase these values, while clouds and aerosols will usually decrease the dose rates.

Typical transmission values for horizontally uniform clouds are given in Table 4. The transmission is a strong function of the cloud optical depth, but only a weak function of the solar zenith angle. For partial cloud cover, the dose is generally reduced relative to clear sky doses. Measured correlations between fractional cloud cover and UV doses are rather scattered, and several different empirical relations have been given, e.g.,

$$D_c/D_o = 1 - 0.56c \quad \text{(in Malaysia; Ilyas, 1987)}$$
$$= 1 - 0.50c \quad \text{(in the U.S.; Cutchis, 1980)}$$
$$= 1 - 0.7c^{2.5} \quad \text{(in Sweden; Josefsson, 1986)}$$

where D_o is the clear sky dose and D_c is the dose when the fractional cloud cover is c. For complete cloud cover, the above dose ratios limit to 0.44, 0.5, and 0.3, while measurements in Australia by Paltridge and Barton (1978) give a fully overcast estimate of 0.2. Generally, these simple relationships must be viewed with some caution because realistic cloud morphology is complex and nonuniform, reflections from the sides of clouds may also contribute, and different types of clouds may be prevailing at different locations.

The effect of changing surface albedo is shown in Fig. 15. The increases are relatively independent of solar zenith angle. The enhancement for large albedo is due not only to the direct local reflections, but also to the atmospheric back-scattering of radiation reflected at the surface. Thus it is important to distinguish between the *local albedo,* which provides direct reflections to the target from surfaces within a few hundreds of meters, and the average *regional albedo* over a zone within approximately a 10-km radius from the target,

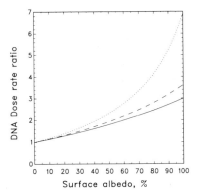

Figure 15. Effect of surface albedo on DNA-damaging dose rates, for a solar zenith angle of 45°. Solid curve is for clear sky, dashed curve is with thin cloud (optical depth = 4), and dotted curve for thick cloud (optical depth = 32). Values are normalized to surface albedo of 0.05.

which may also contribute to the diffuse sky radiation. One interesting aspect is the potential coupling between reflections from the surface and reflections from the base of the cloud, which can lead to effective radiation trapping in the lower atmosphere. This can be seen from Fig. 15, which shows that the effect of albedo increases in the presence of clouds. Note especially the strong enhancement of radiation for the combination of high surface albedo and thick cloud.

Atmospheric pollution can reduce UV levels in certain local areas. The effects of tropospheric ozone are discussed in more detail in the next section. Absorbing pollutant gases, such as NO_2 and SO_2, can decrease UV-B radiation by a few percent in heavily polluted urban areas. Particulate pollution, often in the form of sulfate aerosol haze, can have a significant effect, as shown in Fig. 16. In the figure, the reduction of DNA-damaging daily dose is related to the reduction in ground-level visible range due to the aerosols. The UV

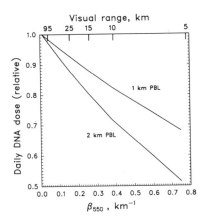

Figure 16. Effect of aerosol on DNA-damaging daily dose, relative to the aerosol-free atmosphere. The lower scale gives the extinction coefficients at 550 nm, corresponding to the visual range shown in the upper scale. Values are shown for different depths of the planetary boundary layer (PBL). From Liu et al. (1991).

reduction also depends on the vertical depth of the boundary layer, where most of the aerosols are contained, and on the chemical composition of the aerosol particles.

Another factor affecting the UV doses is the elevation of the surface above sea level. This is normally a relatively small effect, about 5% to 8% per km. However, higher elevations are often associated with lower pollution levels or less cloud cover, so that the actual increase in UV may be larger.

5.3. Sensitivity to Ozone—Radiation Amplification Factors

Doses for UV-B-induced biological processes are very sensitive to the total overhead atmospheric column of ozone, as illustrated in Fig. 17. For small changes, the sensitivity may be described by a simple proportionality,

$$\Delta D/D = \text{RAF}(\Delta N/N) \qquad \text{(percent rule)} \qquad (28)$$

where D and ΔD are the dose and the dose change, N and ΔN are the ozone column and its change, and the proportionality constant RAF is the *radiation amplification factor*. Values of the RAF, given in Table 6 for various biological processes, are useful for rough estimates of the effects of ozone depletion and also for comparing the sensitivity of different biological processes to UV-B radiation. For example, a 2% ozone depletion will increase DNA-damaging radiation (RAF = 1.9) by 3.8% and erythemal radiation (RAF = 1.1) by 2.2%. Although the RAF is an extremely useful and convenient concept, it suffers from some important limitations and therefore must be used with some caution.

First, the simple proportionality given above must not be applied directly for large ozone changes. In this case it is usually better to integrate the percent rule,

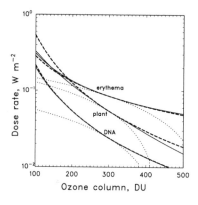

Figure 17. Sensitivity of dose rates to ozone, for three different action spectra, solar zenith angle = 60°, clear skies. Solid lines give the exact radiative transfer calculations. Dashed lines are estimated using the power rule, and dotted lines using the percent rule (see text), from the exact value at 300 DU and assumed RAFs of 2.0 for DNA damage, 2.1 for plant damage, and 1.1 for erythema induction. From Madronich (1992b).

$$D_1/D_2 = (N_1/N_2)^{-RAF} \qquad \text{(power rule)} \qquad (29)$$

where D_1 and D_2 are the doses corresponding to ozone amounts N_1 and N_2, respectively. Thus, for example, an ozone reduction of 20% would imply a 53% increase in DNA-damaging UV according to the power rule, but only a 38% increase according to the percent rule. Figure 17 shows that the power rule describes the changes more accurately over a wider range of ozone column values.

Second, even the power rule is a highly oversimplified representation of the complex atmospheric scattering and absorption processes and the wavelength dependence of the irradiance and various action spectra, for all different environmental conditions such as solar zenith angle, clouds, and albedo. At best, a specific RAF may be used over a small range of conditions with the understanding that if these conditions change substantially, different RAF values may be applicable. RAFs for generalized plant damage and erythema generally increase with increasing solar angle and with increasing ozone, while DNA-damaging RAFs are fortuitously insensitive to these different conditions. Several other atmospheric factors have only negligible effects on RAFs, because they change all wavelengths about equally. These include changes in cloud cover, surface albedo and elevation, atmospheric aerosols, and receiver shape (e.g., scalar versus vector irradiance).

Third, even at constant ozone column, changes in the vertical distribution of ozone can change UV doses (Brühl and Crutzen, 1989). The origin of this effect may be understood as follows. In the stratosphere, most of the UV-B radiation is still contained in the direct solar beam and crosses stratospheric ozone layers with the solar zenith angle θ_o. In the troposphere, most of the radiation has been scattered, and crosses the tropospheric layers at some average angle θ^*, which may be taken as approximately 60° as discussed in Section 4.3. If ΔN ozone molecules are added (or subtracted) in the stratosphere, the additional Beer–Lambert absorption at wavelength λ will be $e^{-\sigma(\lambda)\Delta N/\cos\theta_o}$, while the same number of ozone molecules added (or subtracted) in the troposphere will change the absorption by $e^{-\sigma(\lambda)\Delta N/\cos\theta^*}$. Thus, equal amounts of tropospheric ozone are more effective UV-B absorbers than stratospheric ozone for high sun ($\theta_o < 60°$), but somewhat less effective for low sun ($\theta_o > 60°$). As a result, the RAFs also depend on the altitude of the ozone perturbations.

Finally, serious errors in estimates of dose sensitivity to ozone depletion can be induced by experimental uncertainties in the action spectra determined in laboratory studies. In addition to the usual issues of accuracy and precision, large overestimates of RAFs can be made if the experiments are not carried out over a sufficiently dynamic range. Typically, biological responses fall rapidly with increasing UV-B and UV-A wavelengths. However, surface irradiance

Table 6. Radiation Amplification Factors (RAFs) at 30°N[a]

Effect	RAF[b]		Reference
	January	July	
DNA related			
Mutagenicity and fibroblast killing	[1.7] 2.2	[2.7] 2.0	Zölzer and Kiefer, 1984; Peak et al., 1984
Fibroblast killing	0.3	0.6	Keyse et al., 1983
Cyclobutane pyrimidine dimer formation	[2.0] 2.4	[2.1] 2.3	Chan et al., 1986
(6-4) photoproduct formation	[2.3] 2.7	[2.3] 2.5	Chan et al., 1986
Generalized DNA damage	1.9	1.9	Setlow, 1974
HIV-1 activation	[0.1] 4.4	[0.1] 3.3	Stein et al., 1989
Plant effects			
Generalized plant spectrum	2.0	1.6	Caldwell et al., 1986
Inhibition of growth of cress seedlings	[3.6] 3.8	3.0	Steinmetz and Wellmann, 1986
Isoflavonoid formation in bean	[0.1] 2.7	[0.1] 2.3	Wellmann, 1985
Inhibition of phytochrome-induced anthocyanin synthesis in mustard	1.5	1.4	Wellmann, 1985
Anthocyanin formation in maize	0.2	0.2	Beggs and Wellmann, 1985
Antocyanin formation in sorghum	1.0	0.9	Yatsuhashi et al., 1982
Photosynthetic electron transport	0.2	0.2	Bornman et al., 1984

Effect			Reference
Overall photosynthesis in leaf of *Rumex patientia*	0.2	0.3	Rundel, 1983
Membrane damage			
Glycine leakage from *E. coli*	0.2	0.2	Sharma and Jagger, 1979
Alanine leakage from *E. coli*	0.4	0.4	Sharma and Jagger, 1979
Membrane bound K$^+$-stimulated ATPase inactivity	[0.3] 2.1	[0.3] 1.6	Imbrie and Murphy, 1982
Skin			
Elastosis	1.1	1.2	Kligman and Sayer, 1991
Photocarcinogenesis, skin edema	1.6	1.5	Cole et al., 1986
Erythema reference	1.1	1.1	McKinlay and Diffey, 1987
Skin cancer in SKH-1 hairless mice (Utrecht)	1.4	1.3	de Gruijl (unpubl., 1991)
Eyes			
Damage to cornea	1.2	1.1	Pitts et al., 1977
Damage to lens (cataract)	0.8	0.7	Pitts et al., 1977
Movement			
Inhibition of motility in *Euglena gracilis*	1.9	1.5	Häder and Worrest, 1991
Other			
Immune suppression	[0.4] 1.0	[0.4] 0.8	De Fabo and Noonan, 1983
Robertson–Berger meter	0.8	0.7	Urbach et al., 1974

[a] Adapted from UNEP (1991).
[b] Values in brackets show effect of extrapolating original data to 400 nm with an exponential tail, for cases where the effect is larger than 0.2 RAF units.

increases sharply between 300 and 330 nm, and even small biological responses at longer wavelengths may contribute substantially to the total dose-rate integral. This is illustrated in Table 6, where the RAFs were calculated using the original literature data, as well as with an exponential tail extrapolation to 400 nm. In many instances, significantly different RAFs were found, as shown by the values given in brackets. Generally, this error can be avoided by measuring the biological response over a dynamic range of about two to three orders of magnitude, but this of course represents a formidable challenge to the experimental studies.

6. Trends in UV Reaching the Surface

The interest in environmental UV photobiology has increased considerably in recent years, due to concerns that ambient UV-B levels will increase as a result of stratospheric ozone depletion. Assessment of UV changes which might have already occurred is difficult. At the time of writing, there is no adequate UV-B monitoring network on either a regional or global scale, and so any trends must be inferred through radiative transfer calculations based on measured ozone amounts. While several ground-based stations have been gathering ozone column data since the 1950s, satellite-based instruments have been monitoring global ozone distributions only since late 1978.

The clearest and strongest ozone depletion to date has been taking place during springtime in the southern polar regions, the so-called *ozone hole.* Figure 18 shows the October ozone column values in southern polar regions at the South Pole from 1957 to 1991. Before about 1980, the values scatter between 250 and 325 DU, but have fallen in recent years to between 125 and 175 DU. Values as low as 108 DU have been reported for some days in

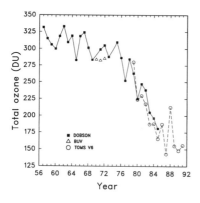

Figure 18. October ozone column measured over Antarctica by different instruments. Squares are ground-based measurements from Halley Bay; circles are from the TOMS instrument aboard the NIMBUS-7 satellite. From WMO (1991).

October 1991 (WMO, 1991). Vertical measurements indicate that this is due to the almost complete disappearance of ozone from the lower stratosphere, and that low ozone correlates very well with high amounts of ozone-destroying compounds such as chlorine monoxide, ClO. The geographical extent of the low-ozone air is confined to the inside of the atmospheric *polar vortex,* which covers most of Antarctica during the winter and early spring. With the spring warming, the vortex breaks up and large parcels of ozone-poor air are transported out of the region, causing significant low-ozone events and enhanced UV-B at midlatitudes (Roy *et al.,* 1990). Although several different explanations have been proposed for the occurrence of the ozone hole, laboratory data and field measurements indicate that reactions occurring on the surface of *polar stratospheric cloud* particles convert inactive chlorine species (such as HCl and $ClONO_2$) to active chlorine (Cl_2 and ClO) during the winter polar night. When the sun rises in the early spring, the active chlorine destroys ozone. As might be expected, the ozone reduction in the Antarctic spring has a significant impact on UV levels. Direct UV-B measurements made at a few locations confirm these recent increases (Lubin *et al.,* 1989, 1992; Lubin and Frederick, 1989, 1990; Stamnes *et al.,* 1990, 1992; Frederick and Alberts, 1991; Beaglehole and Carter, 1992a,b; Smith *et al.,* 1992). The recent measurements by Smith *et al.* (1992) also indicate a decrease of phytoplankton productivity associated with increased UV during the passage of the ozone hole over regions of the Bellingshausen Sea.

Global trends in ozone, observed between 1978 and 1990 by the satellite-based Total Ozone Mapping Spectrometer (TOMS), are much smaller that those associated with the Antarctic ozone hole (Stolarski *et al.,* 1991, 1992; McCormick *et al.,* 1992). Table 7 shows the annually averaged trends in ozone and the corresponding changes in biological yearly doses. The trends also vary with season, as shown in Figs. 19 and 20. Trends are largest in the polar regions, but are still statistically significant at midlatitudes in both hemispheres. The increase between 30°S and 60°S in late spring and early winter is particularly noteworthy, as this is the time of the Southern Hemisphere summer solstice and smallest earth–sun separation, when UV levels are normally at their yearly maximum. No significant trends are seen in the tropics, but clearly even small percentage increases here would translate into large irradiance increments because of the already high natural radiation levels.

The UV increases calculated from the TOMS ozone column measurements may have been offset partly or completely by tropospheric pollutants in some regions of the globe. Tropospheric ozone has been increasing in the Northern Hemisphere at a rate of 1% per year, but it has been steady or even decreasing in the Southern Hemisphere (Oltmans *et al.,* 1989). As only about $\frac{1}{10}$ of the total ozone column is in the troposphere, these trends are insufficient to offset current trends due to stratospheric ozone depletion, but they are

Table 7. Trends in Annual Integral, Percent per Decade, 1979–1989 [a]

Latitude	Total ozone	R–B meter	Erythema induction	Plant damage	DNA damage
85°N	−4.5 ± 3.0	5.1 ± 1.5	4.7 ± 1.5	14.8 ± 4.3	10.1 ± 3.2
75°N	−4.0 ± 2.7	3.9 ± 1.3	4.1 ± 1.4	10.8 ± 3.4	9.0 ± 2.9
65°N	−4.2 ± 1.7	3.4 ± 1.3	4.0 ± 1.5	8.9 ± 3.3	8.1 ± 3.0
55°N	−4.4 ± 1.2	3.3 ± 1.2	4.0 ± 1.5	8.1 ± 3.0	7.7 ± 2.9
45°N	−4.5 ± 1.0	3.1 ± 1.0	4.0 ± 1.4	7.2 ± 2.6	7.2 ± 2.7
35°N	−4.5 ± 1.3	3.0 ± 1.1	4.3 ± 1.6	7.0 ± 2.7	7.5 ± 3.0
25°N	−2.7 ± 1.6	1.8 ± 1.1	2.8 ± 1.9	4.2 ± 2.8	4.8 ± 3.3
15°N	−0.7 ± 1.2	0.5 ± 0.8	0.8 ± 1.4	1.1 ± 1.9	1.3 ± 2.3
5°N	−1.2 ± 1.4	0.8 ± 1.0	1.3 ± 1.6	1.8 ± 2.3	2.2 ± 2.7
5°S	−1.2 ± 1.3	0.8 ± 0.8	1.3 ± 1.4	1.8 ± 2.0	2.1 ± 2.4
15°S	−0.3 ± 1.2	0.2 ± 0.8	0.4 ± 1.3	0.6 ± 1.8	0.7 ± 2.2
25°S	−1.2 ± 1.7	1.0 ± 1.1	1.8 ± 1.7	2.7 ± 2.5	3.3 ± 2.8
35°S	−3.1 ± 1.7	2.3 ± 1.1	3.5 ± 1.6	5.6 ± 2.6	6.3 ± 2.9
45°S	−4.9 ± 1.5	3.7 ± 1.1	5.3 ± 1.6	9.3 ± 2.8	9.9 ± 3.0
55°S	−7.3 ± 1.7	6.0 ± 1.5	7.7 ± 1.9	15.4 ± 4.0	15.1 ± 3.9
65°S	−10.8 ± 2.2	9.5 ± 2.3	11.4 ± 2.8	25.4 ± 6.6	23.4 ± 6.1
75°S	−13.2 ± 2.8	12.8 ± 3.9	15.0 ± 4.8	39.0 ± 13.4	34.4 ± 11.9
85°S	−14.5 ± 3.4	15.6 ± 5.1	16.8 ± 5.7	53.9 ± 20.2	42.7 ± 15.8

Note. Uncertainties are one standard deviation.
[a] From Madronich (1992a).

likely to have persisted over at least several previous decades. According to some estimates, tropospheric ozone in the Northern Hemisphere has roughly tripled during this century, from about 10 ppb to about 30 ppb (Logan, 1985, 1989), which would imply a reduction of DNA-weighted UV doses of about 10% (UNEP, 1991). Similarly, sulfate aerosols have been increasing over the last century National Research Council ([*NRC*] 1986), and recent estimates

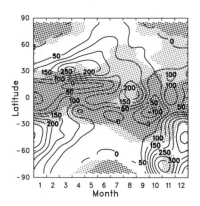

Figure 19. Trends in DNA-damaging daily doses over 1979–1989. Dark shading shows regions where trends differ from zero by less than one standard deviation, light shading by less than two standard deviations. No shading denotes areas with significant trends. Values are in J m^{-2} per decade. From Madronich (1992a).

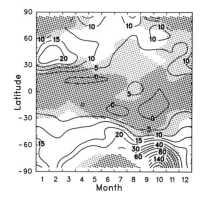

Figure 20. Trends in DNA-damaging daily doses over 1979–1989. Dark shading shows regions where trends differ from zero by less than one standard deviation, light shading by less than two standard deviations. No shading denotes areas with significant trends. Values are in percent per decade. From Madronich (1992a).

suggest a reduction of UV radiation of between 5% and 18% (Liu *et al.,* 1991). However, it should be clear that these reductions apply mostly to urban and relatively polluted regions of industrialized nations, while the depletion of stratospheric ozone is likely to have a global impact.

Clouds remain a major uncertainty in the assessment and prediction of UV trends. Their natural daily, yearly, and longer-term variability can completely obscure small UV trends due to ozone depletion (Frederick and Snell, 1990). There are suggestions that clouds in the Northern Hemisphere may have become more reflective due to sulfate aerosols of anthropogenic origin, which serve as condensation nuclei and therefore might increase the number of small cloud droplets (Intergovernmental Panel on Climate Change [*IPCC*], 1990). Cloud types and distributions may also change in the future, if significant climate change occurs due to anthropogenic emissions of infrared-absorbing gases (such as carbon dioxide, methane, ozone, and others). The direction and magnitude of these changes are currently not known. However, any increase in cloudiness is not likely to offset the large UV-B increases associated with the Antarctic ozone hole, nor the changes in the ratio of UV-B radiation to other potentially useful light (e.g., photosynthetic radiation).

Given the natural geographic and temporal variability of stratospheric ozone, cloud cover, surface albedo, and pollutants such as aerosols and tropospheric ozone, extraction of global UV trends from the limited data currently available is by necessity only tentative, and future measurements will likely require significant revision of the trends given here.

References

Beaglehole, D., and Carter, G. G., 1992a, Antarctic skies 1. Diurnal variations of the sky irradiance, and UV effects of the ozone hole, spring 1990, *J. Geophys. Res.* 97:2589–2596.

Beaglehole, D., and Carter, G. G., 1992b, Antarctic skies 2. Characterization of the intensity and polarization of skylight in a high albedo environment, *J. Geophys. Res.* 97:2597–2600.

Beggs, C. J., and Wellmann, E., 1985, Analysis of light-controlled anthocyanin formation in coleoptiles of Zea mays 1.: The role of UV-B, blue, red and far-red light, *Photochem. Photobiol.* 41:481–486.

Blumthaler, M., and Ambach, W., 1988, Solar UVB-albedo of various surfaces, *Photochem. Photobiol.* 48:85–88.

Bornman, J. F., Björn, L. O., and Åkerlund, H. -E., 1984, Action spectrum for inhibition by ultraviolet radiation of photosystem II activity in spinach thylakoids, *Photobiochem. Photobiophys.* 8:305–313.

Brasseur, G. P., and Solomon, S., 1986, *Aeronomy of the Middle Atmosphere,* 2nd ed., D. Reidel, Dordrecht.

Brühl, C., and Crutzen, P. J., 1989, On the disproportionate role of tropospheric ozone as a filter against solar UV-B radiation, *Geophys. Res. Lett.* 16:703–706.

Caldwell, M. M., Camp, L. B., Warner, C. W., and Flint, S. D., 1986, Action spectra and their key role in assessing biological consequences of solar UV-B radiation chance, in: *Stratospheric Ozone Reduction, Solar Ultraviolet Radiation and Plant Life* (R. C. Worrest and M. M. Caldwell, eds.), Springer-Verlag, Berlin, pp. 87–111.

Chan, G. L., Peak, M. J., Peak, J. G., and Haseltine, W. A., 1986, Action spectrum for the formation of endonuclease-sensitive sites and (6-4) photoproducts induced in a DNA fragment by ultraviolet radiation, *Int. J. Radiat. Biol.* 50:641–648.

Chandrasekhar, S., 1960, *Radiative Transfer,* Dover, New York.

Cole, C. A., Forbes, D., and Davies, R. E., 1986, An action spectrum for UV photocarcinogenesis, *Photochem. Photobiol.* 43:275–284.

Coulson, K. L., and Reynolds, W. D., 1971, The spectral reflectance of natural surfaces, *J. Appl. Meteorol.* 10:1285–1295.

Cutchis, P., 1980, *A Formula for Comparing Annual Damaging Ultraviolet (DUV) Radiation Doses at Tropical and Mid-Latitude Sites,* Federal Aviation Administration Report-FAA-EE 80-21, U.S. Department of Transportation, Washington, DC.

Davidson, J. A., Cantrell, C. A., McDaniel, A. H., Shetter, R. E., Madronich, S., and Calvert, J. G., 1988, Visible-ultraviolet absorption cross section for NO_2 as a function of temperature, *J. Geophys. Res.* 93:7105–7112.

De Fabo, E. C., and Noonan, F. P., 1983, Mechanism of immune suppression by ultraviolet radiation in vivo. I. Evidence for the existence of a unique photoreceptor in skin and its role in photoimmunology, *J. Exp. Med.* 158:84–98.

Dickerson, R. R., Stedman, D. H., and Delany, A. C., 1982, Direct measurements of ozone and nitrogen dioxide photolysis rates in the troposphere, *J. Geophys. Res.* 87:45933–4946.

Doda, D. D., and Green, A. E. S., 1980, Surface reflectance measurements in the UV from an airborne platform, Part 1, *Appl. Opt.* 19:2140–2145.

Doda, D. D., and Green, A. E. S., 1981, Surface reflectance measurements in the UV from an airborne platform, Part 2, *Appl. Opt.* 20:636–642.

Duffett-Smith, P., 1988, *Practical Astronomy with your Calculator,* 3rd ed., Cambridge University Press, Cambridge.

Finlayson-Pitts, B. J., and Pitts, J. N., 1986, *Atmospheric Chemistry,* Wiley-Interscience, New York.

Frederick, J. E., and Alberts, A. D., 1991, Prolonged enhancement in surface ultraviolet radiation during the Antarctic spring of 1990, *Geophys. Res. Lett.* 18:1869–1871.

Frederick, J. E., and Snell, H. E., 1990, Tropospheric influence on solar ultraviolet radiation: The role of clouds, *J. Climate* 3:373–381.

Frölich, C., and Shaw, G. E., 1980, New determination of Rayleigh scattering in the terrestrial atmosphere, *Appl. Opt.* 19:1773–1775.

Goody, R. M., and Yung, Y. L., 1989, *Atmospheric Radiation*, Oxford University Press, New York.

Green, A. E. S., Cross, K. R., and Smith, L. A., 1980, Improved analytic characterization of ultraviolet skylight, *Photochem. Photobiol.* 31:59–65.

Häder, D. -P., and Worrest, R. C., 1991, Effects of enhanced solar ultraviolet radiation on aquatic ecosystems, *Photochem. Photobiol.* 53:717–725.

Hansen, J. E., and Travis, L. D., 1974, Light scattering in planetary atmospheres, *Space Sci. Rev.* 16:527–610.

Haurwitz, B., 1948, Insolation in relation to cloud type, *J. Meteorol.* 5:110–113.

Hoyt, D. V., Kyle, H. L., Hickey, H. R., and Maschhoff, R. H., 1992, The Nimbus 7 solar total irradiance: A new algorithm for its derivation, *J. Geophys. Res.* 97:51–63.

Ilyas, M., 1987, Effect of cloudiness on solar ultraviolet radiation reaching the surface, *Atmos. Environ.* 21:1483–1484.

Imbrie, C. W., and Murphy, T. M., 1982, UV-action spectrum (254–405 nm) for inhibition of K⁺-stimulated adenosine triphosphatase from a plasma membrane of *Rosa damascena, Photochem. Photobiol.* 36:537–542.

Intergovernmental Panel on Climate Change (IPCC), 1990, *Climate Change,* Cambridge University Press, New York.

Josefsson, W., 1986, *Solar Ultraviolet Radiation in Sweden,* National Institute of Radiation Protection in Stockholm, SMHI Report-53, Norrköping, Sweden.

Joseph, J. H., Wiscombe, W. J., and Weinman, J. A., 1976, The delta-Eddington approximation for radiative flux transfer, *J. Atmos. Sci.* 33:2452–2459.

Keyse, S. M., Moses, S. H., and Davies, D. J. G., 1983, Action spectra for inactivation of normal and xeroderma pigmentosum human skin fibroblasts by ultraviolet radiation, *Photochem. Photobiol.* 37:307–312.

Kligman, L. H., and Sayre, R. M., 1991, An action spectrum for ultraviolet induced elastosis in hairless mice: Quantification of elastosis by image analysis, *Photochem. Photobiol.* 53:237–242.

Kondratyev, K. Ya., 1969, *Radiation in the Atmosphere,* Academic Press, New York.

Liu, S. C., McKeen, S. A., and Madronich, S., 1991, Effects of anthropogenic aerosols on biologically active ultraviolet radiation, *Geophys. Res. Lett.* 18:2265–2268.

Logan, J. A., 1985, Tropospheric ozone: Seasonal behavior, trends, and anthropogenic influence, *J. Geophys. Res.* 90:10463–10482.

Logan, J. A., 1989, Ozone in rural areas of the United States, *J. Geophys. Res.* 94:8511–8532.

Lubin, D., and Frederick, J. E., 1989, Measurements of enhanced springtime ultraviolet radiation at Palmer station, Antarctica, *Geophys. Res. Lett.* 16:783–785.

Lubin, D., and Frederick, J. E., 1990, Column ozone measurements from Palmer station, Antarctica: Variations during the austral springs of 1988 and 1989, *J. Geophys. Res.* 95:13883–13889.

Lubin, D., Frederick, J. E., and Krueger, A. J., 1989, The ultraviolet radiation environment of Antarctica: McMurdo station during September–October 1987, *J. Geophys. Res.* 94:8491–8496.

Lubin, D., Mitchell, B. G., Frederick, J. E., Roberts, A. D., Booth, C. R., Lucas, T., and Neuschuler, D., 1992, A contribution toward understanding the biospherical significance of Antarctic ozone depletion, *J. Geophys. Res.* 97:7817–7828.

Madronich, S., 1992a, Implications of recent total atmospheric ozone measurements for biologically active ultraviolet radiation reaching the Earth's surface, *Geophys. Res. Lett.* 19:37–40.

Madronich, S., 1992b, *The Natural Ultraviolet Radiation Environment,* paper presented at the 20th Annual Meeting of the American Society for Photobiology, Marco Island, Florida, June 20–24.

McCartney, E. J., 1976, *Optics of the Atmosphere,* Wiley, New York.

McCormick, M. P., Veiga, R. E., and Chu, W., 1992, Stratospheric ozone profile and total ozone trends derived from the SAGE I and SAGE II data, *Geophys. Res. Lett.* 19:269–272.

McGee, T. J., and Burris, J., Jr., 1987, SO_2 absorption cross section in the near U.V., *J. Quant. Spectros. Radiat. Transfer* 37:165–182.

McKinlay, A. F., and Diffey, B. L., 1987, A reference action spectrum for ultraviolet induced erythema in human skin, in: *Human Exposure to Ultraviolet Radiation: Risks and Regulations* (W. R. Passchler and B. F. M. Bosnajokovic, eds.), Elsevier, Amsterdam.

Meador, W. E., and Weaver, W. R., 1980, Two-stream approximations to radiative transfer in planetary atmospheres: A unified description of existing methods and a new improvement, *J. Atmos. Sci.* 37:630–643.

Molina, L. T., and Molina, M. J., 1986, Absolute absorption cross sections of ozone in the 185- to 350-nm wavelength range, *J. Geophys. Res.* 91:14501–14508.

National Research Council (NRC), 1986, *Acid Deposition Long Term Trends,* National Academy Press, Washington, DC.

Oltmans, S. J., Komhyr, W. D., Franchois, P. R., and Matthews, W. A., 1989, Tropospheric ozone: Variations from surface and ECC ozonesonde observations, in: *Ozone in the Atmosphere,* proceedings of the Quadrennial Ozone Symposium 1988 and Tropospheric Ozone Workshop, Göttingen, Federal Republic of Germany, August 1988 (R. Bojkov and P. Fabian, eds.), Deepak Publishing, Hampton, Virginia.

Paltridge, G. W., and Barton, I. J., 1978, *Erythemal Ultraviolet Radiation Distribution over Australia—the Calculations, Detailed Results and Input Data Including Frequency Analysis of Observed Australian Cloud Cover,* Division of Atmospheric Physics Technical Paper-33, Commonwealth Scientific and Industrial Research Organization, Australia.

Peak, M. J., Peak, J. G., Moehring, M. P., and Webb, R. B., 1984, Ultraviolet action spectra for DNA dimer induction, lethality, and mutagenesis in *Escherichia coli* with emphasis on the UVB region, *Photochem. Photobiol.* 40:613–620.

Pitts, D. G., Cullen, A. P., and Hacker, P. D., 1977, Ocular effects of ultraviolet radiation from 295 to 365 nm, *Invest. Ophthalmol. Visual Sci.* 16:932–939.

Roy, C. T., Gies, H. P., and Graeme, E., 1990, Ozone depletion, *Science,* 347:235–236.

Rundel, R. D., 1983, Action spectra and the estimation of biologically effective UV radiation, *Physiol. Plant.* 58:360–366.

Setlow, R. B., 1974, The wavelengths in sunlight effective in producing skin cancer: A theoretical analysis, *Proc. Natl. Acad. Sci. U.S.A.* 71:3363–3366.

Sharma, R. C., and Jagger, J., 1979, Ultraviolet (254–405 nm) action spectrum and kinetic studies of analine uptake in *Escherichia coli* B/R, *Photochem. Photobiol.* 30:661–666.

Shetter, R. E., McDaniel, A. H., Cantrell, C. A., Madronich, S., and Calvert, J. G., 1992, Actinometer and Eppley radiometer measurements of the NO_2 photolysis rate coefficient during MLOPEX, *J. Geophys. Res.* 97:10349–10359.

Smart, W. M., 1979, *Textbook on Spherical Astronomy,* 6th ed., Cambridge University Press, Cambridge.

Smith, R. C., Prezelin, B. B., Baker, K. S., Bidigare, R. R., Boucher, N. P., Coley, T., Karentz, D., MacIntyre, S., Matlick, H. A., Menzies, D., Ondrusek, M., Wan, Z., and Waters, K. J., 1992, Ozone depletion: Ultraviolet radiation and phytoplankton biology in Antarctic waters, *Science* 255:952–959.

Spencer, J. W., 1971, Fourier series representation of the position of the sun, *Search* 2:172.

Stamnes, K., Tsay, S. C., Wiscombe, W., and Jayaweera, K., 1988, Numerically stable algorithm for discrete-ordinate-method radiative transfer in multiple scattering and emitting layers, *Appl. Opt.* 27:2502–2509.

Stamnes, K., Slusser, J., Bowen, M., Booth, C., and Lucas, T., 1990, Biologically effective ultraviolet radiation, total ozone abundance, and cloud optical depth at McMurdo station, Antarctica, September 15, 1988, through April 15, 1989, *Geophys. Res. Lett.* 17:2181–2184.

Stamnes, K., Jin, Z., Slusser, J., Booth, C., and Lucas, T., 1992, Several-fold enhancement of biologically effective ultraviolet radiation levels at McMurdo station Antarctica during the 1990 ozone "hole," *Geophys. Res. Lett.* 19:1013–1016.

Stein, B., Rahmsdorf, H. J., Steffen, A., Litfin, M., and Herrlich, P., 1989, UV-induced DNA damage is an intermediate step in UV-induced expression of human immunodeficiency virus type 1, collagenase, c-fos, and metallothionein, *Mol. Cell. Biol.* 9:5169–5181.

Steinmetz, V., and Wellmann, E., 1986, The role of solar UV-B in growth regulation of cress (*Lepidium sativum* L.) seedlings, *Photochem. Photobiol.* 43:189–193.

Stolarski, R. S., Bloomfield, P., McPeters, R. D., and Herman, J. R., 1991, Total ozone trends deduced from Nimbus 7 TOMS data, *Geophys. Res. Lett.* 18:1015–1018.

Stolarski, R., Bojkov, R., Bishop, L., Zerefos, C., Staehelin, J., and Zawodny, J., 1992, Measured trends in stratospheric ozone, *Science* 256:342–349.

Toon, O. B., McKay, C. P., Ackerman, T. P., and Santhanam, K., 1989, Rapid calculation of radiative heating rates and photodissociation rates in inhomogeneous multiple scattering atmospheres, *J. Geophys. Res.* 94:16287–16301.

UNEP, 1991, *Environmental Effects of Ozone Depletion: 1991 update* (J. C. Van der Leun, M. Tevini, and R. C. Worrest, eds.), United Nations Environmental Programme, Nairobi, Kenya.

Urbach, F., Berger, D., and Davies, R. E., 1974, Field measurements of biologically effective UV radiation and its relation to skin cancer in man, in: *Proceedings of the Third Conference on Climatic Impact Assessment Program* (A. J. Broderick and T. M. Hard, eds.), U.S. Dept. of Transportation, Washington, DC.

U.S. Standard Atmosphere, 1976, National Oceanic and Atmospheric Administration, National Aeronautics and Space Administration, United States Air Force, Washington, DC.

Warneck, P., Marmo, F. F., and Sullivan, J. O., 1964, Ultraviolet absorption of SO_2: Dissociation energies of SO_2 and SO, *J. Chem. Phys.* 40:1132–1132.

Wellmann, E., 1985, UV-B-Signal/Response-Beziehungen unter natürlichen und artifiziellen Lichtbedingungen, *Ber. Dtsch. Bot. Ges.* 98:99–104.

WMO, 1985, *Atmospheric Ozone 1985,* Global Ozone Research and Monitoring Project—Report No. 16, World Meteorological Organization, Geneva.

WMO, 1989, *Scientific Assessment of Stratospheric Ozone: 1989, Volume I,* Global Ozone Research and Monitoring Project—Report No. 20, World Meteorological Organization, Geneva.

WMO, 1991, *Scientific Assessment of Ozone Depletion: 1991,* Global Ozone Research and Monitoring Project—Report No. 25, World Meteorological Organization, Geneva.

Yatsuhashi, H., Hashimoto, T., and Shimizu, S., 1982, Ultraviolet action spectrum for anthocyanin formation in broom sorghum first internodes, *Plant Physiol.* 70:735–741.

Zölzer, F., and Kiefer, J., 1984, Wavelength dependence of inactivation and mutation induction to 6-thioguanine-resistance in V79 Chinese hamster fibroblasts, *Photochem. Photobiol.* 40:49–53.

2

Simulation of Daylight Ultraviolet Radiation and Effects of Ozone Depletion

Lars Olof Björn and Alan H. Teramura

1. Light Sources for UV-B Photobiology

1.1. Low-Pressure Discharge Lamps

In a gas discharge lamp, when an electric current flows through a gas, the gas emits light, the spectral composition of which depends on the kind of gas used. At low pressure the gas emits a line spectrum, i.e., only light of certain wavelengths is represented (in contrast to the continuous spectrum emitted by an incandescent lamp or a high-pressure discharge lamp).

The basic parts of a gas discharge lamp are a transparent envelope which encloses the gas and two electrodes to conduct current to and from the gas. Other parts may be necessary, such as heating filaments, to release enough electrons to start the current through the gas or to vaporize a metal, e.g., mercury or sodium, when vapors of these metals are used as the emitting gas.

The electrical resistance of an incandescent lamp increases when the current through it increases, since tungsten has a higher resistivity the higher

Lars Olof Björn • Plant Physiology, Lund University, S-220 07 Lund, Sweden.   Alan H. Teramura • Department of Botany, University of Maryland, College Park, Maryland 20742.

Environmental UV Photobiology, edited by Antony R. Young *et al.* Plenum Press, New York, 1993.

the temperature. Thus, an incandescent lamp is self-regulating and burns in a stable way as long as the voltage is constant. In a gas discharge lamp the reverse holds: Its electric resistance decreases with increasing current. Therefore, such a lamp has to be connected to some kind of circuitry limiting the current. In the case of direct current a series resistor is often used; in the case of alternating current, a choke.

Gas discharge lamps containing mercury vapor of very low pressure emit most of the energy as UVR of wavelength 253.7 nm. This wavelength is close to the absorption maximum of nucleic acids, and the radiation is also absorbed by the aromatic amino acids in proteins and many other biological molecules. The photons are also energetic enough to initiate many chemical reactions, and therefore this kind of radiation is very destructive to living matter. Low-pressure mercury lamps with quartz envelopes (which transmit this kind of radiation, in contrast to glass envelopes) are therefore used as sterilization lamps.

Fluorescent lamps are similar, but they have glass envelopes on their inner surfaces that have a fluorescent layer ("phosphor") that converts UV radiation to visible light, or, in lamps such as the Philips TL12 (which is described below), to UV radiation of longer wavelength than the original emission. More will be said about fluorescent lamps in the following sections.

One type of medium-pressure lamp that is important to UV-B photobiology is that which contains mercury together with iodides of iron and cobalt. Typically such a metal halogen UV-B lamp contains about 1 mg cobalt and 2 mg iron per 100 mg mercury. The ratings for the Philips HPA series of lamps run from 400 to 1,920 W, and the lamps are much more compact than fluorescent lamps. Though most of the emission is in the UV-A and visible, UV-B radiance is higher than for fluorescent lamps. A spectrum is shown in Fig. 1.

1.2. Properties of Ultraviolet Fluorescent Lamps

The radiation spectrum from a fluorescent lamp depends on the lamp's fluorescent coating. The type of lamp used in UV-B experiments is the same as that used in solaria. A type manufactured in Europe is the Philips TL12, which comes in 20- and 40-W versions.

The bare lamps emit too much short-wave radiation for a realistic simulation of daylight, and are therefore used with filters. The type of filter used by most researchers is cellulose diacetate.

One property of fluorescent lamps that may cause trouble, especially in field experiments, is that the radiant flux is temperature dependent. Most of the temperature dependence is due to the temperature effect on mercury pressure (temperature also affects the fluorescence yield of the coating). The

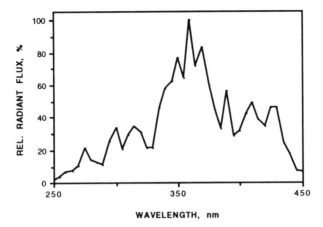

Figure 1. Relative spectral radiant flux from Philips metal halogen UV lamp HPA 1000 S.

pressure of the mercury vapor is determined by the temperature of the coolest spot of the envelope. The optimum pressure is about 0.8 Pa, which corresponds to a cool-spot temperature of 40°C. For lamps in still air that are not enclosed, this corresponds to an ambient temperature of approximately 25°C under normal operating conditions. The radiant flux decreases quickly with lower temperatures, and somewhat more slowly with higher temperatures (Fig. 2).

1.3. Medium- and High-Pressure Gas Discharge Lamps

If the vapor pressure in a mercury lamp is increased, more and more of the emission at 253.7 nm is reabsorbed until very little of this radiation escapes from the vapor. Instead, spectral lines of longer wavelength emerge (medium-pressure mercury lamps). At even higher pressures, the spectral lines are broadened to bands (high-pressure mercury lamps), and finally a continuous spectrum results (super high pressure mercury lamps).

Deuterium (heavy hydrogen) lamps of medium pressure are used as light sources for spectrophotometry in the ultraviolet region and are also potential radiation sources for UV photobiology.

Lamps containing xenon under high pressure are used to obtain a strong, continuous emission from 300 nm up into the infrared. Depending on the composition of the envelope, more or less of the shorter-wavelength ultraviolet also escapes. Xenon lamps come in a great variety of types. We use lamps running on about 24 volts direct current (but ignited, with about 20,000 volts = danger!!) and wattages (rated powers) from 150 to 900 W. Xenon

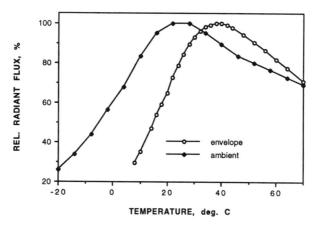

Figure 2. Relative radiant flux from Philips TL12 fluorescent lamps for different tube-wall and ambient temperatures.

lamps of higher wattage are often water-cooled. Electrodeless xenon lamps are also manufactured. They are powered by microwave radiation.

Because xenon lamps with UV-transparent envelopes cause conversion of oxygen to ozone, such lamps must be provided with exhausts to prevent poisoning of plants and people. The same holds for high-powered mercury lamps with UV-transparent envelopes.

1.4. Solar Simulators

1.4.1. Solar Simulators with High-Pressure Xenon Lamps

High-pressure xenon lamps have intense light emission spectrally resembling sunlight. The spectral similarity to sunlight can be further improved using suitable filters. The main drawbacks of xenon lamps are the cost—both in investment (expensive power supplies are required) and in running time (lifetime of a lamp is ca. 1,000 h)—risk of explosion, generation of ozone (except the ozone-free lamps with an envelope absorbing the shortest spectral components), and the instability and inhomogeneous brightness distribution of the arc.

Due to the inhomogeneity of the arc, a "light scrambler" or "light integrator" has to be used to obtain a uniform light field. This is an optical system onto which the arc is focused and which serves as a secondary light source.

Solar simulators are produced by several manufacturers. To describe the system, we will use a solar simulator from Oriel as an example. The optical system is shown in Fig. 3. The air-cooled 300- or 1,000-W xenon lamp is placed at one focus of an ellipsoidal mirror. The radiation is focused onto the light scrambler at the second focus (displaced by a flat mirror). After deflection by another flat mirror the beam is collimated by an ultraviolet transparent lens. All three mirrors are surface metallized.

The spectral composition of the radiation can be manipulated in two ways:

1. A filter between the light scrambler and the final lens can make the spectrum similar to either extraterrestrial solar radiation or direct sunlight at ground level or global radiation at ground level (Fig. 4). Direct sunlight is the radiation coming in from the direction of the sun, while global radiation is the combined daylight from the sun, sky, and ground. We shall return later to the differences in spectral composition between these components.

2. For the first flat mirror in Fig. 3, one can choose a dichroic mirror, which reflects only the ultraviolet part of the xenon spectrum and transmits the long-wavelength components (Fig. 5). Using a dichroic mirror it is possible, on the one hand, to irradiate a sample with very strong UVR without overheating it. The option with a totally reflecting mirror, on the other hand, provides the best simulation of total daylight.

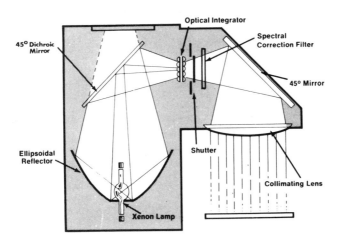

Figure 3. Optical diagram of Oriel solar simulator.

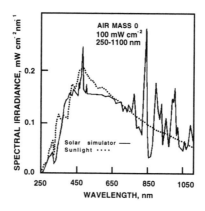

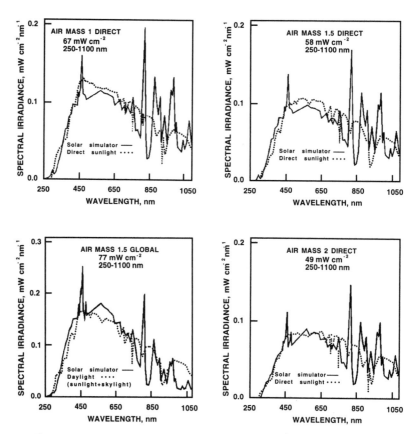

Figure 4. Spectra for Oriel solar simulator, with various spectral correction filters.

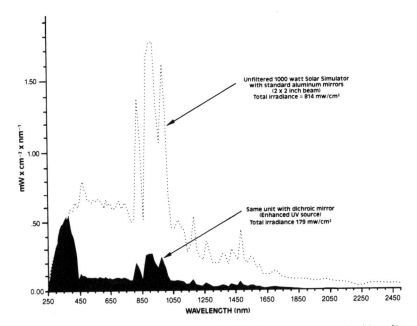

Figure 5. Spectrum for Oriel solar simulator with dichroic mirror, compared with unfiltered spectrum and aluminum mirror.

A different kind of solar simulator is manufactured by Solar Light Company, USA. This solar simulator allows the irradiation of six small areas (e.g., on skin or leaves) via fiber optics.

1.4.2. Solar Simulators with Metal Halide Lamps

Solar simulators with metal halide lamps, aluminum reflectors, and filters are manufactured by Dr. Hönle GmbH, Germany. They can be supplied with filters to mimic either extraterrestrial sunlight or daylight with air mass 1.5, i.e., corresponding to a zenith angle θ such that $\cos \theta = 1.5$. The fit to the natural daylight spectrum is at least as good as that for filtered xenon lamps, and the metal halide lamps are more economical. During the life of the lamp (2,500 h) the intensity decreases by about 20%, slightly more in the UV-B region and less in the visible.

2. Simulation of Ozone Depletion

2.1. How to Determine Suitable UV Irradiations for the Simulation of a Certain Degree of Ozone Depletion

Because different kinds of UV radiation have different quantitative and qualitative effects, it is essential to try to mimic as closely as possible either the total daylight spectrum under normal and ozone-depleted conditions, or the *increase* in UV caused by ozone depletion. Since this match cannot be made exact, one has to do the quantitative comparison of radiation conditions using a *weighting function* (approximating the action spectrum for the radiation effect on whatever is being assessed), as described below.

In laboratory and growth-room experiments, where all the radiation supplied is artificial, one has to simulate the radiation both under normal ozone conditions (control) and under depleted conditions. The best light source for this simulation is the xenon arc lamp combined with short-wave cutoff filters, as described in the previous section.

In field trials, one does not have to mimic the natural UV daylight spectrum, only the change brought about by ozone depletion.

Solarium tubes such as a Philips TL12 (20 or 40 W), in combination with cellulose diacetate filters, preirradiated to decrease the rate of transmission change during the experiment, give a spectrum that deviates considerably from the daylight spectrum, but it matches the spectrum of the *increase* caused by ozone depletion rather well (see Figs. 6 and 7). Figure 8, in which the ordinate is logarithmic, shows clearly, however, that one layer of cellulose diacetate (0.13 mm) is not sufficient to match the short-wavelength tail at higher latitudes.

The next step is to measure the spectral irradiance from the lamp–filter combination, using a spectroradiometer.

Then the spectral irradiance values are entered, nanometer by nanometer, into a computer program for convolution with the weighting function. This is a very simple program with the following program list:

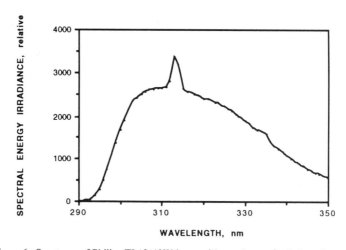

Figure 6. Spectrum of Philips TL12/40W lamp with one layer of cellulose diacetate.

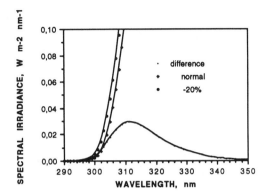

Figure 7. The calculated daylight spectral UV-B irradiance in Lund at noon on June 15, with or without a 20% ozone depletion (cloudless sky), and the increase in UV resulting from the ozone depletion.

```
FOR L=280 TO 360
PRINT L
A=
INPUT B
S1=S1+A*B
LPRINT L,S1:NEXT L
END
```

Before using this program one must fill in the particular formula for A that is to be used (see below). What the program does is to multiply the irradiance values by the mathematical function which approximates the weighting function, and successively adds the products together. This is called "integrated biologically effective irradiance." When the accumulated sum has reached a constant value, the execution can be terminated. The spectroradiometer prints out values in W cm^{-2} nm^{-1}, and by multiplying by 10,000

COMPARISON OF DAYLIGHT CHANGE AND SIMULATION

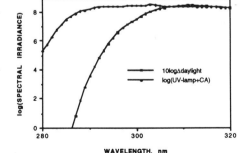

Figure 8. Comparison of the spectrum of a cellulose-diacetate-filtered Philips TL12/40W lamp (upper curve) with the difference between 20% ozone-depleted and normal daylight in Lund at noon on June 15.

we obtain values in W m^{-2} nm^{-1}. After adding the values over the spectrum, the unit will then be W m^{-2}; we call this value X for the present purpose.

To make results obtained with different weighting functions comparable, the functions should always be normalized to unity at 300 nm. Some examples of weighting functions that could be used for A in the program are (L stands for wavelength in nm):

$$A = 2.618*[1 - (L/313.3)^2]* \text{EXP}[-(L - 300)/31.08] \qquad (1)$$

which is an analytic expression for Caldwell's (1971) generalized action spectrum proposed by Green et al. (1974).

$$A = \text{EXP}(-(((265 - L)/21)^2)) \qquad (2)$$

which is an analytic expression for the same spectrum proposed by Thimijan et al. (1978).

$$A = \text{EXP}(88.357 - 0.49*L + 6.23E - 4*L^2) \qquad (3)$$

which is a fit by Rundel (1983) to data obtained by Caldwell for inhibition of photosynthesis in *Rumex patientia;*

$$A = 1 \quad \text{for } L < 292 \text{ nm} \qquad \text{and}$$

$$A = \text{EXP}(1.086 - 0.02028*L) \quad \text{for } L > 292 \text{ nm} \quad (4)$$

which is another fit by Rundel (1983) to data obtained by Caldwell for inhibition of photosynthesis in *Rumex patientia.* A number of other action spectra are listed by Madronich et al. (1991).

The next step is to compute the expected normal daylight spectrum for each hour during the day, and the corresponding values for the ozone depletion being aimed at. For this step one uses a version of the "Daylight" program of Björn and Murphy (1985) that integrates the ultraviolet irradiance over the whole day (see Björn, 1989, and Appendix A for a detailed description), and the same program for multiplying the values for each wavelength and each hour by the same mathematical weighting function as in the previous step. All the values for ozone-depleted atmosphere (one sum) and normal atmosphere (another sum) are then added. Note that one has to use the same form of the mathematical function in both cases, i.e., either

$$\text{EXP}(-(((265 - L)/21)^2)) \qquad (5)$$

(having the value 1 at 265 nm) or

$$\text{EXP}(-(((265 - L)/21)^2))/\text{EXP}(-(((265 - 300)/21)^2)) \qquad (6)$$

(having the value 1 at 300 nm).

The difference is then taken between the sum for ozone-depleted atmosphere and normal atmosphere, and this gives the increase in daily W h m^{-2} of weighted UV-B. We call the value so obtained Y.

The desired daily irradiation time is now obtained as the ratio Y/X. It is probably convenient to convert the value to minutes, or hours and minutes, so timers can be set.

If the values of X and Y are reported separately, it is customary to express Y in the unit kJ m^{-2} (and X could be expressed in kJ m^{-2} h^{-1}) by multiplying by 3.6 (since there are 3.6 ks in 1 h). Make sure not to multiply just one of them by 3.6 before taking the ratio Y/X.

When choosing the time of day, the center of the irradiation period should coincide with *solar* noon. For instance, solar time in Lund runs about 7 min behind the official time.

Usually many ultraviolet lamps are used in a field experiment. Switching them all on and off at the same time would provide a rather unrealistic simulation. A somewhat better and only marginally more expensive solution is to switch every second tube on and off earlier than the rest. Suppose the computation described above results in a scheme where the tubes would be switched on for 2 times x h per day. Instead of switching all on x h before noon and off x h after noon, it is better to switch every second tube on 1.5 times x h before noon and off 0.5 times x h after noon. The other tubes would then be switched on 0.5 times x h before noon and off 1.5 h after noon. This gives the same total fluence as the first scheme, but is more graded, as real daylight UVR is.

2.2. Continuous Temporal Modulation of Lamps for Simulation of Ozone Depletion

A further improvement was introduced by Caldwell *et al.* (1983) by letting natural daylight regulate artificial UVR. In this way the risk for shock effects that could result from the sudden stepping up of UVR is eliminated. More importantly, the ultraviolet addition is decreased when clouds decrease natural daylight. This is important, since plants become more sensitive when visible light (photosynthetically active radiation) is decreased.

However, the irradiation system of Caldwell *et al.* (1983) also has some drawbacks. In that system, visible light is monitored, and this does not change with time in the same way as UV-B radiation. The UV-B is more concentrated at the middle of the day, and the directional distribution is not the same as for visible light. If, for instance, the sun is obscured by a small cloud and the rest of the sky is cloudless, the relative decrease of the visible light is larger than that of the UV-B radiation. Therefore, with the system of Caldwell *et al.* (1983), no accurate simulation of ozone-depleted daylight is possible. A further difficulty is that the sensors for the visible daylight and those for the artificial UV-B in the feedback loop are different and thus have different temperature coefficients. The system therefore should be calibrated differently for various ambient temperatures, which, in practice, is difficult to do.

A system called YMT-UIMS (Fig. 9), designed by Zhilong Yu and Sui-xiang Mai in A. Teramura's group at the University of Maryland, circumvents all these difficulties by monitoring the daylight UV-B and using it for regulation of the artificial UV-B (YMT-UIMS stands for Yu–Mai–Teramura Ultraviolet Irradiation Modulation System). Artificial UV-B is added at a constant fraction of the natural UV-B throughout the day. Although UV-B irradiance varies by orders of magnitude during the day, the fractional increase for a certain ozone depletion remains almost the same. The values in Table 1 were computed for May 1 in Abisko (northern Sweden, latitude 68.3°N, altitude 360 m corresponding to 955 mb, snow-covered ground, no clouds, no aerosol, relative humidity (RH) = 0.5) with natural ozone or 10% depletion.

The YMT-UIMS consists of three major components: the UV-B detectors (one for ambient UV-B and two for ambient plus supplemental UV-B), the

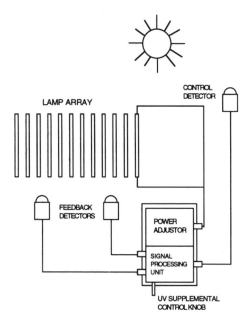

Figure 9. Block diagram of the YMT-6 Ultraviolet Irradiation Modulation System.

Table 1. Comparison of Weighted Irradiance under
Normal Conditions and with 10% Ozone Depletion
from Morning to Noon (Abisko, 68.3° N, May 1)

Hour (A.M.)	Caldwell weighted irradiance, W m^{-2}		Depleted/ normal
	Normal ozone	10% depleted	
4	9.5595e-5	1.1279e-4	1.180
5	2.3165e-4	2.7372e-4	1.182
6	5.97036e-4	7.0492e-4	1.181
7	1.31627e-3	1.5502e-3	1.178
8	2.41133e-3	2.83139e-3	1.174
9	3.7425e-3	4.38266e-3	1.171
10	5.0369e-3	5.8863e-3	1.169
11	5.9799e-3	6.9793e-3	1.167
12 noon	6.3251e-3	7.3791e-3	1.167

control circuitry, and the lamp bank. The system is simple to operate, economical to maintain, has a high power efficiency (60–80%) and a dynamic range of 1:300, is temperature insensitive, and allows a longer effective lifetime of the UV-B lamps than other systems.

The desired supplemental UV-B can be easily adjusted. Separate adjustment of the individual lamps is possible, for greater uniformity of the radiation field than with other systems.

All three detectors in the system are of the same type. The spectral response peaks at 290 nm and is down to 1% of maximum at 260 and 330 nm (Fig. 10). The response to radiation of wavelength exceeding 360 nm is four orders of magnitude below the maximum sensitivity.

2.3. Tests of Weighting Functions

The purpose of this section is to show how action spectra can be used as weighting functions in the study of the biological impact of UVR.

Traditionally, action spectra have been determined by studying how organisms react to monochromatic light. When it comes to effects of UVR on relatively large whole organisms, such as seed plants, it is difficult to determine action spectra in this way. The technical difficulties in irradiating large organisms with monochromatic UVR of sufficient purity and irradiance (or fluence rate) are too great.

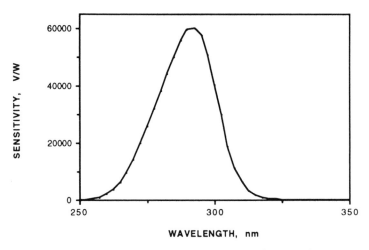

Figure 10. Spectral sensitivity for the sensor used in the YMT-6.

Instead, one has to resort to an "intelligent guess" about the shape of the action spectrum, and then test it using polychromatic radiation, which is also more similar to real daylight. Rundel (1983) has discussed various weighting functions and how to compare them to results of biological experiments with cutoff filters.

We shall demonstrate this principle by using two action spectra which have been employed by "ultraviolet photobiologists," Caldwell's "generalized plant action spectrum" (Fig. 11) and the DNA spectrum (Fig. 12).

The curves labeled "Action" in Figs. 11 and 12 were obtained by multiplying the daylight spectrum, point by point, by the plant action spectrum. The daylight spectrum is expressed as spectral irradiance (unit: $W\ m^{-2}\ nm^{-1}$). The "integral of action" was obtained by summing the values for "Action" for all wavelengths. Note that the value of this "integral of action" depends on the absolute scale of the action spectrum, which is arbitrarily chosen. We have chosen here to approximate the action spectrum with the function

$$EXP(-((265 - L)/21)^2) \qquad (7)$$

which gives it the maximum value 1 at 265 nm and the value 0.06217653 at 300 nm (the wavelength should be expressed in nm). In the literature, spectra are often (but not always) "normalized" to unity at 300 nm. To normalize our spectrum in this way we would instead use the function

$$EXP(-((265 - L)/21)^2)/EXP(-((265 - 300)/21)^2)$$
$$= EXP(-((265 - L)/21)^2)/0.06217653 \qquad (8)$$

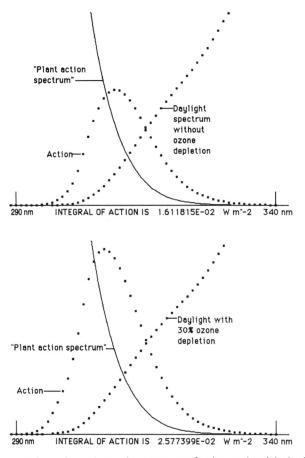

Figure 11. Computation, using a plant action spectrum, of action or ultraviolet load, in Lund at noon on June 15, with 0 and 30% ozone depletion.

We now repeat the computation with a 30% ozone depletion instead of a normal atmosphere. We then obtain the result shown in Fig. 11.

Note that the ratio of the integrals of action in the two cases is 0.02577399/ 0.01611815 = 1.60. This ratio is not dependent on the way we normalize the action spectrum, only on its shape, and on the daylight spectra. The steeper the slope of the action spectrum, the greater the ratio. For a discussion of the so-called radiation amplification factors, see Chapter 1 (this volume).

If the action spectrum is the correct one for a process that we want to study, say, inhibition of the growth rate of a plant, then we would expect to obtain the same result if we expose the plant every day to, e.g., 2 h of radiation

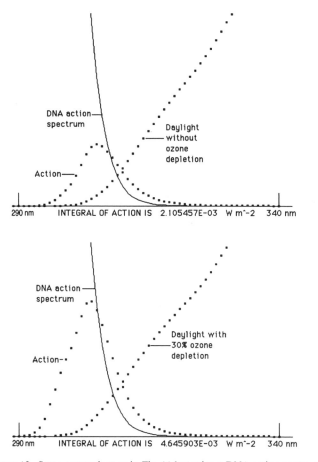

Figure 12. Same comparison as in Fig. 11 but using a DNA action spectrum.

corresponding to the daylight calculated above for a 30% ozone depletion as when we expose the plant every day for 2 times 1.60 = 3.2 h of radiation corresponding to daylight without ozone depletion. Note that this example is simplified, because we assume that the radiation remains constant during the irradiation period, which is not the case with real daylight.

Let us now compare this result to what we obtain with another weighting function, the DNA action spectrum, instead of the plant action spectrum (Fig. 12).

In this case we have the same ozone depletion, but the ratio between the two integrals of action is not 1.60, but 0.004645903/.002105457 = 2.21. The ratio is higher than before, because the DNA action spectrum is steeper than the plant action spectrum.

If our real experiment turns out as we have described, i.e., that 2 h of "ozone-depleted" daylight gives the same effect as 3.2 h of normal daylight, then we conclude that the DNA action spectrum is not appropriate for our case. The plant action spectrum, on the other hand, remains a possibility, but just a possibility, as long as we have only the effects of two different irradiations to compare. If we find, after having treated plants with many different kinds of irradiation, that the results still follow what we would expect for the plant action spectrum, then we can conclude that this spectrum, indeed, is the appropriate one.

Let us turn now to a real series of experiments, a series carried out by Wan Jiangxin in our laboratory in Lund with one of the smallest of all seed-producing plants, *Wolffiella hyalina,* as the experimental material. Petri dishes with quartz lids were inoculated with five plants each and allowed to grow for 14 days, after which time dry and fresh weights of the plants in each dish were determined. During growth the dishes were irradiated by fluorescent UV-B lamps through various Schott cutoff filters, either for 4 h or for 8 h daily. The spectral irradiance was determined for each treatment using a spectroradiometer, and the plant weighted irradiance determined. For the results presented in Fig. 13, the plant action spectrum was normalized to 1 at 300

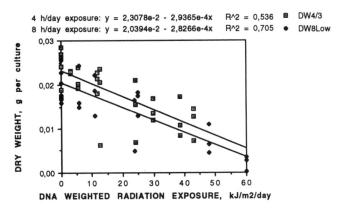

Figure 13. Comparison of regression lines of final dry weight versus weighted UV radiation exposure in experiments in which cultures of the duckweed *Wolffiella hyalina* were grown for 14 days under UV-emitting fluorescent tubes. The exposure was varied in each series by cutting off various portions of the short-wavelength end of the spectrum using glass filters. In one series the cultures were exposed to UV radiation for 4 h per day, in the other for 8 h per day. Within limits of variation, the regression lines should be the same if the correct weighting function has been found (in this case the DNA spectrum was tried). Unpublished experiment by Jiangxin Wan at the Section of Plant Physiology, Lund University.

nm as described above. The dry weight was plotted versus the product (irradiance) × (daily irradiation time). Most people would call this product daily fluence; this term is not correct here since we used a "cosine-corrected" spectroradiometer.

We see from Fig. 13 that we have a lot of scatter in the data. Thus we conclude that almost the same regression lines will be obtained whether we use 4-h or 8-h daily irradiations, and thus the spectrum used is reasonable, at least until we can find a way to decrease experimental uncertainty.

3. Fluorescent Tube Arrays for Homogeneous Light and UV Fields

3.1. Introduction

For many photobiological experiments homogeneous light fields, i.e., areas or spaces with the same irradiance or fluence rate at different points, are required. Tubular fluorescent lamps are often used and are suitable for such experiments provided that high irradiance (or fluence rate) and monochromatic light are not required. The present section is aimed at aiding the experimenter in setting up an optimal geometric arrangement, where the experimental area is not enclosed within highly reflecting walls. The study was initiated to determine the optimal arrangement, in field experiments, of ultraviolet-emitting lamps for simulation of a plant environment with a depleted stratospheric ozone layer. In the first trials it was found that the conventional arrangement with parallel, equidistant lamps does not make optimal use of the lamps.

The experimenter should first decide on the kind of "even illumination" desired. A design optimized for equal irradiance in different parts of the experimental area is, in most cases, not optimized for equal fluence rate, and vice versa. Irradiance is light per unit time and surface area, falling on a surface of specified direction, while fluence rate is light per unit time passing through a sphere of unit cross-sectional area. Irradiance is equal to fluence rate for light perpendicular to the reference plane, but less than fluence rate for all other situations. Irradiance is a relevant quantification of light incident on flat objects with an easily definable reference plane (single leaves, small areas of skin, etc.). Fluence rate is a better quantification for three-dimensional objects and randomly oriented flat objects.

3.2. Theoretical Aspects

The computer program to be described below is based on the following:

1. Fluence rate and irradiance decrease with the square of the distance from a point source.

2. The surface of a lamp is uniformly coated with identical point sources.
3. The brightness of a lamp surface element is the same in all directions.

From this it follows that the irradiance and fluence rate in a certain direction contributed by a slice of a tubular lamp is proportional to the sine of the angle α between the direction and the tube axis.

In Fig. 14 we see a single fluorescent tube suspended at distance z from the plane surface to be irradiated. On the surface we superimpose a coordinate system with its origin opposite the center of the tube, with its y-axis parallel and its x-axis perpendicular to the tube. The tube is of length $2y0$. We consider the contribution to irradiance at point $P = (x1, y1)$ of an infinitesimal length dy of the tube at position y. This contribution is

$$dl = \frac{A \cdot \sin\alpha \cdot \cos\beta \cdot dy}{x1^2 + (y1 - y)^2 + z^2} \tag{9}$$

where A is a constant, $x1^2 + (y1 - y)^2 + z^2$ the square of the distance between P and the tube segment, and β the angle between the light direction and the normal to the plane (incidence angle).

Furthermore,

$$\sin\alpha = \frac{(x1^2 + z^2)^{1/2}}{(x1^2 + (y - y1)^2 + z^2)^{1/2}} \quad \text{and}$$

$$\cos\beta = \frac{z}{(x1^2 + (y - y1)^2 + z^2)^{1/2}} \tag{10}$$

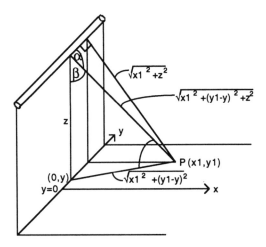

Figure 14. Derivation of expressions for irradiance and fluence rate below a fluorescent tube.

The corresponding expression for fluence rate is the same as for irradiance, except that the factor cos β is left out.

In the program the integration is replaced by a summation over small intervals. The interval size is a compromise between accuracy and computation time. For the examples below, each value of irradiance or fluence rate is a sum of 79 values per tube. A program listing is found in Appendix B.

3.3. Examples

The following are examples of different tube arrangements for irradiation of a square 1.2 m × 1.2 m by a number of 1.2-m-long lamps (Fig. 15).

1. Six tubes are uniformly spaced parallel to each other and with the outermost tubes above the sides of the square. All tubes are 0.45 m above the square to be irradiated. The distances between the tubes are thus 0.24 m. With this arrangement the irradiance is highly nonuniform, with the irradiance at the corners of the square only 42% of the value in the center of the square.

2. Six tubes are arranged parallel to each other as in the first example, but the distance between them is not constant but "cosine distributed": The

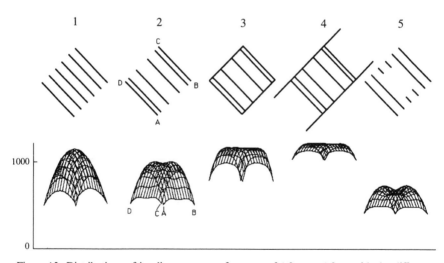

Figure 15. Distributions of irradiance on a surface area of 1.2 m × 1.2 m with the different arrangements of fluorescent lamps shown on top. The distributions are displayed as pseudo three-dimensional diagrams projected "corner on" (see the second example in text for explanation) with the irradiance axis upwards and $x1$ directions (i.e., perpendicular to the "transverse" tubes, A to B and D to C) as well as $y1$ directions (i.e., parallel to the "transverse" tubes, A to D and B to C) horizontally. The bottom line indicates zero irradiance. Irradiance is plotted as the percentage of the irradiance from one tube, on the center normal, at 1-m distance.

Figure 16. Distribution of irradiance compared to distribution of fluence rate on a 1.2-m × 1.2-m surface area at two distances from four fluorescent lamps arranged along the sides of a 1.2-m × 1.2-m square. The bottom line indicates zero irradiance and zero fluence rate, and irradiance as well as fluence rate are plotted in a vertical direction, as percentage of the irradiance from one tube, on the center normal, at 1-m distance.

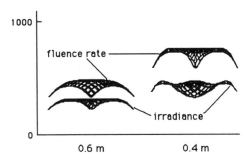

distances vary as if the tubes had been equally spaced on half a cylinder and then projected on a plane (see line 65 in the program listing).

3. Six tubes are arranged as in example 2, 0.45 m above the square to be illuminated. In addition, there are two tubes perpendicular to the first ones, below the ends of the first six tubes. The latter tubes are 0.32 m above two of the sides of the square.

4. As in example 3, except that there are four "perpendicular" tubes instead of two. They are arranged end to end above two of the sides of the square, so that they protrude half a tube length outside the square at each corner.

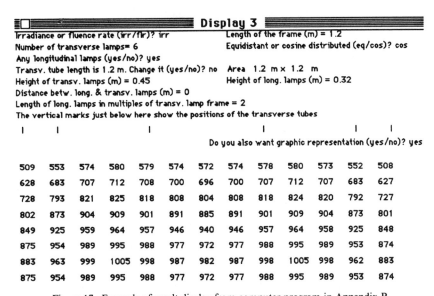

Figure 17. Example of result display from computer program in Appendix B.

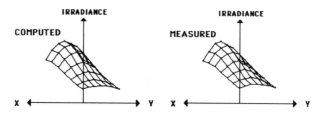

Figure 18. Irradiance distribution below a single fluorescent lamp 1.2 m long, suspended 1 m above the surface to be irradiated. In the x direction (perpendicular to the lamp) the data grid (having 0.2-m by 0.2-m squares) extends from a line just below the lamp to a line at 1.2 m distance. In the y direction (parallel to the lamp) the grid extends the length of the lamp.

5. This arrangement is the same as example 1, except that 70% of the central two tubes is shielded to make the irradiance distribution more uniform.

One final example will demonstrate the fact that irradiance fields and fluence rate fields must be optimized in different ways. In this example, four tubes are arranged at the same height along the four sides of a square. The irradiance field (Fig. 16) is optimally flat at a distance between the tubes and the surface to be irradiated of about 0.6 m (half the tube length), while the optimal fluence rate field is obtained at a tube-to-surface distance of about 0.4 m (one third of the tube length).

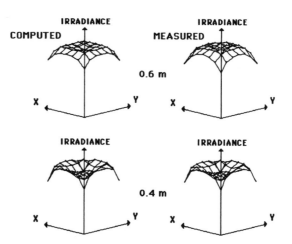

Figure 19. Comparison between computed and measured irradiance distributions below four fluorescent lamps forming a square 0.6 m (top) or 0.4 m (bottom) from the surface to be irradiated. The scale of the top diagram is not the same as that of the bottom diagram, nor are any of the scales the same as for Fig. 16. In each pair of diagrams, however, the scale for the computed and the measured irradiance is the same. The pseudo three-dimensional diagrams have been slightly tilted toward the observer in this figure.

All these examples are readily calculable using the program, which prompts the operator with questions and plots graphical explanations for the terms used. Although we have used graphical presentations here to conserve space, the program prints the results in the form of an array of numbers or, alternatively, as an array of ovals with area and appearance varying with irradiance or fluence rate. An example of a printout is shown in Fig. 17.

3.4. Experimental Verification

To verify the principle of the computations, a frame with fluorescent tubes emitting visible light was used. A LiCor LI-188 Integrating quantum/radiometer/photometer was used together with an LI-190 SE cosine-corrected sensor for checking relative irradiance. Comparisons between computed and measured distributions are made in Figs. 18 and 19. We had no means of checking fluence rate distribution.

ACKNOWLEDGMENTS. This chapter is part of a research program supported by the Swedish Environmental Protection Board (SNV). Dr. J. F. Bornman kindly reviewed the syntax.

Appendix A.

Below is a program listing for a program for calculating the ultraviolet spectral irradiance and the biological load under various daylight conditions. This version of the program computes the physical and biologically weighted spectral irradiance integrated over a day (W h m^{-2}) and can be used for horizontal or tilted planes. Another version of the program can be used for spectral irradiance (W m^{-2}) at a certain time of day. Both versions are available on disk from the author in exchange for an empty disk.

```
      REM: THIS IS THE August 90 VERSION OF THE PROGRAM "DAYLIGHT"
      REM: WRITTEN BY L.O. BJÖRN AND T.M. MURPHY FOR COMPUTATION OF DAYLIGHT
UV
      REM: IT IS DOCUMENTED IN PHYSIOL. VEG. 23:555-561 (1985) AND
"RADIATION MEASUREMENT IN PHOTOBIOLOGY" (ED. B.M. DIFFEY), ACAD. PRESS 1989
      REM: TO RUN THE PROGRAM YOU NEED A MACINTOSH COMPUTER AND A MICROSOFT
BINARY BASIC INTERPRETER FOR IT (VERSION 2.00 OR LATER)
      REM: THE GRAPHICS CAN EASILY BE GREATLY IMPROVED FOR MACINTOSH, BUT WE
WISHED TO KEEP A FORMAT WHICH DOES NOT REQUIRE EXTENSIVE MODIF. FOR OTHER
COMPUTER
```

```
DIM r%(4)
SH=SYSTEM(6)        'Get screen height
SW=SYSTEM(5)        'Get screen width
H%=230
w%=440
WINDOW 10,,((SW-w%)/2,(SH-H%)/3)-((SW-w%)/2+w%,(SH-H%)/3+H%),2
title$="Daylight ultraviolet radiation, tilt numeric version"
TEXTFONT 0
MOVETO (WINDOW(2)-WIDTH(title$))\2,20      'center title$
PRINT title$
TEXTFONT 3
SetRect r%(1),10,30,w%-5,H%-15
M$="        This program was written by L.O. Björn and T.M. Murphy for
computation of the amount of physically measurable and biologically active
ultraviolet radiation in daylight. The principles are described in Physiol.
Vég. 23:555-561 (1985) and"
M$=M$+" the books Radiation Measurement in Photobiology (1989,
B.L.Diffey, ed., ISBN 0-12-215840-7), and Photobiology (L.O. Björn)."
M$=M$+" The program may be duplicated but not sold for profit. If you
use it for a publication, acknowledgement should be given."
M$=M$+" This version can be used for radiation falling on a tilted
plane, and integrates radiation over a day from 5 to 19 at half-hourly
intervals."
TextBox M$,r%(1),0
BUTTON 1,1,"OK", (180,200)-(240,215),1
d%=DIALOG(0)
WHILE d%<>1 AND d%<>6 :d%=DIALOG(0):WEND
WINDOW CLOSE 10
1000 :PRINT "first step: CALCULATE HEIGHT OF SUN"
1003 PI=3.141592
DIM G(121):DIM AC(121):DIM GA(121)
PRINT"ENTER NORTH LATITUDE, MONTH, DATE"
INPUT LA, MO, DA:LPRINT LA,MO,DA
INPUT "TILT ANGLE";TT:LPRINT "TT="TT:TT=TT*PI/180
INPUT "AZIMUTH OF TILT";AZT:LPRINT "AZT="AZT:AZT=AZT*PI/180
LET DN=30.3*(MO-1)+DA
1050 :LET
ED=.398*SIN((DN-80)*2*PI/365+.0335*(SIN(DN*2*PI/365)-SIN(1.3771)))
LET DI=ATN(ED/SQR(1-ED*ED))
DIM O(9,18)
2070 :FOR I=1 TO 18
FOR J=1 TO 9
2090 : READ O(J,I)
NEXT J: NEXT I
2110 : DATA .24, .24, .26, .28, .29, .32, .33, .35, .36
2120 : DATA .24, .24, .26, .27, .29, .32, .34, .36, .36
2130 : DATA .24, .24, .25, .27, .30, .33, .36, .37, .37
2140 : DATA .23, .23, .25, .28, .32, .35, .38, .38, .37
2150 : DATA .23, .23, .26, .29, .33, .37, .39, .38, .37
2160 : DATA .24, .24, .25, .28, .31, .36, .38, .37, .36
2170 : DATA .24, .24, .25, .27, .29, .32, .35, .36, .36
2180 : DATA .24, .24, .25, .26, .28, .31, .33, .36, .36
2190 : DATA .23, .24, .25, .28, .31, .32, .33, .34, .35
2200 : DATA .23, .24, .25, .28, .31, .32, .33, .34, .35
2210 : DATA .24, .23, .24, .27, .29, .32, .33, .34, .35
2220 : DATA .24, .24, .24, .26, .28, .30, .32, .34, .35
2230 : DATA .25, .25, .25, .27, .28, .30, .32, .34, .34
2240 : DATA .26, .26, .27, .27, .28, .31, .32, .33, .34
2250 : DATA .26, .26, .27, .28, .29, .33, .34, .33, .34
```

```
2260 : DATA .25, .25, .27, .28, .32, .38, .36, .34, .34
2270 : DATA .25, .25, .27, .28, .32, .35, .36, .35, .35
2280 : DATA .25, .25, .27, .28, .30, .32, .33, .35, .35
2290 : PRINT "ENTER LONGITUDE (WEST LONGITUDE AS NEGATIVE NUMBER)"
INPUT LO
2310 : LET IA=INT((LA+5)/10)+1
2320 : LET IO=INT((LO+170)/20)+1
LET OO=O(IA,IO)
2340 : IF LA>44 THEN 2370
LET DM=90+(44-LA)*3.1
GOTO 2380
2370 : LET DM=90
2380 : LET AM=.07*(LA+10)/90
LET W3=OO+AM*COS((DN-DM)*2*PI/365)
PRINT "OZONE CONC IS" W3 "atm cm"
2500 : PRINT "third step: CORRECT FOR ENVIRONMENT"
PRINT "WHAT ENVIRONMENT TYPE?","1. RURAL, 2. URBAN, 3. MARITIME"
2610 : INPUT Z1
IF Z1=3 THEN 2680
IF Z1=2 THEN 2660
2640 : KT=.255:K=1.962:Q=.345:L1=.122:L0=439:AZ=.069:BE=1.31:NU=5
2650 : GOTO 2690
2660 : KT=.288:K=2.758:Q=.471:L1=.0827:L0=510:AZ=.363:BE=1.59:NU=.9
2670 : GOTO 2690
2680 : KT=.106:K=3.393:Q=.435:L1=1.049:L0=734:AZ=.032:BE=2.44:NU=5
2690 : PRINT "WHAT GROUND COVER?":PRINT"1. PINE FOREST          ","2.
OPEN OCEAN A"
PRINT "3. OPEN OCEAN B        ","4. GREEN FARMLAND"
PRINT "5. BROWN FARMLAND","6. DESERT SAND"
PRINT "7. BLACK LAVA          ","8. GYPSUM SAND A"
PRINT "9. GYPSUM SAND B       ","10. SNOW COVER"
2720 : INPUT Z2:CLS
2730 : IF Z2=10 THEN 2910
IF Z2=9 THEN 2900
IF Z2=8 THEN 2890
IF Z2=7 THEN 2880
IF Z2=6 THEN 2870
IF Z2=5 THEN 2860
IF Z2=4 THEN 2850
IF Z2=3 THEN 2840
IF Z2=2 THEN 2830
2820 : A0=.0147:DE=.05308:BA=.9181: GOTO 2920
2830 : A0=.0653:DE=.07922:BA=.6636: GOTO 2920
2840 : A0=.0511:DE=.05511:BA=.6013: GOTO 2920
2850 : A0=.0441:DE=-.3948:BA=25.88: GOTO 2920
2860 : A0=.0417:DE=-.2449:BA=24.95: GOTO 2920
2870 : A0=.0387:DE=.08812:BA=2.046: GOTO 2920
2880 : A0=.0186:DE=.03422:BA=1.302: GOTO 2920
2890 : A0=.195:DE=.02156:BA=1.956: GOTO 2920
2900 : A0=.194:DE=.03028:BA=1.764: GOTO 2920
2910 : A0=.289:DE=.01409:BA=.472
2920 : PRINT "WHAT AIR PRESSURE? (MILLIBARS)"
INPUT P
PRINT "WHAT RELATIVE HUMIDITY? (RANGE 0.0-1.0)"
2950 : INPUT RH
PRINT "WHAT AEROSOL LEVEL?"
2970 : INPUT W2
3001 : READ G1, G2, G3, G4, A1, A2, A3, A4
3002 : READ B1, B2, B3, B4, A, KU, T
```

```
3003 : READ MA, MB, PA, PB, QA, QB, VA, VB
3005 : DATA .5346, .6077, 1.0, 0, .8041, 1.437, .2864, 2.662
3006 : DATA .4424, .1, .2797, 3.7, 84.37, .6776, .0266
3007 : DATA 1.389, .5626, 1.12, .878, .8244, .8404, .4166, .1728
FOR KL=5.05 TO 19.05 STEP .5
GI=0:GA=0
M=ED*SIN(LA*PI/180)+COS(DI)*COS(LA*PI/180)*COS((KL-12)*PI/12)
CZ=ED*SIN(LA*PI/180)+COS(DI)*COS(LA*PI/180)*COS((KL-12)*PI/12)
COSAZ=-(ED*COS(LA*PI/180)-SQR(1-ED*ED)*SIN(LA*PI/180)*COS((KL-12)*PI/1
2))/SQR(1-CZ*CZ)
AZ=-ATN(COSAZ/SQR(1-COSAZ*COSAZ))+PI/2
COSTH=CZ*COS(TT)+SQR(1-CZ*CZ)*SIN(TT)*COS(AZ-AZT)
3010 : FOR I=0 TO 40
3025 : L=280+I
H=1-.738*EXP(-(L-279.5)^2/(2*2.96^2))
3040 : H=H-.485*EXP(-(L-286.1)^2/(2*1.57^2))
H=H-.243*EXP(-(L-300.4)^2/(2*1.8^2))
3060 : H=H+.192*EXP(-(L-333.2)^2/(2*4.26^2))
H=H-.167*EXP(-(L-358.5)^2/(2*2.01^2))
H=H+.097*EXP(-(L-368)^2/(2*2.43^2))
3100 : H=.582*((300/L)^5)*(EXP(9.102)-1)*H/(EXP(9.102*300/L)-1)
3110 T1=1.0456*(P/1013)*((300/L)^4)*EXP(.1462*(300/L)^2)
3140
K2=KT*(1+K*EXP(-RH^-3)/(1-RH)^Q*EXP(-(L-300)/(L0*(1+L*RH*(1-RH)^-Q))))
AL=AZ*((300/L)^NU)*EXP(-BE*EXP((RH-1)/.147))
T2=(1-AL)*W2*K2
T3=W3*10.89*1.0355/(.0355+EXP((L-300)/7.15))
T4=AL*K2*W2
3190 TX=.0018
TY=.0003
TZ=.0074
3220 MX=SQR((M*M+TX)/(1+TX))
MY=SQR((M*M+TY)/(1+TY))
MZ=SQR((M*M+TZ)/(1+TZ))
3250 F=1/(1+A*(T3+T4)^KU)
3260 F3=1/(1+A3*T3^QA*W3^VA)
F4=1/(1+A4*T4)
F6=1/(1+B3*T3^QB*W3^VB)
F8=1/(1+B4*T4)
FI=SQR((1+T)/(M*M+T))-1
ML=(A1*T1^MA*F3+A2*T2^PA*(1+A1*(T1^MA)*F3))*F4
3320 RA=(B1*(T1^MB)*F6+B2*(T2^PB))*F8
3330 IF (L-300)/DE>10 THEN 3360
3331 : IF (L-300)/DE<-10 THEN 3362
RG=A0*(1+BA)*EXP((L-300)/DE)/(EXP((L-300)/DE)+BA)
3350 GOTO 3370
3360 RG=A0*(1+BA)
3361 :GOTO 3370
3362 :RG=0
3370 :S=(F+(1-F)*EXP(-T3*FI))*EXP(-FI*(G1*T1+G2*T2))
3372 :DIR=EXP(-T1/MX-T2/MY-T3/MZ-T4/MY)
3373 : IF ATN(M/SQR(1-M*M))<0 THEN H=0
3380 G(I)=H*(COSTH*DIR+S*ML*EXP(-T1-T2-T3-T4))/(1-RA*RG)
4110 AC(I)=EXP(-(((265-I-280)/21)^2))
4130 GA(I)=G(I)*AC(I)
GI=GI+GA(I)
NEXT I
LPRINT "KL="KL,"GI=" GI" W m^-2"
NEXT KL
END
```

Appendix B.

Below is a program listing for a program for calculating the distribution of light from fluorescent tube array.

```
        REM This program is available on disk, in exchange for an empty
Macintosh disk, from L.O. Björn, Dept. of Plant Physiology, Univ. of Lund, Box
7007, S-220 07 Lund, Sweden
        REM It is dedicated to Professor Wolfgang Haupt on the occasion of his
70th birthday
        DIM x5(500):DIM y5(500):DIM c5(500):pi=3.1416:DIM r%(4)
        SH=SYSTEM(6)    'Get screen height:SW=SYSTEM(5)    'Get screen width
        h%=150:w%=400
        WINDOW 10,,((SW-w%)/2,(SH-h%)/3)-((SW-w%)/2+w%,(SH-h%)/3+h%),2
        title$="Fluorescent tube array":TEXTFONT 0
        MOVETO   (WINDOW(2)-WIDTH(title$))\2,20      'center title$
        PRINT title$:TEXTFONT 3
        SetRect r%(1),10,30,w%-5,h%-15
        m$="        This program computes the light distribution from an array
of"
        m$=m$+" fluorescent tubes, either as irradiance or fluence rate. The"
        m$=m$+" values are the percentages of irradiance obtained 1 m from the
centre"
        m$=m$+" of a single tube, on the normal to the tube axis, on a
surface"
        m$=m$+" parallel to the tube. © L.O. Björn 1990.":TextBox m$,r%(1),0
        BUTTON 1,1,"OK",  (170,130)-(230,145),1:d%=DIALOG(0)
        WHILE d%<>1 AND d%<>6 :d%=DIALOG(0):WEND:WINDOW CLOSE 10
        CALL TEXTSIZE(9):FOR mm=1 TO 6:CALL MOVETO(100+50*mm,50)
        CALL LINETO(100+50*mm,170):CALL LINETO(100+50*mm+5,170)
        CALL LINETO(100+50*mm+5,50): CALL LINETO(100+50*mm,50):NEXT mm
        CALL MOVETO(158,25):CALL LINETO(398,25):CALL LINETO(398,30)
        CALL LINETO(158,30):CALL LINETO(158,25):CALL MOVETO(158,190)
        CALL LINETO(398,190):CALL LINETO(398,195):CALL LINETO(158,195)
        CALL LINETO(158,190):CALL MOVETO(10,100):PRINT "transverse lamps  "
        CALL MOVETO(202,100):CALL LINETO(95,98):CALL LINETO(152,120)
        CALL MOVETO(10,30):PRINT "longitudinal lamps":CALL MOVETO(95,30)
        CALL LINETO(160,28):CALL MOVETO(278,195):CALL LINETO(278,190)
        CALL MOVETO(278,25):CALL LINETO(278,30):CALL MOVETO(10,10)
        PRINT "HERE IS A KEY TO THE VOCABULARY USED:":CALL MOVETO(10,175)
        PRINT "distance between transv.":CALL MOVETO(10,185)
        PRINT "and longitudinal lamps":CALL MOVETO(140,170)
        CALL LINETO(140,192):CALL MOVETO(120,180):CALL LINETO(140,180):CALL
MOVETO(152,210):CALL LINETO(402,210):CALL MOVETO(200,230)
        PRINT "length of frame":CALL MOVETO(270,227):CALL LINETO(310,210)
        CALL MOVETO(10,250):INPUT "TO CONTINUE PRESS RETURN!", goon$
        CLS:CALL MOVETO(2,7):INPUT "Irradiance or fluence rate (irr/flr)?",
LJ$
        CALL MOVETO(250,7):INPUT "Length of the frame (m) =",l
        CALL MOVETO(2,20):INPUT "Number of transverse lamps=",n
        CALL MOVETO(250,20)
        INPUT "Equidistant or cosine distributed (eq/cos)?", d$
        CALL MOVETO(2,33):INPUT "Any longitudinal lamps (yes/no)?", LO$
        CALL MOVETO(2,46):LET y0=.6
        INPUT "Transv. tube length is 1.2 m. Change it (yes/no)?",
```

```
changelength$
        IF changelength$="no" THEN 20
        CALL MOVETO(2,46)
        INPUT "Transverse tube length (m) =",length:y0=length/2
        20 :CALL MOVETO(250,46):PRINT "Area "1"m x"2*y0" m"
        CALL MOVETO(2,59):INPUT "Height of transv. lamps (m) =",z
        IF LO$="no" THEN 50
        CALL MOVETO(250,59):INPUT "Height of long. lamps (m) =",z1
        CALL MOVETO(2,72):INPUT "Distance betw. long. & transv. lamps (m)
=",ym
        CALL MOVETO(2,85)
        INPUT "Length of long. lamps in multiples of transv. lamp frame =",ll
        CALL MOVETO(2,98):PRINT "The vertical marks just below here show the
positions of the transverse tubes"
        50 :FOR x1=0 TO 1 +1/12 STEP 1/12: FOR y1=-y0/6 TO y0 STEP y0/6
        c2=0:IF n=0 THEN 179
        IF n=1 THEN 175
        FOR k=1 TO n
        FOR y=-y0+y0/80 TO y0-y0/80 STEP y0/40:IF d$="eq" THEN 70
        IF d$="eq" THEN 70
        LET a=z*z+(x1-.5*1*(1-COS((k-1)/(n-1)*pi)))*
        (x1-.5*1*(1-COS((k-1)/(n-1)*pi)))+(y-y1)*(y-y1)
        65 :LET b=SQR(z*z+(x1-.5*1*(1-COS((k-1)/(n-1)*pi)))*
        (x1-.5*1*(1-COS((k-1)/(n-1)*pi)))):GOTO 100
        70 :LET a=z*z+(x1-1*(k-1)/(n-1))*(x1-1*(k-1)/(n-1))+(y-y1)*(y-y1)
        LET b=SQR(z*z+(x1-1*(k-1)/(n-1))*(x1-1*(k-1)/(n-1))):GOTO 100
        100 :IF LJ$="flr" THEN 150
        c2=c2+100*b*z/a/a/65.42:GOTO 170
        150 :c2=c2+100*b/a/SQR(a)/65.42
        170 :NEXT y:NEXT k:GOTO 179
        175 :FOR y=-y0+y0/80 TO y0-y0/80 STEP y0/40
        LET a=z*z+x1*x1+(y-y1)*(y-y1):LET b=SQR(z*z+x1*x1)
        IF LJ$="flr" THEN 177
        c2=c2+100*b*z/a/a/65.42:GOTO 178
        177 :c2=c2+100/b/a/SQR(a)/65.42
        178 :NEXT y
        179 :c6=0:IF LO$="no" THEN 200
        FOR x2=-1*(ll-1)/2+1/160 TO 1*(ll-1)/2+1-1/160 STEP 1/80
        e1=z1*z1+(x1-x2)*(x1-x2)+(y0+ym-y1)*(y0+ym-y1)
        e2=z1*z1+(x1-x2)*(x1-x2)+(y0+ym+y1)*(y0+ym+y1)
        g1=SQR(z1*z1+(y0+ym-y1)*(y0+ym-y1))
        g2=SQR(z1*z1+(y0+ym+y1)*(y0+ym+y1)):IF LJ$="flr" THEN 180
        c6=c6+100*g1*z1/e1/e1/65.42+100*g2*z1/e2/e2/65.42:GOTO 200
        180 :c6=c6+100*g1/e1/SQR(e1)/65.42+100*g2/e2/SQR(e2)/65.42
        195 :NEXT x2
        200 :i=i+1:x5(i)=x1:y5(i)=y1:c5(i)=INT(c2+c6+.5)
        NEXT y1:NEXT x1:f=i:FOR r=1 TO f
        CALL MOVETO(350*x5(r),265-180*y5(r)):PRINT c5(r):NEXT r
        FOR k=1 TO n:IF n*(n-1)=0 THEN 410
        IF d$="eq" THEN 300
        CALL MOVETO(1*180*(1-COS((k-1)/(n-1)*pi)),115):GOTO 400
        300 :CALL MOVETO(1*(k-1)/(n-1),115)
        400 :PRINT "    I":NEXT k:GOTO 415
        410 : CALL MOVETO(0,115):PRINT "   I"
        415 :CALL MOVETO(230,130):INPUT "Do you also want graphic
representation (yes/no)?", graphic$
        IF graphic$="no" THEN finish
        500 :CLS:INPUT "scale factor=",scf:FOR r=1 TO f
        FOR black = 1 TO c5(r)*c5(r)*scf/10000
```

```
CIRCLE (360*x5(r)+13,260-180*y5(r)+15),black,33,,,250/c5(r):NEXT black
NEXT r:IF (n-1)*n=0 THEN 600
FOR k=1 TO n:IF d$="eq" THEN 300
CALL MOVETO(1*180*(1-COS((k-1)/(n-1)*pi)),140):GOTO 800
600 :CALL MOVETO(1*360*(k-1)/(n-1),140)
800 :PRINT "    I":NEXT k
1000 :CALL MOVETO(125,100)
INPUT "Do you want to repeat the graphic display with another scale
factor (yes/no)?",quit$:IF quit$="yes" THEN 500
finish:END
```

Appendix C.

Below are addresses of companies dealing with ultraviolet research equipment.

Solar Simulators

- Germany: Dr. K. Hönle GmbH, Fraunhoferstrasse 5, D-8033 Martinsried. Telephone: 49-(0)89-85608-0; telefax: 49-(0)89-85608-48.
- United Kingdom: UVALight Technology Ltd., 9th Floor, St. Martins House, 10 The Bull Ring, GB-Birmingham B5 DT. Telephone: 44-(0)21-643-2463/2472.
- Italy: Th. Mohwinckel S.p.A., Via San Christoforo 78, I-20090 Trezzano sul Naviglio. Telephone: (0)1-4452651.
- Switzerland: Anson AG, Friesenbergstr. 108, CH-8055 Zürich. Telephone: (0)1-4614444.
- Sweden: Agaria Industri AB, Box 140, S-18400 Åkersberga. Telephone: 46-(0)764-66085.
- United States: Solar Light Co. Inc., 6655 Lawnton Ave., Philadelphia, Pennsylvania 19126. Telephone: 1-(215)-927-4206. In addition to solar simulators this company markets broadband UV (sunburn) meters. Represented in Europe by Dr. Dieter Kockott UV-Technik, Vogelbergstrasse 27, D-6450 Hanau 7-Steinheim, Germany. Telephone: 49-(0)6181-659162.
- United States: Oriel Corporation, 250 Long Beach Blvd., P.O. Box 872, Stratford, Connecticut 06497. Telephone: 1-(203)-377-0063; telefax: 1((203)-378-2457.
- United Kingdom: Oriel Scientific Ltd. Leatherhead, Surrey. Telephone: 44-(0)372-378822.

- France: Oriel Sarl, Les Ulis. Telephone: (1)-69072020. In addition to solar simulators this company markets other light sources, monochromators, detection systems, lenses, mirrors.

Cellulose Acetate and Other Materials for Filtering UV Radiation

- United Kingdom: Courtaulds Speciality Plastics, P.O. Box 5, Spondon, Derby, DE2 7BP, Telephone: 44-(0)332-661422; telefax: 44-(0)332-660178.
- Austria: Wettlinger Kunststoffe, Neulerchenfelder Strasse 6-8, A-1160 Wien. Telephone: 43-(0)222-439953, 439386, 430638; telefax 43-(0)222-488155. Ask for Ultraphan Diacetatfolie, available in thicknesses from 0.03 to 1 mm.

Lamps

- The Netherlands: Philips Lightning B.V., Building EC 4, P.O. Box 80020, NL-5600 JM Eindhoven.
- Austria: Österreichische Philips Ind. GmbH, Triestert Str. 64, A-1100 Wien.
- Belgium: Philips Lighting, Em. Bockstaellaan 122, B-1020 Brussels.
- France: Philips Eclairage tour Vendome, 204, Rond Point du Pont de Sèvres, F-92516 Boulogne-Bilancourt Ce.
- Germany: Philips Licht GmbH, Steindamm 94, Postfach 104929, D-2000 Hamburg 1.
- Greece: Philips Lighting Hellenique S.A., P.O. Box 15125, GR-10210 Athens.
- Ireland: Philips Electronics Ireland Ltd., Newstead, Clonskeagh, Dublin 14.
- Italy: Philips S.p.A., Piazza IV Novembre, 3, P.O. Box 3992, I-20124 Milano.
- Portugal: PHILIPS ILUMINACAO LDA., 1009 Lisboa Codex, Av. Eng. Duarte Pacheco, 6, P-1000 Lisboa.
- Spain: Iberica de Alumbrado, Martinez Villergas 2, Apartado 2065, E-28027 Madrid.
- Sweden: Philips Ljus AB, P.O. Box 50506, S-11584·Stockholm.
- Switzerland: Philips A.G., Allmendstr. 140, Postfach 670, CH-8027 Zürich.
- United Kingdom: Philips Lighting, City House, London Road 420-430, Croyden, Surrey CR9 3QR.

References

Björn, L. O., 1989, Computer programs for estimating ultraviolet radiation in daylight, in: *Radiation Measurements in Photobiology* (B. L. Diffey, ed.), Academic Press, London, pp. 161–189.

Björn, L. O., & Murphy, T. M., 1985, Computer calculation of solar ultraviolet radiation at ground level, *Physiol. Veg.* 23:555–561.

Caldwell, M. M., 1971, Solar UV irradiation and the growth and development of higher plants, in: *Photophysiology* (A. C. Giese, ed.), Vol. 6, Academic Press, New York, pp. 131–177.

Caldwell, M. M., Gold, W. G., Harris, G., & Ashurst, C. W., 1983, A modulated lamp system for solar UV-B (280–320 nm) supplementation studies in the field, *Photochem. Photobiol.* 37:479–485.

Green, A., Sawada, T., & Shettle, E. P., 1974, The middle ultraviolet reaching the ground, *Photochem. Photobiol.* 19:251–259.

Madronich, S., Björn, L. O., Ilyas, M., and Caldwell, M. M., 1991, Changes in biologically active ultraviolet radiation reaching the earth's surface, in: *Environmental Effects of Ozone Depletion: 1991 update* (J. C. Van der Leun, & M. Tevini, eds.), United Nations Environmental Programme, pp. 1–13.

Rundel, R. D., 1983, Action spectra and estimation of biologically effective UV radiation, *Physiol. Plant.* 58:360–366.

Thimijan, R. W., Carns, H. R., & Campbell, L. E., 1978, Final report (EPA-IAG-D6-0168): *Radiation Sources and Relative Environmental Control for Biological and Climatic Effects of UV Research* (BACER), Environmental Protection Agency, Washington, DC.

Yu, W., Teramura, A. H., & Sullivan, J. H., 1991, *Model YMT-6 UV-B Modulation System. Manual of Operation.* Final report submitted to U.S. Environmental Protection Agency, Corvallis, Oregon.

3

Monitoring Ultraviolet Radiation

Weine A. P. Josefsson

1. Introduction

This chapter will focus on measurements of irradiance in the ultraviolet (UV) part of the solar spectrum. Therefore, in this chapter UV refers to solar ultraviolet radiation. To accurately measure UV is not a simple task, and many complex sources of error may produce spurious data. The present interest in the depletion of the ozone layer and the probable increase of harmful solar UV has initiated the start of many UV-monitoring programs, which will substantially increase the amount of UV data in the near future. There is also an urgent need for an internationally established and accepted standard procedure to measure UV and to calibrate the measurement instruments. The lack of accurate UV standards makes comparisons between different data sets problematic.

Solar radiation is the electromagnetic radiation of the sun. The radiation penetrating the atmosphere is confined to the wavelength range 0.28–3 μm. The short-wavelength part is usually subdivided into ultraviolet-A (UV-A, 315–400 nm) and ultraviolet-B (UV-B, 280–315 nm). American biologists sometimes set the limit between UV-A and UV-B at 320 nm (e.g., Holm-Hansen *et al.*, Chapter 13, this volume). Radiant power or flux per unit area is termed *irradiance*. This is the quantity usually measured by the instruments in a monitoring network. A common unit is W m^{-2}. If the irradiance is

Weine A. P. Josefsson • Research and Development Division, Swedish Meteorological and Hydrological Institute, S-601 76 Norrköping, Sweden.

Environmental UV Photobiology, edited by Antony R. Young *et al.* Plenum Press, New York, 1993.

integrated over time, the quantity is termed *irradiation*. It is the radiant energy per unit area, and J m^{-2} is the recommended unit, but W h m^{-2} can also be used. To avoid confusion the time period of integration should be stated, for example, hourly or daily values.

Although UV is only a minor part of solar radiation, in many respects it is the most active part. UV irradiance may have a large spatial and temporal variation, especially on a short time scale, that is, less than a day. Most UV irradiance is in the UV-A region. This part of the UV range varies more or less with the total solar radiation. Therefore, it can often be adequately estimated from the more frequently available records of global radiation, that is, diffuse plus direct radiation normally measured on a horizontal surface.

The solar irradiance in the UV-B is very small, several orders of magnitude less than that in other parts of the spectrum, but this region is the most harmful. Therefore, much effort has been spent in developing instruments sensitive enough to measure this part of the spectrum. With decreasing wavelength in the UV-B, absorption by ozone and scattering by molecules and aerosols increase, resulting in increasing attenuation. The spectral irradiance varies over several orders of magnitude through the UV-B spectral range and also during the day. To be useful, an instrument must be very sensitive, and it must have a reasonably linear response over a wide range.

2. Instruments

2.1. Spectroradiometers

2.1.1. Description and Problems

A spectroradiometer is an instrument for measuring spectral irradiance as a function of wavelength. It may incorporate either a monochromator or a spectrograph. A monochromator transmits a selected narrow wavelength band. Usually one grating is used to disperse the radiation from an input slit, and the optics will focus the selected wavelength component on the output slit. For UV-B measurements a double monochromator, i.e., two monochromators used in tandem, is often used to reduce the influence of stray light.

A spectrograph disperses the different wavelengths of the radiation on an output plane where, e.g., a CCD (charged coupled detector) or a diode array may be used as detector. The following paragraphs give a brief review of some of the problems connected with spectral measurements of UV.

A spectrograph may record a spectrum almost instantaneously, while a monochromator usually requires several minutes to scan a spectrum. The time for a scan depends on wavelength interval, bandwidth, step length, and the time spent to measure at each position. During the time of the scan, irradiance can change, e.g., due to variable meteorological conditions and changing solar elevation. This is one important source of error if individual spectra are utilized. The width of the measured wavelength interval (bandpass) varies among different instruments. However, it should be no more than about 1 nm to give sufficient spectral information in the UV-B for biological applications (Gibson, 1991). The bandpass of a monochromator is mainly determined by the widths of the slits and dispersion of the grating. Instruments with a smaller bandpass will give a lower signal. This may be compensated for by increasing measurement time. By illuminating the diffuser of a spectroradiometer with a narrow spectral line and varying the wavelength setting around the center of the line, an approximation of the relative transmittance of the monochromator is obtained, often called a *slit function* (Nicodemus, 1979). A plot of the output will be more or less triangular, trapezoidal, or Gaussian, depending on the relative widths of the entrance and exit slits. The full width of these slit functions at half of the peak maximum (FWHM) is called the *spectral slitwidth,* or the *nominal bandwidth.* The spectroradiometer is somewhat responsive to radiation outside the nominal band, an effect that can be reduced by using a double monochromator.

Stray light is an important error source in UV measurements. Stray light is unwanted radiation measured along with desired wavelengths, and is caused by dirty or imperfect optics that cause radiation to be dispersed or reflected in unintended directions. Because the longer-wavelength components in the solar spectrum are several orders of magnitude more intense, even a very small proportion of stray light from the visible or infrared part of the spectrum may distort UV-B measurements. If a grating is used, there are also spectral orders other than that intended that may contribute to the stray light. To reduce this problem, it is very important to have a spectroradiometer specially designed for measurements in the UV. As noted, one way to ensure the purity of the spectrum is to use a double monochromator. Another way to reduce stray light is to combine a single monochromator with a cutoff filter. A third way is to use a detector insensitive to visible and infrared radiation.

The dynamic range of the instrument must be linear or controlled over several decades of irradiance because UV irradiance shows a large variation not only over the spectral range of interest but also with solar elevation.

Measurements are usually made to represent the irradiance on a horizontal surface. The receiving surface of the instrument may be a Teflon or quartz diffuser or an integrating sphere. Error in responsivity, which depends on the direction of incident radiation, is often split into two components, referred to as *cosine* and *azimuthal errors.* Ideally, instrument response should be proportional to the cosine of the angle of incident radiation and independent of the azimuth. This is far from true for most radiometers, especially at large angles of incidence. Although a well-designed integrating sphere often gives a good approximation, it has the disadvantage of low throughput of the radiation.

Sky UV radiation is more or less polarized. The responsivity of a spectroradiometer is often dependent on polarization, because reflection at optical surfaces depends on it. However, measuring the global irradiance using a Teflon diffuser reduces the effect of polarization by approximately 10 times, and an integrating sphere reduces it even more (Liedquist and Werner, 1984). Therefore, this error source should be negligible compared with, e.g., that from cosine error.

Variations in temperature will cause changes in the measured signal of the spectroradiometer. The magnitude of this error should be investigated and corrected for. Ideally the temperature should be stabilized at a prescribed value or at least rapid temperature variations should be avoided. If the instrument is cooled one should be aware of the risk of condensation. Housing the instrument in a container with a desiccant and a humidity indicator that can be easily inspected is good insurance against serious problems.

A spectrograph can record spectra almost instantaneously, while a monochromator has to spend some time scanning the spectrum. If UV radiation varies over time, the spectra measured with a monochromator will be distorted.

Small changes in the monochromator, due to imperfections in moving parts, to changes in temperature, rough handling, and so forth, will affect the position of the spectrum on the detector. Regular control of the wavelength setting is needed to have accurate data. Even small offsets will introduce large errors in the measurements, due to the large wavelength dependence of the solar spectral irradiance in the UV-B range. For example, the relative increase in spectral irradiance by increasing wavelength is about 3% per 0.1 nm at 305 nm and about 10% per 0.1 nm at 295 nm (Saunders and Kostkowski, 1978). These numbers are rough estimates and may vary significantly. Many spectroradiometers have a wavelength uncertainty on the order of 0.1 nm, and they are significantly affected by this source of error. If the wavelength is offset by 1 nm, the error in the spectral irradiance will be much larger than 50% for most of the UV-B. It should be noted that the error in wavelength setting will vary not only along the wavelength range but also depending on the direction of the scan.

2.1.2. Calibration and Accuracy

Good standards are needed to make measurements comparable. Measurements in the UV are based on blackbodies, special hydrogen (deuterium) lamps, synchrotron radiation, or electrical quantities (Liedquist and Werner, 1984). The spectral radiometric quantities of a blackbody of specified temperature can be calculated using Planck's radiation law. The technique to realize the primary standards for UV irradiance is utilized only at large laboratories (e.g., National Institute of Standards and Technology, United States; National Physical Laboratory, United Kingdom; Physikalisch-Technische Bundesanstalt, Germany). Other laboratories have to use secondary standards, usually halogen lamps and deuterium lamps. (Note that the deuterium lamps may change, in steps, between calibrations. Therefore the level of irradiance of the deuterium lamp should be checked before use.) The most stable lamp in the UV-B is the halogen lamp, with an inaccuracy of about 2%. It also has a greater spectral irradiance in the UV-B than the deuterium lamp. Uncalibrated lamps are commercially available from various producers, e.g., the DXW 1000-W quartz tungsten halogen lamp from General Electric.

The calibration of the secondary standard is often transferred to a number of identical lamps used as working standards. The transfer can be realized using the spectroradiometer. Each transfer of calibration will introduce an uncertainty that should be added to the original inaccuracy. When calibrating a spectrometer in the UV-B using a working standard, about 5% has to be added to the total absolute inaccuracy of the instrument. Using the secondary or primary standard will reduce this number. It is important to have good calibration facilities, including a set of calibration lamps. If only one lamp is available, problems will occur when a difference between the most recent and previous calibrations are found. Is it the instrument or the lamp that is malfunctioning? Before calibration the lamps have to be seasoned for at least 30 h. The lamps will age as a function of the burn time. Therefore, a log should be kept indicating the burn time of each lamp. After about 50 h the lamp should be recalibrated.

The absolute irradiance calibration is hard to perform. It may contribute 5–10% to the absolute inaccuracy in UV-B output. However, the precision can probably be controlled within ±5%. The standard lamp must always be properly oriented and aligned to ensure repeatability in the calibration setup. Baffles should be used to remove reflected radiation. The current through the lamp circuit is often used as a measure of the lamp output. Therefore it has to be controlled accurately, as an error of 1% in the lamp current will produce an error of several percent in the UV-B output. The halogen lamp should not be ventilated because this will disturb the transport mechanism between the gas and the coil. To have stable irradiance, at least a 5-min warmup of the lamp is necessary.

Most spectroradiometers used for meteorological and biological measurements are calibrated to an absolute scale of irradiance using a lamp as reference. Intercomparisons between spectroradiometers can also be performed

(Josefsson, 1991; Gardiner and Kirsch, 1992; Seckmeyer *et al.*, 1993). This will give either a relative scale or an absolute one if at least one of the instruments can be regarded as an absolute reference. Usually intercomparisons are done outdoors, preferably at stable atmospheric conditions, i.e., cloud-free sky and close to noon. Different bandwidths, scan times, input optics, and errors in wavelength settings will give discrepancies in the measured spectral irradiances. Some of the complex interactions between different sources of error may give a spurious result that will be missed if only a standard lamp is used to transfer the calibration. Therefore, intercomparisons are necessary if data from different types of spectroradiometers are going to be compared.

The absolute accuracy of spectroradiometric irradiance measurements of the solar UV-B may not seem impressive, especially at the shorter wavelengths and under rapidly varying conditions. Absolute calibration, wavelength error, and, depending on the diffuser, cosine error cause large uncertainties. Summing all possible errors might give a pessimistic estimate of absolute uncertainty close to 20% in the measurement of integrated UV-B with a carefully calibrated, well-operated and well-designed spectroradiometer. However, the precision or random error of well-maintained and regularily calibrated instruments may be about 5% at the longer wavelengths in the UV-B.

2.2. Broadband Radiometers

Because of the high cost of spectroradiometers, several attempts have been made to construct a simple radiometer or dosimeter that directly records UV-B or UV-A or gives a measure of UV radiation. Depending on the type of sensor used, the detectors can be classed as biological, chemical, or physical. Although the biological and chemical sensors often are nonspecific and not suitable for continuous recording, they may have the advantage of giving a direct measure of the desired effect, e.g., erythema, vitamin D synthesis, or photodegradation of polymers. General descriptions of methods, sources of error, and limitations of biological and chemical measurements of UVR can be found in Coulson (1975), Calkins, (1982), Lala (1985), Diffey (1986, 1987), and Berre and Lala (1989).

For UV-A measurements there is a number of broadband-filter instruments. One of these, the Eppley Ultraviolet Pyranometer, has been widely used. It is normally calibrated in such way that the output represents the UV-A irradiance. The author has used a radiometer (Wester, 1983) that measures two spectral bands in the UV in parallel. The combinations of filters and detectors give, for the shorter-wavelength band, a maximum sensitivity close to 306 nm with a half width of 5 nm, and for the longer-wavelength band, a maximum sensitivity at 360 nm and a half width of 10 to 15 nm. It has been shown (Wester, 1983) that output from the 360-nm wavelength band is proportional to UV-A irradiance. This implies that UV-A has a relatively invariant spectral distribution. Another indication of this results from studies of the relation of UV-A to global radiation (Schultze, 1970; Kvifte *et al.*, 1983; Josefsson, 1986; Basher, 1989). It seems that daily values of UV-A can be estimated using global radiation and cloudiness with an accuracy comparable to the uncertainty of the measurement. Because of the problems connected with regular calibration of UV-A radiometers, as well as other instrumental errors, model estimation of UV-A is often a convenient alternative to measurements.

In the 1950s, D. F. Robertson developed an instrument having a sensitivity resembling the erythema action spectrum (Robertson, 1972). The instrument was redesigned (Berger, 1976) and is still used at a number of sites around the world, mostly in the USA. It is often called the Robertson–Berger, or R–B, meter. On top of the instrument there is a transparent Vycor dome. A black UG-11 filter transmits UV radiation and a small proportion of red light. A layer of phosphor deposited on a Corning 4010 filter absorbs UV and emits radiation in the visible region. The greenish Corning 4010 filter transmits much of the emitted radiation of the phosphor and blocks the wavelengths transmitted by the UG-11 filter. The transmitted radiation is detected by a phototube or a photodiode.

Many studies of the R–B meter and of recorded data from these instruments have been published, such as papers on climatology and long-term variations (Berger and Urbach, 1982; Scotto *et al.*, 1976, 1988; Degórska *et al.*, 1992; Frederick and Weatherhead, 1992) and studies of the instrument's characteristics (Blumthaler and Ambach, 1986; Johnsen, 1990; Jokela *et al.*, 1991; Kennedy and Sharp, 1992; DeLuisi *et al.*, 1992).

Deviations from the exact cosine response with angle of incidence can be a source of considerable error for many radiation instruments. The model 500 instrument from the Solar Light Company is similar in principle to the Robertson–Berger meter. The response of the Model 500 instrument increased up to ~20% above what would be expected from the cosine law at about a 60° angle of incidence (Josefsson, 1989; Jokela *et al.*, 1991). At larger angles the response dropped very quickly. The instrument is almost insensitive for radiation with an angle of incidence larger than 80°.

The temperature dependence of the R–B meter and Model 500 instrument relative sensitivity is large but seems to be reasonably linear in the studied ranges of temperature. Assuming a linear dependence between the relative sensitivity and the temperature, Blumthaler and Ambach (1986) found 0.8%/°C, Johnsen (1990) 1.2%/°C, Jokela *et al.* (1991) 0.7–1%/°C for three instruments, and Kennedy and Sharp (1992) measured 0.3%/°C and 0.6%/°C for two instruments.

Using the phosphor-layer principle of the R–B meter, new instruments have been developed. Much effort has been spent on reducing the large temperature dependence and the deviation from true-cosine response. The Solar Light Company, started by Berger in the United States, has introduced a Peltier element that either cools or heats the interior of the instrument to a temperature that can be selected by the operator. The UV radiometer of Yankee Environmental Systems, Inc., is also temperature controlled. A Japanese version manufactured by EKO (Furusawa *et al.*, 1990) has a temperature-compensating circuit and uses a diffuser and interference filter combination instead of the UG-11 filter.

The calibration of broadband-filter instruments is a difficult task. Although the manufacturer tries to build identical instruments, there will be differences in the characteristics of each (e.g., Jokela *et al.*, 1991; DeLuisi, 1992). The spectral responsivity, which varies among instruments, is often temperature dependent, most of which can be eliminated by postcorrection of the data if the air temperature, or preferably the temperature of the instrument, is known. But it is better to have the instrument temperature stabilized. However, due to the remaining difference in spectral responsivity, a ratio of the outputs from two instruments, i.e., roughly the convolution of the radiation source and the spectral responsivity, will vary depending on the spectral distribution of the radiation source. For example, if the sun is used, the measured ratio of two instruments will depend on factors changing the solar spectral distribution, such as the solar elevation and the amount of total ozone. Use of a halogen lamp as a radiation source will introduce radiation of wavelengths shorter than 290 nm, and, correspondingly, the spectral responsivity of the instruments for those wavelengths will affect the result. The ratio of the instruments' outputs may be surprisingly different using the sun and the halogen lamp, respectively.

A practicable way to calibrate a network of broadband instruments is to use a traveling reference instrument of the same type to intercalibrate the UV radiometers. The result will, of course, depend on the factors mentioned. Therefore, one should try to perform the intercomparisons under conditions that are as representative as possible, i.e., high sun, clear sky, and non-extreme ozone amounts. Such a network can provide data suitable for mapping of yearly and daily variations and regional spatial features, and for public information. Because of the large natural variation of UVR at higher latitudes and the low precision of network data, small trends (<1%/year) within a 10-year period will not be detectable with statistical significance.

If a station is equipped with a spectroradiometer, it is possible to take spectral responsivity into account when performing calibrations. This is recommended when calibrating reference broadband instruments.

3. Aspects Concerning Networks

Establishing a network for UV monitoring will often be a balancing act between economy and goals. A network may be established to record data to study the variation and magnitude of UV changes (climatology). The number of recording sites must be sufficient to cover a wide range of latitudes and different climate regimes. Their locations should be carefully selected not only with respect to spatial distribution but also to ensure availability of maintenance and the availability of other meteorological data. Measurements should be of sufficient quality, frequency, duration, and geographic density that spatial and temporal analyses give meaningful results. If the data aquisition is automatic and can be carried out in real time, the network may be used to maintain an alertness for public protection purposes. To reach these goals, broadband radiometers are sufficient. Calibration routines and facilities are required, as well as proper maintenance of the instruments and a data aquisition system to ensure reliable operation.

However, if the goal is to substantially improve understanding of UV variation, high-quality UVR data are essential. A well-maintained spectroradiometer will give a physical quantity as output, e.g., the spectral irradiance as $W\ m^{-2}\ nm^{-1}$. Such a spectrum has a general applicability because it can be convoluted by any action spectrum. Often the calibration is the crucial point to ensure high quality. Tentative specifications and recommendations of spectroradiometric UV measurements and network requirements have been compiled by various groups, e.g., Basher (1989), WMO/UNEP (1990), STEP Project 76 (1991), and Gibson (1991, 1992).

The following are some general aspects of a UV-monitoring program:

- Operation should be secured over a period longer than 10 years.
- Measurements should be taken every day.
- Density of stations should be high enough to separate local effects from regional ones.
- Periodic calibrations and/or intercomparisons should be done.

A UV-monitoring program should not be restricted to UV quantities. If possible, other relevant quantities should be measured in close connection, e.g., global radiation, amount of aerosols, cloudiness, surface reflectance, and total ozone. It is also important to have simple and advanced instruments in parallel operation to compile a data base of broad utility and to ensure that any detected changes are significant, as well as to allow monitoring of performance (internal check). There is also an advantage to having several solar radiation quantities measured at the same site. Redundancy in data can be used for quality control and also for interpolation of missing data.

Usually the sensors of solar radiation instruments are positioned horizontally, and therefore they measure solar irradiance at a horizontal surface. However, in application, e.g., effects on humans and plants, the surface of interest may have various orientations. Therefore, data often have to be recalculated to the surface orientation of interest. If the tilt of the surface is large, the influence of reflected radiation from the adjacent ground must be taken into account. By using a sun tracker, the direct UV irradiance may be measured. Using a model, the horizontal and direct UV data may be combined to give the irradiance on any surface with, in most cases, acceptable accuracy.

At present, most monitoring networks are in an establishment phase (AFEAS, 1992). It will take operators some years to gain experience and establish routines for calibration, and to maintain the calibrations within the networks, which is essential for data quality. New instruments, which will be available in the near future, should be added to the network and not replace existing instruments. A rather long period of overlapping measurements will ensure the homogenity of the whole series.

If there are only small changes in the ozone layer, i.e., of the order a few percent per decade, then it will be hard to detect a significant trend in UV irradiance within a decade with most instruments and networks, due to the large variation of UVR caused by meteorological events and also because of the relatively large inaccuracy inherent in this type of measurement.

A homogeneous and accurate data set is especially needed to make reliable trend analyses. Practical problems such as power failures and frost on the domes of the instruments will cause periods of missing and erroneous data. Regular processing of data with computerized and manual checking will ensure early detection of malfunctioning and inaccurate data. It is well known that bad maintenance may cause errors as large as those caused by irregular calibration. The instruments' characteristics (calibration protocols, etc.) must be documented regularly. The establishment of UV-monitoring networks and calibration routines should be coordinated, preferably internationally, e.g., by the World Meteorological Organization (WMO).

The R–B meter has been used at a number of stations around the world. Some of these stations probably have the longest records of UVR monitoring, and effort has been made to analyze the data and the instruments to extract possible trends in UVR (Berger and Urbach, 1982; Berger, 1987; Scotto et al., 1988; Degórska et al., 1992; Frederick and Weatherhead, 1992).

Despite all their problems, broadband instruments have the advantage of being in more or less continuous operation in all seasons and during all kinds of weather. They are relatively low-cost instruments and can be operated by unqualified personal, which will facilitate a denser network. Recently, there have been several new networks established using broadband radiometers in, e.g., Australia, New Zealand, Japan, Sweden, Finland, Norway, and the United

Kingdom. Several institutions around the world have both spectroradiometers and broadband instruments in operation, some of them even designed for underwater measurements.

The interest in spectroradiometric measurement of UV is increasing rapidly, as indicated by the following presumably incomplete and arbitrary compilation of authors who have published papers on routine spectral measurements. Starting with the important works by Bener (1960, 1972) the number of works has increased rapidly, especially in the 1980s (Kok, 1972; Tarrant and Brock, 1975; Garrison et al., 1978; Wester 1983, 1984a, 1984b, 1987; Hisdal, 1986; Josefsson, 1986; Seckmeyer, 1989; Lubin and Frederick, 1989; Henriksen et al., 1989; Bittar and McKenzie, 1990; Stamnes et al., 1990; McKenz: et al., 1991; Henriksen et al., 1992; McKenzie et al., 1992; Dirmhirn et al., 1993).

A very interesting network is one formed by sites equipped with the Brewer ozone spectrophotometer (Kerr et al., 1984). The instrument was originally designed to monitor total ozone, but it is also able to make UV spectral measurements (Josefsson, 1986; Ito et al., 1991). It is a weatherproof, scanning spectroradiometer that can be operated in an automatic mode. There are over fifty sites around the world equipped with this instrument and thus capable of monitoring UV spectral irradiance (Fig. 1). A network of five stations has been established by the Japan Meteorological Agency (Ito et al., 1991). The site distribution of Brewer instruments is similar to that for UV monitoring, both spectral and broadband, with the exception of Australia and New Zealand, where UV-monitoring programs have been running for a long time.

Another interesting network was established in 1988 by the U.S. National Science Foundation. In this network, three spectroradiometers operate in Antarctica, one operates in southern Argentina, and since 1990, there is also one operating at Pt. Barrow in the Northern Hemisphere. These instruments are manufactured by Biospherical Instruments Inc., which also is responsible for the network. Results obtained from this network have been published by Lubin et al. (1989), Stamnes et al. (1990), Lubin and Frederick (1991), and WMO (1991).

4. Modeling versus Measurements

A model with which to compute UV irradiance, spectral or integrated, is a very useful tool for studying how various parameters affect the UV climate. The variety of models for this purpose range from radiative transfer models to rough empirically based models. The real atmosphere is complex, and

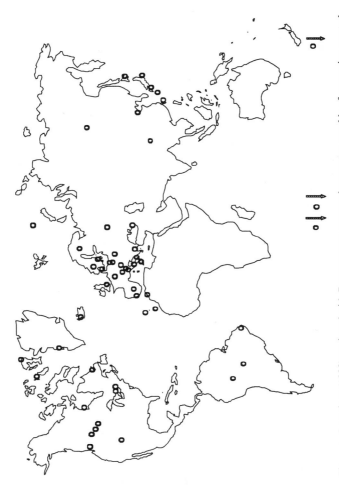

Figure 1. Locations (1992) equipped with Brewer ozone spectrophotometers capable of measuring spectral UV irradiance. The arrows at the bottom indicate Antarctic sites.

some of its characteristics are hard to parameterize into a model. For example, the recent eruption of the volcano Pinatubo in 1991 injected a large amount of aerosols into the atmosphere. Important information necessary for modeling, e.g., amounts, vertical and horizontal distribution, size distributions, and chemical composition of the aerosols, is seldom available. Despite the lack of relevant data, a model can give insight into the nature of the UV dependency of different variables, and calculated results can be compared with measurements.

Atmospheric complexity introduced by aerosols and clouds is perhaps the main argument for having a network measuring UVR, because available information on clouds and aerosols from standard meteorological observations or even from satellites is still so inexact and sparse that accurate calculations are not possible.

References

AFEAS, 1992, *UV-B Monitoring Workshop: A Review of the Science and Status of Measuring and Monitoring Programs,* 10–12 March 1992, Washington, DC, AFEAS, Science and Policy Ass., Inc., Washington, DC.

Basher, R. E., 1989, Perspectives on monitoring ozone and solar ultraviolet radiation, *Transactions of Menzies Foundations,* Vol. 15, pp. 81–88.

Bener, P., 1960, *Investigation on the Spectral Intensity of Ultraviolet Sky and Sun+Sky Radiation (between 297.5 mµ and 370 mµ) under Different Conditions of Cloudless Weather at 1590 m a.s.l.,* Contract AF 61(052)-54, Technical Summary Report No. 1, Davos-Platz, Switzerland.

Bener, P., 1972, *Approximate Values of Spectral Intensity of Natural Ultraviolet Radiation for Different Amounts of Atmospheric Ozone,* Final Technical Report, Contract AF DAJA-68-C-1017, London, Davos-Platz, Switzerland.

Berger, D. S., 1976, The sunburning ultraviolet meter: Design and performance, *Photochem. Photobiol.* 24:587–593.

Berger, D. S., 1987, *Fluctuations and Trends in Environmental UV loads, Human Exposure to Ultraviolet Radiation: Risks and Regulations,* Proc. of a Seminar held in Amsterdam, 23–25 March 1987, W. F. Passchier and B. F. M. Bosnjakovic (eds.), Elsevier, Amsterdam, pp. 213–221.

Berger, D. S., and Urbach, F., 1982, A climatology of sunburning ultraviolet radiation, *Photochem. Photobiol.* 35:187–192.

Berre, B., and Lala, D., 1989, Investigation on photochemical dosimeters for ultraviolet radiation, *Sol. Energy,* 42:405–416.

Bittar, A., and McKenzie, R. L., 1990, Spectral ultraviolet intensity measurements at 45°S: 1980 and 1988, *J. Geophys. Res.* 95:5597–5603.

Blumthaler, M., and Ambach, W., 1986, Messungen der Temperaturkoeffizienten des Robertson-Berger Sunburn Meters und des Eppley UV-Radiometers, *Arch. Meteorol. Geophys. Bioklimatol. Ser. B* 36:357–363.

Blumthaler, M., and Ambach, W., 1990, Indication of increasing solar ultraviolet-B radiation flux in alpine regions, *Science* 248:206–208.

Calkins, J., 1982, Measuring devices and dosage units, in: *The Role of Solar Ultraviolet Radiation in Marine Ecosystems* (J. Calkins, ed.), Plenum, New York.

Coulson, K. L., 1975, *Solar and Terrestrial Radiation, Methods and Measurements*, Academic Press, New York.

Degórska, M., Krzyscin, J., Rajevska-Wiech, B., and Slomka, J., 1992, *Ozone Changes Impact on UV-B Radiation at Belsk, Poland, in 1975–1989*, Institute of Geophysics, Polish Academy of Science, Dept. of Physics of Atmosphere, Warsaw.

DeLuisi, J., Wendell, J., and Kreiner, F., 1992, An examination of the spectral response characteristics of seven Robertson-Berger meters after long-term field use, *Photochem. Photobiol.* 56:115–122.

Diffey, B. L., 1986, Possible errors involved in dosimetry solar UV-B radiation, Stratospheric Ozone Reduction, *Solar Ultraviolet Radiation and Plant Life*, (R. C. Worrest and M. M. Caldwell, eds.), Springer Verlag, Berlin, pp. 75–86.

Diffey, B. L., 1987, A comparison of dosimeters used for solar ultraviolet radiometry, *Photochem. Photobiol.* 46:55–60.

Dirmhirn, I., Sreedharan, C. R., and Venugopal, G., 1993, Spectral ultraviolet radiation instrument and preliminary measurements in mountainous terrain, *Theor. Appl. Climatol.* 46:219–228.

Frederick, J. E., and Weatherhead, E. C., 1992, Temporal changes in surface ultraviolet radiation: A study of the Robertson-Berger meter and Dobson data records, *Photochem. Photobiol.* 56: 123–132.

Furusawa, Y., Suzuki, K., and Sasaki, M., 1990, Biological and physical dosimeters for monitoring solar UV-B light, *J. Radiat. Res.* 31:189–206.

Gardiner, B. G., and Kirsch, P. J., 1992, European Intercomparison of Ultraviolet Spectrometers, 3–12 July 1991, Panorama, Greece, *Air Pollution Research Report 38,* STEP Project 76, Determination of Standards for a UVB Monitoring Network, CEC-DG, Brussels.

Garrison, L. M., Murray, L. E., Doda, D. D., and Green, A. E. S., 1978, Diffuse-direct ultraviolet ratios with a compact double monochomator, *Appl. Opt.* 17:827–836.

Gibson, J. H., 1991, *Justification and Criteria for the Monitoring of Ultraviolet (UV) Radiation,* Report of UV-B measurements workshop, January 22–25, 1991, Denver, Co., Natural Resource Ecology Lab, Colorado State University, Fort Collins, CO.

Gibson, J. H., 1992, *Criteria for Status-and-Trends Monitoring of Ultraviolet (UV) Radiation,* Recommendations of the UV-B monitoring workshop, March 11–12, 1992, Washington, DC, Coordination Office, Natural Resource Ecology Lab, Colorado State University, Fort Collins, CO.

Henriksen, K., Stamnes, K., and Östensen, P., 1989, Measurements of solar U.V., visible and near I.R. irradiance at 78° N, *Atmos. Environ.* 23:1573–1579.

Henriksen, K., Claes, S., Svenøe, T., and Stamnes, K., 1992, Spectral UV and visible irradiance measurements in the Barents Sea and Svalbard, *J. Atmos. Terr. Phys.* V.54:1119–1127.

Hisdal, V., 1986, Spectral distribution of global and diffuse solar radiation in Ny-Ålesund, Spitsbergen, *Polar Research* 5:1–27.

Ito, T., Ueno, T., Kajihara, R., Shitamichi, M., Uekubo, T., Ito, M., and Kobayashi, M., 1991, Development of monitoring technique of ultraviolet irradiance on the ground: An assessment of UV-B increase due to ozone depletion based on spectral observations, *J. Meteorological Research,* 43:213–273.

Johnsen, B. J., 1990, *Måling av ultrafiolett stråling* [Measurements of ultraviolet radiation], Diplomoppgave, Dept. of Physics and Mat., Univ. of Trondheim, Norway (in norwegian).

Jokela, K., Huurto, L., and Visuri, R., 1991, Optical test results for Sunburn UV-Meter Solar Light Model 500, *Proceedings of 1991 Intern. Geoscience and Remote Sensing Symp.,* June 3–6, 1991, Espoo, Finland, pp. 950–954.

Josefsson, W., 1986, *Solar Ultraviolet Radiation in Sweden,* SMHI Reports Meteorology and Climatology, No. 53, Swedish Meteorological and Hydrological Inst., Norrköping, Sweden.

Josefsson, W., 1989, *Testing of the MED-meter and a Proposal of a Solar UV-Network in Sweden*, SMHI, Meteorologi, No. 102, Norrköping, Sweden.

Josefsson W., 1991, *The Intercomparison of Spectroradiometers at SMHI in Norrköping 6-8 of August 1991*, SMHI, Meteorologi, No. 31, Aug. 29, 1991, Norrköping, Sweden.

Kennedy, B. C., and Sharp, W. E., 1992, A validation study of the Robertson–Berger meter, *Photochem. Photobiol.* 56:133–142.

Kerr, J. B., McElroy, C. T., Wardle, D. I., Olafson, R. A., and Evans, W. J. F., 1984, The automated Brewer spectrophotometer, *Proceedings of the Quadrennial International Ozone Symposium*, Halkidiki, Greece, D. Reidel Publishing, Dordrecht, pp. 396–401.

Kok, C. J., 1972, Spectral irradiance of daylight at Pretoria, *J. Phys. D.* 5:1513–1519.

Kvifte, G., Hegg, K., and Hansen, V., 1983, Spectral distribution of solar radiation in the nordic countries, *J. Clim. and Appl. Meteor.* 22:143–152.

Lala, D., 1985, Ultraviolet radiation measurements by photochemical methods, *Bulletin M85: 12*, The National Swedish Institute for Building Research, Gävle, Sweden.

Liedquist, L., and Werner, G., 1984, Noggrannhet och felfaktorer vid optiska strålningsmätningar: 2. Ultraviolett strålning från solarier och sollampor, *Teknisk Rapport SP-RAPP 1984:14*, Borås, Sweden (in swedish).

Lubin, D., and Frederick, J. E., 1989, Measurements of enhanced springtime ultraviolet radiation at Palmer Station, Antarctica, *Geophys. Res. Lett.* 16, 783–785.

Lubin, D., and Frederick, J. E., 1991, The ultraviolet radiation environment of the Antarctic Peninsula: The roles of ozone and cloud cover, *J. Appl. Meteor.* 30:478–493.

McKenzie, R. L., Matthews, W. A., and Johnston, P. V., 1991, The relationship between erythemal UV and ozone, derived from spectral irradiance measurements, *Geophys. Res. Lett.* 18: 2269–2272.

McKenzie, R. L., Johnston, P. V., Kotkamp, M., Bittar, A., and Hamlin, J. D., 1992, Solar ultraviolet spectroradiometry in New Zealand: Instrumentation and sample results from 1990, *Appl. Opt.* 31(30):6501–6509.

Nicodemus, F. E., 1979, Self study manual on optical radiation measurements: Part I-Concepts, Chapter 7, 8 and 9, *NBS Technical Note 910-4*, No. 003-003-02067-2, Washington DC.

Robertson, D. F., 1972, *Solar Ultraviolet Radiation in Relation to Human Sunburn and Skin Cancer*, PhD thesis, No. THE4866, Univ. of Queensland, St. Lucia, Queensland.

Saunders, R. D., and Kostkowski, H. J., 1978, Accurate solar spectroradiometry in the UV-B, *Optical Radiation News*, No.24, NBS, U.S. Dept. of Commerce. Washington, DC.

Scotto, J., Fears, T., and Gori, G., 1976, *Measurements of Ultraviolet Radiation in the United States and Comparisons with Skin Cancer Data*, NIH Publ. No. 76-1029, Bethesda, MD, pp. 1–120.

Scotto, J., Cotton, G., Urbach, F., Berger, D., and Fears, T., 1988, Biologically effective ultraviolet radiation: Surface measurements in the United States, 1974–1985, *Science* 239:762–764.

Seckmeyer, G., 1989, Spectral measurements of global UV-radiation, *Meteor. Rundsch.* 41 Jg, Heft 6, pp. 180–183.

Seckmeyer, G., Blumthaler, M., Fabian, P., Gerber, S., Gugg-Helminger, A., Häder, D.-P., Huber, M., Kettner, C., Köhler, U., Köpke, P., Maier, H., Schäfer, J., Suppan, P., Tamm, E., Thomalla, E., and Thiel, S., 1993, Comparison of spectral UV-radiation measurements systems, (submitted to *Appl. Optics*).

Schultze, R., 1970, *Strahlenklima der Erde*, Wissenschaftliche Forschungsberichte, Band 72, Dr. Dietrich Steinkopff Verlag, Darmstadt.

Stamnes, K., Slusser, J., Bowen, M., Booth, C., and Lucas, T., 1990, Biologically effective ultraviolet radiation, total ozone abundance, and cloud optical depth at McMurdo station, Antarctica, September 15, 1988, through April 15, 1989, *Geophys. Res. Lett.* 17:2181–2184.

STEP Project 76, Minutes of the first meeting of the CEC STEP programme: Determination of Standards for a UVB Monitoring Network, Jan 28th–29th 1991, Coordinator: A. Webb, Dept. of Meteor., Univ. of Reading, U.K.

Tarrant, A. W. S., and Brock, J. R., 1975, Further studies of the spectral power distribution of daylight in the ultraviolet region, *Compte Rendus 18th Session CIE,* Publ. 36, London, pp. 384–392.

Wester, U., 1983, Solar ultraviolet radiation—A method for measuring and monitoring, *Internal report RI 1983-02,* Dept. of Radiation Phys., Karolinska Inst., Stockholm.

Wester, U., 1984a, Solar ultraviolet radiation in Stockholm—Examples of spectral measurements and influences of measurement error parameters, *Internal report RI 1984-03,* Dept. of Radiation Phys., Karolinska Inst., Stockholm.

Wester, U., 1984b, Erythemal efficiency of ultraviolet radiation from the sun and from sunlamps calculated on the basis of measured spectral data, *Internal report RI 1984-05,* Dept. of Radiation Phys., Karolinska Inst., Stockholm.

Wester, U., 1987, Solar ultraviolet radiation on the Canary Islands and in Sweden—A comparison of irradiance levels, in: *Human Exposure to Ultraviolet Radiation: Risks and Regulations,* proceedings of a seminar held in Amsterdam, 23–25 March 1987 (W. F. Passchier and B. F. M. Bosnjakovic eds.), Elsevier, Amsterdam, pp. 275–279.

WMO/UNEP, 1990, Report of the Preparatory Meeting of the Ozone Research Managers of the Parties to the Vienna Convention for the Protection of the Ozone Layer, Geneva, 7–9 February 1990, WMO, Geneva.

WMO, 1991, *Scientific Assessment of Stratospheric Ozone: 1991,* Report No. 25, Global Ozone Research and Monitoring Project, Geneva.

4

Influence of Ozone Depletion on the Incidence of Skin Cancer
Quantitative Prediction

Frank R. de Gruijl and Jan C. Van der Leun

A fake fortuneteller can be tolerated. But an authentic soothsayer should be shot on sight. Cassandra did not get half the kicking around she deserved.

—Lazerus Long, in Robert A. Heinlein's *Time Enough for Love.*

1. Introduction

As with many environmental problems, a quantitative assessment of the biological effects of a stratospheric ozone depletion stretches science beyond the limits of directly verifiable statements. The scientist should take care not to become the modern soothsayer reading the high-tech equivalent of a goat's entrails, providing authoritative answers to those "seeking guidance." The prognoses may take decades to become detectable, and then the forecasted effects can be obscured by many interfering factors. The premise of *ceteris paribus* (other things being equal) will almost surely not be met. To a skeptic such a forecasting exercise may seem scientifically futile, but it need not be if cause (ozone depletion) and effects (e.g., skin cancer) are properly monitored

Frank R. de Gruijl and Jan C. Van der Leun • Institute of Dermatology, University Hospital, 3508 GA Utrecht, The Netherlands.

Environmental UV Photobiology, edited by Antony R. Young *et al.* Plenum Press, New York, 1993.

together with possible modifying factors (e.g., behavioral changes). Moreover, if there are data available which can be pieced together to arrive at quantitative estimates of environmental impacts, the scientist has the moral obligation to speak up and give the expected magnitudes of effects. Thus, governmental officials, administrators, and politicians are provided with the best possible information for a well-balanced environmental policy.

A possible increase in skin cancers in humans was the first quantified biological effect of a stratospheric ozone depletion. The early estimates have repeatedly been updated and refined. The effect on skin cancer has served as a forerunner in developing the methodology for assessing other biological impacts of ozone depletion.

Epidemiological data are, of course, of primary importance for determining the effects of ozone depletion on skin cancer in man: the relationship between ambient solar UV radiation and skin cancer incidence provides a basis for risk assessments. However, epidemiological data may be heavily confounded by uncontrolled factors like migration, human exposure behavior, and the like (IARC, 1992). Animal experiments can be exploited to gain insight into UV carcinogenesis and to arrive at a better understanding of the epidemiological data. For instance, skin cancer in man can only be correlated to sunlight; it is impossible to deduce which part of the spectrum causes skin cancer. One needs *a priori* knowledge to interpret the data in terms of solar UV exposure. This knowledge is most directly derived from animal experiments. A complementary combination of epidemiological and experimental data provides an optimum basis for risk assessments. In this chapter, we will make a comparison between the human and animal data, and look for parallels that may firm up the risk calculations, or differences that should restrain us in our risk assessments.

1.1. UV Carcinogenesis

In the late nineteenth century, physicians already suspected that regular, extensive exposure to sunlight caused degenerative skin changes ("aging") and skin cancers (excluding malignant melanomas). Unna (1894) made such an inference from his observations on sailors in his clinic in Hamburg, and Dubreuilh (1896) from the high frequency of keratoses and skin cancers among people working in the vineyards in the Bordeaux area. Later epidemiological studies substantiated the relationship of skin cancers with sunlight and, further, with the susceptibility to sunburn (Silverstone and Searle, 1970; Fears *et al.,* 1977; Vitaliano and Urbach, 1980; Scotto and Fears, 1981).

The evidence for a causal relationship between skin cancer and UV radiation followed in the twenties, when Findlay (1928) reported on his experiments in which mice were exposed to radiation from a mercury arc. Later on, Roffo (1939) provided more specific spectral data by showing that the carcinogenic effect disappeared when the radiation (either from a mercury arc or from the sun) was filtered through window glass. Hence, radiation of wavelengths below 315 nm (filtered out by the glass) was carcinogenic, and radiation above 315 nm seemed to do nothing or at most very little. In a period of about forty years after these early experiments there was little progress in the knowledge on the wavelength dependency of carcinogenesis. Interest in the subject was revived in the seventies with the first concerns about a potential ozone depletion, increased UV loads, and the possible increase in skin cancer.

1.2. Types of Skin Cancers and Linkages with UV Radiation

Skin cancers in humans can be divided into two main classes: the cutaneous malignant melanomas (CMMs) and the nonmelanoma skin cancers (NMSCs). In this chapter we will briefly discuss the CMMs, but concentrate on the NMSCs.

1.2.1. Cutaneous Malignant Melanomas

The CMMs stem from the melanocytes (pigment cells) in the skin. These tumors can grow extremely aggressively and metastasize very rapidly. Their relationship with UV radiation is complex (De Gruijl, 1989; IARC, 1992); e.g., these tumors are not preferentially located on skin areas with the highest sun exposure, and office workers (indoor profession) appear to run a higher risk than farmers (outdoor profession). Studies on immigrants (mainly from Great Britain) in (sunny) Australia show that the risk (especially for superficial spreading melanoma) increases markedly for people who arrived before adolescence (Holman and Armstrong, 1984), suggesting that sun exposure in early life is important. Recent data on the experimental induction or growth promotion of melanocytic tumors by UV radiation have made the "UV connection" more credible (Ley et al., 1989; Setlow et al., 1989; Donawho and Kripke, 1991; Husain et al., 1991). Apparently, the risk of melanoma in humans is not simply related to the accumulated UV dose, with the exception of lentigo maligna melanoma.

The steady, dramatic increase in mortality due to melanomas in the Western world since the beginning of this century is associated with an increasing mortality in successive birth cohorts (Magnus, 1973), a trend which has been reported to flatten out (Venzon and Moolgavkar, 1984) and, more recently, has even been reported to reverse in younger birth cohorts (Scotto et al., 1991). These trends, which are not at all well understood, are often attributed to increased sunbathing, fashions in bathing suits, and so forth (no firm quantitative data on exposure trends available).

We consider the UV pathogenesis of melanomas to be riddled with question marks. It is still unclear how the animal data may serve in interpreting the epidemiological data and in constructing a proper quantitative model of the UV-related risk. Based on a correlation between melanoma incidences and ambient UV loads (Scotto and Fears, 1987), one can venture to estimate increases in melanoma after a certain ozone depletion (De Gruijl, 1989), but we will not go into that in this chapter.

1.2.2. Nonmelanoma Skin Cancers

The NMSCs consist of two main subclasses: basal cell carcinomas (BCCs) and squamous cell carcinomas (SCCs). These skin cancers stem from epithelial cells that form the outermost layer of the skin, the epidermis. This part of the skin absorbs most of the carcinogenic UV radiation ($\lambda < 315$ nm). A BCC appears to be composed of undifferentiated cells from the germinal, basal layer of the epidermis and from appendages of the epidermis. An SCC is a neoplasm of epidermal cells that differentiate toward keratin formation, and in advanced stages it will lose any form of structural organization and the cells may become spindle shaped.

Although NMSCs grow slower than CMMs and are far less dangerous, they are genuine cancers: they are invasive, and, if neglected long enough, they can even metastasize, an extremely rare event with BCCs, but not that rare with neglected SCCs. Because NMSCs grow slowly, they are easily sighted and often cosmetically annoying. They are usually removed when they are small and in an early stage of development.

There is a large body of data, both experimental and epidemiological, that substantiates a causal relationship between the accumulated UV dose to the skin and SCCs (e.g., Vitaliano and Urbach, 1980; Forbes et al., 1978). For BCCs there are hardly any experimental data (a few were induced in rats [Hueper, 1942] and nude mice [Anderson and Rice, 1987]). In UV experiments the majority of induced tumors are SCCs and fibrosarcomas (perhaps misdiagnosed SCCs; see Morison et al., 1986) developing at a much higher rate than the occasionally observed BCCs—much like the increased tumor incidence in patients treated with psoralen plus UV-A (PUVA) photochemotherapy (Stern et al., 1984) or in immunosuppressed renal transplant patients (Hardie et al., 1980; Harteveld et al., 1990). Normally, BCCs occur more frequently (3 to 6 times) among Caucasians than SCCs, but in the PUVA-treated and renal transplant patients the situation is dramatically reversed.

The epidemiology of SCC and BCC is broadly speaking, the same: people with a fair complexion (especially those of Irish origin) and those who work outdoors run a higher risk than their respective counterparts, and the tumor sites show a clear preference for regularly exposed skin areas (especially head and neck area). On closer inspection, there are, however, also notable differences: relatively many of the BCCs occur on the trunk (9–12%), and virtually none on the hands (<1%), whereas relatively many SCCs occur on the hands (12–17%), and fewer on the trunk (about 5%) (see Scotto and Fears, 1981). If we assume the tumor frequency to be directly proportional to exposed skin area, then frequency ratios of 3.8 (females) and 6.3 (males) for head/neck versus hand areas for SCCs are not unreasonable, but ratios of 140 (females) and 100 (males) are found for BCCs. The virtual absence of BCCs on the hands suggests that this part of the body lacks a prerequisite for the development of BCCs: could it be that BCCs develop predominantly in sebaceous skin areas (chest and upward)?

As in other parts of Europe, there has been an upward trend in the incidence of BCC in the Netherlands from 1975 through 1988, amounting to a 20% increase in males and a 50% increase in females; for SCC there has been no clear trend (Coebergh et al., 1991). Similar results have been reported for the white population in the USA by Scotto and Fears (1981), who compared incidence data acquired in 1971–72 with those acquired in 1977–78 and found a 15–20% increase in BCC. Strong upward trends in both BCC and SCC have, however, been reported by others (e.g., Gallagher et al., 1990). Standard cancer registries may show variable percentages of underscoring NMSC incidences; time trends in underscoring cannot be ruled out. Why there would be an upward trend in BCC and more variation in SCC is an open question.

Later on we will discuss the possible quantitative differences between SCC and BCC in their dependencies on UV exposure. In a case-control study, Vitaliano and Urbach (1980) reported that the risk of SCC increases more strongly with increasing hours of sun exposure than the risk of BCC. In a cross-sectional prevalence study among 808 white male watermen, Vitasa et al. (1990) found that SCC cases had, on average, received more sunlight between 30 and 60 years of age than their tumor-free colleagues; surprisingly, BCC cases got increasingly less sunlight from 30 to 60 years of age. These latter results on BCC incidence are rather difficult to reconcile with a suspected UV pathogenesis. BCCs probably have lower average growth rates than SCCs. A long, UV-independent latency period after the tumor-initiating UV event could explain why there need not be significantly higher sun exposure in a BCC patient's recent past; a BCC patient may have moderated his sun exposure after finding a first small tumor or because of his sunburn susceptibility. Tumor patients and controls were apparently only age-matched, and surprisingly, the authors found that ease of sunburning was associated with incidence of BCC but not SCC. Although difficult to interpret, the study on the watermen does suggest a difference between incidence of SCC and BCC in UV dose dependency.

1.3. Stratospheric Ozone Depletion

In the early seventies the prospect of large-scale supersonic, stratospheric commercial transport raised the first worries about the lasting integrity of the stratospheric ozone layer. Atmospheric scientists were aware of the fact that the ozone layer constituted a vital, protective atmospheric filter against harmful solar UV radiation. Degradation of this filter would, therefore, pose a direct threat to our biosphere. McDonald (1971) presented the first crude assessment to inform the U.S. Senate of the possible severity of an ozone depletion. The biological impact he anticipated was an increase in NMSCs.

Analogous to "elasticy" in econometry, a dimensionless ratio of a (small) relative change in X (say ozone concentration) over a causal (small), relative disturbance in Y (say water concentration) was employed in atmospheric chemistry to describe reaction schemes. McDonald extended this description to quantify the relationship between NMSC incidence, $I(=X)$, and the amount of stratospheric ozone, $O_3 (=Y)$. Thus, he defined the so-called amplification factor as

$$AF = -(dI/I)/(dO_3/O_3) \tag{1}$$

where dI denotes a small change in I. McDonald estimated the AF to equal 6 by attributing half of the North–South gradient (an estimated overall doubling per 5 degrees latitude, a doubling per 10 degrees due to ozone differences) in NMSCs among the white U.S. population to the differences in the path-lengths for sunlight through stratospheric ozone, i.e., a 1% decrease in ozone would ultimately lead to a 6% increase in NMSC.

Later on, Molina and Rowland (1974) drew attention to the anthropo-genic emission of chlorofluorocarbons into the atmosphere as a potential threat to the stratospheric ozone layer. Around 1980 the public concern about the problem slackened, and criticism on reliability of the assessments increased because the scientists were "only speculating" and could not measure anything. Then the unexpected Antarctic ozone hole was measured in the mideighties, which gave a new impetus to the study of this global environmental problem. Besides skin cancer, other potential health effects, such as cataracts, immuno-suppression, and virus activation, have now come to light. These effects are not well quantified (yet), but they could have a grave impact (Longstreth *et al.*, 1991).

2. Definitions

First we will concisely treat some concepts we will use in analyzing the skin cancer data and in the estimate of the increased skin cancer incidence after a certain ozone depletion.

2.1. Amplification Factors

In McDonald's (1971) approach, the change in the amount of ozone with latitude was directly correlated with a corresponding change in the NMSC incidence: an overall amplification factor, AF, quantified this relationship (Eq. 1). The causal relationship between UV exposure and NMSC was only implicitly present. Possible latitudinal trends in UV exposure that were not related to ozone were discussed and, by necessity, only roughly accounted for.

A better method is to ascertain step by step (1) how the ambient UV load depends on ozone and (2) how the NMSC incidence depends on the UV load. To this end, Van der Leun and Daniels (1975) have rewritten Eq. 1 as

$$AF = RAF \cdot BAF \tag{2}$$

where

$$RAF = -(dUV_{Am}/UV_{Am})/(dO_3/O_3) \tag{3}$$

and

$$BAF = (dI/I)/(dUV_{Am}/UV_{Am}). \tag{4}$$

RAF is the radiation (or optical) amplification factor, BAF is the biological amplification factor, and UV_{Am} is ambient (annual) UV load. The contribution of UV radiation is thus made explicit. In doing so, however, one has brought to light new problems. For instance, what is the proper definition of the UV load; that is, how should the carcinogenic UV dose be measured? To do this, climatological data on carcinogenic UV loads should be acquired and correlated with NMSC incidences.

Ambient UV loads have been measured on a large scale with so-called Robertson–Berger meters. The spectral responses of these R–B meters roughly resemble the skin's spectral response for sunburn; that is, wavelengths below 315 nm are most effective. The RAF for the R–B meter (spectral response from Eq. 1 in Rundel and Nachtwey, 1983) is given in Table 3.

When the carcinogenic UV loads have been measured or calculated, they can be correlated with corresponding NMSC incidences to estimate the BAF. As we will discuss, the RAF can be most directly estimated from data on SCCs induced in mice; such data can also assist in determining the BAF.

2.2. Measures of Tumor Induction

Let us consider a birth cohort of individuals who are regularly exposed to UV radiation and the skin tumors that will subsequently develop.

The *yield, Y(t)*, of tumors at a certain point in time, *t*, is defined as the average number of tumors per individual at risk (including tumors that have been removed). As the majority of skin carcinomas induced by UV radiation are not acutely lethal, an individual may contract multiple primary tumors in the course of time.

The time- or age-specific *incidence, I(t)*, is the rate at which new tumors occur at time *t*, that is, the increase in the yield per unit time:

$$I(t) = dY(t)/dt \tag{5}$$

The *chance, P(t)*, of having contracted a tumor before time *t* equals the percentage of individuals that ever contracted a tumor up to time *t* (i.e., the prevalence of tumor-bearing mice in experiments). Here we assume that no deaths have occurred which were unrelated to tumor induction. But if such deaths do occur there are statistical methods available to correct for censoring effects and to reconstruct the chance in the absence of deaths (a nonparametric, actuarial method and a parametric maximum-likelihood method; e.g., see De Gruijl and Van der Leun, 1991).

If the tumors occur randomly (in the absence of deaths), then

$$P(t) = 1 - \exp[-Y(t)] \tag{6}$$

The *age distribution, n(a)*, defines what fraction of a population is of a certain age *a*; $\int n(a)da = 1$. The overall tumor incidence, *I*, in a population in a certain year is given by an integration of the age-specific incidences weighted by the age distribution.

$$I = \int I(a)n(a)da \tag{7}$$

Here we have assumed that the age-specific incidence, *I(a)*, is the same for every birth cohort in the population (i.e., a stationariness in the induction of tumors).

In epidemiology, the incidence I is usually given per 100,000 persons per year (or per 100,000 personyears), and $I(a)$ as the incidence in a certain age group (say, 40–49 years of age). Equation 7 then simply becomes a summation in which $I(a)$ is weighted by the fraction of the population in that age group.

The relationships in Eqs. 5 and 6 follow from probability theory, where $I(t)$ would be the named hazard function, $Y(t)$ the cumulative hazard function, and $P(t)$ the cumulative probability distribution (e.g., see formula 6.4.3 in Kalbfleisch, 1979).

3. Animal and Human Data on Tumor Incidence

In the sixties, a hairless mouse model became available in the USA for experiments on UV carcinogenesis. Large skin areas of these mice could be exposed to UV radiation without any shaving as with haired mice. Moreover, the hairless mice appeared to be more appropriate as a model for humans because no fibrosarcomas were induced, in contrast with shaved mice. Only SCCs and precursors were induced in the hairless mice. For the sake of brevity we will limit the following discussion to the experimental results our group obtained with SKH:HRI hairless mice, and we will mainly use the epidemiological data on NMSC from a specially designed survey done from 1977 to 1978 in the USA (Scotto and Fears, 1981). In addition, ambient UV measurements were determined with R–B meters. Hence, the data on tumor incidences could be directly correlated with *measured* ambient UV loads.

To simplify the discussion, we will only consider the induction of tumors in terms of yield or incidence (the two are directly related, see Eq. 5). The relevant epidemiological data are given as incidences, and in the animal experiments prevalences as well as yields have been measured simultaneously. It turns out that the statistically expected relationship between prevalence and yield (Eq. 6) holds fairly well, albeit small differences in the susceptibilities or treatments of the individual mice can cause slight disturbances (i.e., a certain tumor yield may be reached somewhat earlier than expected from the prevalence; see De Gruijl *et al.,* 1983; De Gruijl and Van der Leun, 1991).

To simplify the discussion further, we will describe the tumor induction with Weibull statistics; a description with lognormal statistics is also possible, but it is less easy. The cumulative hazard function (the yield) of a Weibull distribution is directly proportional to the stochastic variable (time or UV dose) to a certain power. In simple multi-hit or multi-event theories, this power represents the number of (time- or dose-dependent) "hits" or "events" leading up to a tumor. This interpretation is, however, too naive. We could consider a tumor to be the clonal expansion of a transformed cell, and such a transformation to be an extremely rare event. But we also have to recognize that multitudes of cells in each individual run the risk of such a transformation. Based on the theory of the "probability of extremes," we can then expect a host of probabilistic models to asymptotically approach Weibull statistics for tumor induction in individuals (Pike, 1966; De Gruijl, 1982). In spite of this reservation, we do think that this power is directly related to stages in the tumorigenesis (De Gruijl and Van der Leun, 1991).

3.1. Time or Age Dependence

If people persist in a certain level of solar exposure throughout their lives they will steadily accumulate their lifetime dose of UV radiation. The age or time dependency of tumor occurrences is, therefore, related to an increasing total UV dose.

When the backs of 24 SKH:HRI mice were exposed daily to 1.5 kJ/m^2 UV radiation from a Westinghouse FS40 sunlamp, the first small tumors (diameter < 1 mm) were observed after 6 to 7 weeks. A yield of 1 for 1-mm tumors was reached in about 73 days; for 4-mm tumors, in about 135 days. Tumors smaller than 4 mm showed a wide variation in growth speed, some stopped growing, and some even regressed. With lower daily doses it took longer for the yield to build up: with 0.05 kJ/m^2/day of UV it took about 500 days to reach a yield of 1 for 1-mm tumors, and about 650 days for 4-mm tumors. The time difference between the yield of 1-mm and 4-mm tumors was, therefore, not simply a constant time delay introduced by a steady growth of tumors; some 1-mm tumors did not make it to the 4-mm size (see De Gruijl et al., 1983; De Gruijl and Van der Leun, 1991).

For all tumor sizes up to 4 mm in diameter and all daily doses, the yield as a function of time is well described by

$$Y(t) = (t/t_1)^p \tag{8}$$

with $t = 0$ on the day of first exposure, $Y(t_1) = 1$, and t_1 depending on tumor size; p is the power mentioned above—it determines the steepness of $Y(t)$. The mean of p for tumors with diameters smaller than 2 mm is 7.2 (see Table 1 under SCC), and the range is 4.4 to 9.6. Because of the relatively small group sizes (24–48 mice), the errors in the values of p are rather large (10–15%).

If we look at a human population at a certain moment in time we get a cross-sectional view of the age-specific incidence. If we subsequently assume stationariness in tumor induction (as with Eq. 7), then this cross-sectional age-specific incidence should be equal to the longitudinal time-specific incidence in each birth cohort. As discussed in Section 1.2.2, upward trends have, however, been reported and may have affected the cross-sectional age dependency.

The incidence of both SCC and BCC in humans increases strongly with age. We assume that every person exposes himself or herself regularly to sunlight, and we take the age to equal the number of years of exposure to sunlight, i.e., $t = a$ in Eq. 8 and t_1 becomes a_1. According to Eq. 5, $I(a)$ is then found by differentiating $Y(a)$ with respect to a:

$$I(a) = (p/a_1)(a/a_1)^{p-1} \qquad (9)$$

The incidence of both SCC and BCC can be described by Eq. 9 for ages 30 to 80 years. In Table 1 we give the values of p for SCC and BCC as we determined them by fitting a linear relationship to $\ln[I(a)]$ versus $\ln(a)$ by the method of error-weighted least squares to data from the Netherlands (Coebergh

Table 1. Time- or Age-Specific Tumor Induction: Values of p

		SCC	BCC
mice (SKH:HRI)	m+f	7.2 ± 0.8	—
Iceland	m	7.5 ± 0.9	5.5 ± 0.3
	f	—	4.5 ± 0.3
Netherlands	m	6.6 ± 0.4	5.4 ± 0.1
	f	8.9 ± 0.7	4.8 ± 0.1
Switzerland	m	7.6 ± 0.2	5.0 ± 0.4
	f	6.0 ± 0.5	4.1 ± 0.2
Seattle	m	6.1 ± 0.8	4.5 ± 0.2
	f	4.6 ± 0.6	4.7 ± 0.3
Minneapolis	m	6.4 ± 0.1	5.0 ± 0.2
	f	6.5 ± 0.6	5.0 ± 0.2
Detroit	m	5.8 ± 0.5	4.7 ± 0.3
	f	6.3 ± 0.3	4.7 ± 0.3
Utah	m	5.2 ± 0.3	4.5 ± 0.3
	f	5.3 ± 0.3	4.7 ± 0.3
San Francisco	m	5.8 ± 0.3	4.6 ± 0.3
	f	4.7 ± 0.2	4.7 ± 0.3
Atlanta	m	5.7 ± 0.4	4.6 ± 0.3
	f	5.3 ± 0.2	4.7 ± 0.3
New Orleans	m	5.4 ± 0.5	4.7 ± 0.3
	f	4.5 ± 0.3	4.6 ± 0.3
New Mexico	m	4.7 ± 0.5	4.7 ± 0.3
	f	5.3 ± 0.3	4.8 ± 0.3
New Zealand	m	4.9 ± 0.4	4.5 ± 0.2
(North Island)	f	5.4 ± 0.6	4.2 ± 0.2

Note. ± standard error; m = males, f = females.

et al., 1991), Switzerland (Levi *et al.,* 1988), Iceland (McKnight and Mag-nussen, 1979), several locations in the USA (Scotto and Fears, 1981), and New Zealand (Freeman *et al.,* 1982). The values of p for SCC at high latitudes are strikingly close to the 7.2 found in mice (Table 3a in De Gruijl, 1982). As we go closer to the equator (New Zealand, and the USA), we see a drop in the value of p for SCC to about 5. We see practically no trend (perhaps a very small one) in the value of p for BCC; it varies around 5.

The values of p for BCC could well be affected by the observed increase in the incidence over the years. This could be due to a gradual increase in BCC in successive birth cohorts, similar to the increase in CMM, only less strong. Such a gradual increase in successive birth cohorts will reduce the steepness of the age-specific incidence in a cross-sectional view at a certain moment in time.

The value of p will also decrease as the "center of gravity" of lifetime UV exposure is shifted more toward early life (De Gruijl and Van der Leun, 1991).

Incidences in humans were recorded as the number of *cases* with *newly* diagnosed carcinomas per 100,000 personyears. More than one new tumor on a patient was counted as a single new case. Approximately 10% of the cases had multiple skin cancers, 2.5% both BCC and SCC (Scotto and Fears, 1981). Note that this makes these reported incidences slightly different from our theoretical definition in Eq. 5.

The percentage of cases with multiple tumors seems rather high, not what one would expect if the tumors occur randomly. This may indicate that people with an NMSC run an exceedingly high risk of contracting more NMSCs; together with the mere fact of their advanced age, this higher risk can be caused by their genetic background, by their (professional) behavior and related UV exposure pattern, perhaps by the mere fact of their having contracted a first NMSC (a conditioning, perhaps, of an immunologic nature), or a combination of these possibilities. But there is also a much simpler, quantitative reason to consider as the direct cause of the tumor multiplicity: the delay between the first appearance of a tumor and consultation with a physician. During this patient delay there is a possibility of contracting a second tumor. The probability of such an event will increase as the delay becomes longer and as the incidence of tumors increases (either with age or with increasing solar UV load). We have hardly any data on this delay, but we expect it to be quite substantial: perhaps 2 months to 5 years for BCCs on the face and 2 to 10 years for BCCs on the trunk. With the U.S. incidence of BCCs, say 1,000 to 1,500 per 100,000 personyears at 60 years of age, with a delay of 5 years, the percentage of cases with multiple tumors in this age group would be on the order of 5% to 8%. In combination with the presence of high-risk groups (outdoor workers), this delay may easily explain the high percentage of cases with multiple tumors. Interestingly, substantial delays to consultation will also tend to lower the value of p. This might explain a decrease in p with decreasing latitudes, increasing UV loads, and increasing incidences.

3.2. Dose Dependence

As mentioned in Section 3.1, the rate of tumor occurrence in mice is lower with lower daily doses: it takes longer to reach a certain yield. For tumors up to 1 mm in diameter we can simply rewrite Eq. 8 as

$$Y(t) = (D/D_0)^{p1}(t/t_0)^p \tag{10}$$

where D is the daily dose, $p1$ is a constant power of D, and D_0 and t_0 are mutually dependent constants. By combining Eqs. 8 and 10 we find that

$$t_1 D^r = \text{constant} \tag{11}$$

with $r = p1/p$. The relationship in Eq. 11 is easily determined from our experiments, and we find that $r = 0.61$ (De Gruijl $et\ al.$, 1983). Hence, $p1 = 4.4 (=0.61\ p)$. For 4-mm tumors r can become 0.5, but the mean of p increases to about 8.5, and $p1$ becomes 4.3, which is consistent with the p for 1-mm tumors (Table 3a in De Gruijl, 1982; De Gruijl and Van der Leun, 1991). In a direct multiparameter fit of the data, $p1$ varies from 3.2 to 4.1, and p varies between 5.8 and 7.0 (Table 5 in De Gruijl, 1982). In all, the value of $p1$ ranges from 3.2 to 4.4 for SCC in mice, but we think it is closer to the latter value.

Following Eq. 10 for the dose dependency of the yield, one would expect the incidence in humans to be of the form

$$I(a) = (p/a_0)(\text{UV}_{\text{Am}}/\text{UV}_0)^{p1}(a/a_0)^{p-1} \tag{12}$$

where UV_{Am} is the annual ambient UV dose. Substituting Eq. 12 into Eq. 7, we would expect to find that

$$I \propto \text{UV}_{\text{Am}}^{p1} \tag{13}$$

The U.S. data on incidences can be correlated with UV_{Am} as measured by R–B meters: $\ln(I)$ versus $\ln(\text{UV}_{\text{Am}})$ can be fitted with a linear relationship by the method of error-weighted least squares (see Fig. 1). The results for $p1$ are given in Table 2. Figure 1 clearly shows that much of the variation in I is not explained by Eq. 13. Apparently, local factors (exposure habits, genetic com-

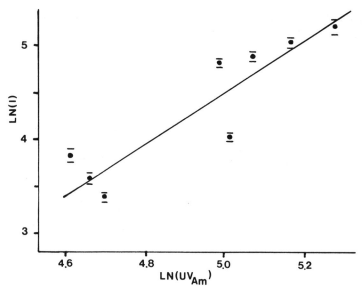

Figure 1. Ln(incidence of SCC) versus ln(UV$_{Am}$) for white, U.S. male populations, UV$_{Am}$ measured by R–B meters. Data from Scotto and Fears (1981).

position of the population, etc.) introduce large deviations (up to 50%) from the fitted relationship. This is reflected in the rather large errors in $p1$.

In animal experiments, the animals receive well-controlled UV exposures, whereas epidemiological data for humans reflect a wide variation in exposure habits. We could, however, assume a human population to consist of sub-populations with certain UV susceptibilities and certain annual UV loads. We assume that the white U.S. populations at different geographic locations have similar subpopulations in which personal UV exposures are proportional to UV$_{Am}$. Each of the subpopulations at a certain location has a different rate of tumor induction (i.e., a different set of UV$_0$, a_0 values), but we could assume

Table 2. Dose Dependence: Values of $p1$

		SCC	BCC
SKH:HRI mice	f+m	3.2–4.4	—
U.S. whites	m	2.7 ± 0.7	1.4 ± 0.5
	f	2.8 ± 0.7	1.4 ± 0.4

Note. ± standard error; m = males, f = females.

the induction kinetics, as expressed by $p1$ and p, to be the same in all of these subpopulations. The overall yield at a location is a weighted average of all the subyields, and it is of the form given in Eq. 10. The overall incidence is then properly represented by Eq. 12.

The fact that $p1 < p$ appears to signify that the age dependency is stronger than the dependency on the annual UV load; that is, there are probably important rate-limiting steps in the tumorigenesis which are not related to UV radiation (De Gruijl and Van der Leun, 1991).

Instead of directly establishing the parallels between the dose and time dependencies in mouse and human (Eq. 10 versus Eq. 12), one can give Eq. 11 a pivotal role by assuming that it is a fundamental relationship between dose fraction (per day or per year) and tumor induction time, as Blum (1959) did. We can use Eqs. 10 and 11 for humans (i.e., $t = a$) to find the value of $p1$ as $r \times p$, with $r = 0.5$ to 0.6 as determined in animal experiments. In an earlier study this approach yielded a dose dependency consistent with the one measured for the NMSC data from a U.S. survey conducted in 1971–72 (De Gruijl and Van der Leun, 1980). Here we can use the new values of p in Table 1 and take $r = 0.5$. For SCC in the USA we then find values of $p1$ ranging from 2.3 to 3.3, which is nicely in line with the results given in Table 2. For BCC we get values for $p1$ ranging from 2.3 to 2.5, which appear to be too high in comparison with the results in Table 2. The dose–time relationship in Eq. 11 with $r = 0.5$ is apparently not valid for BCC in the U.S. population.

The value of $p1$ found for SCC in the white U.S. population appears to be smaller than $p1$ in the mice (see Table 2), although the ranges of possible values overlap. A reduction in $p1$ could simply be caused by different exposure habits (e.g., people in sunnier areas may, on average, spend less time in sunlight), or by a relative decrease in UV-susceptible people in sunnier areas. As discussed in Section 3.1, patient delays will tend to increase the percentage of cases with multiple tumors, and this will also tend to decrease $p1$. Migration will also reduce $p1$. If a substantial portion of SCCs in humans are unrelated to solar UV exposure, the UV dose dependence, and possibly the age dependence, too, could become less steep. Or if the solar UV radiation competes with another carcinogen in only one of several steps in the carcinogenic process, the UV dose dependency would be lessened. Following this line of thought, it is imaginable that the $p1$ and p for BCC are about 1 to 1.5 units smaller than those for SCC because the genesis of BCC lacks a UV-related step in comparison to that of SCC.

The factors mentioned above could explain a reduced $p1$ in humans, but there could, of course, also be a more fundamental reason: a physiological difference between mouse and human. The absence of pigment in the albino skin of the mice is, of course, a factor to reckon with, but this animal model seems appropriate for the most susceptible, fair-skinned people (note that pigmented mice have mainly pigment in the dermis, not so much in the epidermis, in contrast to humans). It is known that the epidermis thickens both in humans and mice under chronic UV exposure. In mice this thickening has been found to develop rapidly with initial exposure and more gradually at a later stage (De Gruijl and Van der Leun, 1982). How the human epidermis behaves during an annual cycle of UV exposure, and in the course of a lifetime, is not exactly known. Because the epidermis protects its germinal, basal layer by filtering out much of the harmful UV radiation, a thickening of the epidermis could offer more protection against UV tumorigenesis. Differences in the kinetics of this epidermal thickening in mouse and human could introduce differences in the dose and time dependencies of UV tumorigenesis. Differences in DNA-repair proficiencies between mouse and human may introduce differences in susceptibility to UV tumorigenesis, but not necessarily in the induction kinetics as expressed by $p1$ and p; the process could simply be speeded up by a higher mutability of cells in mice.

It is important to note here that the dose dependency measured in the U.S. survey hinges on UV loads as measured by R–B meters. If the spectral response of the R–B meter is dramatically different from the wavelength dependency of the carcinogenic UV dose, the measured dose dependency will be incorrect. We will see later on that corrections will tend to make $p1$ smaller.

3.3. Wavelength Dependence: The Action Spectrum

The wavelength dependence of UV carcinogenesis is crucial in an assessment of the carcinogenic risk of ozone depletion. Here, the human data on skin cancer cannot provide a direct answer; this spectral information has to come from other sources. The crude knowledge on the wavelength dependence of UV carcinogenesis (see Section 1.1) has led some scientists to suggest that an erythemal (sunburning) UV dose may be used as a carcinogenic dose. Others have criticized the soundness of such a substitution, and have suggested that DNA damage (Setlow, 1974) or mutations (Kubitschek *et al.,* 1986) induced by UV radiation, after correction for epidermal transmission, provide a mechanistically better basis for the assessment of the carcinogenic UV dose. These approaches yield substitutes for or, at best, plausible constructions of the wavelength dependence.

Animal experiments provide the most direct way of ascertaining the wavelength dependence of UV carcinogenesis. Ideally one would like to irradiate the mice monochromatically—vary the wavelength in steps of 10 nm throughout the UV range, check the dose–time relationship at each wavelength, and ascertain how this relationship shifts as the effectiveness changes from one wavelength to the next. In reality this is near to impossible because the mice have to be irradiated chronically (or at least for a substantial length of time), and using a single expensive monochromator would create even more expensive routine handwork for technicians. Moreover, the monochromator has to be powerful enough to be effective at each wavelength throughout the UV, and getting enough power (especially at >340 nm) may conflict with the demand of monochromaticity. A more practical approach is to use an "ensemble" of different broadband spectra from various types of UV sources. Data acquired with such an ensemble of spectra implicitly contain the information on the wavelength dependence. Our group and the photobiological group of the former Skin and Cancer Hospital in Philadelphia have accumulated such data.

Cole *et al.* (1986) found that the edemal (MEE48) dose in SKH:HRI mice yielded a fair estimate of the carcinogenic dose for these mice. In 1987 our group made a first attempt to extract the wavelength dependence from the animal data (Sterenborg and Van der Leun, 1987; Slaper, 1987). With more recent data added, we found that none of the presently available wavelength dependencies described our data on experimental UV carcinogenesis adequately (De Gruijl and Van der Leun, 1992). We therefore derived the wavelength dependency anew (De Gruijl *et al.*, 1993) and represented it as a so-called action spectrum, dubbed SCUP. The action spectrum is a set of factors to weight the contributions of different wavelengths according to their respective effectivenesses; the sum of these weighted contributions equals the carcinogenic UV dose. The SCUP action spectrum is depicted in Fig. 2 together with the earlier experimental action spectra for UV carcinogenesis.

4. Effect of Ozone Depletion

We have now gathered the required data for an assessment of the impact of ozone depletion on NMSC incidence.

4.1. Radiation Amplification Factor

In Table 3 some RAFs are given for various carcinogenic UV doses. Note that the RAFs for the R–B meter responses are substantially lower than the other RAFs, which is mainly due to the fact that longer wavelengths (>315 nm) are relatively more effective in the R–B meter response, and the radiation at these wavelengths is hardly affected by ozone depletion.

Because the SCUP action spectrum is inferred directly from experimental data on UV tumorigenesis, we will use this RAF of 1.4 in our assessment. It should be noted, however, that other plausible RAFs are not much different from 1.4; they range from 1.3 to 1.6.

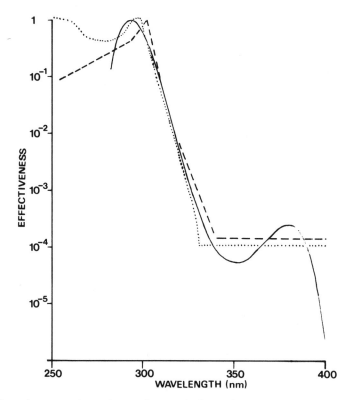

Figure 2. Action spectra for carcinogenesis: the solid line depicts the SCUP action spectrum, the dotted line the edemal action spectrum MEE48, and the dashed line the "Sterenborg–Slaper" action spectrum.

It should be noted that these RAFs are based on a small change (-1%) in ozone. Because the carcinogenic dose may not follow larger changes in ozone according to a power relation with a constant RAF, the change in the carcinogenic UV dose can best be calculated directly for a large change in ozone.

4.2. Biological Amplification Factor

The BAF is equal to $p1$ in Eqs. 12 and 13 (substitute Eq. 13 into Eq. 4), if the carcinogenic UV_{Am} is correct in these equations. The presented values of $p1$ for the human data are based on UV_{Am} doses as measured by R–B meter. These values of $p1$ have to be divided by 1.106 to adjust them to values of $p1$ for the SCUP-effective UV doses (using the empirical method of Green

Table 3. Radiation Amplification Factors
for Various Effective UV Doses

	RAF	
Type of UV dose:	30° N lat.	0–65° N lat.
R–B meter		
[Rundel and Nachtwey (1983)]	0.98	0.96–1.06
Erythema in man[a]		
[24 h; Parrish et al. (1982)]	1.31	1.25–1.33
IEC[b] erythemal dose[a]	1.15	1.08–1.18
Mutations, through epidermis[a]	1.56	1.54–1.60
Edema in mice[a]		
[MEE48; Cole et al. (1986)]	1.54	1.50–1.56
Carcinogenesis in mice[a]		
[Sterenborg and Van der Leun (1987);		
Slaper (1987)]	1.59	1.58–1.59
Carcinogenesis in mice		
[SCUP; de Gruijl et al., (1993)]	1.40	1.39–1.41

Note. UV radiation through 3.2 atm.mm ozone.
[a] From Kelfkens et al. (1990).
[b] International Electrotechnical Committee.

and Hedinger, 1978). Hence, the values of $p1$ in Table 2 are lowered by 10%, which is a small change considering the errors.

The value of $p1$ for SCC in humans (2.5 ± 0.6 after correction) is lower than that found in mice (4.3, and maybe as low as 3.2), whereas the values for p for humans at high latitudes are in the range found in the animal experiments. This might indicate that the $p1$ from the human data has been estimated too low. Several factors have been discussed that could have reduced the estimate of $p1$ from ambient UV loads at different geographic locations. The U.S. data on SCC do, however, appear to obey Eq. 11 with $r = 0.5$ as for large tumors in mice (i.e., $p1/p \approx 0.5$). We are not yet able to carry out further corrections on the human data and will, therefore, use the values of $p1$ for humans as estimates of BAFs; for BCC we have to use these values, because there are no experimental data on BCC available.

The present BAFs apply to the white U.S. population because they are based on values of $p1$ derived from that population. Considering the higher values of p for SCC at higher latitudes, it is imaginable that $p1$ also increases, especially if $p1/p \approx 0.5$ holds. This could increase the BAF by as much as 40%. However, this expectation appears to be nullified by the results obtained by Moan et al. (1989) with Norwegian data. They correlated incidence data from registries with computed annual ambient UV loads (measured either as erythemal or mutation-effective doses), and found a lower BAF for SCC, 1.6–1.8, and, surprisingly, a higher BAF for BCC, 2.1–2.3. The range of ambient UV loads was rather limited: UV_{Am} varied by a factor of 1.6. In the U.S. data the range was somewhat larger: the R–B meter dose varied by a factor of 2.0, and the SCUP dose by a factor of 2.1.

Note that the BAF is based on a small change in UV_{Am}. Large changes in UV_{Am} should be dealt with through Eq. 13.

4.3. Overall Amplification Factor

Having established the best estimates of RAF and BAF, it is simple arithmetic to find the overall amplification factors (AFs). Based on the U.S. data, we find that AF = 1.4 × 2.5 = 3.5 (±0.8) for SCC, and AF = 1.4 × 1.4 = 2.0 (±0.6) for BCC. Taking BCC and SCC together in a ratio of 4:1, we find the overall AF = 2.3 (±0.5) for NMSC as the weighted average of the AFs of BCC and SCC.

Again, it has to be mentioned that these AFs pertain to a small change in ozone; they concisely quantify the sensitivity to an ozone depletion. Larger changes should be dealt with as suggested in Sections 4.1 and 4.2.

5. Conclusions

In looking over the body of data summarized above, it is clear that we cannot claim complete certainty about the UV dose dependencies of BCC and SCC incidences in human populations. There is still a need to improve on the epidemiological data and to find out more about BCC and melanoma experimentally. Consequently, the impact of ozone depletion on these incidences can only be assessed with some reservations. Nevertheless, the data offer enough information to make some quantitative assessments, albeit with rather large margins of error.

The wavelength dependence for UV carcinogenesis has been determined in animal experiments. These data provide the most direct information on this wavelength dependence, which cannot be derived from human data. The UV doses measured by R–B meters can be adjusted to carcinogenic doses, but these corrections are relatively small with regard to the errors in the dose dependencies.

Both the human and mouse data on SCC incidence can be described by Weibull statistics, where the UV dose fraction (a daily or yearly dose) and exposure time (age for humans) are the stochastic variables. The time dependency in animals appears to be equal to that of humans at high latitudes (i.e., equal values of p), and the dose–time relationship in animals could be equal to that in humans (i.e., equal values of $r = p1/p \approx 0.5–0.6$). However, the time and dose dependencies in white U.S. populations appear to be somewhat less than those in mice (if incidences are correlated with ambient UV loads, p and $p1$ lower than in mice by 1.5–2.0). As discussed, the epidemiological data could be affected by many factors that could have reduced these dose and time dependencies. However, there are no data available to make corrections for these potentially modifying effects. Estimates of the dose dependence in humans should, of course, preferentially be based on human data. The animal data can be used to substantiate and refine the Weibull model for the human data. The animal data indicate that the dose dependency of SCC extracted from epidemiological data could be too small. Epidemiology is the sole source of quantitative information for BCC.

For the U.S. white population we find that a 1% decrease in ozone would ultimately yield a $3.5\% \pm 0.8\%$ increase in SCC, a $2\% \pm 0.6\%$ increase in BCC, and an overall $2.5\% \pm 0.5\%$ increase in NMSC.

This assessment gives numbers on expected differences between two stationary situations. An assessment of the effects during a change in ozone is much harder; it requires additional information on tumor responses in the period after changes in dosages. There are no human data available, but animal experiments can assist in such an assessment for SCC (De Gruijl and Van der Leun, 1991).

ACKNOWLEDGMENT. The comments by Bruce Armstrong, Ph.D., were greatly appreciated.

References

Anderson, D. E., and Rice, J. M., 1987, Tumorigenesis in athymic nude mice skin by chemical carcinogenesis and ultraviolet light, *J. Nat. Cancer Inst.* 78:125–134.

Blum, H. F., 1959, *Carcinogenesis by Ultraviolet Light,* Princeton University Press, Princeton, NJ.

Coebergh, J. W. W., Neumann, H. A. M., Vrints, L. W., Van der Heijden, L., Meijer, W. J., and Verhagen-Teulings, M. Th., 1991, Trends in the incidence of non-melanoma skin cancer in the SE Netherlands 1975–1988: A registry-based study, *Br. J. Dermatol.* 125:353–359.

Cole, C., Forbes, P. D., and Davies, R. E., 1986, An action spectrum for photocarcinogenesis, *Photochem. Photobiol.* 43:275–284.

De Gruijl, F. R., 1982, *The Dose-Response Relationship for UV Tumorigenesis,* Ph.D. thesis, University of Utrecht, pp. 2–5, 51–63.

De Gruijl, F. R., 1989, Ozone change and melanoma, in: *Atmospheric Ozone and Its Policy Implications* (T. Schneider *et al.,* eds.), Elsevier, Amsterdam, pp. 813–821.

De Gruijl, F. R., and Van der Leun, J. C., 1980, A dose-response model for skin cancer induction by chronic UV exposure of a human population, *J. Theor. Biol.* 83:487–504.

De Gruijl, F. R., and Van der Leun, J. C., 1982, Effect of chronic UV exposure on epidermal transmission in mice, *Photochem. Photobiol.* 36:433–438.

De Gruijl, F. R., and Van der Leun, J. C., 1991, Development of skin tumors in hairless mice after discontinuation of ultraviolet irradiation, *Cancer Res.* 51:979–984.

De Gruijl, F. R., and Van der Leun, J. C., 1992, Action spectra for carcinogenesis, in: *The Biologic Effects of UVA Radiation* (F. Urbach, ed.), Valdemar, Overland Park, Kansas, pp. 91–98.

De Gruijl, F. R., Van der Meer, J. B., and Van der Leun, J. C., 1983, Dose-time dependency of tumor formation by chronic UV exposure, *Photochem. Photobiol.* 37:53–62.

De Gruijl, F. R., Sterenborg, H. I. C. M., Forbes, P. D., Davies, R. E., Cole, C., Kelfkens, G., Van Weelden, H., Slaper, H., and Van der Leun, J. C., 1993, Wavelength dependence of skin cancer induction by ultraviolet irradiation of albino hairless mice, *Cancer Res.* 53:53–60.

Donawho, C. K., and Kripke, M. L., 1991, Evidence that the local effect of ultraviolet radiation on the growth of murine melanomas is immunologically mediated, *Cancer Res.* 51:4176–4181.

Dubreuilh, W., 1896, Des hyperkeratoses circonscriptes, *Ann. Dermatol. Syphiligr.* (Paris) 7:1158–1204.

Fears, T. R., Scotto, J., and Schneiderman, M. A., 1977, Mathematical models of age and ultraviolet effects on the incidence of skin cancer among whites in the United States, *Am. J. Epidemiol.* 105:420–427.

Findlay, G. M., 1928, Ultraviolet light and skin cancer, *Lancet* 2:1070–1073.

Forbes, P. D., Davies, R. E., and Urbach, F., 1978, Experimental ultraviolet photocarcinogenesis: Wavelength interactions and time-dose relationship, in: *NCI Monograph 50* (M. L. Kripke, and E. R. Sass, eds.), NCI, Bethesda, MD, pp. 31–38.

Freeman, N. R., Fairbrother, G. E., and Rose, R. J., 1982, Survey of skin cancer incidence in the Hamilton area, *N. Z. Med. J.* 95:529–533.

Gallagher, R. P., Ma, B., MacLean, D. I., Yang, C. P., Ho, V., Carruthers, J. A., and Warshawki, L. M., 1990, Trends in basal cell carcinoma, squamous cell carcinoma, and melanoma of the skin from 1973 through 1987, *J. Am. Acad. Dermatol.* 23:413–421.

Green, A. E. S., and Hedinger, R. A., 1978, Models relating ultraviolet light and non-melanoma skin cancer incidences, *Photochem. Photobiol.* 28:283–291.

Hardie, I. R., Strong, R. W., Hartley, L. C. J., Woodruff, P. W. H., Clunie, G. J. A., 1980, Skin cancer in Caucasian renal allograft recipients living in a subtropical climate, *Surgery* 87:177–180.

Harteveld, M. M., Bouwes Bavinck, J. N., Kootte, A. M. M., Vermeer, B. J., and Vandenbroucke, J. P., 1990, Incidence of skin cancer after renal transplantation in the Netherlands, *Transplantation* 49:506–509.

Holman, C. D. J., and Armstrong, B. K., 1984, Cutaneous malignant melanoma and indicators of total accumulated exposure to the sun: An analysis separating histogenic types, *J. Natl. Cancer Inst.* 73:75–82.

Hueper, W. C., 1942, Morphological aspects of experimental actinic and arsenic carcinomas in the skin of rats, *Cancer Res.* 2:551–559.

Husain, Z., Pathak, M. A., Flotte, T., and Wick, M. M., 1991, Role of ultraviolet radiation in the induction of melanocytic tumors in hairless mice following 7,12-dimethylbenz(a)anthracene application and ultraviolet irradiation, *Cancer Res.* 51:4964–4970.

IARC, 1992, *Solar and Ultraviolet Radiation. Monographs on the Evaluation of Carcinogenic Risks to Humans,* Vol. 55, International Agency for Research on Cancer, Lyon, France.

Kalbfleisch, J. G., 1979, *Probability and Statistical Inference I*, Springer Verlag, New York.

Kelfkens, G., De Gruijl, F. R., and Van der Leun, J. C., 1990, Ozone depletion and increase in annual carcinogenic ultraviolet dose, *Photochem. Photobiol.* 52:819–823.

Kubitschek, H. E., Baker, K. S., and Peak, M. J., 1986, Enhancement of mutagenesis and human skin cancer rates resulting from increased fluences of solar ultraviolet radiation, *Photochem. Photobiol.* 43:443–447.

Levi, F., Vecchia, C. L., Te, V. C., and Mezzanotte, G., 1988, Descriptive epidemiology of skin cancer in the Swiss canton of Vaud, *Int. J. Cancer* 48:811–816.

Ley, R. D., Applegate, L. A., Padilla, R. S., and Stuart, T. D., 1989, Ultraviolet radiation-induced malignant melanoma in *Monodelphis domestica, Photochem. Photobiol.* 50:1–5.

Longstreth, J. D., De Gruijl, F. R., Takizawa, Y., and Van der Leun, J. C., 1991, Human health, Chap. 2 in UNEP report *Environmental Effects of Ozone Depletion: 1991 Update* (J. C. Van der Leun, and M. Tevini, eds.), UNEP, Nairobi, Kenya, pp. 15–24.

Magnus, K., 1973, Incidence of malignant melanoma of the skin in Norway, 1955–1970, variations in time and space and solar radiation, *Cancer* 32:1275–1286.

McDonald, J. E., 1971, Statement submitted at hearings before the House Subcommittee on Transportation, in: *Cong. Rec.* 117(39), p. 3493.

McKnight, C. K., and Magnussen, B., 1979, Tumours in Iceland, *Acta Pathol. Microbiol. Scand. Sect. A* 87:37–44.

Moan, J., Dahlback, A., Hendriksen, T., and Magnus, K., 1989, Biological amplification factor for sunlight-induced nonmelanoma skin cancer at high latitudes, *Cancer Res.* 49:5207–5212.

Molina, M. J., and Rowland, F. S., 1974, Stratospheric sink for chlorofluoromethanes: Chlorine atom-catalysed destruction of ozone, *Nature* 249:810–812.

Morison, W. L., Jerdan, M. S., Hoover, T. L., and Farmer, E. R., 1986, UV radiation-induced tumors in haired mice: Identification as squamous cell carcinomas, *J. Natl. Cancer Inst.* 77: 1155–1162.

Parrish, J. A., Jaenicke, K. F., and Anderson, R. R., 1982, Erythema and melanogenesis action spectra of normal skin, *Photochem. Photobiol.* 40:485–494.

Pike, M. C., 1966, A method of analysis of a certain class of experiments in carcinogenesis, *Biometrics* 22:142–161.

Roffo, A. H., 1939, Über die physikalische Aetiologie der Krebskrankheit, *Strahlentherapie* 66: 328–350.

Rundel, R. D., and Nachtwey, D. S., 1983, Projections of increased non-melanoma skin cancer incidence due to ozone depletion, *Photochem. Photobiol.* 38:577–591.

Scotto, J., and Fears, T. R., 1981, *Incidence of Nonmelanoma Skin Cancer in the United States,* publ. no. *NIH 82-2433,* U.S. Dept. of Health and Human Services, Washington, DC.

Scotto, J., and Fears, T. R., 1987, The association of solar ultraviolet and skin melanoma incidence among caucasians in the United States, *Cancer Investig.* 5:275–283.

Scotto, J., Pitcher, H., and Lee, J. A. H., 1991, Indications of future decreasing trends in skin melanoma mortality among whites in the United States, *Int. J. Cancer* 49:490–497.

Setlow, R. B., 1974, The wavelengths in sunlight effective in producing skin cancer: a theoretical analysis, *Proc. Natl. Acad. Sci. U.S.A.* 71:3363–3366.

Setlow, R. B., Woodhead, A. D., and Grist, E., 1989, Animal model for ultraviolet radiation-induced melanoma: Platyfish–swordtail hybrid, *Proc. Natl. Acad. Sci. U.S.A.* 86:8922–8926.

Silverstone, H., and Searle, J. H. A., 1970, The epidemiology of skin cancer in Queensland: Influence of phenotype and environment, *Br. J. Cancer* 24:235–252.

Slaper, H., 1987, Action spectra for photocarcinogenesis, Chap. 7 in: *Skin Cancer and UV Exposure: Investigations on the Estimation of Risk,* Ph.D. thesis, University of Utrecht, pp. 147–152.

Sterenborg, H. J. C. M., and Van der Leun, J. C., 1987, Action spectra for tumorigenesis by ultraviolet radiation, in: *Human Exposure to Ultraviolet Radiation: Risks and Regulations* (W. F. Passchier and B. F. M. Bosnjakovic, eds.), Elsevier, Amsterdam, pp. 173–190.

Stern, R. S., Laird, N., Melski, J., Parrish, J. A., Fitzpatrick, T. B., and Bleich, H. L., 1984, Cutaneous squamous cell carcinoma in patients treated with PUVA, *N. Engl. J. Med.* 310: 1156–1161.

Unna, P., 1894, *Histopathologie der Hautkrankheiten,* August Hirschwald, Berlin.

Van der Leun, J. C., and Daniels, F., 1975, Biological effects of stratospheric ozone decrease: A critical review of assessments, in: *CIAP Monograph* 5 Part 1, Chapter 7, Appendix B (D. S. Nachtwey, ed.), Dept. of Transportation, Washington, DC, pp. 105–124.

Venzon, D. J., and Moolgavkar, S. H., 1984, Cohort analysis of malignant melanoma in five countries, *Am. J. Epidemiol.* 119:62–70.

Vitaliano, P. P., and Urbach, F., 1980, The relative importance of risk factors in nonmelanoma carcinoma, *Arch. Dermatol.* 116:454–456.

Vitasa, B. C., Taylor, H. R., Strickland, P. J., Rosenthal, F. S., West, S., Abbey, H., Ng, S. K., Munoz, B., and Emmett, E. A., 1990, Association of nonmelanoma skin cancer and actinic keratosis with cumulative solar ultraviolet exposure in Maryland watermen, *Cancer* 65:2811–2817.

5

UV-Induced Immunosuppression

Relationships between Changes in Solar UV Spectra and Immunologic Responses

Frances P. Noonan and Edward C. De Fabo

1. Introduction

In mammals, ultraviolet-B radiation impacts on the skin and on the eyes (Zigman, Chapter 6 this volume). Because of effective absorption by skin components, very little of the incident UV-B penetrates beyond the epidermis. In spite of this, UV-B radiation is intimately linked with the mammalian immune system. In this chapter, we will describe a specific immunosuppression initiated by UV-B and discuss the mechanism of its generation. UV-B-induced immunosuppression may be a fundamental biological regulatory mechanism, designed to protect against autoimmune attack on sunlight-altered skin, the adverse effect of which is to permit the outgrowth of UV-induced skin cancers. In addition to this specific immunosuppression, irradiation with UV-B also releases a number of cytokines and induces the expression on keratinocytes of cell-surface immunologic receptors, which may be of significance in UV-related dermatologic disorders.

Frances P. Noonan and Edward C. De Fabo • Laboratory of Photoimmunology and Photobiology, Department of Dermatology, The George Washington University Medical Center, Washington, DC 20037.

Environmental UV Photobiology, edited by Antony R. Young *et al.* Plenum Press, New York, 1993.

The chief purpose of this chapter, namely, to discuss the effect on immunologic responses of increases in solar UV-B, can be addressed in a quantitative fashion for UV-induced immunosuppression since the action spectrum for this effect has been established. In contrast, the action spectra and in most cases even the *in vivo* UV dose responses for release of cytokines and for the induction of cell-surface receptors are not known and thus a less rigorous discussion is possible of these UV-B effects.

2. Historical Background

Although it has been well known for many years (Blum, 1959) that UV-B exposure is critical to the development of skin cancers (and UV carcinogenesis is discussed elsewhere in this volume), the idea that UV irradiation alters immune responses has been of more recent genesis. In 1963, Hanisko and Suskind reported that contact hypersensitivity responses in guinea pigs were significantly lowered if the animals were UV irradiated (Hanisko and Suskind, 1963). Graffi *et al.* reported in 1964 that UV-induced tumors in mice were resistant to transplantation, indicating that they were highly antigenic, i.e., were readily recognized and eliminated by the immune system (Graffi *et al.*, 1964). This group found also that each UV tumor had individually specific antigens (Pasternak *et al.*, 1964). These findings, although they raised the very interesting question of how such antigenic tumors could escape immunologic destruction in the mouse in which the tumor arose, were not pursued until 1974. Kripke then confirmed the antigenicity of UV-induced tumors in mice (Kripke, 1974). The observations which gave impetus to the study of photoimmunology were those of Fisher and Kripke and of Daynes and co-workers (Daynes and Spellman, 1977; Fisher and Kripke, 1977), who demonstrated that highly antigenic UV-induced tumors could be transplanted not only into mice immunosuppressed by X-irradiation or by chemical immunosuppressants, but also into mice given a subcarcinogenic dose of UV irradiation, thus suggesting that UV radiation compromised the immune system of the host.

2.1. UV Suppression in UV Carcinogenesis

The effect of UV irradiation on immunity to UV-induced tumors was extensively investigated. Mice were irradiated with UV, usually three times weekly for between 1 and 3 months, and their ability to accept a transplanted UV tumor, which had arisen in another mouse of the same strain, was assessed. These UV tumors did not grow when transplanted into normal mice but did grow when transplanted into immunosuppressed animals of the same strain. It was found that a regimen of UV irradiation which did not produce any tumors nevertheless conferred on the UV-irradiated animals the ability to accept a UV tumor graft (Daynes and Spellman, 1977; Fisher and Kripke, 1977). The tumors transplanted to an unirradiated area of a UV-treated mouse also grew, and the effect could be transferred from one mouse to another in a parabiotic pair, indicating that the suppressive effect of UV was systemic (Fisher, 1978). The immunosuppressive effect did not appear to be initiated through the eyes but rather through the UV-irradiated skin (Fisher, 1978). Remarkably, it was demonstrated that susceptibility to transplanted UV tumors could be transferred to unirradiated animals with T lymphoid cells from the irradiated host (Kripke and Fisher, 1976; Daynes and Spellman, 1977; Fisher, 1978; Fisher and Kripke, 1978). Thus, UV irradiation initiated the formation in the lymphoid tissues of a population of suppressor T cells which down regulated the immunologic rejection response, permitting tumor outgrowth. The specificity of these suppressor T cells was for UV-induced tumors as a group. There was no effect on the growth of spontaneous tumors or on the growth of tumors induced by chemical carcinogens (Fisher and Kripke, 1978). Most importantly, Fisher and Kripke showed that transfer of these suppressor cells not only was effective in permitting outgrowth of transplanted UV tumors, but also acted in a primary carcinogenesis experiment, both increasing the number of UV tumors formed and decreasing the latent period of tumor appearance (Fisher and Kripke, 1982). De Fabo and Kripke showed that chronic irradiation with UV (UV doses three times weekly for 1 to 3 months) was not necessary, and that susceptibility to transplanted UV tumors could be initiated in a dose-dependent manner with a single UV treatment with as little as 15-min exposure from the UV source (De Fabo and Kripke, 1979). They also showed that the wavelengths responsible for suppression of tumor rejection were in the UV-B (Section 3.2.2) (De Fabo and Kripke, 1980). All of these experiments were done with haired mice of various inbred strains which were shaved before UV irradiation. These animals develop fibrosarcomas in response to UV irradiation. A systemic effect of UV irradiation was also described in hairless mice, a strain which develops papillomas and squamous cell carcinomas in response to UV irradiation, suggesting that UV suppression is operative in this model also (De Gruijl and Van der Leun, 1982).

The role of UV suppression in UV carcinogenesis is envisaged as facilitating the outgrowth of neoplastically transformed cells (Fig. 1). The photobiologic characteristics of immunosuppression differ from those of carcinogenesis (Section 3.2 and De Gruijl and Van der Leun, Chapter 4, this volume), suggesting that the initial photoreceptor(s) differed for these two events. In this scheme, UV radiation performs at least three functions in UV

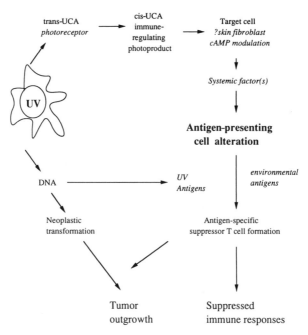

Figure 1. Working hypothesis for the mechanism of UV suppression and its role in UV carcinogenesis. UV photons interact with *trans*-urocanic acid (*trans*-UCA), causing it to isomerize to *cis*-UCA, which, by a series of unidentified steps possibly involving factors from skin fibroblasts, causes an alteration to antigen-presenting cell (APC) function. Because of this alteration to APCs, antigens introduced to UV-irradiated mice—either external environmental antigens or UV antigens formed in skin cells by the interaction of UV photons with DNA—lead to the production of antigen-specific suppressor T cells. In UV carcinogenesis, these suppressor T cells prevent the immunologic rejection of UV tumors, permitting their outgrowth.

carcinogenesis: (1) the events leading to neoplastic transformation of a skin cell, which occur via DNA; (2) immunosuppression, which occurs via a different photoreceptor, urocanic acid (UCA) (Section 4); and (3) the expression of UV antigens on skin cells. Steps (2) and (3) interact to produce suppressor cells specific for UV antigens. The overall resultant is tumor outgrowth. Recent reviews of the role of tumor immunity in UV carcinogenesis include Roberts *et al.* (1989), Kripke (1990), and Ward and Schreiber (1991).

2.2. UV Effects on Immunity

Once the immunosuppressive effect of UV radiation on the tumor rejection response had been described, the effect of UV on other immune responses was investigated. It was found that irradiation with UV did not cause a generalized immunosuppression (Fisher, 1978; Kripke *et al.,* 1977; Spellman *et al.,* 1977), consistent with the observation that UV-irradiated mice, even animals irradiated with chronic, large doses of UV, did not die from wasting disease or from infection as did other heavily immunosuppressed animals, e.g., congenitally athymic (nude) mice or animals treated with large doses of X-irradiation. In UV-irradiated animals, antibody responses were normal (Fisher, 1978; Kripke *et al.,* 1977; Spellman *et al.,* 1977), a finding which has since been confirmed and which indicates that the immunosuppressive effects of UV radiation are selective. This selectivity may reside in the antigen presentation arm of the immune response (Section 3.1.1). Mitogen responses and phagocytic responses of leukocytes were also normal (Norbury *et al.,* 1977). No effect was observed on the survival of skin allografts (Fisher, 1978; Kripke *et al.,* 1977). More recent studies have since shown, however, that UV and its mediators can in fact prolong transplants from one strain of animal to another (allogeneic transplants) (Section 4.1). Graft versus host response (the immune attack of allogeneic lymphoid cells on the recipient) was decreased by prior UV irradiation of the recipient (Glazier *et al.,* 1984).

Most importantly, contact hypersensitivity responses were suppressed by prior UV irradiation (Jessup *et al.,* 1978). This last observation has been most useful in investigating the mechanism of UV suppression, since contact hypersensitivity (CHS) responses, in which the antigen is applied topically to the skin, and delayed-type hypersensitivity (DTH) responses, in which the antigen is injected subcutaneously, represent a major class of cell-mediated immune responses (Brostoff *et al.,* 1991). They are mediated by a subclass of T lymphocytes carrying the cell-surface marker CD4. These cells are stimulated by antigen together with antigen-presenting cells carrying Class II molecules (IA and IE in the mouse and HLA-D in man), which are products of the major histocompatibility genes. The best-known examples of CHS responses are the poison ivy reaction, common in North America, and the skin response to metals such as nickel. Exposure to the antigen stimulates the proliferation of antigen-specific T cells. On reexposure to the same antigen, not necessarily at the original site, these already-primed T cells release mediators at the site of antigen application, initiating a cascade of inflammatory reactions resulting in a skin lesion. The term *delayed-type hypersensitivity* is used since, in contrast to other types of hypersensitivity reactions, a time delay of 1–2 days after reexposure to antigen is required before the skin reaction reaches its maximum. The significance of this class of reaction is not limited to the skin, since the DTH response is also considered to play an important role in the immune response to many infectious agents and in transplantation.

2.2.1. Contact and Delayed-Type Hypersensitivity Reactions

The initial studies of Jessup and colleagues demonstrated that a chronic regimen of UV irradiation could suppress the DTH response to the sensitizer dinitrochlorobenzene (DNCB) injected into the footpad (Jessup et al., 1978). This suppression was systemic since neither the site of sensitization nor the site of challenge was exposed to UV radiation. Subsequently, it was shown that the CHS response to DNCB and to trinitrochlorobenzene (TNCB), applied in a more conventional way to the skin in acetone solution, was suppressed in a dose-dependent manner with a single UV treatment (Noonan et al., 1981a). This effect was systemic, the sensitizer being applied to the skin of the shaved abdomen and the CHS response elicited on the ears, which were protected from UV. Suppression was initiated via UV irradiation of the skin of the shaved back. Interestingly, there was a time delay of between 1 and 3 days after UV exposure in manifestation of suppression; mice contact sensitized on the abdomen within 24 h after a single dose of UV on the back exhibited normal CHS responses (Noonan et al., 1981c). A similar time delay in manifestation of susceptibility to a UV tumor transplant after a single UV dose had previously been found (De Fabo and Kripke, 1980). Suppression of CHS after a single UV dose persisted for 14–21 days after UV irradiation (Noonan et al., 1981c). Suppression of CHS or DTH response to a wide variety of antigens has since been demonstrated, including chemical sensitizers (Greene et al., 1979; Harriott-Smith and Halliday, 1988; Mottram et al., 1988; Reeve et al., 1989), viruses (Howie et al., 1986; Yasumoto et al., 1987), bacteria (Jeevan and Kripke, 1989), parasites (Giannini, 1986), and alloantigens (Ullrich, 1986; Molendijk et al., 1987; Mottram et al., 1988). UV suppression is most effective prior to sensitization, but also appears to suppress the expression or elicitation of these responses (Howie et al., 1986). In many of these examples of UV suppression of DTH and CHS responses, transfer of the suppressive effects of UV irradiation to untreated animals with T lymphoid cells from a UV-irradiated and sensitized donor has been demonstrated (Greene et al., 1979; Harriott-Smith and Halliday, 1988; Noonan et al., 1981a; Norval et al., 1989), analogous to the suppressor T cells observed which suppress the rejection of UV tumors (Fig. 1). In the case of suppression of DTH and CHS responses, exogenous antigen must be applied to the UV-irradiated animal to cause the formation of suppressor T cells, whereas in the case of suppression of tumor immunity it is postulated that the relevant tumor antigen is formed endogenously in the skin by UV radiation (Noonan et al., 1981b).

2.2.2. Cytotoxic Responses

Cytotoxic immune responses are mediated by a class of T cells different from that which mediates DTH responses (Brostoff et al., 1991). Cytotoxic

T cells carry CD8 surface markers, and their function is to kill cells which express on the surface antigen, e.g., viral antigen as well as Class I molecules (H2-K and -D in the mouse and HLA-A, -B, and -C in man), products of the major histocompatibility complex. Cytotoxic cells are particularly important in viral and in tumor immunity.

Cytotoxic T-cell responses are also decreased in UV-irradiated mice (Jensen, 1983; Thorn, 1978). These experiments were done by priming mice with antigen after UV irradiation and measuring the cytolytic activity *in vitro* of lymphoid cells from these animals against appropriate target cells bearing the same antigen. This *in vitro* assay measures a secondary cytolytic response, since the primary response to antigen has already taken place *in vivo.* Under these circumstances, cytolytic activity of cells from UV-treated mice was depressed. Transfer of suppressor T cells from UV-irradiated mice into the donor mouse prior to priming with antigen also depressed the cytolytic response, analogous to the experiments described above with DTH and CHS responses (Jensen, 1983).

2.2.3. Antibody Responses

Antibody responses to a variety of antigens have been investigated in UV-irradiated mice and, with a single exception, have been shown to be unaffected (Kripke *et al.,* 1977; Spellman *et al.,* 1984; Spellman *et al.,* 1977). A decreased antibody response was observed in UV-irradiated mice when antigen was delivered by scarification (scratching) of the skin at the UV-irradiated site (Spellman *et al.,* 1984). This unusual method of antigen delivery may have forced antigen presentation by a pathway different from that normally employed for initiation of antibody responses.

2.2.4. Macrophage Functions

Early studies indicated that macrophage functions, e.g., phagocytosis, were normal in UV-irradiated mice (Norbury *et al.,* 1977). Macrophages are, however, activated by factors released from T cells in immunized animals, and it is possible that more careful scrutiny may reveal alterations to macrophage function in UV-irradiated animals.

3. Mechanism of UV Suppression

The mechanism by which UV irradiation causes immunosuppression is of great interest, not only because it appears to be an unusual example of a biological regulatory mechanism linking environmental UV and mammals, but also because alterations to the balance between immunity and suppression may be an important factor in various UV-related diseases. A number of mechanisms—which may not all be exclusive—have been put forward to explain the mechanism of UV suppression. Before entering into a discussion of these mechanisms, we will first describe the immunologic and photobiologic parameters of UV suppression which must be accommodated within any hypothesis for its mechanism.

3.1. Immunologic Aspects of UV Suppression

The central immunologic problem in UV-irradiated mice, first described by Greene *et al.* in 1979, appears to be a functional alteration to antigen-presenting cells. These authors suggested that, because of this alteration, antigen given to a UV-irradiated animal results in the stimulation of suppressor rather than effector T cells. A centralized hypothesis for the action of UV irradiation on the immune system can be formulated as shown in Fig. 1.

3.1.1. Antigen Presentation Alteration

Antigen-presenting cells may be macrophages, dendritic cells, epidermal Langerhans cells, B lymphocytes, or occasionally other specialized cells (Brostoff *et al.*, 1991). The function of these cells is to internalize and process antigens and to reexpress antigenic fragments on their surface in a manner which can be recognized by responding lymphocytes. Interaction between antigen-presenting cells and lymphocytes is a highly orchestrated process requiring the interaction of cell-surface proteins and factors produced by the antigen-presenting cell and by the lymphocyte. Different subclasses of T and B lymphocytes appear to have differing antigen-presenting cell requirements (Mosmann and Coffman, 1989; Mosmann and Moore, 1991; Bloom *et al.*, 1992). The selectivity of UV suppression (e.g., DTH responses are depressed but antibody responses are unaffected) may possibly be explained by a selective effect of UV irradiation on some classes of antigen-presenting cells.

The experiments of Greene and co-workers showed that UV-irradiated mice could be successfully immunized *in vivo* if the antigen was presented together with normal antigen-presenting cells (Greene *et al.*, 1979; Noonan *et al.*, 1981c). This was done by immunizing normal or UV-treated mice with trinitrophenylated (TNP) spleen cells from normal or UV-irradiated donors and measuring the subsequent DTH response. UV-treated mice gave a positive DTH response to immunization with TNP spleen cells from normal mice, but not to TNP spleen cells from UV-treated mice. In the unirradiated animals a positive DTH response was produced to TNP cells from either normal or

UV-irradiated mice, albeit a lower response in the latter case, since the normal antigen-presenting cells of the UV-irradiated mice could reprocess the antigen. Further, suppressor T cells were found only in spleens from UV-treated mice immunized with TNP–UV-treated cells.

These investigators later demonstrated a functional alteration to antigen-presenting spleen cells from UV mice in a number of *in vitro* assays (Letvin *et al.*, 1980) and postulated that a redistribution of antigen-presenting cells away from the spleen in response to UV irradiation may be responsible. Highly purified dendritic cells (potent antigen-presenting cells which play a critical role in CHS responses and in transplantation) from UV-treated mice were found, however, to be less effective at stimulating proliferation of lymphocytes *in vitro,* indicating some change to the antigen-presenting cell itself (Noonan *et al.*, 1988). Thus the decreased antigen-presentation ability of spleen cells could not be explained solely by a redistribution of cells in response to UV irradiation.

3.1.2. Suppressor T-Cell Formation

As indicated above, one of the earliest observations in the investigations of UV-mediated suppression was that this effect could be transferred to unirradiated animals with T lymphocytes from UV-treated donors. Suppressor T cells have since been described in a wide variety of systems in response to UV irradiation and antigen treatment as described above.

Suppressor cells, while also a consistently reproducible phenomenon in many systems other than UV suppression, have been a controversial topic in immunology. In spite of the advances in molecular immunology, suppressor cells have yet to be cloned and isolated, let alone their activity characterized in molecular terms. Of the many described suppressor factors, none has as yet been molecularly characterized. It has long been known, however, that there is an inverse relationship *in vivo* between cell-mediated immunity and antibody formation in many immune responses, suggesting some form of antagonistic cross-regulation (Parish, 1972). More recently, at the molecular level, T cells (T_H2 CD4) which help B cells produce antibody have been shown to produce a cytokine, IL-10, which down regulates the activity of T_H1 CD4 cells which produce DTH responses (Mosmann and Moore, 1991). Bloom has recently reviewed the topic of suppressor cells and has described a scheme in which suppression results from stimulation of an antagonistic T-cell subset rather than of a class of unique suppressor cells and that this scheme is controlled at the level of the antigen-presenting cell (Bloom *et al.*, 1992). This concept is consistent with the working hypothesis shown in Fig. 1.

3.2. Photobiologic Parameters of UV Suppression

The question of the photobiologic characteristics of UV suppression was addressed after the immunologic investigations were initiated. Photobiologic approaches have been of great assistance in establishing the mechanism by which UV initiates suppression.

3.2.1. UV Dose Responses and Reciprocity

De Fabo and Kripke established that a chronic regimen of UV treatments (three times weekly for 1 to 3 months) as had been used previously was not necessary to enable a mouse to accept a transplanted UV tumor. Using single doses of UV, they found dose-dependent UV suppression, as determined by acceptance of UV tumors transplanted to an unirradiated site on a UV-treated animal. Interestingly, however, a time delay of 2–3 days after UV irradiation was necessary for suppression to be detected. By varying the UV dose rate over a 10-fold range, it was found that Bunsen–Roscoe reciprocity held, i.e., suppression was dependent only on total UV dose delivered, and was independent of dose rate over this range. Further, by varying the number of treatments between 1 and 12, at a constant dose rate, used to deliver a given UV dose, it was shown that UV suppression was also independent of dose-fractionation; i.e., the amount of suppression was dependent only on the total UV dose regardless of how many treatments were used to deliver that dose (De Fabo and Kripke, 1979, 1980). These photobiologic characteristics differed greatly from those observed for UV-induced carcinogenesis, in which reciprocity does not hold and more total UV is required to form tumors at higher dose rates than at lower dose rates. Thus, it was postulated that the events by which UV initiated suppression and tumor formation differed. These two events interact, however, in tumor outgrowth, as shown in Fig. 1.

We investigated the photobiologic characteristics of UV suppression of contact hypersensitivity and found that this form of suppression also showed Bunsen–Roscoe reciprocity and was independent of dose-fractionation (Noonan et al., 1981a). These observations suggested that suppression of tumor rejection and suppression of contact sensitivity may be initiated by the same mechanism, but with differing UV dose requirements.

3.2.2. Wavelengths Responsible

The wavelengths of UV responsible for initiating suppression were found to be in the UV-B (280–315 nm) and UV-C (200–280 nm) ranges (De Fabo and Kripke, 1980; Noonan et al., 1981a). The UV source used in these experiments, the FS40 sunlamp, emits radiation in the UV-C, UV-B, and UV-

A (315–400 nm) ranges, with a peak at 313 nm in the UV-B range (Noonan *et al.*, 1981b). Filtration of this source with Mylar, which removes >99% of wavelengths shorter than 315 nm, prevented UV suppression. Subsequent experiments have confirmed that UV-A of wavelengths greater than 320 nm does not initiate immunosuppression, at least at the doses used (Aubin *et al.*, 1991; De Fabo *et al.*, 1992). However, the question of any contribution to suppression of UV-A in solar irradiation, where the UV-A radiation levels are very high, is one which has not been thoroughly investigated. In particular, given the action spectrum for UV suppression (Section 3.2.3), it may be predicted that sufficient UV-A II, i.e., UV-A in the shorter part of the spectrum (315–340 nm), could be immunosuppressive.

3.2.3. Action Spectrum

Having established that the wavelength band responsible for immune suppression lay within the UV-B and UV-C regions of the solar spectrum, it was possible to analyze which wavelengths within this waveband were the most efficient at initiating immune suppression. Efficiency in this case means how many photons of each wavelength are needed to produce a constant level of response, e.g., 50% suppression. The most efficient wavelengths will require the least number of photons.

Known as an "action spectrum" (De Fabo, 1980), such a wavelength-by-wavelength analysis is extremely useful in analyzing photobiologic effects. According to action spectrum theory, the shape of an action spectrum is exactly congruent with the *in vivo* absorption spectrum of the unknown, or putative, photoreceptor. This is because in order for a photon to do "work," in this case initiate immunosuppression, it must first be absorbed. Thus, not only does an action spectrum establish which wavelengths are effective in producing a light-driven response, it also produces a relative wavelength analysis of the efficiency of each wavelength. Further, this relative wavelength analysis is indicative of the identity of the photoreceptor by revealing its absorption spectrum.

3.2.3.1. Deriving an Action Spectrum.

Action spectroscopy is based on several assumptions, one of the most important of which is that the photoreceptor is not screened by compounds external to it. Ideally it should be superficially located. Significant screening by such compounds would distort the true shape of the action spectrum and make its interpretation difficult, if not impossible. Thus, not only is it necessary to derive an action spectrum very carefully, but one has also to be aware of those parameters which may interfere with its interpretation (De Fabo, 1980).

Another important aspect of action spectroscopy is wavelength resolution. The optical dispersion system should be able to produce as narrow a band of radiation as possible so as to be able to resolve in detail the true absorption spectrum of the photoreceptor. The radiation should be dispersed over as wide an area as possible. In the past, this was a formidable problem. For example, using a grating-type monochromator, very narrow bands of radiation can be produced, but increasing the area of irradiation by opening the slit width causes the broader spectrum wavelengths to be dispersed, thereby greatly reducing the resolution of any action spectrum obtained. However, this problem can now be solved by coupling interference or bandpass filters to a high-power source of radiant energy such as a xenon arc lamp (see Chapter 2, this volume). These filters must, however, be sufficiently blocked to eliminate the transmission of radiation outside the bandpass of interest. With this type of interference filter, wavelengths from the UV-C (250 nm) through to the visible (800 nm) range can be produced with half band widths of 2.5 nm in 5-nm steps. The half band width is the width of the transmitted waveband at 50% maximum transmission. By mounting these filters to intersect a beam of light from a 2.5-kW xenon arc lamp, an irradiation area of approximately 50–60 cm^2 can be produced. The production of such narrow-band radiation over an area of 50 cm^2 enables the irradiation of relatively large objects such as mice (De Fabo and Noonan, 1983). Thus, detailed action spectra can be derived *in vivo* on mammals with radiation across the solar spectrum.

3.2.3.2. Action Spectrum for Immune Suppression. The first step in determining an action spectrum is to derive dose-response curves at each of the wavelengths to be tested. In the case of immune suppression by UV radiation, for each point in the dose response, twelve mice were divided into groups of three. Three animals were UV irradiated and contact sensitized, three were contact sensitized only, and the remaining six animals were used as controls (see De Fabo and Noonan, 1983, for more complete details). The CHS response was measured in each of these groups of mice. UV irradiation with each of these narrow bands of UV was found to suppress this immune response to the chemical sensitizer in a dose- and wavelength-dependent fashion. Dose-response determinations were completed using each of 10 UV wavelengths. To derive the action spectrum, the UV dose for 50% suppression was calculated from the dose response for each wavelength. This UV dose was quantum corrected to account for energy differences between photons of different wavelengths, and the number of photons needed at each wavelength was calculated. To plot the action spectrum, the reciprocal of the number of photons which produced 50% suppression was plotted versus wavelength. Since the most effective wavelength at causing immunosuppression was 270 nm, the values derived at the other wavelengths were normalized against this value.

The shape of the action spectrum for immune suppression (Fig. 2) indicated a wavelength maximum between 270 and 275 nm, a shoulder at 280 nm, and a gradual decline to about 3% of maximum at 320 nm (De Fabo and Noonan, 1983). The absorption spectra of a number of UV-absorbing compounds normally found in mammalian skin were plotted on the same scale as the action spectrum. For example, the action spectrum for DNA effects is shown in Fig. 2. It could then be determined that urocanic acid was the compound in the skin the absorption spectrum of which best matched the action spectrum (Fig. 2) (De Fabo and Noonan, 1983).

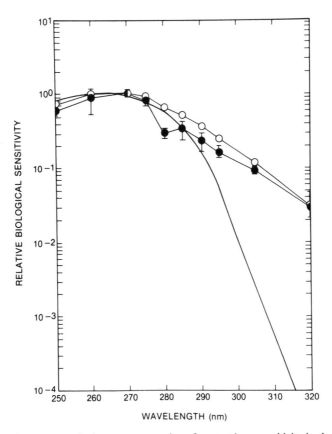

Figure 2. Action spectrum for immune suppression of contact hypersensitivity in the BALB/c mouse (closed circles). The open circles represent the absorption spectrum of urocanic acid, and the solid line the DNA action spectrum (Setlow, 1974). From De Fabo and Noonan (1983). Reprinted from *Journal of Experimental Medicine,* 1983, 158:84–98, by copyright permission of the Rockefeller University Press.

Urocanic acid undergoes a reversible UV-induced isomerization from the *trans* to the *cis* isomer *in vitro* and in mammalian, including human, skin (Baden and Pathak, 1967; Morrison *et al.,* 1980; Norval *et al.,* 1989; Pasanen *et al.,* 1990) (Fig. 3), reaching a photostationary state or photoequilibrium. UCA was originally considered to be a natural sunscreen (Baden and Pathak, 1967). Isomerization of UCA from the *trans* to the *cis* isomer is accompanied by a small change in the λ_{max} of <5 nm to shorter wavelengths (Morrison *et al.,* 1980). UCA also forms photoadducts with DNA (Farrow *et al.,* 1990). Most of the UCA of the skin is found in the stratum corneum, the outermost skin layer. The stratum corneum consists of terminally differentiated, dead epidermal cells which contain keratin, a protein complex, and a histidine-rich protein, fillagrin. Urocanic acid, which is produced in the *trans* configuration, is formed by deamination of histidine—possibly the histidine in fillagrin—by the enzyme histidase (histidine ammonia-lyase, E.C. 4.3.1.3.) (Taylor *et al.,* 1991). The activation of histidase is thought to be a consequence of the terminal differentiation process of epidermal cells. We found that removal of the stratum corneum prevented the generation of UV suppression (De Fabo and Noonan, 1983), further supporting the hypothesis that UCA was the photoreceptor for UV suppression. Considerable experimental evidence has subsequently been produced supporting the hypothesis that UCA is the photoreceptor for UV suppression (Section 4).

3.2.4. Biologically Effective Doses of UV for Immunosuppression in Sunlight

Artificial UV sources emit different spectral distributions of radiation, and the spectral emissions of artificial UV sources differ from that of the sun. Because, as just discussed, different wavelengths of UV have differing effectiveness at initiating UV effects, this means that UV dose responses derived in the laboratory do not necessarily mimic UV dose responses in natural sunlight. In order to make such comparisons, knowledge of the action spectrum

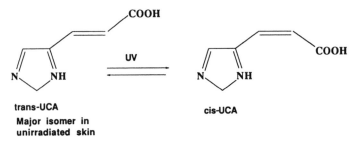

Figure 3. Structure and UV-induced isomerization of urocanic acid.

for an effect is critical. By using the action spectrum to "weight" the spectral output of a UV source, e.g., the sun, a biologically effective dose rate (irradiance) can be determined. For example, by multiplying the spectral output of solar UV by the action spectrum for UV suppression, the output of each wavelength in the solar UV is adjusted for its relative effectiveness at causing immunosuppression. In the laboratory, the most effective wavelength for immune suppression is 270 nm, which does not exist in terrestrial sunlight, and thus the biologically effective irradiance for immune suppression at 270 nm is zero. On the other hand, 320-nm UV, which has a relative effectiveness of only 3% in the laboratory, is one of the most effective wavelengths at causing immune suppression in sunlight because it is so abundant. A simple mathematical expression makes this clear.

$$\mathrm{BEI}_{(\mathrm{IMS})} = \int A(\lambda)I(\lambda)d(\lambda) \tag{1}$$

where $\mathrm{BEI}_{(\mathrm{IMS})}$ is the biologically effective or weighted irradiance for immune suppression, $A(\lambda)$ is the biological action spectrum, and $I(\lambda)$ is the irradiance for each wavelength. Thus, at any given wavelength, the product of the action spectrum and the irradiance gives the number of effective photons within a small region of wavelengths $d(\lambda)$. The integral is then simply the sum of the products over these wavelength ranges and represents the area under the product curve. To calculate the biologically effective dose, or BED, the biologically effective irradiance, or dose rate, is multiplied by time of exposure:

$$\mathrm{BED}\ (\mathrm{J/m^2}) = \mathrm{BEI}\ (\mathrm{W/m^2}) \times \mathrm{time(s)} \tag{2}$$

remembering that $1\mathrm{W} = 1\ \mathrm{J/s}$.

We have calculated the biologically effective dose rate for immune suppression in noonday sunlight as a function of latitude and season (De Fabo et al., 1990). For this calculation, a radiative transfer model for solar irradiance was used, which assumes clear conditions. The action spectrum was extrapolated to 330 nm, assuming that the contribution of wavelengths longer than 330 nm was negligible because of the very low absorption of UCA at these wavelengths. Figure 4 shows the action spectrum together with the calculated solar irradiance spectrum at 40°N in July. The product curves derived from multiplying the action spectrum by the solar irradiance spectrum (for January and July at 40°N) are shown in Fig. 5. Integration of the area under each of these curves yielded the biologically effective irradiance "weighted" for UV-induced immunosuppression at July and January, respectively, at this latitude. Figure 6 shows the biologically effective irradiance for immunosuppression ($\mathrm{BEI}_{\mathrm{IMS}}$) calculated for latitudes from 80°N to 80°S

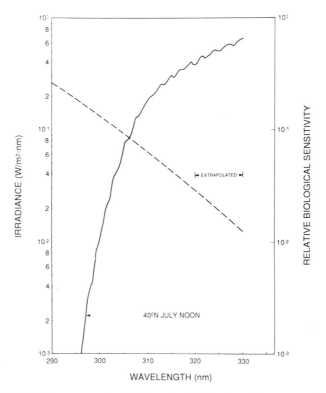

Figure 4. The dashed line represents the UV-B portion of the *in vivo* action spectrum for immune suppression in the BALB/c mouse given as relative biological sensitivity (adapted from De Fabo and Noonan, 1983). The solid line is the solar irradiance in W/m²/nm calculated from the radiation transfer model of Frederick and Lubin (1988) for local noon in July at 40°N latitude. Reproduced from De Fabo *et al.* (1990).

for both January and July. A biologically effective irradiance for immune suppression (BEI_{IMS}) of between 0.2 and 0.3 W/m² was found in summer between approximately 40°N and 40°S latitude. This means that a BALB/c mouse at these latitudes would need between 21 and 26 min of noonday summer sunlight exposure on a clear day in summer for 50% suppression of CHS. Using actual ground-based measurements of UV radiation taken at Rockville, Maryland (39°N latitude), an exposure time for the same effect was predicted to be 47 min, consistent with the fact that totally clear conditions are not normally present. In general, as latitude decreases, the BEI_{IMS} increases and BEI_{IMS} values are higher in summer than in winter.

Another advantage of calculating biologically effective irradiances is that radiation amplification factors (discussed in Chapters 1 and 4, this volume)

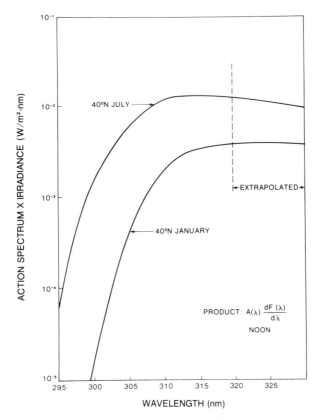

Figure 5. Product curves resulting from the convolution of the action spectrum for immune suppression with the predicted solar irradiance spectrum at 40°N latitude in summer and winter. The integrated area under each curve is the biologically effective irradiance for immune suppression, BEI_{IMS}. The equation for this convolution is shown in the figure. The values for BEI_{IMS} from these curves were 2.7×10^{-1} W/m^2 for July and 0.7×10^{-1} W/m^2 for January at 40°N. Reproduced from De Fabo et al. (1990).

can be calculated, and thus the predicted effect on biological responses can be calculated for predicted conditions of ozone depletion. Stratospheric ozone depletion will cause changes both in irradiance and in wavelengths of UV radiation at the earth's surface. The radiation amplification factor is defined as the change in the biologically effective irradiance relative to the change in total column ozone. For UV-induced immunosuppression, the radiation amplification factor is between 0.6 and 0.9, depending on latitude and season (De Fabo et al., 1990). This means that for a 1% decrease in column ozone there is a 0.6% to 0.9% increase in biologically effective solar UV-B irradiance

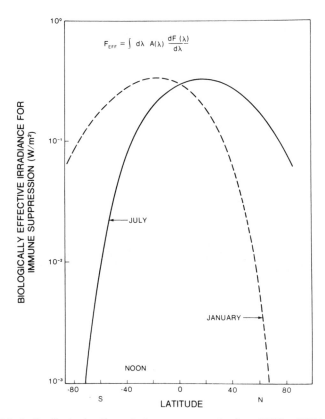

Figure 6. Biologically effective irradiance for immune suppression from 80°N to 80°S in summer and winter. The BEI_{IMS} values were calculated at each latitude using the equation near the top of the figure. F_{EFF} is the biologically effective or "weighted" irradiance of sunlight for immune suppression, A (λ) is the biological action spectrum for immune suppression, and dF (λ)/$d(\lambda)$ is the unweighted predicted solar flux in W/m²/nm. Reproduced from De Fabo et al. (1990).

for immune suppression. Tabulation of predicted increases in BEI_{IMS} at selected latitudes with varying levels of ozone depletion is given in De Fabo et al. (1990, p. 815). The biological consequences of these predicted increases for humans (the biological amplification factor) is presently unknown.

3.3. Postulated Mechanisms of UV Suppression

A number of differing hypotheses have been put forward to explain the mechanism by which UV initiates immunosuppression. Some of the early hypotheses have been disproven by subsequent experimentation. Furthermore, some which are currently presented as distinct mechanisms may in fact be different aspects of the same process. Although the photoreceptor which initiates UV suppression has been identified, the entire sequence of events leading to the generation of the alteration to antigen-presenting cells and to suppressor T cells is not known.

3.3.1. Direct UV Irradiation of Circulating Immune Cells

The action spectrum for UV suppression shows a peak at 270 nm (Fig. 2), indicating that the wavelengths of UV which are most effective at immunosuppression are also those wavelengths which are most heavily absorbed by the epidermis, i.e., the least penetrating, shortest wavelengths. This observation is consistent with a mechanism for initiation of immunosuppression by a superficial photoreceptor rather than a mechanism in which subepidermal circulating blood cells are directly exposed to UV radiation.

Many experiments have investigated the effects of UV radiation *in vitro* in depressing antigen presentation and other functions of blood monocytes, and of peritoneal and splenic macrophages (e.g., Krutmann *et al.*, 1990). It must be borne in mind, however, that *in vitro* experiments may not necessarily mimic the *in vivo* situation. Under normal conditions, such cells would not be exposed to UVB *in vivo*. In pathologic conditions, however, in which infiltrating cells enter into skin lesions, such cell types may be susceptible to UV-B.

3.3.2. UV Irradiation of Skin Cells

In contrast to cells within the circulation, skin cells are within range of UV-B. Various cell types within the skin have been postulated to be the target cell initiating UV suppression, and these will be discussed in turn.

Epidermal Langerhans cells (LCs) are antigen-presenting cells located above the basal layer of the epidermis (Stingl, 1990). They are present as a network, with a pattern of dendritic processes, and can be readily visualized in epidermis by a number of (immuno)histological staining procedures. These cells present antigen *in vitro* and are believed to play an important role in initiating immune responses to antigens which enter the body via the skin (e.g., contact sensitizers). Langerhans cells have, for example, been found in the lymph nodes draining the area of contact sensitization, but they do not appear to be essential for initiation of CHS responses, since animals could still be contact sensitized through skin from which the epidermis (including LCs) had been removed (Streilein, 1989). In this case other dendritic antigen-presenting cells of the dermis or within the lymph node may be responsible.

After UV irradiation of the skin, LCs lose their dendritic appearance, and their numbers are decreased in a dose-dependent manner (Noonan *et al.,* 1984; Toews *et al.,* 1980). They may migrate out of the skin in response to UV, but this has not to date been definitively shown. It was initially postulated that this UV-induced loss of epidermal LCs was responsible for UV suppression by reducing the number of antigen-presenting cells available to initiate contact sensitivity (Toews *et al.,* 1980). This explanation does not appear to be the case, however, since UV suppression can occur without detectable loss of epidermal LC numbers, and loss of LCs can occur without detectable suppression (Lynch *et al.,* 1981; Noonan *et al.,* 1984; Streilein and Bergstresser, 1988). It is also difficult to explain systemic suppression in terms of loss of LC, since LC numbers are normal at unirradiated sites on UV-treated animals (Lynch *et al.,* 1981; Noonan *et al.,* 1984). Using narrow bands of UV radiation, we were able to deplete mouse epidermis of LCs in a dose-dependent manner with 270- or 290-nm UV. Significantly, depletion of LCs occurred at doses considerably lower than those needed to cause suppression. Furthermore, using 320-nm UV, even at doses which caused 50% suppression, no alterations to LC numbers or morphology could be detected (Noonan *et al.,* 1984).

Although the loss of epidermal LCs *per se* does not appear to be the critical event initiating UV suppression, this does not indicate that LCs have no role in UV suppression. If, as is postulated in Fig. 1, UV-induced mediators alter the function of antigen-presenting cells, then it would be anticipated that LCs, along with antigen-presenting cells in the lymph node and spleen, could have their function altered as a result of exposure to UV-released factors.

In mice, the epidermis also contains a small population of Thy1+ cells which bear the gamma-delta T-cell receptor (Stingl, 1990). The function of these cells in the skin has not been clearly established. Evidence has been presented that, if UV-irradiated Thy1+ cells are injected intravenously, suppression results (Cruz *et al.,* 1989). The Thy1+ cell is not a good candidate for the target cell initiating UV suppression, however, since it is not repopulated in the skin after UV irradiation. This is inconsistent with the UV dose-response curves which show that repeated UV exposure causes increases in UV suppression.

Epidermal keratinocytes release a number of factors in response to UV radiation, both known immunologic mediators and as yet unidentified factors. The role of these agents in UV suppression is currently an area of considerable investigation and is discussed further in the next section. It may be well to note, however, that under many circumstances in which UV suppression is initiated in a dose-dependent manner in the mouse model, the UV doses used are lethal to keratinocytes (Noonan *et al.,* 1989).

3.3.3. Release of Factors from UV-Irradiated Keratinocytes

Keratinocytes are the major cell type in the epidermis. These cells differentiate as they move outward from the basal layer, eventually becoming keratinized to form the outer dead cell layer, or stratum corneum. Keratinocytes are very potent cells with the ability to secrete a wide range of cellular growth factors which are important in regulating their growth and differentiation (McKenzie and Sauder, 1990; Roberts *et al.,* 1988). Keratinocytes are also a source of cytokines, soluble factors which play a major role in the immune response, regulating the differentiation of immune cells and interactions between them. The production of some of these factors by keratinocytes is increased by UV irradiation. Keratinocytes have also been shown to produce suppressor factors in response to UV irradiation (Kim *et al.,* 1990). Finally, UV irradiation of keratinocytes induces the expression on the cell surface of molecules, e.g., SS-A/Ro antigen (Furukawa *et al.,* 1990), which is important in photosensitive lupus or ICAM-1 (Tang and Udey, 1991), a cellular adhesion molecule.

The role of these UV effects on keratinocytes in the generation of UV suppression is difficult to assess because of the lack of detailed quantitative photobiologic analyses; e.g., no detailed UV dose responses or wavelength-dependence data are available. It is also not clear whether *in vivo* release of such factors is caused by direct interaction of UV with the keratinocyte or by indirect mechanisms. Nevertheless, these factors must be considered as candidates for the systemic factor(s) postulated to alter antigen-presenting cell function in Fig. 1. A very interesting aspect of these studies, which is beyond the scope of this review, is the role of the keratinocyte-derived cytokines, growth factors, and cell-surface receptors in dermatologic disorders such as psoriasis, cutaneous T-cell lymphoma, atopic dermatitis, photosensitive lupus, and other diseases (Cooper, 1990; McKenzie and Sauder, 1990). It may be possible that disorders in UV suppression contribute to some of these diseases, but there is to date no evidence to support this hypothesis.

The identified factors induced by UV irradiation from keratinocytes include interleukins 1, 3, and 6, contra-interleukin-1, prostaglandins, granulocyte macrophage colony-stimulating factor (GM-CSF), tumor necrosis factor (TNF) alpha, tumor growth factor alpha, nerve growth factor, and basic fibroblast growth factor (McKenzie and Sauder, 1990; Roberts *et al.,* 1988). Interleukin-1 and contra-interleukin-1 would appear unlikely candidates for involvement in UV suppression since their actions are antagonistic. Prostaglandin E_2 is known to be immunosuppressive and *in vivo* administration of indomethacin, which antagonizes prostaglandin activity, has been shown to reverse UV suppression (Chung *et al.,* 1986).

These studies do not appear to have been pursued further. Most recently, it has been postulated from genetic studies on susceptibility to UV suppression (Section 6) that TNF alpha is a mediator of UV suppression (Yoshikawa and Streilein, 1990).

A number of suppressive factors have been described in serum from UV-irradiated mice (Swartz, 1984; Harriott-Smith and Halliday, 1988), leading to investigations of suppressive factors released from UV-irradiated keratino-cytes (Kim *et al.*, 1990; Schwartz *et al.*, 1986). To date, however, none of these suppressor factors has been molecularly characterized.

3.3.4. Two Forms of Suppression: A Misnomer

One of the most confusing topics in the area of UV suppression has been the concept that there are two differing forms of UV-induced suppression. These were postulated to be initiated respectively by antigen applied directly to a site receiving a "low" dose of UV (local suppression) or by antigen applied to an unirradiated site on an animal treated with a "high" dose of UV radiation (systemic suppression) (Bergstresser, 1986).

The terms "low dose" and "high dose" are intrinsically unsatisfactory since they have never been accurately defined and the actual values for UV dose differ from study to study. The terms "local" and "systemic" are more descriptive, but even in this case local suppression is somewhat of a misnomer since, although the antigen is applied "local" to the area of UV irradiation, the subsequent immune response, i.e., CHS or DTH, is elicited distal to the site of UV irradiation (the ear or the footpad). No evolutionary advantage to the organism has been postulated which could account for the necessity to evolve two separate mechanisms for carrying out UV suppression.

In order to clarify this confusing situation, we have recently (Noonan and De Fabo, 1990) quantitatively assessed directly the question of local and systemic suppression by UV radiation in two strains of inbred mouse, BALB/c and C57BL/6. In both strains we determined detailed UV dose responses and kinetics of suppression of CHS responses. Sensitizer was applied either directly to the UV-irradiated site or to the unirradiated site. We found that, within a mouse strain, the UV dose-responses were identical for local and systemic suppression, but that there was a time delay in the generation of systemic suppression. Local suppression (i.e., antigen applied directly to a UV site) could be detected immediately after UV irradiation. The UV dose-response curves were parallel for both strains. These observations are consistent with a single mechanism of UV suppression which is initiated at the site of irradiation and takes between 1 and 3 days to be exhibited systemically. Further, and perhaps more interesting, more than 6 times the UV dose was re-quired to suppress BALB/c mice than to suppress C57BL/6 mice, suggesting a genetic susceptibility to UV suppression. The idea that "low" doses of UV initiate local suppression and "high" doses of UV initiate systemic suppression thus arose because the original experiments on local suppression were done with C57BL/6 mice which were sensitized immediately after UV irradiation (Toews *et al.*, 1980), and the original experiments on systemic suppression were done with BALB/c mice which were sensitized 3 days after UV irradiation (Noonan *et al.*, 1981a).

3.3.5. Generation of Suppressor-Inducer Cells in Human Skin in Response to UV Irradiation

Cooper and co-workers have identified in UV-irradiated human skin a particular class of macrophages ($OKM5^+CD1^-DR^+$) which are suppressor-inducer cells, i.e., they stimulate the proliferation of suppressor CD4+ T cells (Baadsgaard *et al.*, 1988, 1989). This group has carried out extensive investigations of this cell, which appears in human skin 3 days after UV-B irradiation. There are to date no reports of an equivalent cell in mouse skin, which has hampered comparison with animal investigations. Nevertheless, this cell appears to be an interesting candidate for playing a role in UV suppression.

3.3.6. Initiation via a Specific Photoreceptor

As discussed above, the question of whether UV suppression is initiated by interaction between UV and a specific photoreceptor in the skin was addressed by carrying out a detailed *in vivo* action spectrum for UV suppression in the mouse. We concluded from that action spectrum that the most likely candidate for the photoreceptor was urocanic acid (De Fabo and Noonan, 1983), based on the congruence between its absorption spectrum and the action spectrum, on its superficial location in the skin (removal of the stratum corneum had prevented the generation of suppression), and on its *trans*-to-*cis* isomerization, a common means of photobiologic regulation. The evidence which has subsequently been produced supporting this hypothesis will be detailed in Section 4.

The action spectrum for UV suppression deviated from the DNA action spectrum in the longer wavelengths of the UV-B by more than two orders of magnitude (Fig. 2), making it unlikely that DNA could be the photoreceptor. Nevertheless, recent evidence has been presented supporting a role for DNA lesions in initiating UV suppression. UV suppression of the CHS response has been investigated (Applegate *et al.*, 1989) in the opossum *Monodelphis domestica,* a marsupial (nonplacental mammal) which exhibits photorepair. In photorepair, DNA lesions, in particular, thymine dimers—the most frequent UV lesion in DNA—are repaired by exposure to long-wave ultraviolet light (UV-A), using an endogenous enzyme, DNA photolyase (reviewed in De Gruijl and Roza, 1991). Photorepair appears not to exist in adult mouse skin. UV suppression in the opossum showed photorepair; i.e., suppression was reversed by subsequent exposure to UV-A light. Further, the UV-induced loss of epidermal Langerhans cells also showed photorepair. Photoreversal of these effects did not occur in mice. These experiments have been interpreted as evidence for a central role for thymine dimers in initiating UV suppression (Applegate *et al.*, 1989). The strategy of equating photoreversal unequivocally with reversal of thymine dimer formation has, however, been criticized (De Gruijl and Roza, 1991). This idea has been further supported, however, in the mouse model by use of liposomes encapsulating the repair enzyme T4 endonuclease, which nicks DNA at thymine dimers, permitting the cell's own enzymes to repair the DNA. Application of endonuclease-containing liposomes to mouse skin after UV irradiation prevented UV suppression and partially reversed the formation of thymine dimers (Kripke *et al.*, 1992). An interesting note is that UCA forms photoadducts with DNA, and the effect of photoreversal on these adducts is unknown (Farrow *et al.*, 1990).

4. *cis*-UCA Effects on Immunity

The conclusion from the action spectrum for UV suppression as discussed above was that urocanic acid was the most likely candidate for the photoreceptor. It was further postulated that UCA initiated suppression by isomerizing from the *trans* to the *cis* isomer. Considerable evidence has since accumulated from a number of experimental approaches to support this hypothesis.

4.1. Investigations of UCA *In Vivo*

We first tested the hypothesis that UCA initiated immune suppression by UV-irradiating histidinemic mice. These animals are deficient in the enzyme histidine ammonia-lyase and, as a result, accumulate histidine in the blood and tissues. Important to our investigations, these animals are also deficient in skin UCA, since UCA is the end product of the histidine ammonia-lyase reaction. In contrast to their wild-type counterparts, histidinemic mice showed a greatly impaired UV-induced immune suppression of the CHS response (De Fabo *et al.*, 1983). This finding has since been confirmed in inbred mice congenic for this defect (De Fabo, unpublished observations).

In the reciprocal experiment, Reilly and De Fabo increased skin levels of UCA by increasing histidine levels in the diet (Reilly and De Fabo, 1991).

Compared to animals fed a control diet, histidine-fed animals showed significantly greater UV-induced immune suppression, and more of the *cis* isomer of UCA was formed in response to a given UV dose.

Ross, Norval, and associates demonstrated that administration of the *cis* isomer of UCA could mimic the immunosuppressive effects of UV radiation by suppressing the DTH response to herpesvirus Type I (Norval *et al.*, 1989; Ross *et al.*, 1986). They found that a *cis/trans*-UCA mixture, but not *trans*-UCA alone, given topically, subcutaneously, or intravenously to mice, suppressed the DTH response with the formation of antigen-specific suppressor T cells. This effect of *cis*-UCA was indistinguishable from the suppressive effects of UV radiation. In these experiments, *cis*-UCA acted in a dose-dependent manner and was active at doses as low as 1 μg/mouse.

Given *in vivo, cis*-UCA, but not *trans*-UCA, caused an antigen-presenting cell defect in splenic dendritic cells, indistinguishable from that caused by UV radiation (Noonan *et al.*, 1988). These experiments were done by preparing splenic dendritic cells from normal and UCA-treated mice and testing their ability to stimulate, in the presence of antigen, the *in vitro* proliferation of purified T cells from immunized mice. Interestingly, addition of *cis*-UCA *in vitro* directly to the proliferation assay had no effect, suggesting that the action of *cis*-UCA on antigen-presenting cells *in vivo* occurred indirectly via an intermediate step(s).

More recently, *cis*-UCA has been found to be suppressive in a number of experimental transplantation systems. The initial evidence indicated that UV radiation did not suppress the immunologic rejection of allografts (transplants from one strain of animal to another). These early experiments were done using skin allografts, which, for reasons related to the immunology of transplantation, are the most difficult grafts to prolong. Mottram *et al.* (1988) were able, however, to prolong heart allografts in mice with a single dose of UV radiation. Corneal transplants in rabbits were also prolonged by prior UV irradiation of the corneal graft and of the recipient eye (Williams *et al.*, 1987; Young *et al.*, 1989). *Cis*-UCA but not *trans*-UCA administration has been shown to prolong survival of corneal allografts in rabbits (Williams *et al.*, 1990) and of heart and skin allografts in rats and mice (Gruner *et al.*, 1990, 1992). *Cis*-UCA has also been shown to alleviate graft versus host disease in mice, i.e., the immune attack of lymphocytes on an allogeneic recipient (Gruner *et al.*, 1992).

As indicated in Fig. 1, from the original studies on UV suppression of tumor rejection, it would be predicted that *cis*-UCA should play a role in UV carcinogenesis, facilitating the outgrowth of UV tumors by its immunosuppressive action. Evidence for such an effect has been found (Reeve *et al.*, 1989). UCA was topically applied to hairless mice during a UV carcinogenesis experiment. Four times as many tumors per mouse were found in the UV-

and UCA-treated group than in the group exposed to UV only, and a higher percentage of tumors were of the more aggressive squamous carcinoma type in the UV- and UCA-treated animals.

Most recently, preliminary findings have indicated that *cis*-UCA is active in a mouse model of intra-abdominal abscess formation. In this model, CD8+ T cells, not antibody, mediate abscess formation. Prior UV irradiation of the host or treatment with *cis*-UCA, but not with *trans*-UCA, depressed abscess formation (Finlay-Jones *et al.*, 1992).

4.2. Cellular and Molecular Targets of Action of UCA

The most critical current area of investigation is the identification of the cellular and molecular targets of action of *cis*-UCA. *Cis*-UCA is formed in the skin in a dose-dependent manner in response to UV irradiation and can be readily quantitated in skin extracts by high-performance liquid chromatography (De Fabo *et al.*, 1989; Norval *et al.*, 1989; Pasanen *et al.*, 1990). It has not, however, been possible to identify *cis*-UCA systemically, suggesting that it does not accumulate in the lymphoid organs (De Fabo, unpublished observations). The idea that the antigen-presenting cell itself is not the target for *cis*-UCA is supported by the experiments described above in which *cis*-UCA was shown to be inactive *in vitro* in an antigen-presenting cell assay, suggesting that a number of intermediates are formed in response to *cis*-UCA *in vivo* (Noonan *et al.*, 1988). In earlier studies, transplantation of a large area of skin had enabled the recipients to accept a UV tumor (Palaszynski and Kripke, 1983). Together, these findings suggest that the cellular target of action for *cis*-UCA is in the skin. Two lines of evidence have been pursued. Norval and co-workers have demonstrated that *cis*-UCA can decrease the number of epidermal Langerhans cells in the skin (Norval *et al.*, 1990). We have found that UCA can regulate the cyclic-AMP activity of skin fibroblasts (Palaszynski *et al.*, 1992). Further research is necessary to identify unequivocally the cellular target(s) of action of *cis*-UCA.

UCA has structural analogy to histamine. Both UCA and histamine are derived from histidine, UCA by deamination and histamine by decarboxylation. Norval and colleagues investigated the immunosuppressive activity of a number of structural analogues of UCA and found that methylation of the imidazole ring removed suppressive activity but that saturation of the side chain did not (reviewed in Norval *et al.*, 1989). They also observed that histamine was immunosuppressive. Both Norval *et al.* (1990) and Matheson and Reeve (1991) found that administration of histamine antagonists prevented immunosuppression. These observations suggested that UCA acts via a histamine receptor. In our laboratory it was observed that both histamine and *trans*-UCA upregulated, in a dose-dependent manner, the activity of cyclic

AMP (a critical cellular signaling molecule) in human skin fibroblasts (Palaszynski *et al.*, 1992). Most interestingly, *cis*-UCA prevented this up regulation, as did cimetidine, a histamine-receptor antagonist. These findings suggest that UCA may bind via histamine (like) receptors, but they do not exclude the possibility that there are cellular receptors specific for UCA isomers. It can be postulated that *cis*-UCA regulates the production of suppressive factors from skin fibroblasts. More research is necessary to establish if this is the case and to identify the factor(s) involved.

5. Role of UV Suppression in Infectious Disease

An area of current interest is the role of UV suppression in infectious disease. Given the effects of UV on immunity described above, it might be predicted that irradiation with UV-B plays a role in infectious disease, particularly in diseases of the tropics where the UV radiation levels are high. In these studies, however, it must be remembered that the suppressive effect of UV is selective and that antibody responses are unaffected. As described above, a depressed DTH response to herpesvirus Type I has been found on UV irradiation, which may be related to the clinically observed recrudescence of herpes lesions caused by sunlight (Norval *et al.*, 1989). The DTH response to herpesvirus Type II is also suppressed by UV irradiation (Yasumoto *et al.*, 1987). Most interestingly, Giannini observed that UV irradiation suppressed the formation of skin lesions caused by the parasite *Leishmania*, but did not decrease the number of parasites in the skin and may have promoted parasite dissemination (Giannini, 1986; Giannini and De Fabo, 1988). Similar results have been found with skin lesions caused by mycobacteria (Jeevan and Kripke, 1989). As indicated above, preliminary studies in a mouse model indicated that UV irradiation decreased intra-abdominal abscess formation (Finlay-Jones *et al.*, 1992). Further research in these areas will be of considerable interest.

6. UV Suppression in Man

A critical issue is whether the immunosuppressive effects of UV radiation described in animal models also occur in humans. In 1983 it was first demonstrated that contact and delayed hypersensitivity responses in humans were suppressed by exposure to UV (Hersey *et al.*, 1983; Kalimo *et al.*, 1983). More recently, dose-dependent suppression of the CHS response has been described in humans (Cooper *et al.*, 1991). Suppression of CHS has been observed both in caucasians and in blacks (Vermeer *et al.,* 1991), indicating that in humans, as in the mouse model (Section 7), UV suppression is independent of pigmentation. CHS responses in UV-irradiated skin cancer patients were lower than in UV-irradiated controls, suggesting that skin cancer patients may be more susceptible to UV suppression (Yoshikawa *et al.,* 1990).

In terms of mechanism, the suppressor-inducer macrophage has been described in human skin (Baadsgaard *et al.,* 1988, 1989). UCA also occurs in human skin, for the most part above the pigment layer, and may explain why suppression is independent of pigmentation. UCA in human skin isomerizes in response to irradiation with UV (Pasanen *et al.,* 1990), reaching a photostationary state of about 1:1 *cis/trans* in heavily exposed areas such as the face (De Fabo, unpublished). UCA isomers also regulate cyclic-AMP activity in human skin fibroblasts, suggesting UCA is biologically active in these cells (Palaszynski *et al.,* 1992).

7. Genetic Susceptibility to UV Suppression

An important factor which has to be taken into account in assessing the biological effects of increased UV radiation is the underlying susceptibility of the organism. There are significant differences in susceptibility to UV suppression between mouse strains (Streilein and Bergstresser, 1988; Noonan and De Fabo, 1990; Yoshikawa and Streilein, 1990). BALB/c mice, for example, require more than 6 times the UV dose to cause 50% suppression of CHS than do C57BL/6 mice (Noonan and De Fabo, 1990). Pigmentation does not appear to be a critical factor. Differences in susceptibility to UV suppression of the CHS response also appear to occur in humans (Vermeer *et al.,* 1991; Yoshikawa *et al.,* 1990). Genetically determined differences in susceptibility to UV suppression may be a significant factor in UV-related diseases such as skin cancer (Yoshikawa *et al.,* 1990).

Streilein and co-workers have concluded, based on a study of UV suppression in a number of inbred mouse strains and hybrids between these strains, that susceptibility to UV suppression in the mouse is regulated by at least two genes, the *Lps* gene on chromosome 4 in the mouse, which controls immune responsiveness to lipopolysaccharide, and the *TNFα* gene, which is located in the major histocompatibility complex on chromosome 17 of the mouse (Streilein and Bergstresser, 1988; Yoshikawa and Streilein, 1990). Our own preliminary studies have not, however, supported a role either for the *Lps* gene or for a gene within the major histocompatibility complex in controlling susceptibility to UV suppression in the mouse (Noonan *et al.,* 1991). Further studies are necessary to resolve this question.

8. Effect of Sunscreens on UV Suppression

Given that UV suppression is caused by wavelengths in the UV-B range, it might be anticipated that sunscreens which are applied topically to the skin to prevent UV-B-induced sunburn would prevent the generation of UV-induced immunosuppression. This has not proven to be the case, however, for reasons which are not yet clear. A number of studies have found that application of sunscreens containing PABA (para-aminobenzoate) or its commercially used derivatives did not prevent the generation of UV suppression (Fisher *et al.,* 1989; Gurish *et al.,* 1981; Reeve *et al.,* 1991). In these studies, suppression was measured either as the acceptance of UV-induced tumors or as suppression of contact hypersensitivity. In the studies of Fisher and colleagues, sunscreens of SPF 6 or 15 containing Padimate O and oxybenzone were shown to prevent erythema, but they did not prevent suppression of CHS responses in hairless mice irradiated with either an FS40 sunlamp or solar-simulating radiation (Fisher *et al.,* 1989). In another study with hairless mice, a sunscreen containing 6.5% ortho-PABA and 4.5% benzophenone (a UV-A absorber) did not protect hairless mice from suppression of CHS responses, whereas a sunscreen containing 7.5% EHMC (a cinnamate derivative) and 4.5% benzophenone did protect (Reeve *et al.,* 1991). A more recent study, however, using the same sunscreens and the same mouse strain but a different UV source found protection from UV-induced loss of epidermal Langerhans cells and Thy1+ cells by the sunscreens, but no protection of UV-induced suppression of CHS responses (Ho *et al.,* 1992). In humans, two commercial sunscreen products, of SPF 6 and 15, containing a variety of active ingredients, did not protect against the UV-induced loss of alloactivating capacity in human skin *in vivo* (van Praag *et al.,* 1991). Alloactivating capacity is the ability to stimulate the proliferation of allogeneic lymphocytes, i.e., lymphocytes from another individual. This property is largely due to the expression of Class II molecules of the major histocompatibility complex which, in the skin, are expressed predominantly in epidermal Langerhans cells. The same investigators had previously established that application of a film of these sunscreens to a quartz cover above a tissue culture dish containing peripheral blood monocytes did protect against UV suppression of alloactivation by these cells (Mommaas *et al.,* 1990).

These studies clearly raise a number of important issues. The explanation for lack of protection against UV suppression in most of the studies cited above is unknown. It would appear unlikely to have a spectral explanation since the absorption spectra of the sun-protecting agents show major overlap with the action spectrum for immunosuppression at the wavelengths emitted by the UV sources used for these studies (De Fabo and Noonan, 1983; Reeve *et al.,* 1991; van Praag *et al.,* 1991; De Fabo *et al.,* 1992). Also, sunscreens do protect against erythema and against the aging effects of UV irradiation (Harrison *et al.,* 1991). There are no reports to date of the effect of sunscreens on the UV isomerization of UCA, which occurs superficially in the stratum corneum and may not be completely prevented by sunscreens. Alternatively, there may be other explanations. In any case, these findings raise questions about the wisdom of prolonged sun exposure using sunscreens, which protect against erythema (sunburn) but do not protect against UV suppression.

9. Conclusions

In conclusion, we have described a selective, immunosuppressive effect of ultraviolet-B radiation. This immunosuppression appears to be initiated by the interaction between UV radiation and a unique photoreceptor in mammalian skin, urocanic acid. UV suppression may be a protective mechanism, evolved to protect against autoimmune attack on sun-damaged skin but when stressed by high UV doses has the adverse consequence of permitting the outgrowth of skin cancers. The UV doses necessary to initiate UV suppression are readily available in natural sunlight. Increased solar UV-B predicted to result from stratospheric ozone depletion will increase UV suppression. The consequences of increased UV suppression to human and animal health are predicted to include increases in skin cancer. The effects of increased UV suppression on other UV-related diseases are at present uncertain.

ACKNOWLEDGMENTS. These studies were supported by grants from the National Institutes of Health, USA, the U.S. Environmental Protection Agency, the American Cancer Society, The Anti-Cancer Foundation of the Universities of South Australia, and the National Health and Medical Research Council, Australia.

References

Applegate, L. A., Ley, R. D., Alcalay, J., and Kripke, M. L., 1989, Identification of the molecular target for the suppression of contact hypersensitivity by ultraviolet radiation, *J. Exp. Med.* 170:1117–1131.
Aubin, F., Dall'Acqua, F., and Kripke, M. L., 1991, Local suppression of contact hypersensitivity in mice by a new bifunctional psoralen, 4,4,5'-trimethylazapsoralen, and UVA radiation, *J. Invest. Dermatol.* 97:50–54.
Baadsgaard, O., Fox, D. A., and Cooper, K. D., 1988, Human epidermal cells from ultraviolet light-exposed skin preferentially activate autoreactive CD4+2H4+suppressor-inducer lymphocytes and CD8+ suppressor/cytotoxic lymphocytes, *J. Immunol.* 140:1738–1744.

Baadsgaard, O., Lisby, S., Wantzin, G. L., Wulf, H. C., and Cooper, K. D., 1989, Rapid recovery of Langerhans cell alloreactivity without induction of autoreactivity after in vivo ultraviolet A, but not ultraviolet B exposure of human skin, *J. Immunol.* 142:4213–4218.

Baden, H. P., and Pathak, M. A., 1967, The metabolism and function of urocanic acid in skin, *J. Invest. Dermatol.* 48:11–17.

Bergstresser, P. R., 1986, Ultraviolet B radiation induces "Local immunosuppression," *Curr. Prob. Dermatol.* 15:205–218.

Bloom, B. R., Salgame, P., and Diamond, B., 1992, Revisiting and revising suppressor T cells, *Immunol. Today* 13:131–136.

Blum, H. F., 1959, *Carcinogenesis by Ultraviolet Light,* Princeton University Press, Princeton.

Brostoff, J., Scadding, G. K., Male, D., and Roitt, I. M., 1991, *Clinical Immunology,* Gower Medical Publishing, London.

Chung, H. T., Burnham, D. K., Robertson, B., Roberts, L. K., and Daynes, R. A., 1986, Involvement of prostaglandins in the immune alterations caused by the exposure of mice to ultraviolet radiation, *J. Immunol.* 137:2478–2484.

Cooper, K. D., 1990, Psoriasis: Leukocytes and cytokines, *Dermatol. Clinics* 8:737–746.

Cooper, K. D., Oberhelman, L., Hamilton, T. A., Baadsgaard, O., Terhune, M., LeVee, G., Anderson, T., and Koren, H., 1992, UV exposure reduced immunization rates and promotes tolerance to epicutaneous in humans—relationship to dose, CD1a-DR+ epidermal macrophage induction and Langerhans cell depletion. *Proc. Natl. Acad. Sci. U.S.A.* 89:8497–8501.

Cruz, P. D. J., Nixon-Fulton, J., Tigelaar, R. E., and Bergstresser, P. R., 1989, Disparate effects of in vitro low dose UVB irradiation on intravenous immunization with purified epidermal subpopulations for the induction of contact hypersensitivity, *J. Invest. Dermatol.* 92:160–165.

Daynes, R. A., and Spellman, C. W., 1977, Evidence for the generation of suppressor cells by ultraviolet radiation, *Cell. Immunol.* 31:182–187.

De Fabo, E. C., 1980, On the nature of the blue light photoreceptor; still an open question, in: *The Blue Light Syndrome* (H. Senger, ed.), Springer-Verlag, Berlin, pp. 187–197.

De Fabo, E. C., and Kripke, M. L., 1979, Dose-response characteristics of immunologic unresponsiveness to UV-induced tumors produced by UV irradiation of mice, *Photochem. Photobiol.* 30:385–390.

De Fabo, E. C., and Kripke, M. L., 1980, Wavelength dependence and dose-rate independence of UV radiation induced suppression of immunologic unresponsiveness of mice to a UV-induced fibrosarcoma, *Photochem. Photobiol.* 32:183–188.

De Fabo, E. C., and Noonan, F. P., 1983, Mechanism of immune suppression by ultraviolet irradiation in vivo. I. Evidence for the existence of a unique photoreceptor in skin and its role in photoimmunology, *J. Exp. Med.* 158:84–98.

De Fabo, E. C., Noonan, F. P., Fisher, M. S., Burns, J., and Kacser, H., 1983, Further evidence that the photoreceptor mediating UV-induced systemic immune suppression is urocanic acid, *J. Invest. Dermatol.* 80:319 (Abstract).

De Fabo, E. C., Reilly, S. K., and Noonan, F. P., 1989, Kinetics and UVB dose-response of cis urocanic acid formation in mouse skin parallels UVB-induced systemic suppression, *J. Invest. Dermatol.* 92:418 (Abstract).

De Fabo, E. C., Noonan, F. P., and Frederick, J. E., 1990, Biologically effective doses of sunlight for immune suppression at various latitudes and their relationship to changes in stratospheric ozone, *Photochem. Photobiol.* 52:811–817.

De Fabo, E. C., Reilly, D. C., and Noonan, F. P., 1992, Mechanism of UVA effects on immune function. Preliminary studies, in: *Biological Effects of UVA,* (F. Urbach, ed.), Valdenmar, Overland Park, KS, pp. 227–237.

De Gruijl, F. R., and Roza, L., 1991, Photoreactivation in humans, *J. Photochem. Photobiol. B: Biol.* 10:367-371.

De Gruijl, F. R., and Van der Leun, J. C., 1982, Systemic influence of pre-irradiation of a limited skin area on UV-tumorigenesis, *Photochem. Photobiol.* 35:379-383.

Farrow, S. J., Mohammad, T., Baird, W., and Morrison, H., 1990, Photochemical covalent binding of urocanic acid to polynucleic acids, *Chem. Biol. Interact.* 75:105-118.

Finlay-Jones, J. J., Spencer, L. K., Farmer, L., Kenny, P. A., Cox, K. O., McDonald, P. J., and Noonan, F. P., 1992, Ultraviolet (UV) irradiation and the pathogenesis of infection: The influence of UV irradiation on intra-abdominal abscess formation in mice, *Photochem. Photobiol.* 55:7S (Abstr.).

Fisher, M. S., 1978, A systemic effect of ultraviolet irradiation and its relationship to tumor immunity, *Nat. Cancer Inst. Monogr.* 50:185-188.

Fisher, M. S., and Kripke, M. L., 1977, Systemic alteration induced in mice by ultraviolet light irradiation and its relationship to ultraviolet carcinogenesis, *Proc. Natl. Acad. Sci. U.S.A.* 74:1688-1692.

Fisher, M. S., and Kripke, M. L., 1978, Further studies on the tumor-specific suppressor cells induced by ultraviolet radiation, *J. Immunol.* 121:1139-1144.

Fisher, M. S., and Kripke, M. L., 1982, Suppressor T lymphocytes control the development of primary skin cancers in ultraviolet irradiated mice, *Science* 216:1133-1134.

Fisher, M. S., Menter, J. M., and Willis, I., 1989, Ultraviolet radiation-induced suppression of contact hypersensitivity in relation to Padimate O and oxybenzone, *J. Invest. Dermatol.* 92: 337-341.

Frederick, J. E., and Lubin, D., 1988, The budget of biologically active ultraviolet radiation in the earth-atmosphere system, *J. Geophys. Res.* 93:3825-3832.

Furukawa, F., Kashihara-Sawami, M., Lyons, M. B., and Norris, D. A., 1990, Binding of antibodies to the extractable nuclear antigens SS-A/Ro and SS-B/La is induced on the surface of human keratinocytes by ultraviolet light (UVL): Implications for the pathogenesis of photosensitive cutaneous lupus, *J. Invest. Dermatol.* 94:77-85.

Giannini, M. S. H., 1986, Suppression of pathogenesis in cutaneous Leishmaniasis by UV irradiation, *Infect. Immun.* 51:838-843.

Giannini, M. S. H., and De Fabo, E. C., 1988, Abrogation of skin lesions in cutaneous Leishmaniasis by ultraviolet B irradiation, in: *Leishmaniasis: The First Centenary (1885-1985). New Strategies for Control* (D. T. Hart, ed.) Plenum, New York, pp. 65-73.

Glazier, A., Morison, W. L., Bucana, C., Hess, A. D., and Tutschka, J., 1984, Suppression of epidermal graft-versus-host disease with ultraviolet radiation, *Transplantation* 37:211-213.

Graffi, A., Pasternak, G., and Horn, K. -H., 1964, Die Erzeugung von Resistenz gegen isologe Transplantate UV-induzierter Sarkomen der Maus, *Acta Biol. Med. Ger.* 12:726-728.

Greene, M. I., Sy, M. S., Kripke, M. L., and Benacerraf, B., 1979, Impairment of antigen-presenting function by ultraviolet radiation, *Proc. Natl. Acad. Sci. U.S.A.* 76:6592-6595.

Gruner, S., Stoppe, H., Eckert, R., Sonnichsen, N., and Diezel, W., 1990, Verlangerung der Transplantatuberlebenszeit durch eine PUVA-Behandlung des Transplantatempfangers. Bedeuting von cis-Urocaninsaure, *Dermatol. Monatsschr.* 176:49-54.

Gruner, S., Diezel, W., Stoppe, H., Oesterwitz, H., and Henke, W., 1992, Inhibition of skin allograft rejection and acute graft-versus-host disease by cis-urocanic acid, *J. Invest. Dermatol.* 98:459-462.

Gurish, M. F., Roberts, L. K., Krueger, G. G., and Daynes, R. A., 1981, The effect of various sunscreen agents on skin damage and the induction of tumor susceptibility in mice subjected to ultraviolet irradiation, *J. Invest. Dermatol.* 76:246-251.

Hanisko, J., and Suskind, R. R., 1963, The effect of ultraviolet radiation on experimental cutaneous sensitization in guinea pigs, *J. Invest. Dermatol.* 49:183-191.

Harriott-Smith, T. G., and Halliday, W. J., 1988, Suppression of contact hypersensitivity by short-term ultraviolet irradiation. II. The role of urocanic acid, *Clin. Exp. Immunol.* 72:174–177.

Harrison, J. A., Walker, S. L., Plastow, S. R., Batt, M. D., Hawk, J. L., and Young, A. R., 1991, Sunscreens with low sun protection factor inhibit ultraviolet B and A photoaging in the skin of the hairless albino mouse, *Photodermatol. Photoimmunol. Photomed.* 8:12–20.

Hersey, P., Hasic, E., Edwards, A., Bradley, M., and Haran, G., 1983, Immunological effects of solarium exposure, *Lancet* 1:545–548.

Ho, K. K-L., Halliday, G. M., and Barnetson, R. St. C., 1992, Sunscreens protect epidermal Langerhans cells and Thy1+ cells but not local contact sensitization from the effects of ultraviolet light, *J. Invest. Dermatol.* 98:720–724.

Howie, S. E. M., Norval, M., and Maingay, J., 1986, Exposure to low-dose ultraviolet radiation suppresses delayed-type hypersensitivity to *Herpes Simplex* in mice, *J. Invest. Dermatol.* 86: 125–128.

Jeevan, A., and Kripke, M. L., 1989, Effect of a single exposure to ultraviolet radiation on *Mycobacterium bovis Bacille Calmette Guerin* infection in mice, *J. Immunol.* 143:2837–2843.

Jensen, P. J., 1983, The involvement of antigen-presenting cells and suppressor cells in the ultraviolet radiation-induced inhibition of secondary cytotoxic T cell sensitization, *J. Immunol.* 130:2071–2074.

Jessup, J. M., Hanna, N., Palaszynski, E., and Kripke, M. L., 1978, Mechanisms of depressed reactivity to dinitrochlorobenzene and ultraviolet-induced tumors during ultraviolet carcinogenesis in BALB/c mice, *Cell. Immunol.* 38:105–115.

Kalimo, K., Koulu, L., and Jansen, C. T., 1983, Effect of a single UVB or PUVA exposure on immediate and delayed skin hypersensitivity reactions in humans, *Arch. Dermatol. Res.* 275: 374–378.

Kim, T. Y., Kripke, M. L., and Ullrich, S. E., 1990, Immunosuppression by factors released from UV-irradiated epidermal cells: Selective effects on the generation of contact and delayed type hypersensitivity after exposure to UVA or UVB radiation, *J. Invest. Dermatol.* 94:26–32.

Kripke, M. L., 1974, Antigenicity of murine skin tumors induced by ultraviolet light, *J. Natl. Cancer Inst.* 53:1333–1336.

Kripke, M. L., 1990, Effects of UV radiation on tumor immunity, *J. Natl. Cancer Inst.* 82:1392–1396.

Kripke, M. L., and Fisher, M. S., 1976, Immunologic parameters of ultraviolet carcinogenesis, *J. Natl. Cancer Inst.* 57:211–215.

Kripke, M. L., Lofgreen, J. S., Beard, J., Jessup, J. M., and Fisher, M. S., 1977, *In vivo* immune responses of mice during carcinogenesis by ultraviolet irradiation, *J. Natl. Cancer Inst.* 59: 1227–1230.

Kripke, M. L., Cox, P., and Yarosh, D. B., 1992, Effect of T4N5 liposome-enhanced DNA repair on UV-induced immune suppression in the mouse, *Proc. Natl. Acad. Sci. U.S.A.* 89:7516–7520.

Krutmann, J. K., Kammer, G. M., Toossi, Z., Waller, R. L., Ellner, J. J., and Elmets, C. A., 1990, UVB radiation and human monocyte accessory function. Differential effects on premitotic events in T cell activation, *J. Invest. Dermatol.* 94:204–209.

Letvin, N. L., Greene, M. I., Benacerraf, B., and Germain, R. N., 1980, Immunologic effects of whole-body ultraviolet irradiation: Selective defect in splenic adherent cell function *in vitro,* *Proc. Natl. Acad. Sci. U.S.A.* 77:2881–2885.

Lynch, D. H., Gurish, M. F., and Daynes, R. A., 1981, Relationship between epidermal Langerhans cell density, ATPase activity and the induction of contact hypersensitivity, *J. Immunol.* 126: 1892–1897.

Matheson, M. J., and Reeve, V. E., 1991, The effect of the antihistamine cimetidine on ultraviolet-radiation-induced tumorigenesis in the hairless mouse, *Photochem. Photobiol.* 53:639–642.

McKenzie, R. C., and Sauder, D. N., 1990, Keratinocytes cytokines and growth factors: Functions in skin immunity, *Dermatol. Clinics* 8:649–662.

Molendijk, R. J. H., van Gurp, L. M., Donselaar, I. G., and Benner, R., 1987, Suppression of delayed type hypersensitivity to histocompatibility antigens by ultraviolet radiation, *Immunology* 62:299–305.

Mommaas, A. M., Marinus, C. G., van Praag, J. N., Bouwes Bavinck, J. N., Out-Luyting, C., Vermeer, B. J., and Claas, F. H. J., 1990, Analysis of the protective effect of topical sunscreens on the UVB-radiation induced suppression of the mixed lymphocyte reaction, *J. Invest. Dermatol.* 95:313–316.

Morrison, H., Avnir, D., Bernasconi, C., and Fagan, G., 1980, Z/E photoisomerization of urocanic acid, *Photochem. Photobiol.* 32:711–714.

Mosmann, T. R., and Coffman, R. L., 1989, Th1 and Th2 cells: Different patterns of lymphokine secretion lead to different functional properties, *Annu. Rev. Immunol.* 7:145–173.

Mosmann, T. R., and Moore, K. W., 1991, The role of IL-10 in cross regulation of Th1 and Th2 clones, *Immunoparasitol. Today* 12:A49–A53.

Mottram, P. L., Mirosklavos, A., Clunie, G. J. A., and Noonan, F. P., 1988, A single dose of UV radiation suppressed delayed type hypersensitivity responses to alloantigens and prolongs heart allograft acceptance in mice, *Immunol. Cell Biol.* 66:377–385.

Noonan, F. P., and De Fabo, E. C., 1990, Ultraviolet-B dose-response curves for local and systemic immunosuppression are identical, *Photochem. Photobiol.* 52:801–810.

Noonan, F. P., De Fabo, E. C., and Kripke, M. L., 1981a, Suppression of contact hypersensitivity by UV radiation and its relationship to UV-induced suppression of tumor immunity, *Photochem. Photobiol.* 34:683–690.

Noonan, F. P., De Fabo, E. C., and Kripke, M. L., 1981b, Suppression of contact hypersensitivity by ultraviolet radiation: An experimental model, *Spring. Sem. Immunopathol.* 4:293–304.

Noonan, F. P., Kripke, M. L., Pedersen, G. M., and Greene, M. L., 1981c, Suppression of contact hypersensitivity in mice by ultraviolet irradiation is associated with defective antigen presentation, *Immunology* 43:524–533.

Noonan, F. P., Bucana, C., Sauder, D. N., and De Fabo, E. C., 1984, Mechanism of immune suppression by UV irradiation in vivo. II. The UV effects on number and morphology of epidermal Langerhans cells and the UV-induced suppression of contact hypersensitivity have different wavelength dependencies, *J. Immunol.* 132:2408–2416.

Noonan, F. P., De Fabo, E. C., and Morrison, H., 1988, Cis-urocanic acid, a product formed by ultraviolet B irradiation of the skin initiates an antigen presentation defect in splenic dendritic cells in vivo, *J. Invest. Dermatol.* 90:92–99.

Noonan, F. P., De Fabo, E. C., and Elgart, M. E., 1989, Histologic evidence that inflammation and epidermal cell killing are not necessary for UV-induced systemic suppression, *J. Invest. Dermatol.* 92:492 (Abstract).

Noonan, F. P., Reilly, S. K., Hoffman, H., and De Fabo, E. C., 1991, Susceptibility to UVB-induced immune suppression: Effect of genetic factors and of diet, *Photochem. Photobiol.* 53:89S (Abstract).

Norbury, K. C., Kripke, M. L., and Budmen, M. B., 1977, In vitro reactivity of macrophages and lymphocytes from UV-irradiated mice, *J. Natl. Cancer Inst.* 59:1231–1235.

Norval, M., Simpson, T. J., and Ross, J. A., 1989, Urocanic acid and immunosuppression, *Photochem. Photobiol.* 50:267–275.

Norval, M., Gilmour, J. W., and Simpson, T. J., 1990, The effect of histamine receptor antagonists on immunosuppression induced by the cis-isomer of urocanic acid, *Photodermatol. Photoimmunol. Photomed.* 7:243–248.

Palaszynski, E. W., and Kripke, M. L., 1983, Evidence for an interaction between the immune system and UV light in the skin of mice undergoing carcinogenesis, *Transplantation* 36:465–467.

Palaszynski, E. W., Noonan, F. P., and De Fabo, E. C., 1992, *Cis*-urocanic acid down regulates the induction of cyclic AMP by either *trans* urocanic acid or histamine in human dermal fibroblasts *in vitro, Photochem. Photobiol.* 55:165–171.

Parish, C. R., 1972, The relationship between humoral and cell-mediated immunity, *Transplant. Rev.* 13:35–66.

Pasanen, P., Reunala, T., Jansen, C. T., Rasanen, L., Neuvonen, K., and Ayras, P., 1990, Urocanic acid isomers in epidermal samples and suction blister fluid of nonirradiated and UVB-irradiated human skin, *Photodermatol.* 7:40–42.

Pasternak, G., Graffi, A., and Horn, K. -H., 1964, Der Nachweis individualspezifischer Antigenitat bei UV-induzierten Sarkomen der Maus., *Acta Biol. Med. Ger.* 13:276–279.

Reeve, V. E., Greenoak, G. E., Canfield, P. J., Boehm-Wilcox, C., and Gallagher, C. H., 1989, Topical urocanic acid enhances UV-induced tumour yield and malignancy in the hairless mouse, *Photochem. Photobiol.* 49:459–464.

Reeve, V. E., Bosnic, M., Boehm-Wilcox, C., and Ley, R. D., 1991, Differential protection by two sunscreens from UV radiation induced immunosuppression, *J. Invest. Dermatol.* 97:624–628.

Reilly, S. K., and De Fabo, E. C., 1991, Dietary histidine increases mouse skin urocanic acid levels and enhances UVB-induced immune suppression of contact hypersensitivity, *Photochem. Photobiol.* 53:431–438.

Roberts, L. K., Smith, D. R., Seilstad, K. H., and Jun, B. -B., 1988, Photoimmunology: The mechanisms involved in immune modulation by UV radiation, *J. Photochem. Photobiol. B. Biol.* 2:149–177.

Roberts, L. K., Lynch, D. H., Samlowski, W. E., and Daynes, R. A., 1989, Ultraviolet radiation and modulation, *Immunol. Ser.* 48:167–215.

Ross, J. A., Howie, S. E., Norval, M., Maingay, J., and Simpson, T. J., 1986, Ultraviolet irradiated urocanic acid suppresses delayed type hypersensitivity to herpes simplex virus in mice, *J. Invest. Dermatol.* 87:630–633.

Schwartz, T., Urbanska, A., Gschnait, F., and Luger, T. A., 1986, Inhibition of the induction of contact hypersensitivity by a UV-mediated epidermal cytokine, *J. Invest. Dermatol.* 87:289–291.

Setlow, R. B., 1974, The wavelengths in sunlight effective in producing skin cancer: A theoretical analysis. *Proc. Natl. Acad. Sci. U.S.A.* 71:3363–3365.

Spellman, C. W., Woodward, J. G., and Daynes, R. A., 1977, Modification of immunologic potential by ultraviolet radiation. I. Immune status of short-term UV-irradiated mice, *Transplantation* 24:112–119.

Spellman, C. W., Anderson, W. L., Bernhard, E. J., and Tomasi, T. B., 1984, Suppression of antibody responses to topically applied antigens by ultraviolet light irradiation, *J. Exp. Med.* 160:1891–1900.

Stingl, G., 1990, Dendritic cells of the skin, *Dermatol. Clinics* 8:673–679.

Streilein, J. W., 1989, Antigen-presenting cells in the induction of contact hypersensitivity in mice: Evidence that Langerhans cells are sufficient but not required, *J. Invest. Dermatol.* 93:443–448.

Streilein, J. W., and Bergstresser, P. R., 1988, Genetic basis of ultraviolet B effects on contact sensitivity, *Immunogenetics* 27:252–258.

Swartz, R. P., 1984, Role of UVB-induced serum factor(s) in suppression of contact hypersensitivity in mice, *J. Invest. Dermatol.* 83:305–307.

Tang, A., and Udey, M. C., 1991, Inhibition of epidermal Langerhans cell function by low dose ultraviolet B radiation. Ultraviolet B radiation selectively modulates ICAM-1 (CD54) expression by murine Langerhans cells, *J. Immunol.* 146:3347–3355.

Taylor, R. G., Levy, H. L., and McInnes, R. R., 1991, Histidase and histidinemia. Clinical and molecular considerations, *Mol. Biol. Med.* 8:101–116.

Thorn, R. M., 1978, Specific inhibition of cytotoxic memory cells produced against UV-induced tumors in UV-irradiated mice, *J. Immunol.* 121:1920–1926.

Toews, G. B., Bergstresser, P. R., and Streilein, J. W., 1980, Epidermal Langerhans cell density determines whether contact hypersensitivity or unresponsiveness follows skin painting with DNFB, *J. Immunol.* 124:445–449.

Ullrich, S. E., 1986, Suppression of the immune response to allogeneic histocompatibility antigens by a single exposure to ultraviolet radiation, *Transplantation* 42:287–291.

van Praag, M. C., Out-Luyting, C., Claas, F. H., Vermeer, B. J., and Mommaas, A. M., 1991, Effect of topical sunscreens on the UV radiation induced suppression of the alloactivating capacity in human skin *in vivo, J. Invest. Dermatol.* 97:629–633.

Vermeer, M., Schmieder, G. J., Yoshikawa, T., van den Berg, J. -W., Metzman, M. S., Taylor, J. R., and Streilein, J. W., 1991, Effects of ultraviolet B light on cutaneous immune responses of humans with deeply pigmented skin, *J. Invest. Dermatol.* 97:729–734.

Ward, P. L., and Schreiber, H., 1991, MHC class I restricted T cells and immune surveillance against transplanted ultraviolet light induced tumors, *Semin. Cancer Biol.* 2:321–328.

Williams, K. A., Ash, J., Mann, T., Noonan, F. P., and Coster, D. J., 1987, Cells infiltrating inflamed and vascularized corneas, *Transplant. Proc.* 19:2889–2891.

Williams, K. A., Lubeck, D., Noonan, F. P., and Coster, D. J., 1990, Prolongation of rabbit corneal allograft survival following systemic administration of urocanic acid, in: *Ocular Immunology Today* (M. Usui, S. Ohno, and K. Aoki eds.), Elsevier, Amsterdam, pp. 103–106.

Yasumoto, S., Hayashi, Y., and Aurelian, L., 1987, Immunity to Herpes simplex virus Type 2. Suppression of virus-induced immune responses in ultraviolet B-irradiated mice, *J. Immunol.* 139:2788–2793.

Yoshikawa, T., and Streilein, J. W., 1990, Genetic basis of the effects of ultraviolet B light on cutaneous immunity. Evidence that polymorphism at *Tnfa* and *Lps* loci governs susceptibility, *Immunogenetics* 32:398–405.

Yoshikawa, T., Rae, V., Bruins-Slot, W., van den Berg, J. -W., Taylor, R., and Streilein, J. W., 1990, Susceptibility to effects of UVB radiation on induction of contact hypersensitivity as a risk factor for skin cancer in humans, *J. Invest. Dermatol.* 95:530–536.

Young, E., Olkowski, S. T., Dana, M., Mallette, R. A., and Stark, W. J., 1989, Pretreatment of donor corneal epithelium with ultraviolet B radiation, *Transplant. Proc.* 21:3145–3146.

6

Ocular Damage by Environmental Radiant Energy and Its Prevention

Seymour Zigman

1. Introduction

Sunlight is very rich in the ultraviolet wavelengths of radiant energy. The deepest penetration of ultraviolet energy into the body is through the eye. The cornea absorbs ultraviolet energy of wavelengths shorter than 300 nm and can be damaged by it (photokeratitis). Another target of UV damage is the DNA in the nuclei of corneal epithelial cells. Repair of the DNA is quite rapid, however, as is the replacement of damaged cells (Pitts *et al.*, 1986). Ultraviolet energy reaching the lens is in the wavelength range of 300 to 400 nm. Nearly all of this ultraviolet energy is absorbed by the numerous light-absorbing, -scattering, and -fluorescing chemical species present in the human lens. These include RNA and DNA (absorption maxima, λ max, at 254 nm), proteins and amino acids (λ max at 280 nm and 260 nm, respectively), and numerous chromophores that absorb not only this shorter-wavelength UV energy, but also 300- to 400-nm UV and blue light. The chromophore concentration increases in humans with aging, which partially explains the darkening of the lens in the elderly (Lerman and Borkman, 1978).

Seymour Zigman • Department of Ophthalmology and Biochemistry, University of Rochester School of Medicine and Dentistry, Rochester, New York 14642.

Environmental UV Photobiology, edited by Antony R. Young *et al.* Plenum Press, New York, 1993.

Lens light scattering and autofluorescence are two optical processes that adversely influence vision, especially because of 300- to 400-nm radiation and blue light absorption (Zigman, 1990). Large-particle scattering of short-wavelength light by the lens is mostly dependent on lens proteins and their state of aggregation or de-aggregation and also on the physical state of lens fiber-cell membranes. The autofluorescence that is excited by 300- to 400-nm radiation is due mainly to lens chromophores and oxidized protein side chains, mostly of tryptophan oxidation products and protein carbonyl groups and cross-links (Dillon *et al.*, 1976).

Thus, our consideration of UV damage to the lens is twofold:

1. Photochemical production of toxic photoproducts that adversely influence cell metabolism and photochemical damage to structural and informational macromolecules.
2. Excessive scattering and fluorescence of 300- to 400-nm radiation and blue light that interfere with the unencumbered passage of images through the lens, leading to poor vision.

In this chapter, details of the mechanism by which 300- to 400-nm (and blue) radiant energy adversely affects the structure and the function of the lens are presented. Several reviews have already covered this material (Lerman, 1980; Zigman, 1986a,b; 1987a,b). This chapter will allude briefly to them, but will also introduce newer material.

Germane to the discussion of UV damage to the lens is the quality and quantity of the radiant energy in the environment, both outdoors and indoors. Figure 1 depicts the relative and measured irradiances in the range of light commonly encountered by humans. Sunlight radiant energy provides both UV-A (315–400 nm) and UV-B (280–315 nm) wavelengths. Light in the visible and infrared ranges are the most prominent types of radiant energy in sunlight. Fluorescent lamps are of several types, but only the daylight fluorescent spectrum matches that of sunlight qualitatively. The cool-white and warm-white spectra have higher relative irradiances of blue light than sunlight, but lack sunlights longer visible wavelengths. Incandescent lamps emit mainly longer wavelengths, but are deficient in the shorter-wavelength ranges.

As for damage to the lens, the shorter wavelengths are the most efficient. UV-B and UV-A wavelengths together represent most of the environmental or artificial lighting that can damage the lens. Black light bulb (BLB) fluorescent lamps, used in our laboratory experiments, provide mainly UV-A wavelengths, while fluorescent sunlamps provide mainly UV-B. Unless monochromatic light is used to expose living tissues for experimental purposes, both UV-A and UV-B are represented in the energy to which the tissue is exposed, and one cannot totally separate the effects of UV-A from UV-B. In the BLB lamp,

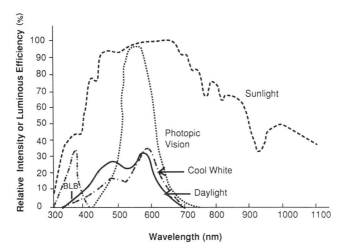

Figure 1. Emission spectra of terrestrial solar radiation, two fluorescent light lamps (daylight and cool-white), and a black light bulb (BLB). The human photopic spectral sensitivity curve is also shown.

only about 3% of the emitted energy is in the UV-B range; much greater irradiances of UV-A than of UV-B are also present in sunlight, since ozone preferentially removes UV-B from sunlight before it reaches the earth.

While the focus of this chapter is mainly on the 300- to 400-nm UV energy range, one cannot ignore visible light as a hazard to the lens as well. In circumstances where an individual has a sensitivity to light, perhaps due to the excessive use of vitamins such as riboflavin or the antibiotic tetracycline or the tranquilizer chlorpromazine (and others), blue light can become detrimental (Dayhaw-Barker and Barker, 1986). The greater concern about visible light damage is to the retina (Sperling *et al.,* 1980; Tso, 1989), which is known to sustain damage from blue light; this will be referred to later. It should be stated, however, that 300- to 400-nm radiation damage to the human retina is only a major concern in aphakic eyes, as the normal lens would absorb 300- to 400-nm UV and, in aged eyes, a large portion of blue light as well (Collier and Zigman, 1989).

In order to discuss generally the role of short-wavelength radiant energy in vision impairment due to lens damage, it is important to outline the manner in which light interacts with the ocular lens. How light is handled by the ocular lens depends on the species, the age of the lens, its state of clarity, the history of the individual, and the states of protein and pigments. Essentially, light in the visible range is transmitted nearly totally by young human and most animal lenses (i.e., 400 to 700 nm is transmitted). Some animal and human lenses contain chromophores that absorb light below 450 nm (e.g., the gray squirrel lens; the elderly human lens). With aging, many components that interfere with the transmission of short-wavelength light are increased. Thus, elderly humans and those with early cataracts have lenses with accumulated chromophoric material that causes the yellowing and browning of the lens and an enhancement of 300- to 400-nm UV and short-wavelength blue light—stimulated autofluorescence. Also increasing with aging are structural changes leading to enhanced light scattering in the lens (crystallin aggregation, membrane damage, and refractive-index discontinuities). Figure 2 illustrates the influence of human lens light-scattering and light absorption properties on an image viewed through a few types of cataractous lenses.

Details of lens light transmission and thresholds for UV-induced damage can be found in Zigman (1987a), and will not be repeated here. Zigman (1986a,b) describes in detail the chemistry of chromophores in the human lens, the photochemical inhibition of several enzymes whose functions are essential to maintaining lens transparency, and evidence of photochemically induced protein cross-linking and aggregation. Newer information relative to the above areas, however, will follow.

2. Light Interaction with the Lens

The lens has only limited capability to directly modify the radiant energy from sunlight that is absorbed by its biological molecules. Such absorption can, however, cause the molecules to become electronically excited and more reactive. Often the peak absorption capability of a biomolecule is not at the wavelength of maximum emission of a light source. For example, DNA absorbs maximally at a wavelength of 254 nm, protein at 280 nm, and aromatic amino acids at 280 or 260 nm, but all have a tail of absorbance that extends weakly up to 400 nm. Present in the lens and aqueous humor are low concentrations of chromophores that do absorb energy in the wavelength range of 300 to 500 nm. Examples are riboflavin and kynurenines, but these molecules not only absorb 300- to 500-nm light, but are also photosensitizers that transfer absorbed energy to macromolecules that can be altered indirectly in this way.

When oxygen and metal ions are also present in the system, as they are in the eye (aqueous humor), excitation of molecules is paralleled by the formation of oxygen-excited states. These include singlet oxygen, hydroxyl radicals, and superoxide. These molecules are highly reactive and can alter structural and enzymatic proteins. Even the structure of DNA can be altered by interactions of excited sensitizer molecules with it. The significance of oxygen radicals to ocular tissue damage has been provided in great detail (see Varma and Lerman, 1984).

Several investigators have shown that one toxic oxygen product, singlet oxygen, forms due to UV exposure and causes lens protein cross-linking (Goosey *et al.,* 1983; Zigman *et al.* 1988). Other free radicals and excited states are also present in the lens (Zigman, 1983b). They are oxidation products of the amino acid tryptophan, but their function is not known.

2.1. New Basic Findings on UV-Induced Lens Damage: *In Vivo* Findings with Squirrels

The lenses of gray squirrels housed in an environment of high UV-A irradiance (6 mW/cm^2—about ½ that of sunlight; 12 h/day; 400 days) developed obvious anterior cortical opacities (see Fig. 3). When histological preparations were examined, numerous deficient epithelial cells were seen in the central area of the lens. There were swollen and vacuolated cells, ballooned cells, absent cells, and proliferative cells (multilayers) present in the central region of the epithelium. Other damage to the central anterior cortical fibers midway between the capsule and the nuclear border was observed in the form of vacuolated and swollen fibers with lakes of fluid and irregular spacings (see Fig. 4). When the content of aggregated insoluble lens proteins was measured, it was found that a significant increase in these high-molecular-weight proteins had occurred throughout the lens. The above structural and gross chemical changes created light-scattering and refractive-index changes.

The biochemical changes observed provide evidence that the entire chemical machinery of the lens had been compromised by UV exposure. For example, there was a loss of specific crystallin components in the cortex and an increase in low-molecular-weight peptides in the nucleus (see Fig. 5). Many other subtle changes in lens crystallins were observed. The chemistry of sulfur compounds in the lens determines the oxidation–reduction state of the tissue. Damaged or sick lenses have generally shown losses of reducing capacity in the form of sulfhydryl losses and disulfide increases. This is a sign that in the affected lenses, the oxidation–reduction balance has been offset by a combination of the loss of reducing agents (i.e., ascorbate and glutathione in the reduced state) and the decreased activity of reducing enzymes such as glutathione reductase and catalase (Zigman, 1986b). This study also found that a major shift in the levels of -SH groups toward the formation of mixed protein–thiol disulfides had occurred throughout the lens due to UV exposure.

A brief summary of the *in vivo* 300- to 400-nm radiation effects on the lens of a diurnal animal, the gray squirrel, is provided in Table 1. Details of the findings can be found in Zigman *et al.* (1991a).

Figure 2. Photographically simulated views seen through various human lenses to show the influence of light scattering and pigmentation. (A) through a young lens; (B) through an older lens; (C) through a cortical cataractous lens; (D) through a nuclear cataractous lens.

2.2. *In Vitro* Studies on Squirrel and Rabbit Lenses

Due to the difficulty of elucidating specific mechanisms of action of 300- to 400-nm UV radiation *in vivo,* an *in vitro* system has been developed. The lens is especially suitable for *in vitro* studies, because it survives well in physiological media for long time periods (i.e., days) and its metabolism continues

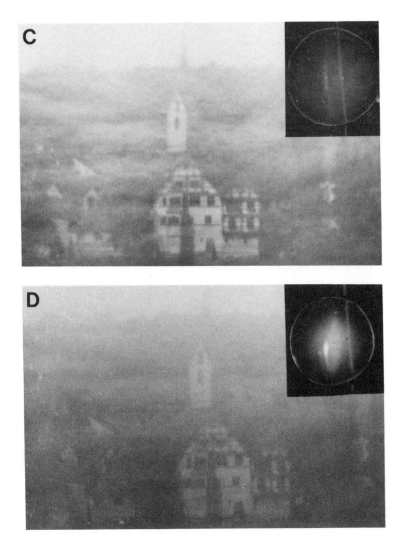

Figure 2. (*Continued*)

almost normally during this time as long as there is sufficient glucose for energy, a buffering system to maintain the pH, and maintenance of an isosmotic pressure to prevent lens swelling or shrinkage.

2.2.1. UV-Induced Damage

When lenses of diurnal gray squirrels were placed anterior side up in medium 199 with the appropriate glucose content and salt mixture (Earle's

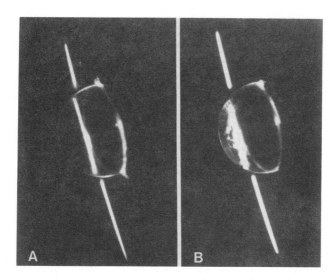

Figure 3. Slit-lamp view of a squirrel lens after 1 year of 300- to 400-nm (BLB lamp) exposure (6 mW/cm^2 for 12 h/day). (A) control animal; (B) exposed.

salts), at an osmolarity of 290 mOsmols, and a temperature of 27°C, they remained clear for at least 24 h. When the lenses were exposed from above to BLB (UV-A-emitting) lamps in this medium at an irradiance of 2 to 4 mW/cm^2 (at 365 ± 50 nm), they developed an anterior cortical opalescence in 2 to 16 h that did not appear in the controls (Fig. 6). This was observed histologically as lens epithelial abnormalities (swollen epithelium, pyknotic nuclei) and damaged subepithelial outer fiber cells (loss of regularity, formation of vacuoles and fluid lakes; Fig. 7). The lens epithelial cells were found to have diminished Na-K-ATPase activity (for controlling water balance), aggregation of crystallins, and also degradation of crystallins (Torriglia and Zigman, 1988; Fig. 8).

Another important type of lens epithelial-cell damage was to the cytoskeletal element actin. The shape and integrity of these cells depends in part on the structural integrity of actin, a macromolecule that is part of the cell membrane system but also is present in polygonal arrays of filaments (Rafferty and Scholz, 1989). In the epithelia of several different species, specific fluorescent staining probes (Phalloidin or Rhodamine) for actin, as well as electron microscopy, revealed that the actin structure was broken down within a matter of a few hours due to exposure to UV radiation (Fig. 9). Chemical studies have also shown that purified actin degrades partially when it is exposed to 300- to 400-nm irradiation (Table 2).

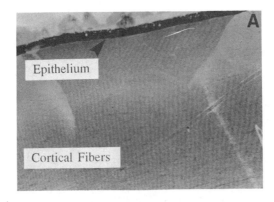

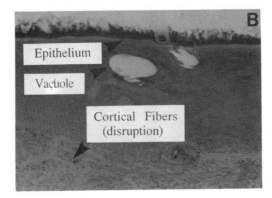

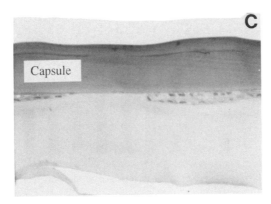

Figure 4. Histological observations of the damage to the squirrel lens induced by *in vivo* exposure to 300- to 400-nm radiation for 400 days (6 mW/cm^2 for 12 h/day). Anterior central epithelial-cell and mild cortical fiber-cell swelling and disruption are evident. (A) control; (B) UV exposed; (C) UV exposed.

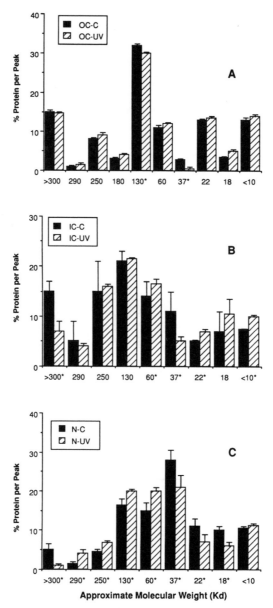

Figure 5. Alterations in lens crystallins of squirrel eye after 300- to 400-nm exposure for 400 days (light bars) compared to control (dark bars). Analyses of total soluble lens protein separated by high-pressure liquid chromatography. (A) outer cortex (OC); (B) inner cortex (IC); (C) nucleus (N). Asterisks denote statistically significant differences from control with $p \leq 0.05$ ($n = 4$). Note: UV-induced changes are described in text.

Table 1. *In Vivo* Effects of Exposure to Black Light on the Squirrel Lens

Gross changes:
    Anterior superficial opacities.
Histology:
    Damaged epithelial cells, showing swelling and then disintegration in small numbers.
    Hydropic areas in the anterior superficial cortex.
    Destruction of the outer cortex fiber cells.
Proteins:
    Enhanced aggregated proteins toward nucleus.
    Altered distribution of soluble proteins show more aggregation and protein degradation,
        especially inner cortex and nucleus.
    Altered protein chains of 30 to 32 kd—beta-crystallins.
Oxidation–reduction system:
    Glutathione drops in outer cortex and epithelium of lens.
    -SH groups drop in the outer cortex and epithelium.
    Increase of mixed disulfides (cysteine and glutathione in outer portion of the lens; lesser but
        definite similar changes in the inner cortex and nucleus.
Conclusion:
    Exposure to 300- to 400-nm, ambient radiation for long time periods at subsolar levels
        causes superficial opacities and biochemical defects throughout the squirrel lens.

Note. Exposure was at 6–7 mW/cm^2 for 12 h/day for 1 year.

Altered states of the crystallins of squirrel lenses have also been observed in the *in vitro* studies described above. As the 300- to 400-nm radiation first strikes the lens epithelium, so that the greatest radiant energy is absorbed there, the major changes observed occur. Figure 10 shows that with time, there is a buildup of aggregated crystallins, which, with further irradiation, begin to yield crystallin breakdown products. This finding is in line with those of the *in vivo* studies, which show both aggregation and breakdown of lens epithelial cell crystallins in the same species.

As discussed earlier, oxidation via excitation of molecules is thought to be responsible for the changes in the lens that lead to damage. However, first the body's natural defense mechanisms against oxidative damage must be overcome. These defenses include catalase and glutathione peroxidase, which destroy hydrogen peroxide (H_2O_2); superoxide dismutase, which destroys superoxide anion; glutathione and ascorbate, which serve as reducing agents; glutathione reductase, which preserves reduced -SH groups; ascorbic acid, which serves as a soluble antioxidant; and alpha-tocopherol and beta-carotene, which protect membrane (lipid-soluble) components. The fact that UV-induced damage does occur proves that these protectors can be overcome.

In vitro damage to lenses and lens cultures is caused by specific oxidants and free radicals (Zigman, 1991). The H_2O_2 damage to lens epithelial cells was increased by adding catalase inhibitors, and when BCNU was added to

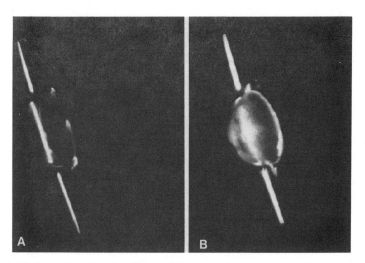

Figure 6. Slit-lamp view of *in vitro* 300- to 400-nm exposed squirrel lens incubated with pho-tosensitizer (0.1 mM tetracycline) (B) in comparison with control (A). Anterior opacities are obvious in irradiated lens.

inhibit glutathione reductase (an important enzyme of the redox cycle), dam-age was increased. When substances that would prevent singlet-oxygen for-mation (DABCO), superoxide formation (TEMPO), and metal-catalyzed ox-idation (desferrioxamine) were tested, they did diminish toxicity. Hightower and McCready (1991) have recently reported an inhibition of Na-K-ATPase and catalase activity in rabbit lens epithelium due to UV-exposure at 313 nm.

2.2.2. Vitamin E and Protective Substances

Antioxidants and free-radical scavengers appear to reduce lens damage leading to cataracts. Epidemiology data have upheld vitamin E, vitamin C, and beta-carotene as being effective cataract retardants (Mares-Perlman *et al.,* 1991; Wu *et al.,* 1991; Vitale *et al.,* 1991). Studies on rats *in vitro* have also shown that both vitamin E and vitamin C can retard oxidation-induced cataracts when added to the media (Varma *et al.,* 1979, 1982). Vitamin C delays the onset of selenium cataracts in mice (Devamanoharan *et al.,* 1991). A summary of naturally occurring UV-related oxidants and protectors is pro-vided in Fig. 11.

Another process to be considered here is the formation of UV-induced photoproducts in the physiological fluids used to incubate lenses *in vitro.* The UV-induced photoproducts of tryptophan have been characterized (Sun and

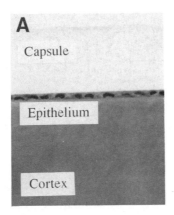

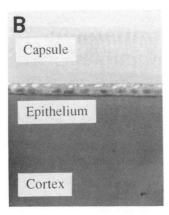

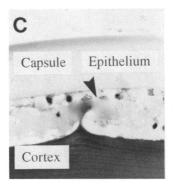

Figure 7. Histological sections of anterior segment of squirrel lens irradiated *in vitro* as described in Fig. 6. Vacuolation of epithelial cells occurs first (after 2 h), and swelling of the interepithelium-capsule space occurs later. (A) control; (B) 2-h exposure; (C) 16-h exposure.

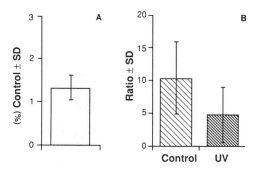

Figure 8. Increase in squirrel lens weight (i.e., water content) and decrease of epithelial cell Na-K-ATPase activity after *in vitro* 300- to 400-nm exposure for 16 h ($n = 4$). (A) Change in lens weight as percentage of control; (B) Change in rate of ATP breakdown by epithelial versus cortical Na-K-ATPase activity.

Zigman, 1979). These indole compounds, in conjunction with UV radiation, are capable of binding to lens proteins photochemically, and also of inhibiting the activity of several of the important protective enzymes listed above (i.e., glutathione reductase, catalase, and Na-K-ATPase; Zigman, 1986b). *In vitro* exposure of lenses to 300- to 400-nm radiation enhances the production of oxidants and toxic photoproducts that not only attack the crystallins and the cell membrane components but also inhibit the protective enzymes.

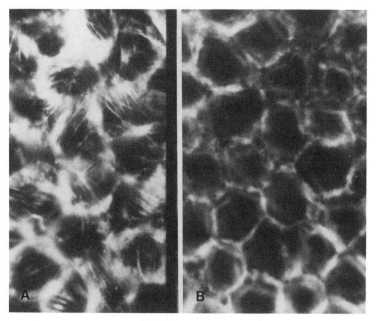

Figure 9. Flat-mount histological preparation of dogfish lens epithelium stained with fluorescent actin-specific dye (Phalloidin plus Rhodamine). (A) control lens; (B) exposed lens. Note the loss of fluorescent actin filaments in UV-exposed lens.

Table 2. Actin Degradation by 300- to 400-nm Radiation

	% polymer		% monomer	
0 h	86.3%		13.7%	

	UV irradiation		Dark control	
	% polymer	% monomer	% polymer	% monomer
3 h	84	16.0	86.1	13.9
8 h	83.6	16.4	85.8	14.2
18 h	68.1	31.9	88.3	11.7

Of great significance to the above discussion are new findings that metabolic energy production in the squirrel lens is inhibited during *in vitro* exposure to 300- to 400-nm radiation (Thomas *et al.*, 1991). Nuclear magnetic resonance measurements show that the ATP level in UV-exposed lenses is markedly depleted in a period of a few hours compared to unexposed lenses. After 4 h of UV exposure, a marked (30%) decrease in ATP level was observed, and within 22 h the decrease was 67%. This finding implies that nearly all

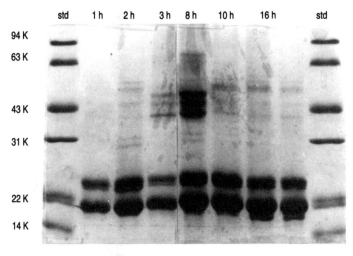

Figure 10. Polyacrylamide gel electrophoresis of squirrel lens epithelial-cell soluble proteins after *in vitro* exposure to 300- to 400-nm radiation. With increasing exposure time there is an accumulation of aggregated protein in the upper part of the gel, and, subsequently, new bands of peptides at the bottom of the gel are observed (K = kilodaltons; std = protein standards).

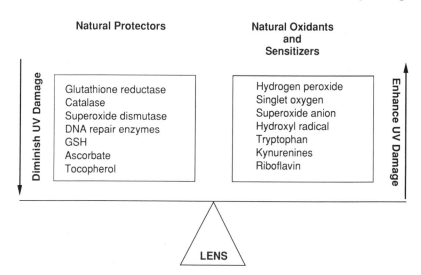

Figure 11. Endogenous UV-damage-protecting agents and UV sensitizers in the lens and aqueous humor.

biochemical processes requiring metabolic energy could be inhibited by such UV exposure.

Table 3 summarizes the *in vitro* effects of BLB lamps on squirrel lenses.

2.3. 300- to 400-nm Radiation Effects on Lens Epithelial Cells in Culture

The changes due to 300- to 400-nm exposure were also observed in lens epithelial cells in culture. The damaging effects on the cells and their cytoskeletal component actin were observed by microanatomical and histochemical methods and by biochemical analyses of the following macromolecules: Na-K-ATPase, catalase, glutathione reductase, and superoxide dismutase. Attempts to moderate the damage observed were made by adding alphatocopherol (vitamin E) to the culture media during UV exposure. The significance of vitamins C, E, and A to cataract formation is discussed later.

Tissue-cultured rabbit lens epithelial cells in given UV exposure for 2 h (6 J/cm^2) had pyknotic nuclei and some loss of cell-to-cell contact. Cells were protected when maintained in medium containing 10 μg/ml of alphatocopherol (Fig. 12).

Phalloidin–Rhodamine fluorescence of the UV-exposed cells revealed the loss of filamentous actin in the cells and the presence of prominently clumped actin near the nucleus, as compared with the regular filaments in the controls. When vitamin E was added to the media at 10 μ/ml, protection of the filamentous actin was achieved (Fig. 13).

Table 3. *In Vitro* **Effects of Exposure to Black Light on the Squirrel Lens**

Gross changes:
  Anterior superficial opacities
Histology:
  Swollen and disrupted lens epithelium.
  Hydropic area between epithelium and cortex.
Proteins:
  Altered crystallins in epithelium; both aggregation and breakdown observed.
  Little if any cortical or nuclear change in crystallins.
Enzyme changes:
  Drop in activity of Na-K-ATPase in epithelium.
Cytoskeleton:
  Actin filaments in epithelium disrupted.
Metabolism:
  Decrease in lens ATP level.
Conclusion:
  In vitro exposure to 300- to 400-nm radiation damages the lens epithelium principally
  through Na-K-ATPase enzyme inhibition, which causes salt–water imbalance, and
  through structural damage relative to cytoskeletal defect.

The enzymes Na-K-ATPase and catalase of lens epithelial cells in culture were rapidly inhibited by UV exposure. Vitamin E strongly protected their activities, as shown in Fig. 14.

Lens DNA is present nearly exclusively in the single layer of epithelial cells located in the most anterior aspect of the tissue. As stated above, the highest irradiances of UV energy reach this layer (Table 4). Besides this, the major synthetic machinery for manufacturing nearly all macromolecules required for cell division, growth, and differentiation occurs in the epithelium (Zigman, 1985a). The future state of the lens depends strongly on how well epithelial-cell DNA, whose information guides these processes, survives environmental UV insult.

Most studies to date on DNA damage caused by UV radiation have been done on a variety of human cells (Peak *et al.,* 1985; Peak *et al.,* 1984). UV-A and UV-B have both been shown to cause DNA damage in these cells. Generally, UV radiation causes DNA lesions such as thymine dimers and single-strand breaks. Cultured normal human cells were shown to contain efficient repair enzymes (Peak *et al.,* 1990), but in xeroderma pigmentosum cells repair enzymes had reduced activity.

The above discussion implies that lens epithelial-cell DNA exposed to the damaging effects of UV radiation can be repaired. Cell culture is the best system for determining the interaction of epithelial-cell DNA with environmental radiation. Workers have been successful in culturing the lens epithelium of cows,

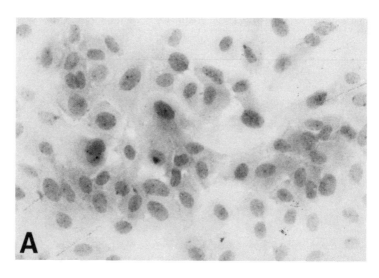

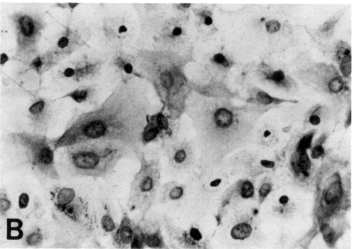

Figure 12. Morphology of squirrel lens epithelial cells in minimum essential medium (MEM). (A) control—no UV exposure; (B) exposed to 300–400 nm for 120 min; (C) control plus 10 μg/ml of vitamin E; (D) exposed to 300–400 nm for 120 min plus vitamin E at 10 μg/ml.

rabbits, and humans (Reddan *et al.*, 1980), which has made it possible to study the direct effect of UV radiation on the DNA of lens epithelium using new analytical methods that allow the observation of DNA damage.

Low doses of UV-B (280–315 nm) exposure were shown to cause single-strand breaks and thymine dimers in tissue-cultured bovine lens epi-

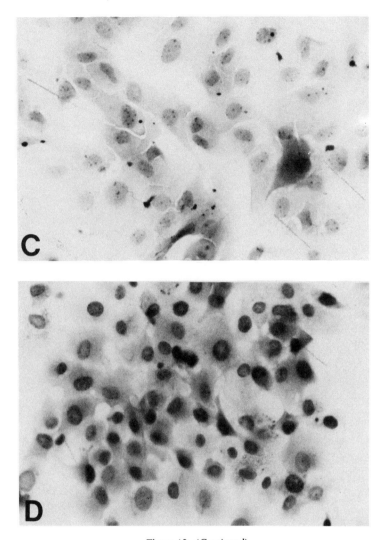

Figure 12. *(Continued)*

thelial-cell DNA (Spector *et al.*, 1990). The irradiance needed was very low (<1 mW/cm^2), and the exposure times were very short (only a few minutes). UV-A was less damaging and produced mainly DNA single-strand breaks, but not thymine dimers. These findings parallel those of other researchers who have observed the same types of damage in other types of cultured cells.

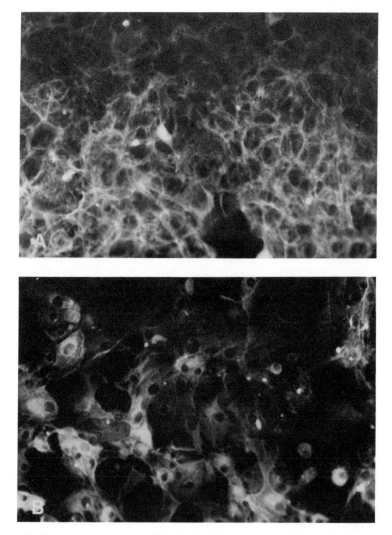

Figure 13. Fluorescence microscopy of Rhodamine–Phalloidin-stained filamentous actin in squirrel lens epithelial cells in MEM culture. (A) control cells with no UV; (B) UV-exposed cells for 2 h; (C) control cells plus 10 μg/ml vitamin E; (D) UV-exposed cells with vitamin E in the media.

In our laboratory, cultured rabbit lens epithelial cells (from J. Reddan) were exposed for up to 2 h with BLB fluorescent lamps (GTE-Sylvania) that emit at 350 ± 50 nm at an irradiance of 2 to 4 mW/cm^2. Cell survival studies indicated a cell-killing threshold dose of 3 J/cm^2. At different time intervals up to 2 h, (exposure of 6 to 12 J/cm^2), cells were harvested, and the single-

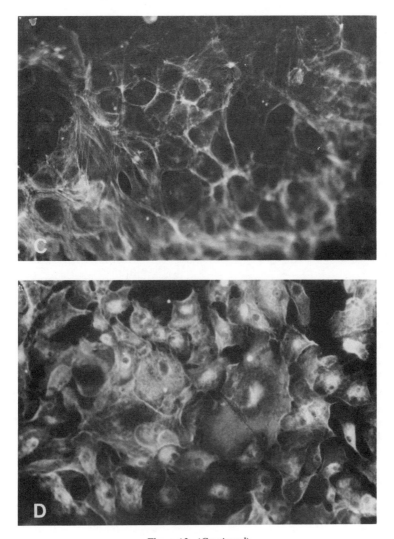

Figure 13. (*Continued*)

strand breaks in DNA were estimated by the alkaline elution technique. It was found that by 30 min of exposure, a large number of single-strand breaks had been induced. The degree of strand breakage was directly related to the dose of radiation (Fig. 15A). When the cells that had been UV exposed for 2 h were returned to darkness (Fig. 15B), they repaired the single-strand breaks by only 80 to 90% within 4 h (Sidjanin and Zigman, 1993). This finding

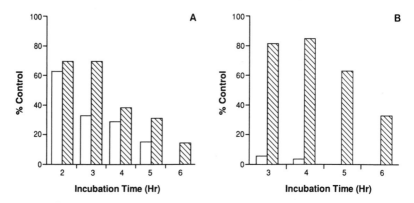

Figure 14. Effects of 120-min UV exposure on squirrel lens epithelial cells incubated for varying times in MEM culture (light bars) and the influence of vitamin E at 2.5 μg/ml (dark bars). (A) Na-K-ATPase activity; (B) catalase activity. Data expressed as percentage of control (i.e., nonirradiated) values.

indicates that the lens epithelial cells have DNA repair enzymes that can restore DNA structure and function.

Studies of the effects of the oxidant H_2O_2 on lens epithelial-cell cultures show that there are many similarities between H_2O_2 damage and UV-induced damage but some crucial differences as well. Hydrogen peroxide does cause DNA single-strand breaks and thymine dimers on lens epithelial-cell cultures (Spector *et al.*, 1990). When the ability of lens epithelial cells to repair single-stranded DNA breaks was assessed, those induced by H_2O_2 were quickly repaired, whereas repair of single-strand breaks caused by UV radiation took much longer.

Table 4. Penetration of UV (365 nm)
Radiation through Squirrel Lens Layers

Lens layer	% of UV reaching the layer
Anterior cortex	100
Inner cortex (ant)	78
Outer nucleus (ant)	19
Inner nucleus	6
Outer nucleus (post)	1.4
Inner cortex (post)	0.3
Posterior cortex	0.06

Note. Figures reflect an average of 4 lenses.

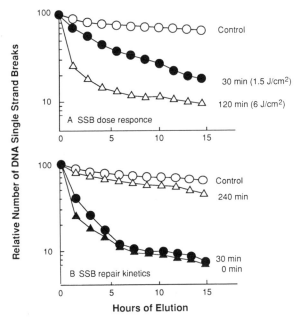

Figure 15. Effect of UV-A exposure on rabbit lens epithelial cell survival and DNA in minimum essential medium (MEM). (A) dose-response curves for DNA single-strand breaks (SSB) assessed by alkaline elution; (B) SSB repair kinetics assessed by alkaline elution at 0, 30, and 240 min in darkness after cells were exposed to a UV-A dose of 6.2 J/cm² (SSB yields from mean of 4 experiments).

This remarkable finding means that the lens epithelium is protected against catastrophic injury to its genetic material by having the ability to repair it very efficiently. The significance of lens DNA repair was first suggested by Jose in 1978. It is probable that this mechanism protects animals and humans from cataractous lens changes that involve DNA damage from UV exposure.

The 300- to 400-nm UV effects on cultured lens epithelial cells are summarized in Table 5.

Direct exposure of purified human lens crystallins in buffer to UV-A radiation resulted in gamma-crystallin aggregation, as determined by polyacrylamide-gel electrophoresis (PAGE), and major shifts in isoelectric points from pH 6.8 to 7.5 down to below pH 5.0. No such changes occurred in human alpha-crystallins under the same conditions. We have found similar results with gamma-crystallins isolated from human cataracts. The isoelectric points of both catalase and superoxide dismutase (purified enzymes) were altered markedly in both buffer and culture medium. Loss of catalase activity was related to the isoelectric point (pI) change observed. However, in the photochemical studies, alpha-tocopherol did not protect catalase or superoxide dismutase from UV damage (see Fig. 16).

2.4. Evidence of Human Cataract Formation from UV Radiation in Sunlight

Several papers and reviews lead us to the conclusion that exposure to sunlight UV radiation enhances cataract formation in humans (Hiller *et al.*, 1983; Zigman, 1987b; Zigman *et al.*, 1979). These were mainly retrospective studies of populations in various geographic locations. Taken as a whole, the findings strongly support a role for sunlight in cataract formation. The most complete and recent study of sunlight-related cataract is by Taylor *et al.* (1988), who studied a group of Maryland fisherman. Degree of exposure to sunlight was estimated from interview and by measuring the irradiance in the geographic area being sampled. The components of sunlight in the UV-A and UV-B ranges were quantitated, and the cataracts were graded in the eye as cortical and/or nuclear. Statistical analysis of the data showed a positive, highly significant relationship between the amount of exposure to UV-B radiation and the presence of cortical cataracts in the fishermen. No relationship was found between cataracts and UV-A. In a study of 400 elderly patients who had cataract extractions or who were treated for retinal diseases in Florida outpatient clinics, a statistically significant relationship between years of living in Florida and formation of cataracts was found. Cortical cataract was statistically significantly related to living in Florida 4.5–9.5 years. Leske *et al.* (1991) have provided detailed epidemiological data on the reduction of human cataract risk factors with antioxidant vitamin usage.

Protein profiles of selected cataracts with specifically located opacities (nuclear only, cortical only, etc.) were studied in detail. Sodium dodecyl sulfate

Table 5. 300- to 400-nm Radiation Effects on Tissue-Cultured Lens Epithelial Cells

Damaged cells, blebbing, death of cells.
Reduced cell growth rate.
Reduced colonization of explanted cells.
Disruption of actin filaments.
Reduction of Na-K-ATPase activity.
Reduction of catalase and glutathione reductase activities.
Single-strand breaks in DNA.

Note. Exposure was to black light at 1.5 to 2 mW/cm^2 for up to 2 h.
Note added in proof. α-Tocopherol protected each of the active targets except strand breaks in DNA.

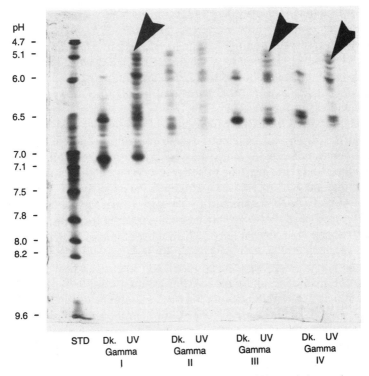

Figure 16. The use of isoelectric focusing to show photochemical changes in human lens gamma-crystallins due to UV exposure. The arrows identify the increased acidic components of each gamma-crystallin due to UV exposure (Dk = dark; STD = protein standards).

PAGE revealed few differences in lens soluble proteins between cataractous and normal cortices or nuclei. High performance liquid chromatography showed that the proteins of cataractous cortices and their nuclei differed very little from those of age-matched controls. The cortical proteins of nuclear cataracts appeared normal. More cross-linked high-molecular-weight proteins and more peptides with molecular weights < 20,000 daltons were found in cataractous nuclei than in the nuclei of age-matched controls. Opacity in the cortex seemed to be unrelated to crystallin distribution changes, which do occur in opaque nuclei. Cortical opacities seemed related to structural changes in the fiber cells, while nuclear opacities were related to altered states of the crystallins, such as aggregation and breakdown.

It is important to compare these findings with other reports of the distribution of UV-A and UV-B wavelengths in sunlight and their effects on the lens. A study of cataract types in active people living in three latitudinally distinct locations indicated that the brown nuclear cataract was statistically significantly related to outdoor occupations (Weale, 1973; Zigman et al., 1979). Also significant is the finding that the lower the latitude, the higher the incidence of nuclear plus brunescent cataracts. The relative irradiances of UV-A and UV-B in this study showed that UV-A predominated in all three latitudes (by 5 to 8 times), and that the increase of irradiance in lower latitudes was nearly equivalent for UV-A and UV-B.

The relative depths of penetration of UV-A and UV-B wavelengths through the cornea and lens can be used to determine where each range causes photochemical change. As stated above, only environmental light < 300 nm passes through the cornea. The majority of UV-B radiation is absorbed in the lens epithelium and anterior cortex, whereas the majority of UV-A penetrates farther so as to be 94% absorbed by the time it reaches the lens nucleus. It could be speculated that UV-B preferentially damages the epithelium, whereas UV-A damages the cortical region. Such a hypothesis is compatible with the results of the *in vivo* experiments on squirrels described earlier; that is, along with the superficial opacity seen, epithelial and anterior cortical morphological and histological damage are observed, suggesting osmotic-type insult to the lens (epithelial-cell swelling, outer fiber-cell membrane disintegration). However, *in vivo* crystallins were altered not only in the superficial region of the lens, but also in the nucleus. Because sunlight contains a majority of UV-A and a minority of UV-B wavelengths, both deep and superficial effects would be expected.

Certain other chemical components of the eye that are modified upon exposure to UV-A or UV-B may play a part in cataract generation. For example, the epithelium of the lens is in contact with the aqueous humor, which contains sensitizers, such as riboflavin, and higher levels of tryptophan than the inner parts of the lens. Also, antioxidants and free-radical scavengers are at higher concentrations in the epithelium and cortex than in the nucleus. UV-B would more efficiently create tryptophan products and oxygen derivatives in the aqueous humor, but UV-A effects would be enhanced preferentially by riboflavin sensitization. This is because riboflavin absorbs mainly in the short-wavelength blue and UV-A wavelength ranges, and not in the UV-B range. Additionally, the greater penetration of UV-A to the lens nucleus and the lower antioxidant and free-radical scavenging activity there favors UV-A as the wavelength range that damages the nucleus. Many questions concerning the mechanism of lens damage from UV radiation still need to be answered.

2.5. Lens–Light Interactions That Diminish Vision

The lenses of older humans, in particular, absorb 300- to 400-nm and blue light strongly. In cataractous lenses which have excessive pigmentation, absorption reduces the visual usefulness of these shorter wavelengths of environmental radiant energy (Taylor *et al.,* 1988). Lenses of elderly and cataractous humans also scatter light excessively due to their highly aggregated crystallin proteins and altered fiber-cell membranes, which increase in aged and cataractous human lenses (Zigman *et al.,* 1976). Irregular refractive-index gradients also contribute to scattering, and the shorter wavelengths are scattered the most. Another parameter of the loss of lens transmittance with aging is autofluorescence, which is excited mainly by UV-A and blue wavelengths (Weale, 1985). Zuclich *et al.* (1992) has shown that lens autofluorescence strongly enhances veiling glare, which adversely influences visual function. Figure 17 provides some data on the level of light scattering and autofluorescence in human cataractous lenses (see Zigman *et al.,* 1991b).

The above discussion leads to the hypothesis that vision loss in aged and cataractous humans is greatly related to optical aberrations in the use of blue and 300- to 400-nm wavelengths of light. Such defects would lead to hazy vision, poor acuity, and loss of contrast sensitivity. This hypothesis has been tested by preventing the shorter wavelengths of light in the environment from entering the eye, and then testing visual function as contrast sensitivity.

The efficacy of short-wavelength-induced light scattering and autofluorescence in obscuring a visual image has been demonstrated photographically (Zigman, 1990). Figure 18 illustrates how a photograph can be altered by adding 300- to 400-nm UV and blue light to the normal light used to produce a photograph and passing the illuminating light through solutions that have both light-scattering and fluorescence properties similar to those of the aged human lens. A filter that cuts out the shorter-wavelength light improves image quality. Figure 19 was photographed using natural outdoor illumination first without and then with a short-wavelength cutoff filter which absorbs light with wavelengths < 480 nm. The improvements in image quality result from reduced scattered light and fluorescence.

The influence of short-wavelength light on vision in humans and the enhancement of the contrast sensitivity function has been shown by Zigman (1992). Table 6 shows the enhancement of contrast sensitivity in human subjects with varying optical characteristics and impairments by including a short-wavelength cutoff filter (the SEE MORE® lens) in the visual path, which cuts off light with wavelengths < 450 nm. Line directions on a chart were described by each patient in order to determine their threshold for discerning the lines. In sum, short-wavelength radiant energy contributes greatly to poor vision in elderly and visually impaired people, and its elimination could improve their vision.

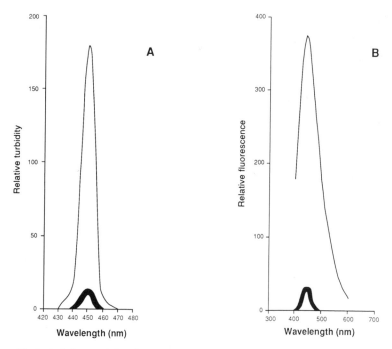

Figure 17. Optical characteristics of a 56-year-old normal human lens (bold line) and a cataractous human lens shown by (A) light scattering (450 nm exc); (B) autofluorescence (360 nm exc). Both determined with light beam entering the anterior side of the lens.

3. Practical Considerations

The previous discussion strongly supports the concept that 300- to 400-nm radiation damages the lens. The *in vitro* animal data indicate that the major direct damage is to the epithelium and outer portion of the lens. This result is compatible with the absorption characteristics of the lens as well, since nearly all of the UV energy is absorbed anteriorly. However, the *in vivo* effects show that cortical and even nuclear defects can occur. Actual time and lens differentiation and growth thus come into play in the chronic exposure of the lens to the light environment. As UV-damaged lens epithelial cells are displaced more internally, they eventually display the acute damage that they sustained during each exposure.

This brings up another point to consider in relation to sunlight UV radiation damage to the human lens that is described earlier in this chapter and in epidemiological studies. Many elderly humans residing in the northern USA retire at ages 60–70 and relocate to a warmer and sunnier climate. Here

Figure 18. Photographs taken through a solution with short-wavelength light-scattering and fluorescence properties similar to those of the human lens. (A) without a filter; (B) with a cutoff filter (<480 nm). Note the enhanced contrast with the filter.

they are more active and have leisure time to spend in the sun swimming, gardening, fishing, golfing, or just walking outdoors. Several changes in the ocular tissues of the elderly have by this time occurred. The activities of certain protective enzymes have diminished with aging, including catalase and glutathione reductase. Lens protein susceptibility to aggregation and degradation has increased with age, and there is additional protein unfolding that increases the potential for reduced transparency and enhanced light scattering. The elderly human lens also absorbs short-wavelength light more efficiently because of its denser pigmentation. Thus, through increased exposure to sunlight and the above physiological changes, susceptibility to UV damage increases. These facts underscore the need for eye protection for the elderly in the form of glasses or contact lenses.

Further, because the ozone layer of the atmosphere seems to be thinning, it is likely that human eyes will receive and suffer from the increased UV-B irradiance that is anticipated due to this environmental change. Just as the incidence of skin cancer is projected to increase with ozone thinning, so will the prevalence of cataracts in humans. Again, eye-protective devices are strongly indicated.

The rise in atmospheric UV radiation may increase the risk of using extrinsic photosensitizers, in the form of drugs or chemicals used for various medical conditions. A registry of ocular side effects of medications has been instituted by Fraunfelder (1982), some of which relate to photosensitization. Table 7 includes a list of several known photosensitizers and an indication of their relative effectiveness as lens protein sensitizers (Roberts and Dillon, 1984–85). This area of research is not well established

Figure 19. Photographs of a flower bed taken in sunlight. (A) with a scattering filter only; (B) also with a cutoff (<480 nm) filter. The brightness of the picture is enhanced by eliminating the UV and short-wavelength blue light, which are scattered most.

Table 6. Average Contract Sensitivity Thresholds for Human Subjects

Subjects	n	Average (yrs)	Contrast Thresholds (× 1,000)[a]				
			Line frequencies				
			1.5	3.0	6.0	12.0	18.0
Normal controls	(14)	63.5					
Without filter			34	23	25	74	200
With filter			28	19	21	50	125
Cataractous	(15)	76.5					
Without filter			96	74	175	300	+++
With filter			80	56	125	240	287
Retina degenerate	(10)	75.0					
Without filter			83	60	80	187	263
With filter			48	30	50	163	264
Aphakic	(9)	75.8					
Without filter			39	32	39	43	200
With filter			25	25	23	27	86

[a] All figures are averages of the number of eyes measured with the conditions stated.

for humans and, in fact, has not even been intensively studied in living animals. Nevertheless, heterocyclic-ring-containing chemical agents provided for human use should be considered as potential photosensitizers. Again, eye protection is indicated when photosensitizing drugs are being used.

Table 7. Relative Rates of Photopolymerization of Lens Protein[a]

Sensitizer	Rate[b]
None	1
3-hydroxy-kynurenine glucoside	1.5
Chlorpromazine	24
Tetracycline	39
Fluorescein	7,800
Rose Bengal	11,400
Hematoporphyrin	14,100

[a] From Roberts and Dillon, 1984–85.
[b] Relative to lens protein without added sensitizer.

4. The Lens Protects the Retina from Light Damage

4.1. The Importance of Pigment

Human and squirrel lenses contain yellow pigments that strongly absorb 300- to 400-nm and short-wavelength blue light (Zigman and Paxhia, 1988). Generally, diurnal animals have lenses with such pigments, while nocturnal animals do not. Two important functions of this pigment are apparent. First, it enhances vision by diminishing chromatic aberrations and scattering and enhancing contrast sensitivity. Second, it absorbs light that is known to damage the retina. Figure 20 shows squirrel retina exposed to 300- to 400-nm UV radiation through a normal lens and with the lens surgically removed prior to exposure. Only in the aphakic eye was the retina damaged (Collier and Zigman, 1987).

4.2. Intraocular Lens Implants

The original intraocular lens (IOLs) implants were of polymethyl methacrylate (PMM), which transmits light with wavelengths longer than 310–320 nm (Zigman, 1982). Two problems are associated with this property of PMM. First, these IOLs transmit short-wavelength light that is known to damage the retina. Second, they allow much higher irradiances of short-wavelength light to reach the retina than do normal eyes. This means that much glare,

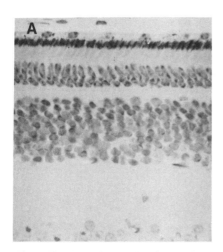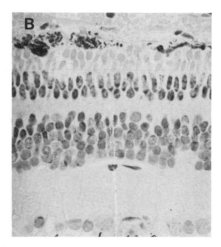

Figure 20. Photomicrographs of the squirrel retina exposed to UV. (A) a normal eye; (B) photoreceptor damage in aphakic eye.

light scattering and chromatic aberration interfere with sharp vision. More recent IOLs have added pigment that absorbs wavelengths of light below 400 nm. While this protects against retinal damage from UV radiation and eliminates some optical anomalies, the 400- to 450-nm UV that does penetrate can be a hazard to retinal structure and function (Mainster, 1978). It would seem to be worthwhile to provide IOLs that absorb light below 450 nm to prevent the transmission of 300- to 400-nm and short-wavelength blue light (Zigman, 1985b).

ACKNOWLEDGMENTS. I enthusiastically thank the members of the Ophthalmic Biochemistry Laboratory for their scientific labor, my research administrator for her editorial input, and the NIH (EY 00459) and Research to Prevent Blindness, Inc., for research support.

References

Collier, R., and Zigman, S., 1987, The gray squirrel lens protects the retina from near-UV radiation damage, in: *Degenerative Retinal Disorders: Clinical and Laboratory Investigations* (J. G. Hollyfied, R. E. Anderson, and M. M. Lavail, eds.), Alan R. Liss, New York, pp. 571–585.

Collier, R. J., Zigman, S., 1989, Comparison of retinal photochemical lesions after exposure to near-UV or short-wavelength visible radiation, in: *Inherited and Environmentally Induced Retinal Degenerations* (J. G. Hollyfield, R. E. Anderson, and M. M. Lavail, eds.), Alan R. Liss, New York, pp. 569–575.

Dayhaw-Barker, P., and Barker, F. M., 1986, Photoeffects on the eye, in: *Photobiology of the Skin and Eye* (E. M. Jackson, ed.), Marcel Dekker, New York, pp. 117–147.

Devamanoharan, P. S., Morris, S. M., Medhat, G. H., and Varma, S. D., 1991, Prevention of selenite cataract by vitamin C, *Invest. Ophthalmol. and Vis. Sci.* (suppl.) 32:724.

Dillon, J., Spector, A., and Nakanishi, K., 1976, Identification of beta carbolines isolated from fluorescent human lens proteins, *Nature* 295:422–423.

Fraunfelder, T., 1982, *Drug-Induced Ocular Side Effects and Drug Interactions,* Lea and Febiger, Philadelphia.

Goosey, J. D., Zigler, S. T., and Kinoshita, J., 1983, Crosslinking of lens crystallins in a photodynamic system: A process mediated by singlet oxygen, *Science* 208:1278–1280.

Hightower, K., and McCready, J., 1991, U.V. cataract in cultured rabbit lenses, *Invest. Ophthalmol. and Vis. Sci.* 32(suppl.):748.

Hiller, R., Sperduto, R. D., and Ederer, F., 1983, Epidemiological association with cataract. in *The 1971–1972 National Health and Nutrition Examination Survey. Am. J. Epidemiol.* 118:239–249.

Jose, J. G., 1978, The role of DNA damage, its repair and its misrepair in the etiology of cataract: A review, *Ophthalmic Res.* 10:52–62.

Lerman, S., 1980, *Radiant Energy and the Eye,* Chapters 2 and 3, Macmillan, New York.

Lerman, S., and Borkman, R. F., 1978, Photochemistry and lens aging, in: *Interdisciplinary Topics in Gerontology: Gerontological Aspects of Eye Research* (E. von Hahn, ed.), Vol. 13, S. Karger, Basel, pp. 154–183.

Leske, M. C., Chylack, T., Jr., Suh-Yuh, P. N., Wu, M. A., 1991. The lens opacities case-control study: Risk factors for cataract, *Arch. Ophthalmol.* 109:244–251.

Mainster, M., 1978, Solar retinitis, photic maculopathy, and the pseudopakic eye, *Am. Intraocular Lens Impant Soc. J.* 4:84–86.

Mares-Perlman, J. A., Klein, B., Klein, R., Ritter, L., Linton, K., and Luby, M. H., 1991, Relationship between diet and cataract prevalence, *Invest. Ophthalmol. and Vis. Sci.* (suppl.) 32:723.

Peak, J. G., Peak, M. J., and MacCoss, M., 1984, DNA breakage caused by 334 nm UV light is enhanced by naturally occurring nucleic acid components and nucleotide enzymes, *Photochem. Photobiol.* 39:713–716.

Peak, M. J., Peak, J. G., and Jones, C. A., 1985, Different (Direct and indirect) mechanisms for the induction of DNA protein crosslinks in human cells by far and near-UV radiations (290 and 405 nm), *Photochem. Photobiol.* 42:141–146.

Peak, J. G., Pilas, B., Peak, M. J., 1990, Repair of single-strand breaks induced by 365 nm UVA and H_2O_2 in human P3 and xeroderma pigmentosum cells, *Photochem. Photobiol.* (suppl.) 51:325.

Pitts, D. G. L., Chu, W. -F., Bergmanson, J. P. G., 1986, Damage and recovery in the UV-exposed cornea, in: *Hazards of Light, Myths and Realities, Eye and Skin* (J. Cronly-Dillon, E. S. Rosen, and J. Marshall, eds.), Pergamon Press, Oxford, pp. 209–219.

Rafferty, N. S., and Scholz, D. L., 1989, Comparative study of actin filament patterns in lens epithelial cells, *Curr. Eye Res.* 8:569–579.

Reddan, J. R., Friedman, T. B., Mostafapor, M. K., 1980, Establishment of epithelial cell lenses from individual rabbit lenses, *J. Tiss. Culture Methods* 6:57–60.

Roberts, J. E., and Dillon, J., 1984–85, A comparison of the photodynamic effects of photosensitizing drugs, *Lens Res.* 2:733–744.

Sidjanin, D., and Zigman, S., 1993, Single-strand breaks in lens epithelial cells DNA induced by UV-A, *Curr. Eye Res.* (submitted).

Spector, A., Kleinman, N. J., and Wang, R. -R., 1990, UV light induced DNA damage and repair in bovine lens epithelial cell cultures, *Invest. Ophthalmol. and Vis. Sci.* (suppl.) 31:436.

Sperling, H. G., Johnson, C., and Harwerth, R. S., 1980, Differential spectral photic damage to primate cases, *Vis. Res.* 20:1117.

Sun, M., and Zigman, S., 1979, Isolation and identification of tryptophan photoproducts from aqueous solutions of tryptophan exposed to near-UV light, *Photochem. Photobiol.* 29:893.

Taylor, H. R., West, S. K., Rosenthal, F. S., and Munoz, B., 1988, Effect of ultraviolet radiation on cataract formation, *New Engl. J. Med.* 319:1429–1433.

Thomas, D. M., Papadopoulou, O., Mahendroo, P. P., and Zigman, S., 1993. Phosphorous-31 NMR study of the effects of UV on squirrel lenses, *Invest. Ophthalmol. and Vis. Sci.* (vol. 56, in press).

Torriglia, A., and Zigman, S., 1988, The effect of near-UV light on Na-K-ATPase of the rat lens, *Curr. Eye Res.* 6:539–548.

Tso, M. O. M., 1989, Experiments on visual cells by nature and man: In search of treatment for photoreceptor degeneration, *Invest. Ophthalmol. and Vis. Sci.* 12: Friedenwald Award and Lecture. 2421–2454.

Varma, S. D., and Lerman, S. (eds.), 1984, Proc. 1st int'l symp on light and oxygen effects on the eye, *IRL Press,* Washington, DC.

Varma, S. D., Kumar, S., and Richards, R. D., 1979, Light-induced damage to ocular lens cation pump: Prevention by vitamin C, *Proc. Natl. Acad. Sci. U.S.A.* 76:3504–3506.

Varma, S. D., Beachy, N. A., and Richards, R. D., 1982, Photoperoxidation of lens lipids, *Photochem. Photobiol.* 36:623–626.

Vitale, S., West, S., Hallfirsch, J., Miller, D., and Singh, V. N., 1991. Plasma vitamin C, E, and β-carotene levels and risk of cataract, *Invest. Ophthalmol. and Vis. Sci.* (suppl.) 32:723.

Weale, R. A., 1973, The effects of the aging lens on vision, *Ciba Foundation Symp. 19 (new series).* 5–24.

Weale, R. A., 1985, Human lens fluorescence and transmissivity and their effects on vision, *Exp. Eye Res.* 41:457–473.

Wu, S. -Y., Leske, M. C., Chylack, L. T., Jr., Hyman, L., Underwood, B., Sperduto, R., and Khu, P., 1991. The lens opacities control study: II. Biochemical risk factors, *Invest. Ophthalmol. and Vis. Sci.* (suppl.) 32:723.

Zigman, S., 1982, Tinting of intraocular lens impants, *Arch. Ophthalmol.* 100:98.

Zigman, S., 1983, Role of tryptophan oxidation in ocular tissue damage, Proc of Max Planck Institute of Biochemistry, Conference on Tryptophan, in: *Progress in Tryptophan and Serotin Research* (H. G. Schlossberger, W. Kochen, B. Unzer, and A. Steinhart, eds.), W. de gruyter, Berlin, pp. 449–468.

Zigman, S., 1985a, Selected aspects of lens differentiation, *Biol. Bull.* 168:189–213.

Zigman, S., 1985b. UV-absorbing intraocular lenses, *Am. Intraocular Lens Implant Soc. J.* 11: 386–387.

Zigman, S., 1986a. Photobiology of the lens, in: *The Ocular Lens* (H. Maisel, ed.), Marcel Dekker, New York, pp. 301–347.

Zigman, S., 1986b. Recent research on near UV radiation and the eye, in: *Biological Effects of UVA Radiation* (F. Urbach and R. W. Gange, eds.), Praeger, New York, pp. 251–261.

Zigman, S., 1987a, Light damage to the lens, in: *Clinical Light Damage to the Eye* (D. Miller, ed.), Springer-Verlag, New York, pp. 65–78.

Zigman, S., 1987b. Recent research on near UV radiation and the eye, in: *The Biological Effects of UVA radiation* (F. Urbach and R. W. Gange, eds.), Praeger, Westport, CT, pp. 252–265.

Zigman, S., 1990, Vision enhancement using a short-wavelength light-absorbing filter, *Optometry and Vis. Sci.* 67:100–104.

Zigman, S., 1991. Effects of radiant energy on the ocular tissues. in: *Photobiology* (E. Riklis, ed.), Plenum Press, New York, pp. 769–785.

Zigman, S., 1992, Light filters to improve vision, *Invest. Ophthamol. and Vis. Sci.* 69:325–328.

Zigman, S., and Paxhia, T., 1988, The nature and properties of squirrel lens yellow pigment, *Exp. Eye Res.* 47:819–829.

Zigman, S., Groff, J., Yulo, T., and Griess, G., 1976, Light extinction and protein profiles in human lenses, *Exp. Eye Res.* 23:555–567.

Zigman, S., Datiles, M., and Torczynski, E., 1979, Sunlight and human cataracts, *Invest. Ophthalmol.* 18:462–471.

Zigman, S., Paxhia, T., and Waldron, W., 1988, Effects of near UV radiation on the protein of the grey squirrel lens, *Curr. Eye Res.* 6:531–537.

Zigman, S., Paxhia, T., McDaniel, T., 1991a, Effect of chronic near-UV exposure on the grey squirrel lens in vivo, *Invest. Ophthalmol. and Vis. Sci.* 32:1723–1732.

Zigman, S., Sutliff, G., and Rounds, M., 1991b, Relationships between human cataracts and environmental radiant energy: *Cataract formation, Light scattering and Fluorescence. Lens and Eye Tox. Res.* 8:259–280.

Zuclich, J., Glickman, R. D., and Menendez, A. R., 1992, In situ measurements of lens fluorescence and its interference with visual function, *Invest. Ophthalmol. and Vis. Sci.* 33:410–415.

7

Vitamin D Synthesis under Changing UV Spectra

Ann R. Webb

1. Introduction

Ultraviolet radiation gains most attention from its detrimental effects on health. However, there are many actions in everyday life which can be constructive or destructive, depending upon the circumstances and context in which they are done: jogging is good for the cardiovascular system but in excess can cause shin splints. In the same way, casual exposure to solar UV radiation has some benefits, whereas excess sunbathing not only produces the discomfort of erythema but also increases the long-term risk of damage to the skin and eye. One of the major beneficial effects, not as evident as erythema, is the cutaneous synthesis of vitamin D, the vitamin essential for the growth and maintenance of a healthy skeleton. Conversion from the vitamin's precursor, 7-dehydrocholesterol (7DHC), in epidermal cells to its active form 1,25-dihydroxyvitamin D ($1,25(OH)_2D_3$), in the blood involves a chain of events triggered by UV-B exposure. To be more specific, cutaneous vitamin D synthesis requires radiation at wavelengths less than 315 nm, the same wavelengths which are most detrimental in other aspects. However, vitamin D synthesis occurs at doses far below those of a minimal erythemal dose (MED). Besides sounding a positive note for UV-B, the photochemistry of

Ann R. Webb • Department of Meteorology, University of Reading, Reading RG6 2AU, England.

Environmental UV Photobiology, edited by Antony R. Young *et al.* Plenum Press, New York, 1993.

vitamin D also provides a fascinating example of a complex control system that prevents vitamin D toxicity after prolonged exposure to sunlight, and illustrates a delicate balance between skin color, latitude, lifestyle, and health of skeleton and skin.

Vitamin D (which is actually a hormone and not a vitamin) is essential to build and maintain a healthy skeleton. By maintaining blood calcium levels within the required range, vitamin D enables bone mineralization and neuromuscular actions to function normally. Calcium metabolism is regulated by the active form of vitamin D, 1,25(OH)$_2$D, which enhances the intestinal absorption of dietary calcium and phosphorous (Holick and Potts, 1980). It is also responsible for encouraging the formation of osteoclasts, the bone cells which remodel and mobilize bone calcium stores. Thus, for a healthy skeleton, vitamin D is required throughout life. In children, whose skeleton is growing and forming, a deficiency of vitamin D leads to rickets, which is observable as curvature of the spine and thighs, enlarged joints of the rib cage and long bones, and weakness of the leg muscles. The adult form of rickets is osteomalacia (Frame and Parfitt, 1978), a disease which even in the early stages (before clinical diagnosis) may increase the risk of fractures, particularly in elderly people who have already lost some bone mass in the aging process. Rickets and osteomalacia can be cured by ensuring that the sufferer receives and maintains an adequate level of vitamin D.

There are two sources and two forms of vitamin D: vitamin D$_2$ (ergocalciferol) is the result of UV irradiation of a yeast sterol; vitamin D$_3$ (cholecalciferol) comes from the irradiation of a sterol present in skin cells, 7-dehydrocholesterol. One natural source of the vitamin is therefore its synthesis in skin exposed to sunlight. The second source is the diet. Vitamin D (D$_2$ and D$_3$) is not common in many foods, but it is present in fatty fish and fish liver oil, and a small amount is also found in eggs. However, some foods are fortified with the vitamin—e.g., margarine in Britain, milk in America—or vitamin D can be taken as a supplement and is often included in multivitamins. Once in the blood the action of the vitamin is the same regardless of source or form, the difference in supply is only apparent in control of vitamin D toxicity: symptoms of excess include nausea and vomiting, and if prolonged can lead to kidney stones and eventually death (Department of Health and Social Security, 1980). Vitamin D taken orally can produce high levels of the circulating vitamin, but even repeated, long-term exposure to sunlight does not result in toxic levels in the blood.

The role of sunlight in vitamin D synthesis has been recognized, if not understood, since the early nineteenth century, when a Polish physician in Warsaw recommended that rachitic children be carried into the sunlight (Sniadecki, 1840). A century later in New York, in 1921, sunlight was proven as a cure for rickets when Hess and Unger (1921) exposed seven rachitic children to various amounts of sunshine and observed radiological evidence of the disease being resolved. Two years earlier, Huldchinsky (1919) had shown that radiation from a mercury vapor quartz lamp could be used to treat rickets, and that the effect was not localized: when only one arm was exposed to the radiation both arms were healed. At about the same time, a dietary antirachitic factor was discovered (Mellanby, 1919). Initially thought to be vitamin A, it was later identified as an independent substance and named vitamin D. Since that time, understanding of the vitamin D story, from synthesis to biological action, has advanced tremendously, revealing a complex and carefully regulated system. A historical account of vitamin D research is given by Holick *et al.* (1982).

2. The Synthesis of Vitamin D

Figure 1 illustrates the sequence of events that begins with the action of sunlight on 7DHC in the skin and ends with the active form of vitamin D_3 circulating in the body. Control mechanisms for vitamin D_3 synthesis within the skin are also indicated. The influence of sunlight on the vitamin D status of an individual is limited to the initial stages of cutaneous synthesis. Thereafter, bound to blood-borne vitamin D binding protein (DBP), it is no longer exposed to solar UV-B radiation, and further events depend on metabolic (and physiological) functions. However, vitamin D status *in vivo* is usually assessed by the measurement of circulating levels of 25OHD (Hadad and Chyu, 1971), an intermediate stage between the skin synthesized vitamin D_3 and the active $1,25(OH)_2D_3$. This allows for a number of individual-specific factors to influence the correlation between sunlight exposure and vitamin D status. In the skin the controlling factors for synthesis of vitamin D are duration, time and place of exposure, and skin pigmentation. The resulting circulating concentrations of 25OHD also depend on the skin area exposed. Vitamin D synthesis begins with the photoconversion of 7DHC (also known as provitamin D_3) to previtamin D_3 (pre-D_3). All skin cells contain 7DHC, but the epidermal layers have the highest concentration of 7DHC per unit area of skin, and therefore the most potential for pre-D_3 synthesis (Holick *et al.*, 1980). A slow, heat-induced isomerization converts pre-D_3 to vitamin D_3 at a temperature-dependent rate. Results from Holick *et al.* (1979a,b) showed that at body temperature (37°C) it takes 28 h for 50% pre-D_3 to convert to vitamin D_3; equilibrium (at 80% conversion) is achieved after 4 days. This means that after the initial exposure to UV radiation, vitamin D_3 synthesis can occur for several days without further exposure. At lower temperatures the rate of isomerization is slower, taking 48 h to reach 50% conversion at

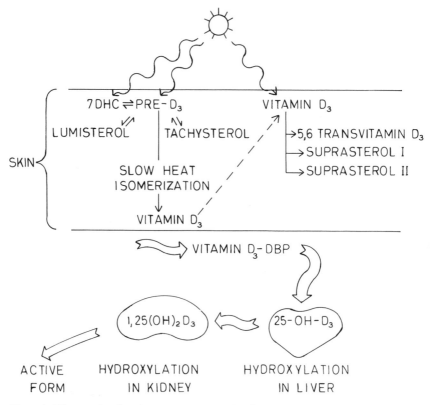

Figure 1. Diagram showing the cutaneous synthesis of vitamin D_3 and its removal from, or degradation in, the skin.

25°C, and attaining only 2% conversion after 7 days at -20°C. Whereas UV irradiation is not necessary for the second stage of vitamin D_3 synthesis, the heat isomerization process must compete with other reactions of pre-D_3, described below, if the skin remains exposed to sunlight.

2.1. The Photoisomers of Previtamin D_3

UV irradiation of pre-D_3 promotes three possible photoreactions: reversal of the initial reaction and restoration of the parent 7DHC, or the formation of either of two further photoisomers, lumisterol and tachysterol (Fig. 1). As in the case of 7DHC, the exchange between these two photoisomers and pre-D_3 is a reversible process (Velluz *et al.*, 1949).

The photoisomers have different absorption spectra (MacLaughlin *et al.*, 1982). The rate of each reaction and the resulting equilibrium conditions will depend on the relationship of the irradiating spectrum and the compounds' absorption spectra. For example, if the solution of 7DHC is irradiated with monochromatic radiation (e.g., 295 nm), the equilibrium mixture is different from that produced under simulated sunlight (Fig. 2, Holick *et al.*, 1982). This equilibrium can be changed by altering the irradiating spectrum, as occurs when the sun is the source of radiation. For example, a shift of the irradiating spectrum toward longer wavelengths ($\lambda > 300$ nm), such as occurs early and late in the day, increasingly favors the photoconversion of tachysterol to pre-D_3 rather than the reverse. The results of irradiating solutions of 7DHC with sunlight at different latitudes and times are presented in Section 2.3.

The extrapolation of these laboratory results to the formation of pre-D_3 and its photoisomers in human skin is further complicated by wavelength-dependent epidermal attenuation of UV-B radiation. The greatest attenuation occurs at the shortest wavelengths (Anderson and Parrish, 1982), which are the most effective for the photoconversion of 7DHC to pre-D_3. The action spectrum for any pre-D_3 (and thus vitamin D_3) formation *in vivo* is therefore a convolution of the *in vitro* action spectrum and the attenuation spectrum of the skin. By no means a constant, the latter depends on skin thickness (a function of age) and pigmentation.

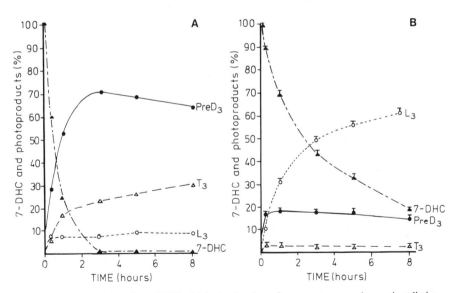

Figure 2. Photoproducts from 7DHC with increasing time of exposure to monochromatic radiation centered at 295 nm (half band width 10 nm) (A) and simulated tropical solar radiation (B). Reproduced with permission from Holick *et al.*, 1981. Copyright 1981 by AAAS.

2.2. Regulating Factors in the Skin

The two major factors governing the cutaneous production of pre-D_3 are the intensity and spectral quality of the incident UV radiation, and the concentration of the parent compound 7DHC. Most vitamin D_3 is produced in the epidermis, whose 7DHC concentration decreases linearly with age, concentration at age 80 being half that at age 20 (MacLaughlin and Holick, 1985). While the thickness of the dermis decreases with time after the age of 20, the same is not true for the epidermis (Tan et al., 1982). Thus, skin samples of young subjects exposed to simulated sunlight produced more than twice the amount of pre-D_3 generated in the skin of elderly subjects. Aging therefore reduces the skin's capacity to synthesize pre-D_3.

The intensity and spectrum of radiation reaching the target 7DHC (rather than the surface of the skin) can be altered by pigmentation. Melanin is a strong absorber of UV radiation and competes with 7DHC for UV photons, providing a limiting factor for pre-D_3 synthesis at low-irradiance levels. However, increased melanin content does not prevent pre-D_3 formation (Holick et al., 1981), it just changes the radiation requirements. A pigmented skin irradiated with the relevant wavelengths at a higher intensity, or for a longer period of time, can produce amounts of pre-D_3 comparable to those produced by unpigmented skin. This has been shown in vivo by exposing Asians and white-skinned people to 1.5 MED and measuring the resulting increase in serum vitamin D (Lo et al., 1986). The Asians, with higher MEDs, received more total radiation than the white-skinned people, but the comparable erythemal doses resulted in similar increases in serum vitamin D. Increasing a single dose of UV radiation from 1 to 4 MED caused more than a 4-fold increase in serum vitamin D (Adams et al., 1982).

Pigmentation is only a limiting factor to pre-D_3 production if the dose of irradiation is insufficient relative to melanin content of the skin (e.g., for highly pigmented people living at high latitudes). A more universal regulator is the photoequilibrium discussed in Section 2.1. The maximum amount of pre-D_3 produced in sunlight is 15% of the original 7DHC (Holick et al., 1981), regardless of pigmentation or exposure time. Continued exposure results only in increasing amounts of lumisterol and tachysterol. Given sufficient time after the initial irradiation, some of the pre-D_3 will undergo heat isomerization to vitamin D_3, upsetting the quasi-equilibrium between photoisomers. A second irradiation may then restore the balance by the photoconversion of lumisterol, tachysterol, or 7DHC to pre-D_3. Lumisterol and tachysterol have no known biologic actions (MacLaughlin et al., 1982) other than regulating the amount of pre-D_3 that can accumulate in human skin. If not returned by irradiation to their parent compound, they are probably lost by epidermal desquamation.

All the above are natural regulators of cutaneous pre-D_3 production. One common artificial regulator is the use of sunscreens, which preferentially attenuates the UV-B portion of solar radiation. Because the action spectra for erythema and vitamin D_3 synthesis overlap, sunscreen use can inhibit vitamin D_3 synthesis. Previtamin D_3 production occurs at suberythemal doses, and a sunscreen with a sun protection factor of 8 has been shown to prevent pre-D_3 production both *in vivo* and *in vitro* (Matsuoka *et al.*, 1987).

2.3. Control by Latitude and Season

Irrespective of skin type, cutaneous pre-D_3 synthesis is regulated by solar exposure in two ways. For clinical purposes, when circulating levels of the active metabolite are paramount, the skin area and frequency of exposure are important factors. Thus, lifestyle can play an important role in determining vitamin D status. The effectiveness of exposure for a given time is dependent on the irradiating spectrum, which is determined by time and place of exposure.

In general, solar irradiance decreases as zenith angle increases: when the sun is close to the horizon, radiation travels a longer path through the atmosphere, and is attenuated more, than when the sun is directly overhead. This effect is particularly noticeable with UV-B, where absorption by stratospheric (and tropospheric) ozone is strongly wavelength dependent (Molina and Molina, 1986), and Rayleigh scattering ($\alpha \lambda^{-4}$) is most effective. At $\lambda <$ 290 nm, all solar radiation is absorbed in the upper atmosphere by oxygen and ozone. With increasing UV-B wavelength, the steeply declining ozone absorption spectrum results in an increase in surface irradiance of several orders of magnitude from 300 to 320 nm. Ozone concentration and zenith angle determine the short-wavelength limit of the solar spectrum: at low zenith angles the spectrum at the ground is shifted toward shorter wavelengths— those most efficient for pre-D_3 synthesis.

Zenith angle is a function of latitude, season, and time of day. Ozone concentration too has a latitudinal dependency and a seasonal cycle on which are superimposed natural long- and short-term changes (plus recent changes induced by man). It is the natural changes in potential pre-D_3 production which will be discussed below, changes dominated by the factors which determine zenith angle.

As latitude increases, the seasonal changes in zenith angle become more pronounced. At the equator the solar declination (the latitude at which the sun is directly overhead) is never far away, and in the middle of the day the zenith angle is always small. The opposite extreme is found in polar regions, where the sun never rises above the horizon during the winter months and does not set during the summer months, although it is always low in the sky. Moving from equatorial to polar latitudes, the summer to winter contrast becomes more pronounced while mean irradiance decreases. These seasonal changes in zenith angle (path length) result in annual cycles of both irradiance and short-wavelength limit of solar UV-B radiation, and hence a seasonal dependence on the ability to produce vitamin D_3 in the skin. At high latitudes cutaneous vitamin D_3 synthesis may only be possible for a few months each year, and this is reflected in circulating levels of 25OHD, which are seen to increase during the spring and summer and decline during the autumn and winter (Stamp and Round, 1974; Beadle *et al.*, 1980; Devgun *et al.*, 1981; Lamberg-Allardt, 1984).

To investigate the latitudinal and seasonal dependency of pre-D_3 photosynthesis, a model system was developed by Webb *et al.* (1988) for exposure to sunlight at different locations and times of year. This was a solution of 7DHC in methanol, contained in a quartz test tube. Aliquots of the solution could be taken at intervals throughout a given day and analyzed by high-performance liquid chromatography (HPLC) to reveal the percentage of original 7DHC which had undergone single or multiple photoconversion to pre-D_3 and its isomers. The results are discussed in Section 2.3.1.

2.3.1. Results of Seasonal and Latitudinal Exposures

In initial studies designed to identify the months of the year when solar radiation would convert 7DHC to pre-D_3, the model solutions were exposed for 3 h from 1130 to 1430 local standard time with aliquots taken every hour. Maximum hourly UV dose was thus received in the first hour of exposure (around local noon) and declined thereafter. Exposures were on clear days as close to the 15th of the month as possible, to represent the average for the month. Monthly experiments took place in Boston (42°N) and Edmonton (51°N), but trials were intermittent in Los Angeles (34°N) and Puerto Rico (18°N). Results showed both a seasonal and latitudinal dependency on pre-D_3 production, as expected.

In Boston, no pre-D_3 was detected in solutions exposed to sunlight in the months of November to February, with this nonproductive period being extended by a month on either side at 10° further north in Edmonton. By contrast, a January exposure in Puerto Rico gave a pre-D_3 content in the exposed solution equivalent to that of a midsummer exposure in Boston. Solutions located in Los Angeles indicated pre-D_3 production throughout the year but at low levels during the winter months.

Inspection of the seasonal change in solar spectrum at Boston establishes the reason for the lack of pre-D_3 production during winter. The most effective wavelengths for the photoconversion of 7DHC to pre-D_3 (wavelengths < 300 nm; Fig. 3) are not present in the solar spectrum at this time of year. Figure 4 shows the irradiance at four wavelengths at different months, measured

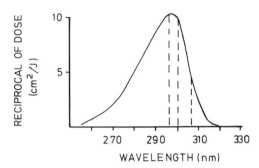

Figure 3. The action spectrum for the formation of pre-D₃ from 7DHC in human epidermis showing the four wavelengths represented in Fig. 4: 296, 300, 306, and 316 nm.

with an Optronics 742 spectroradiometer at noon on clear days in Boston (Webb *et al.*, 1988). The wavelengths of 296, 300, 306, and 316 nm have relative efficiencies for 7DHC conversion of 1, 0.92, 0.45, and zero, respectively. Irradiances at all wavelengths change by orders of magnitude throughout the year, but seasonal change is most pronounced at the shortest, most effective wavelengths. Indeed, 296-nm radiation cannot be detected during the winter months, and radiation at 300 nm is dramatically reduced. At lower latitudes there is less seasonal change in zenith angle, and UV-B irradiance at all wavelengths remains high enough to initiate pre-D₃ synthesis throughout the year, but at lower levels during the winter months.

Once in solution, pre-D₃ is able to undergo photoconversion to any of its three photoisomers. Photoreversed products of the initial reaction could

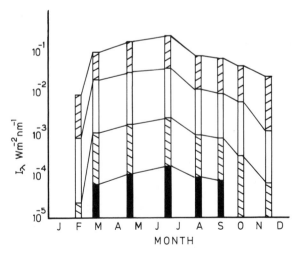

Figure 4. Noon, Boston. The annual change in clear sky irradiance at four UV wavelengths: 296 nm (■), 300 nm (▨), 306 nm (□), and 316 nm (▨).

not be distinguished from the original 7DHC (exposure of a pure pre-D_3 solution showed that this reaction does occur in sunlight). Lumisterol in increasing amounts and small amounts of tachysterol were observed in exposed solutions. Figure 5 shows the photoproduct production from 7DHC for an exposure period beginning at 1130 hours during June in Boston. Noon at this time of year provides the maximum irradiance available at a given latitude, and it can be seen that the initial photoconversion of 7DHC is very rapid: pre-D_3 was detected after only 2 min in sunlight. It was 30 min before lumisterol was detected in the solution, and 1 h before tachysterol appeared. As time progressed the rate of net accumulation of isomers originating from 7DHC decreased, and the quasi-equilibrium mixture also changed. Following its rapid appearance and increase during the first hour, the net pre-D_3 production slowed considerably in the next 2 h, although the steadily increasing lumisterol depended on 7DHC conversion and an intermediate pre-D_3 state. This plateau effect, a maximum of pre-D_3 in equilibrium, was also observed at other exposure sites and under artificial sunlight (Holick *et al.*, 1981; Webb *et al.*, 1988) and is not the result of the particular solar spectrum in Boston or the decreasing irradiance throughout the experimental period. An equilib-

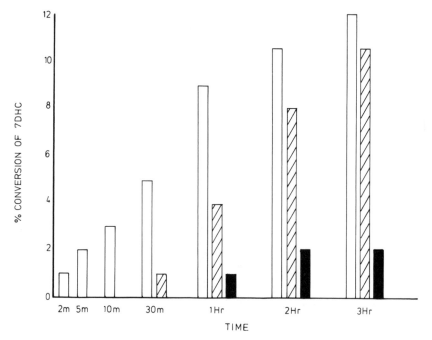

Figure 5. The accumulation of photoproducts of 7DHC in solution during exposure to June sunlight in Boston: pre-D_3 (□), lumisterol (▨), tachysterol (■).

rium mixture of photoisomers originating from 7DHC will not support more than approximately 12% pre-D_3 under solar irradiation.

At other times of year, or at other latitudes, where solar irradiance is lower and pre-D_3 production much slower, the appearance of lumisterol is correspondingly slower. Tachysterol is often not detected in the mixture at low UV irradiances and is only evident during the midsummer months in Boston. This is a result of tachysterol's absorption spectrum, which extends to longer wavelengths than any of the other isomers—wavelengths at which solar irradiance is nearly always sufficient to cause photoreversal. A starting solution of tachysterol produces photoproducts at all seasons of the year. Thus it is only when the short- to long-wavelength UV ratio is sufficiently high that net production of tachysterol becomes possible.

2.3.2. Relation of Photochemistry in the Model to That in Skin

The experimental solution exposed to unattenuated UV-B indicates the maximum conversion of 7DHC that could occur at any time. To relate the model to humans, freshly excised skin was exposed to sunlight alongside the solutions in Boston (Webb *et al.*, 1988). After exposure of the *ex vivo* skin, the isomers of pre-D_3 were extracted from the epidermis and analyzed by HPLC in the same way as the model solutions. Comparison of photoproducts found in the skin and model solution indicated similar photochemistry in both systems, but UV-B-absorbing epidermal chromophores attenuated the radiation reaching the 7DHC in skin and hence reduced the rate of formation of photoproducts. At latitudes where pre-D_3 formation is limited by season, this additional UV-B attenuation in the skin extends the period when no pre-D_3 production occurs. This can be illustrated by the first springtime observation of pre-D_3 in Boston. In the model solution, 2% of pre-D_3 was detected in the 7DHC solution in February, but it was almost a month later in mid-March that pre-D_3 was first detected in skin samples exposed to sunlight (Webb *et al.*, 1988). In the following months the trend of pre-D_3 production in skin followed that of the model solution, increasing to a maximum in June and July and then declining. Quantifying the difference between the two systems under both natural and simulated sunlight, it was found that percentage conversion in skin was about half that seen in the clear solution in a quartz test tube.

2.3.3. Further Influences on Vitamin D Synthesis and Status

Skin is a constant attenuator of UV radiation, but the effects of cloud are transient. All results presented thus far have been for clear sky conditions. UV-B radiation, like visible radiation, is attenuated when the sky is cloud covered, the reduction in irradiance being dependent on the type and amount of cloud. As cloud cover is continually changing, its effect is difficult to quantify, but it will slow the rate of reactions as illustrated in Fig. 6, which shows the accumulation of pre-D_3 and lumisterol in Boston on two consecutive days in September, one clear and one completely overcast.

Metabolic factors may influence vitamin D status, but in vitamin D depleted individuals, an increase in circulating 25OHD concentrations has shown a direct relationship with short periods of sunlight exposure (Webb *et al.*, 1991). When there is neither an equilibrium limit to cutaneous pre-D_3 production nor a metabolic limit to the subsequent production of 25OHD, UV-B exposure benefits vitamin D status. However, extended exposure to UV-B will become progressively less productive for pre-D_3 (as seen in *ex vivo* skin and the model solution). In individuals with a high degree of circulating 25OHD, the uptake of vitamin D from the skin is limited (Stanbury *et al.*, 1980), and sunlight degrades the unused vitamin (Section 3). Optimal exposure times are best explored without physiological complications, using the test tube solution model.

Further work with the 7DHC model solution has extended the scope of the initial studies, using a different protocol, at the original and additional locations—Bergen (61°N), Johannesburg (26°S), Buenos Aires (34°S), and Cape Town (35°S). Aliquots of the solution were exposed for consecutive 1-h periods throughout the day and analyzed to give the hourly percentage

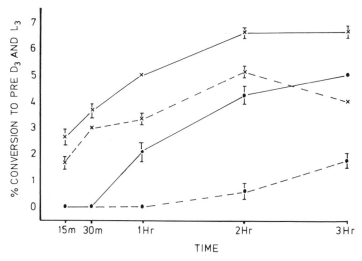

Figure 6. Pre-D_3 (×) and lumisterol (●) observed in solution after exposure in Boston on two September days, one cloud-free (——) and one overcast (- - -).

conversion of 7DHC to pre-D_3. Results (Holick, personal communication) illustrate a diurnal dependency for pre-D_3 synthesis, and further control by zenith angle in addition to that of season and latitude dependency. In Boston, pre-D_3 synthesis of more than 1% is detected after 1-h exposure from 0600 to 1800 (EST) in the month of June, with the maximum hourly conversion rate of 6.5% at around noon. A month later, in July, the day length for pre-D_3 synthesis has become 0700–1700 (local standard time), while 19° further north in Bergen this productive period does not begin until 0900, and reaches a maximum hourly conversion rate of only 2.5%.

During the winter months, no significant pre-D_3 is observed in solution in Boston, but closer to the equator there is sufficient solar UV-B irradiance during the middle hours of the day to produce pre-D_3. In July in Johannesberg (southern hemisphere winter), the noontime hourly conversion rate is 6.5%, equivalent to a Boston summer, and the effective conversion hours are from 0900 to 1600. Cape Town, latitudinally between Boston and Johannesberg, gives the expected intermediate result; pre-D_3 synthesis is detected from 1100 to 1400 local time, with a maximum hourly rate of 1.5%. Adequate exposure times in terms of attaining and maintaining a healthy vitamin D status must therefore be assessed for each locality and time of year. As pre-D_3 synthesis is a rapid process, given the right spectrum, the necessary exposures should be suberythemal. As discussed, continued exposure becomes counterproductive, and successive long exposures can destroy vitamin D.

3. Photodegradation of Vitamin D

It is well documented that vitamin D_3 exposed to high levels of UV radiation produces several overirradiation products (Dauben and Bauman, 1961; Havinga, 1973), but the effect of sunlight on vitamin D_3 in the skin has only recently been studied. The model solution of pre-D_3 formation developed in Boston was adapted to investigate the irradiation of vitamin D_3 by sunlight (Webb *et al.*, 1989). Using a solution of vitamin D_3 in methanol contained in the quartz test tube, and following a similar exposure, sampling, and analysis procedure, solar-induced photoproducts of vitamin D_3 were identified. The three main photolysis products were 5,6-transvitamin D_3 (Holick *et al.*, 1976) and the suprasterols I and II (Dauben and Bauman, 1961) as indicated in Fig. 1.

5,6-transvitamin D_3 can be hydroxylated in rats to 25-hydroxy-5,6-transvitamin D_3 (Kumar *et al.*, 1986), which is biologically active and can substitute for 1,25(OH)$_2$D$_3$ (Holick *et al.*, 1972). The antiproliferative properties of 1,25(OH)$_2$D$_3$ have been used to treat the skin disease psoriasis by inhibiting activity of the epidermis (Holick *et al.*, 1987), raising interesting implications about the generation of 5,6-transvitamin D_3 in the skin. The suprasterols have no known biologic activity.

From a photobiologic standpoint, the important finding was that vitamin D_3 was degraded throughout the year in Boston sunlight. Vitamin D_3 was reduced by 30% after a 3-h exposure around noon in December, and by 90% after a similar exposure in June (Webb *et al.*, 1989). This would indicate that vitamin D_3 can be photodegraded by wavelengths longer than those required to initiate its production. To confirm this, the action spectrum for the breakdown of vitamin D_3 was determined and found to extend from UV-C to 335 nm (pre-D_3 production requires $\lambda < 315$ nm). Sufficient irradiance at wavelengths between 315 and 335 nm is present in the solar spectrum incident in Boston to degrade the photolabile vitamin D_3 throughout the year. These longer wavelengths are also able to penetrate further into the skin than wavelengths < 315 nm, and may reach the dermis or even the dermal capillary bed (Anderson and Parrish, 1982), raising the possibility that winter sunlight may destroy vitamin D_3 in the circulation while not producing it in the skin. To test the hypothesis that there may be protective factors in the skin or blood that prevent this photodestruction, vitamin D_3 was generated in *ex vivo* skin and exposed to sunlight. Human serum containing vitamin D_3 bound to DBP was similarly exposed, and in both cases photodegradation of the vitamin was observed (Webb *et al.*, 1989).

In considering the paradox that an essential vitamin should be destroyed more readily than it is produced (with both reactions being instigated by natural sunlight), the overall context in which the exposure takes place must not be forgotten. At those latitudes where no cutaneous pre-D_3 production occurs during the winter months, it is unlikely that any significant skin area would be exposed to sunlight for an appreciable amount of time anyway. Not only is the day length short and the amount of solar radiation often further attenuated by cloud, but the temperature is such that significant skin exposure is unlikely. Climate and behavior greatly modify the potential effects of sunlight. Where the climate does encourage year-round exposure, the solar spectrum is able to initiate both the production and destruction of vitamin D_3 throughout the year. However, it must be remembered that there is a time lag of hours to days between the initiation process (7DHC to pre-D_3) and the appearance of vitamin D_3 in the skin. Exposure to sunlight on one day will result in vitamin D_3 production over the next 2–3 days (Holick *et al.*, 1979b), and it is probably only vitamin D_3 not removed from the skin by the time a subsequent exposure occurs which may be photodegraded. However, during photodegradation of existing vitamin D_3, a new sequence of production is beginning. As with isomerization to lumisterol and tachysterol to limit the amount of pre-D_3 in the skin, the photodegradation of vitamin D_3 serves to prevent unused vitamin D_3 from accumulating. Thus, photocontrol at the cutaneous level helps to prevent vitamin D_3 toxicity.

A further control for toxicity may result from another form of vitamin D, 24-dehydrovitamin D_3, which has been identified from mammalian skin (Holick *et al.*, 1985). *In vitro,* 24-dehydrovitamin D_3 is an efficient inhibitor of the 25-hydroxylation of vitamin D (Bolt *et al.*, 1987), blocking the metabolism of 25OHD, and may thus help prevent excessive sunlight exposure from causing vitamin D intoxication.

4. The Potential Effects of Ozone Depletion

Depletion of stratospheric ozone would lead to an increase in UV-B irradiance at the earth's surface. Thus, with ozone depletion, the initial rate of pre-D_3 production would increase or become possible at latitudes and times where cutaneous synthesis of the vitamin is currently prevented by lack of UV-B. In this case ozone depletion would be beneficial, extending the period of the year when vitamin D synthesis in skin is possible and shortening the exposure time necessary for adequate pre-D_3 formation. Sections of the population who spend little time outdoors, have pigmented skin, or have no dietary source of vitamin D would benefit from an increased ability to synthesize the vitamin and a reduced risk of deficiency.

In populations where the probability of maintaining sufficient vitamin D levels from casual exposure to sunlight already exists, a decrease in ozone and its attendant increase in UV-B irradiance would produce less obvious changes. Reducing ozone would alter the shape of the incident UV-B spectrum in favor of the shorter wavelengths and would therefore change the equilibrium state of pre-D_3 photoisomers formed in the epidermis. An extreme example of such a change is shown in Fig. 2, where equilibrium mixtures generated under short-wavelength UV-B (295 ± 5 nm) and simulated tropical sunlight are shown. However, this change in photochemical balance is unlikely to affect concentrations of vitamin D and its metabolites in the circulatory system. Heat isomerization of pre-D_3 to vitamin D_3 is a slow reaction independent of incident radiation. With increased UV-B irradiance, more 7DHC may undergo conversion to pre-D_3, but in full sunlight at current mid- to low-latitude UV-radiation levels, the photoisomer equilibrium supports less than 20% pre-D_3 (Section 2.3.1, Fig. 2). A moderate change in ozone, equivalent to moving several degrees of latitude toward the equator, is not expected to substantially alter this upper limit of pre-D_3 available for isomerization to vitamin D_3. As now, any excess vitamin D remaining in the skin would be photodegraded on the next exposure to sunlight.

A potentially greater influence on vitamin D synthesis in skin may come indirectly from ozone depletion. If public awareness of the dangers of UV-B increases sufficiently to change exposure habits and significantly increases the use of sunscreens, then the natural synthesis of pre-D_3 will also be prevented or reduced. Taken to extreme this behavior could be detrimental to skeletal health. It will remain true that vitamin D_3 synthesis can be initiated at UV-B doses far below those that will damage the skin, and there would still be beneficial aspects to brief exposure to solar radiation.

5. Summary

Exposure to sunlight (or other UV-B radiation sources) is necessary for the cutaneous synthesis of vitamin D_3. For much of the world's population, natural sunlight provides for an adequate vitamin D status, although this source is subject to all the factors which control the amount of solar UV radiation reaching the earth's surface. Seasonal cycles of vitamin D status have been observed in the populations of many mid- to high-latitude countries where the irradiance and spectral properties of sunlight are limiting factors for vitamin D_3 production in winter: the seasonal potential for the initiation of pre-D_3 synthesis by sunlight has been investigated using a solution of 7DHC.
After the conversion of 7DHC to pre-D_3, sunlight is no longer required for the formation of vitamin D_3 and its active metabolites, and excessive exposure becomes counterproductive. The relationship between dose and pre-D_3 formation in the epidermis is not linear since pre-D_3 itself is photoisomerized to two inert products (lumisterol and tachysterol) with the equilibrium mixture of isomers subtly dependent on the irradiating spectrum. Under solar radiation conditions, the amount of pre-D_3 in the mixture is limited, and prolonged exposure serves only to prevent the accumulation of pre-D_3 in skin. In the event of repeated exposure day after day, further control is provided by the photodegradation of any vitamin D_3 that remains in the skin at a time when further synthesis is being initiated.
Sunlight, through complex photochemistry, is thus the primary regulator of vitamin D_3 synthesis in skin at climatic extremes. It limits the season of production at high latitudes, and regulates the net productivity where there is no limit imposed by the spectrum and period of irradiation.

References

Adams, J. S., Clemens, T. L., Parrish, J. A., and Holick, M. F. 1982, Vitamin D synthesis and metabolism after ultraviolet irradiation of normal and vitamin D deficient subjects, *N. Engl. J. Med.* 306:722–725.

Anderson, R. R., and Parrish, J. A., 1982, Optical properties of human skin, in: *The Science of Photomedicine* (J. D. Regan, and J. A. Parrish, eds.), Plenum Press, New York, pp. 147–194.

Beadle, P. C., Burton, J. L, and Leach, J. F., 1980, Correlation of seasonal variation of 25-hydroxycalciferol with UV radiation dose, *Br. J. Dermatol.,* 102:289–293.

Bolt, M. J. G., Holick, M. F., Holick, S. A., and Rosenberg, I. H., 1987, 24-Dehydrocholecalciferol is a potent inhibitor of rat liver microsomal vitamin D-hydroxylase, *Fed. Proc.* 46:603.

Dauben, W. G., and Baumen, P., 1961, Photochemical transformations IX. Total structure of suprasterol II, *Tetrahedron Lett.* 16:565–572.

Department of Health and Social Security, 1980, *Report on Health and Social Security Subjects 19. Rickets and Osteomalacia.* Report of the working party on fortification of food with vitamin D. HMSO, London.

Devgun, M. S., Patterson, C. R., Johnson, B. E., and Cohen, C., 1981, Vitamin D in relation to season and occupation, *Am. J. Clin. Nutr.* 34:1501–1504.

Frame, B., and Parfitt, A. M., 1978, Osteomalacia: Current concepts, *Ann. Intern. Med.* 89:966–982.

Hadad, J. G., and Chyu, K., 1971, Competitive protein binding radioassay for 25-hydroxychalciferol, *J. Clin. Endocrinol. Metab.* 33:992–995.

Havinga, E., 1973, Vitamin D, example and challenge, *Experentia* 29:1181–1193.

Hess, A. F., and Unger, L. J., 1921, Cure of infantile rickets by sunlight. *J. Am. Med. Assoc.* 77: 39.

Holick, M. F., and Potts, J. T., Jr., 1980, Vitamin D, in: *Harrison's Principles of Internal medicine* (K. J. Isselbacher, R. D. Adams, E. Braunwald, R. G. Petersdorf, and J. D. Wilson, eds.), 9th ed., McGraw-Hill, New York, pp. 1843–1849.

Holick, M. F., Garabidian, M., and Deluca, H. F., 1972, 5,6-Trans-25-hydroxycholecalciferol: Vitamin D analog effective on intestine of anephric rats, *Science* 176:1247–1248.

Holick, M. F., Garabedian, M., and Deluca, H. F., 1976, 5,6-Transisomers of cholecalciferol and 25-hydroxycholecalciferol in anephric animals. *Biochemistry* 11:2715–2719.

Holick, M. F., Holick, S. A., McNeill, S. C., Richtand, M., Clark, M. B., Potts Jr., J. T., 1979a, in: *Vitamin D: Basic Research and its Clinical Applications* (A. W. Norman, K. Schaefer, D. von Herath, H. G. Grigolet, J. W. Coburn, M. F. DeLuca, E. B. Mawer, and T. Suda, eds.), Gruyter, Berlin/New York, pp. 173–176.

Holick, M. F., McNeill, S. C., MacLaughlin, J. A., Holick, S. A., Clark, M. B., Potts Jr., J. T., 1979b, The physiologic implications of the formation of previtamin D_3 in skin, *Trans. Assoc. Am. Physicians* 92:54–63.

Holick, M. F., MacLaughlin, J. A., Clark, M. A., Holick, S. A., Potts, J. T., Jr., 1980, Photosynthesis of previtamin D_3 in human skin and the physiologic consequences, *Science* 210:203–205.

Holick, M. F., MacLaughlin, J. A., and Doppelt, S. H., 1981, Regulation of cutaneous previtamin D_3 photosynthesis in man: Skin pigment is not an essential regulator, *Science* 211:590–593.

Holick, M. F., MacLaughlin, J. A., Parrish, J. A., and Anderson, R. R., 1982, The photochemistry and photobiology of vitamin D_3, in: *The Science of Photomedicine* (J. D. Regan and J. A. Parrish, eds.), Plenum, New York, pp. 195–218.

Holick, M. F., Smith, E., and Pincus, F., 1987, Skin as the site of vitamin D synthesis and target tissue for 1,25 dihydroxyvitamin D_3. Use of calcitriol (1,25 dihydroxyvitamin D_3) for treatment of psoriasis. *Arch. Dermatol.* 123:1677–1683.

Holick, S. A., St. Lezin, M., Young, D., Malaika, S., and Holick, M. F., 1985, Isolation and identification of 24-dehydroprovitamin D_3 and its photolysis to 24-dehydroprevitamin D_3 in mammalian skin. *J. Biol. Chem.* 260:12181–12184.

Huldchinsky, K., 1919, Heiling von rachitis durch funtsliche hohensonne, *Dtsch. Med. Wochenschr.* 14:712–713.

Kumar, R., Nagubandi, S., Jardine, I., Londowski, J. M., and Bollman, S., 1986, The isolation and identification of 5,6-trans-25-hydroxyvitamin D_3 from the plasma of rats dosed with vitamin D_3, *J. Biol. Chem.* 256:9389–9391.

Lamberg-Allardt, C., 1984, Vitamin D intake, sunlight exposure and 25-hydroxy-vitamin D levels in the elderly during one year, *Ann. Nutr. Metab.* 28:144–150.

Lo, C. W., Paris, P. W., and Holick, M. F., 1986, Indian and Pakistani immigrants have the same capacity as Caucasians to produce vitamin D in response to ultraviolet radiation, *Am. J. Clin. Nutr.* 44:683–685.

MacLaughlin, J. A., and Holick, M. F., 1985, Aging decreases the capacity of human skin to produce vitamin D_3, *J. Clin. Invest.* 76:1536–1538.

MacLaughlin, J. A., Anderson, R. R., and Holick, M. F., 1982, Spectral character of sunlight modulates photosynthesis of previtamin D_3 and its photoisomers in human skin, *Science* 216:1001–1003.

Matsuoka, L. Y., Ide, L., Wortsman, J., MacLaughlin, J. A., and Holick, M. F., 1987, Sunscreen supresses cutaneous vitamin D_3 synthesis, *J. Clin. Endocrinol. Metab.* 64:1165–1168.

Mellanby, E., 1919, An experimental investigation on rickets, *Lancet* 1:407–412.

Molina, L. T., and Molina, M. J., 1986, Absolute absorption cross sections of ozone in the 185–350 nm wavelength range, *J. Geophys. Res.* 91:14501–14508.

Sniadecki, J., 1840, Cited by J. Mozolowski, 1939, Jedrzej Sniadecki (1768–1883) on the cure of rickets, *Nature* 143:121.

Stamp, T. C. B., and Round, J. M., 1974, Seasonal changes in human plasma levels of 25-hydroxyvitamin D, *Nature* 247:563–565.

Stanbury, S. W., Mawer, E. B., Taylor, C. M., and deSilva, P., 1980, The skin, vitamin D and the control of its hydroxylation. An attempted integration, *Min. Elec. Metab.* 3:51–60.

Tan, C. Y., Stratton, B., Marks, R., and Payne, P. A., 1982, Skin thickness measurement by pulsed ultrasound: Its reproducibility, validation, and variability, *Br. J. Dermatol.* 106:657–667.

Velluz, L., Amiard, G., and Petit, A., 1949, Le precalciferols relations d'equilibre avec le calciferol, *Bull. Soc. Chim. Fr.* 16:501–508.

Webb, A. R., Kline, L., and Holick, M. F., 1988, Influence of season and latitude on the cutaneous synthesis of vitamin D_3: Exposure to winter sunlight in Boston and Edmonton will not promote vitamin D_3 synthesis in human skin, *J. Clin. Endocrinol. Metab.* 67:373–378.

Webb, A. R., DeCosta, B. R., and Holick, M. F., 1989, Sunlight regulates the cutaneous production of vitamin D_3 by causing its degradation, *J. Clin. Endocrinol. Metabol.* 68:882–887.

Webb, A. R., Steven, M. D., Hosking, D. J., and Campbell, G. A., 1990, Correction of vitamin D deficiency in elderly long-stay patients by sunlight exposure, *J. Nutritional Med.* 1:201–208.

8

Can Cellular Responses to Continuous-Wave and Pulsed UV Radiation Differ?

Tiina Karu

1. Introduction

Interest in the action of UV radiation (UVR) on mammalian cells stemmed initially from the need to interpret the effects of this radiation on humans. This interest was a focus at Finsen's Light Institute in Copenhagen, where some of the first basic studies on the effects of UVR on animal cells were carried out at the beginning of the century (Giese, 1964). During recent decades, investigation into UVR-induced effects on cells has been one of the most active and extensively studied areas in photobiology. For review see, e.g., Rauth (1970), Painter (1970), Peak and Peak (1989), and Moan and Peak (1989).

For most biological effects of UVR, such as lethality, mutation, and transformation, DNA is considered to be the main target. For example, action spectra for lethality and mutagenicity, which are similar for both prokaryotes and eukaryotes, follow the spectra for DNA absorption and pyrimidine dimer induction (Rothman and Setlow, 1979; Coohill, 1986; see also Fig. 16a).

Tiina Karu • Laser Technology Center of the Russian Academy of Sciences, 142092 Moscow Region, Troitzk, Russia.

Environmental UV Photobiology, edited by Antony R. Young *et al.* Plenum Press, New York, 1993.

With the advent of lasers, new UVR sources emitting pulsed radiation with various pulse durations and peak intensities have become available. It would be of interest to compare cellular responses to classical continuous-wave (CW) and pulsed UV radiation for two reasons. First, during evolution, cells have adapted to CW (solar) radiation, and it is not clear *a priori* that their responses to CW radiation and pulsed UVR at the same wavelengths are exactly similar. Second, in the case of very short and intense laser pulses, it is possible to excite singlet and triplet electronic states (two-quantum excitation) of the photoacceptor molecule (Fig. 1b,c). It is known from experiments with DNA bases and viruses (Letokhov, 1983; Nikogosyan and Letokhov, 1983; Nikogosyan, 1990) that two-quantum excitation produces photoproducts different from those produced by single-photon UVR excitation.

The aim of our experiments (Karu *et al.*, 1981, 1982, 1983a,b; 1984a,b; 1988; Karu, 1986; Karu and Kalendo, 1987) was to find a way to damage quiescent (resting) cells. The high resistance of quiescent tumor cells to various damaging agents (e.g., ionizing radiation, chemotherapeutic drugs) is one of the factors decreasing the efficiency of tumor therapy (Baserga, 1971; Valeriote and van Putten, 1975; Zubrod, 1978; Epifanova *et al.*, 1983). The information available indicates that cultures of tumor cells in the plateau phase of growth, as identified by the content of cellular subpopulations, can serve as models

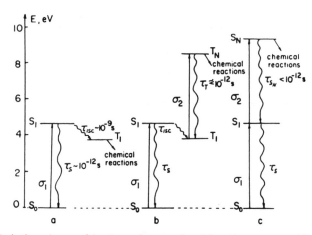

Figure 1. Excitation schemes of the electronic states of nucleic acid components: (a) one-quantum excitation of a singlet state with subsequent conversion to a triplet state with relaxation time τ_{isc}, (b) two-quantum excitation via intermediate triplet (T) states, and (c) two-quantum excitation via intermediate singlet (S) states. τ marks relaxation times of respective excited states, and σ_1 and σ_2 are cross sections of absorption of first and second photons.

for solid tumors (Baserga, 1971; Hahn and Little, 1972; Hahn, 1975; Grdina *et al.,* 1977).

To damage cells, two laser UVR sources and one conventional UVR source were used (see details in Fig. 2). The wavelengths of all three UVR sources are very close and fit to the maximum of the first absorption band of DNA (see Fig. 16a for the absorption spectrum). Powerful picosecond pulses (PPP) at 266 nm can cause two-quantum excitation of DNA (Fig. 1b,c), and high-repetition-rate pulses (HRRP) at 271 nm, as well as CW radiation at 270 nm, excite DNA by the single-quantum mechanism following classical laws of photobiology (Fig. 1a). The action of PPP on DNA bases, viruses, and

POWERFUL PICOSECOND PULSES (PPP), λ = 266 nm

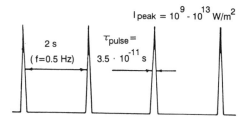

HIGH REPETITION RATE PULSES (HRRP), λ = 271.2 nm

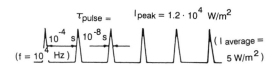

CONTINUOUS WAVE (CW) UV RADIATION, λ =270 nm,

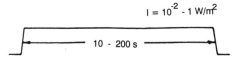

Figure 2. Parameters of UV pulses used in present work. The 4th harmonic of a Nd^{3+}:YAG laser emitting single powerful picosecond pulses (PPP) at 266 nm, and the 2nd harmonic of a Cu-vapor laser emitting high-repetition-rate nanosecond pulses (HRRP) at 271.2 nm with high peak and low average intensity (I) were the laser UVR sources used. A mercury lamp with a mono-chromator, λ = 270 nm, was the conventional UVR source used. (I_{peak} = peak intensity; τ_{pulse} = pulse duration; $I_{average}$ = average intensity.)

bacteria (but not on mammalian cells) has been extensively studied (Niko-gosyan and Letokhov, 1983; Nikogosyan, 1990; Schulte-Frohlinde *et al.*, 1990; Cadet and Vigny, 1990). Laser sources emitting HRRP are new to photo-biologic studies.

In this chapter, I will describe two ways to damage plateau-phase HeLa cells (as compared to exponentially growing cells) by pulsed laser UVR. First, it will be shown that certain pulse rates and fluence ranges of HRRP enhance colony-forming ability and induce replicative synthesis of DNA in a subpop-ulation of plateau-phase cells. Second, two intense picosecond pulses with a 4- to 6-s interval between them enhance the transport processes of cellular membranes without causing any changes in the viability of cells or in the rate of DNA synthesis.

It is well known that exposure to solar radiation and UVR under labo-ratory conditions can induce skin tumors. Recent experimental data suggest that an inappropriate expression of protooncogenes due to point mutation or gene amplification, deletion, or rearrangement may be involved in UV car-cinogenesis (Ananthaswamy and Pierceall, 1990). The experimental data on UV-induced replicative synthesis in plateau-phase HeLa cells presented here may be of interest to those concerned with mechanisms of UV carcinogenesis, despite the fact that a HeLa culture is in itself transformed.

The increase in the permeability of cellular membranes can probably be used to introduce allogenic material or chemotherapeutic agents into quies-cent cells.

2. Induction of Replicative DNA Synthesis in Plateau-Phase Cells

2.1. Basic Information about Cells and Experimental Technique

HeLa cell cultures grown as monolayers using nutrient medium 199 supplemented with 10% bovine serum and antibiotics were irradiated 3 days (exponentially growing, or log-phase, cells) or 10 days (plateau-phase cells) after plating. The cells were grown without changing the nutrient medium (unfed culture) and without CO_2 supply. The HeLa culture spread well and did not form multilayers (see Karu *et al.*, 1990, for details). The labeling index of the log-phase population was 19.1% $\pm$ 3.0%, and that of the plateau-phase culture was 5.0% $\pm$ 0.8%. The mitotic indices of log-phase and plateau-phase populations were 1.1% $\pm$ 0.1% and 0.1%, respectively.

Cultivation of cells, radiometric techniques of pulse labeling or continuous labeling, and autoradiography are described in detail elsewhere (Karu *et al.*,

1983b, 1984a,c, 1987, 1990). These methods are rather standard and follow the description given by Hauschka (1973).

There are, however, two methodological points in our experiments which differed from usual experimental technique. First, cells were cultivated and irradiated in special chambers with quartz windows allowing irradiation of the monolayer without removal of the nutrient medium (Fig. 3). Under these conditions, changes in the partial pressure of oxygen and pH jumps in the nutrient medium could be avoided. Second, in contrast to the classical Puck technique for evaluation of clonogenicity (i.e., individual cells, either in suspension or attached to substate, exposed to radiation and then allowed to form colonies), we irradiated confluent monolayers and then plated the cells into a fresh nutrient medium to form colonies.

2.2. Specific Responses of Plateau-Phase Cells to Various Sources of UVR

Monolayers of plateau-phase HeLa cells were exposed to high-repetition-rate pulses (HRRP) at 271 nm, continuous-wave (CW) UVR at 270 nm, or powerful picosecond pulses (PPP) at 266 nm at various fluences. After 3 h the cells were removed from the culture chamber in maximal mild conditions (Versene instead of trypsin, temperature at 37°C), and plated in scintillation vials. Two experiments were performed. In the first, 100 cells were plated into every flask—control (nonirradiated) and test—and incubated for 14 days. Then the colonies were fixed with methylene blue, their numbers determined and their diameters measured under a high-power dissecting microscope. The results of this experiment, which reflects the colony-forming ability of plated cells, are presented in Figs. 4a and 5. In the second experiment, 100 cells were again plated into every flask, but the number of cells in three separate flasks was determined every day. The result of this experiment (growth curves) is presented in Fig. 4b.

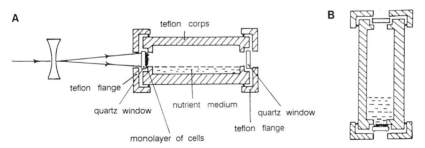

Figure 3. Cell culture chamber for UV irradiation of monolayers: (A) position during irradiation; (B) position during cultivation.

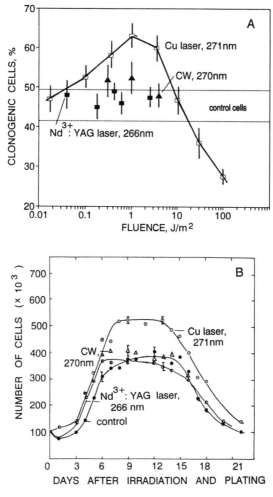

Figure 4. Colony formation and multiplication of HeLa cells exposed in the plateau phase of growth to various UVR sources and plated 3 h after irradiation into fresh medium. ($\triangle$) = CW; ($\mathbf{O}$) = Nd:YAG; ($\bullet$) = control; ($\odot$) = Cu laser. (A) number of colonies formed as a function of fluence after 14-day incubation period; (B) growth curves of cells given a fluence of 2 J/m²; cells per vial counted daily.

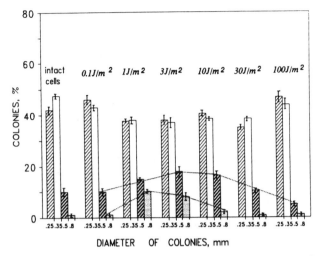

Figure 5. Distribution of colonies by diameter (d) after 271.2-nm HRRP at various fluences. Dashed lines mark the dependence of the percentage of large ($0.5 < d < 0.8$ mm) and intermediate-size ($0.35 < d < 0.5$ mm) colonies on the fluence.

The percentage of clonogenic cells in the nonirradiated culture was 45.2% ± 0.4%. As can be seen in Fig. 4a, irradiation with HRRP at fluences from 0.1 to 3 J/m² caused an increase in the number of colonies. At higher fluences (from 10 to 100 J/m²), the number of colonies decreased as compared to the unexposed control. Under the same experimental conditions, PPP and low-intensity CW UVR in the fluence range 0.1–3 J/m² did not cause statistically significant changes from the control level.

A comparison of the growth curves indicates that PPP and CW UVR did not have much effect on cell growth (Fig. 4b). Irradiation with HRRP at 2 J/m² caused an increase in cell proliferation during the exponential phase of growth and the formation of a plateau with a higher number of cells than that of the control (Fig. 4b).

Investigation of colony size distributions after irradiation of plateau-phase cells with various fluences of HRRP (Fig. 5) indicated that in the cultures exposed to stimulating fluences (0.1–3 J/m², as seen in Figure 4a), the number of intermediate-size ($0.35 < d < 0.5$ mm) and large ($0.5 < d < 0.8$ mm) colonies increased, and the number of abortive ($d < 0.25$ mm) and small ($0.25 < d < 0.35$ mm) colonies decreased. At fluences above 3 J/m², however, the percentage of large and medium-size colonies decreased, whereas the action of light on the abortive and small colonies had no effect.

The data presented in Figs. 4 and 5 suggest that at certain low fluences (0.1–3 J/m^2) of HRRP, irradiation not only results in an increase in the number of colonies but also induces increased proliferation of a subpopulation of the irradiated cells.

2.3. Specific Responses of Plateau-Phase Cells to High-Repetition-Rate UVR

Measurement of [^{3}H]thymidine (a DNA precursor) incorporation into DNA is frequently done to estimate the effects of DNA-damaging agents. There are two different ways to perform such measurements.

In one method, the radioactive precursor is added to cell culture for a certain time period, e.g., 20 min, after which time the amount of radioactive DNA is determined, allowing changes in DNA synthesis rate to be evaluated (radiometric pulse-labeling). In the second method, the pulse-labeled cells are fixed and developed with photographic emulsion (autoradiography). This technique makes it possible to count the number of DNA-synthesizing cells, and by determining the number of silver grains above each nucleus, to evaluate the rate of DNA synthesis in individual cells.

Figure 6 presents the fluence-dependent changes in DNA synthesis rate measured by radiometric pulse-labeling in exponentially growing and plateau-phase HeLa cells 2.5 h after exposure to HRRP, PPP, or CW low-intensity UVR. As seen in Fig. 6a, the log-phase HeLa cells responded to HRRP and CW UVR in a similar way: DNA synthesis was inhibited by increasing the fluence. In the case of PPP, some stimulation was observed in the fluence range 10^{-2}–10 J/m^2, and inhibition of DNA synthesis occurred at fluences higher than 10 J/m^2.

The responses of the plateau-phase cells (Fig. 6b) were different from those of log-phase cells. PPP did not affect DNA synthesis in the fluence range used. HRRP and CW low-intensity UVR had pronounced but opposite effects on DNA synthesis rate: CW UVR inhibited DNA synthesis in a fluence-dependent manner; HRRP stimulated it (optimal fluences near 0.5–1 J/m^2), and no inhibition of DNA synthesis was observed by increasing the fluence of HRRP.

The stimulation of DNA synthesis by HRRP was specific for plateau-phase cells (Fig. 7). In this series of experiments, measurement of DNA synthesis rate as a function of the fluence of HRRP was done on the 8th, 10th, 11th, and 13th days of cultivation of the cells. In other words, the cells were irradiated at the end of the log-phase and at the beginning, in the middle, or at the end of the plateau-phase, as illustrated by the schemes on the right of Fig. 7. As seen in Fig. 7, all plateau-phase cultures (10th, 11th, and 13th days of cultivation) responded to irradiation with increasing [^{3}H]thymidine incor-

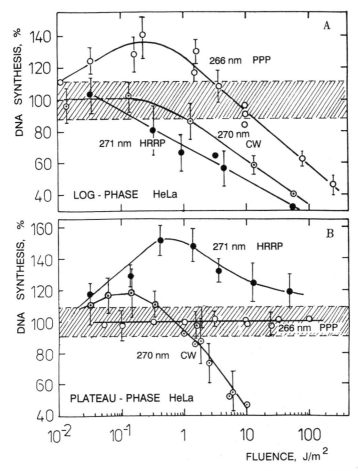

Figure 6. Incorporation of [³H]thymidine during 20-min pulse labelings measured 2.5 h (starting point of the pulse label) after the exposure of (A) log-phase or (B) plateau-phase HeLa monolayers to HRRP at 271 nm, PPP at 266 nm, or CW UVR at 270 nm. Control cells are denoted by the shaded area.

poration. No inhibition of DNA synthesis was observed as the fluence was increased.

The stimulation of DNA synthesis also depended on the pulse repetition rate of the radiation (Fig. 8). Two bands of pulse repetition rates (near 8–12 kHz and 19–25 kHz, with maxima at 10 kHz and 21 kHz) had an effect on DNA synthesis. In this experiment we used a Nd³⁺:YAG laser emitting at 266 nm, but the parameters of the pulses (their peaks, average intensities,

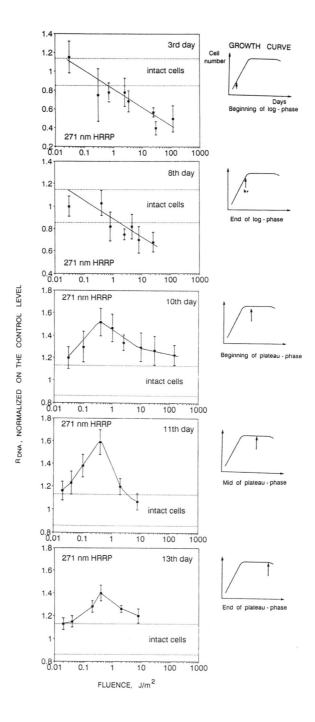

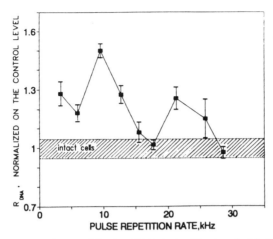

Figure 8. Incorporation of [³H]thymidine into DNA during 20-min pulse labelings measured 2.5 h (starting point of the pulse label) after the exposure of plateau-phase HeLa cells to HRRP at a fluence of 5 J/m² as a function of pulse repetition rate.

and duration) were very close to the respective parameters of the 2nd harmonic of the Cu-vapor laser.

This result indicates that the action of HRRP on plateau-phase HeLa cells is connected with DNA damage specific and characteristic of this type of radiation.

The [³H]thymidine experiments described above were performed 2.5 h after irradiation of the cells. The kinetics of [³H]thymidine incorporation measured by pulse-labeling at various time points after irradiation varied with the type of radiation used (Fig. 9a,c). After irradiation with CW UVR (Fig. 9c), DNA synthesis was inhibited during the first hours and then returned to the control level. After irradiation with HRRP, DNA synthesis remained at the control level in the first 2 h of postirradiation and then increased rapidly (Fig. 9a).

We also measured [¹⁴C]uridine incorporation into RNA after both types of irradiation. With CW UVR, no statistically significant changes were found in the first 5 h as compared to the nonirradiated cells (Karu *et al.,* 1983b).

Figure 7. Incorporation of [³H]thymidine during 20-min pulse labelings measured 2.5 h (starting point of the pulse label) after the exposure of HeLa monolayers to HRRP at different days of cultivation. The schemes on the right illustrate the growth phase of the population at the moment of irradiation. Ordinate is defined as the ratio of experimental and control.

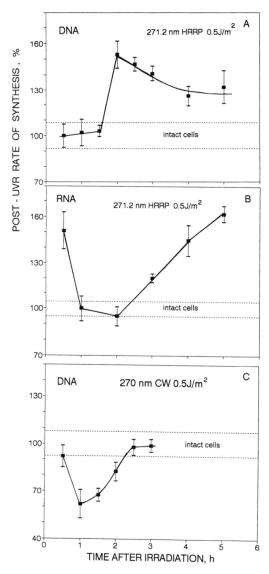

Figure 9. Incorporation of (A,C) [³H]thymidine or (B) [¹⁴C]uridine during 20-min pulse labelings measured at different time points (time in the abscissa marks the beginning of the pulse labeling) after the exposure of plateau-phase HeLa cells to (a,b) HRRP or (c) CW UVR.

When the plateau-phase cells were irradiated with HRRP, [^{14}C]uridine incorporation increased 30 min after irradiation, then returned to the control level, and later increased again (Fig. 9b). Similar kinetic changes in RNA synthesis rate have been described for cells overcoming the $G_0 \rightarrow S$ transition (Epifanova, 1977). The data presented in Fig. 9b are clearly insufficient to answer the question of whether or not the increased RNA synthesis found in our experiments (Fig. 9b) is connected with an activation of proliferation.

This series of experiments suggests that the increase of [^{3}H]thymidine incorporation after irradiation with HRRP is a specific response of plateau-phase HeLa cells under our experimental conditions.

2.4. Does High-Repetition-Rate UVR Induce Replicative or Reparative DNA Synthesis?

One way to distinguish between replicative and reparative synthesis of DNA is by counting silver grains in autoradiographs. This method is used to determine the percentage of DNA-synthesizing cells (labeling index). Also, by counting the number of grains above every nucleus, the proportions of replicative and reparative DNA synthesis can be determined (cells undergoing replicative synthesis are heavily labeled; cells undergoing reparative DNA synthesis are only weakly labeled).

Figure 10 presents the labeling index 2.5 h after irradiation of plateau-phase HeLa cells with HRRP at various fluences. The number of cells incorporating [^{3}H]thymidine increased and exceeded the control level by about 3

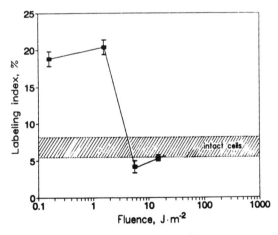

Figure 10. Labeling index (percentage of cells incorporating [^{3}H]thymidine) 2.5 h after the exposure of plateau-phase HeLa cells to 271-nm HRRP at various fluences (radioautographic measurements).

times at fluences of 0.16 or 1.6 J/m². After increasing the fluence (5.6 and 16 J/m²), the number of DNA-synthesizing cells decreased and dropped to near the control level. The stimulative fluences here (0.16 and 1.6 J/m²) coincide with those determined by radiometric measurements of [³H]thymidine incorporation (Figs. 6b and 7).

Figure 11 shows the distribution of silver grains above the nuclei of DNA-synthesizing cells. The shift of the cumulative curves to the left from the control curve (intact cells) when the cells were irradiated at 5.6 and 16 J/m², points to a reduced number of cells with heavily labeled nuclei, i.e., to suppression of replicative DNA synthesis. With lower fluences (0.16 and 1.6 J/m²), the cumulative curves shift to the right, i.e., there is a greater number of cells with heavily labeled nuclei, which indicates activation of replicative DNA synthesis.

When the cell population immediately after the irradiation is treated with [³H]thymidine and continuously incubated with it for several hours, the fraction initially labeled represents cells in the S phase at the moment of irradiation, while the subsequent increase in the labeling index (percentage of DNA-synthesizing cells) reflects the flow of cells into the S phase during the interval studied. It is obvious from Fig. 12a that the percentage of labeled cells increased in both control and irradiated cultures; however, the irradiated cells underwent a rapid increase in the number of labeled cells during the first hour after treatment.

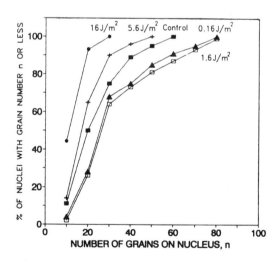

Figure 11. Silver grain counts in control and 271.2-nm HRRP-irradiated plateau-phase HeLa cells after 20-min pulse labeling with [³H]thymidine (radioautographic measurements). The respective labeling indices are shown in Fig. 10.

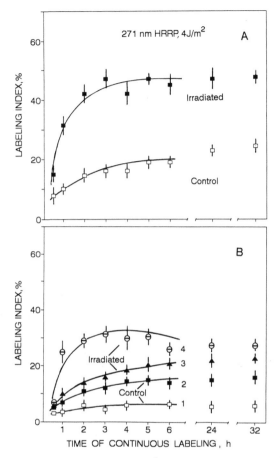

Figure 12. Labeling index (percentage of DNA-synthesizing cells) after the exposure of plateau-phase HeLa cells to 271-nm HRRP at 4 J/m², measured by the autoradiographic continuous-labeling technique: (A) percentage of labeled cells in irradiated and control cultures, and (B) percentage of cells with highly labeled (curves 2 and 3) and weakly labeled nuclei (curves 1 and 4) in control (curves 1 and 2) and irradiated (curves 3 and 4) cultures.

By counting the number of silver grains above every nucleus it is possible to draw some conclusions about the rate of DNA synthesis. We divided the cells into three groups: unlabeled cells (less than 8 grains per nucleus), weakly labeled cells (8 to 29 grains), and heavily labeled cells (more than 30 grains). The third group includes HeLa cells undergoing normal replicative synthesis of DNA and the second group includes the cells undergoing reparative synthesis of DNA (Paribok and Semyonova, 1970).

As seen in Figure 12b, the percentage of labeled cells in the irradiated cultures increased mainly due to the number of weakly labeled cells (curve 4). A lesser increase was seen in the number of heavily labeled, irradiated cells (curve 3).

The data obtained show that [³H]thymidine incorporation into DNA of plateau-phase HeLa cells after irradiation with HRRP in a certain fluence range (0.1–3 J/m² (Figs. 6b, 7, 9a) and at certain pulse rates (near 10 and 21 kHz, Fig. 8) is mainly due to reparative DNA synthesis (Fig. 12b). A lesser [³H]thymidine uptake is accomplished via replicative DNA synthesis (Figs. 11 and 12b). Under our experimental conditions, HRRP is able to induce both types of DNA synthesis, replicative as well as reparative.

2.5. Is the Induction of Replicative Synthesis by DNA-Damaging Agents a Common or a Unique Phenomenon?

A stream of literature data suggest that UVR suppresses replicative synthesis of DNA and induces reparative synthesis of DNA in cellular cultures of mammalian origin (for review see Painter, 1970; Rauth, 1970; Moan and Peak, 1989). Most experimental data concern mammalian cells in the exponential phase of growth (log phase). A few investigations of UVR effects on plateau-phase or growth-arrested mammalian cells (Little, 1969; Hahn, 1975; Chan and Little, 1979; Wechelbaum et al., 1980; Kantor et al., 1980; Kantor, 1986; Gill and Coohill, 1987) suggest (1) that quiescent cells have extreme resistance to high doses of UVR, (2) that lethal action of UVR is (as in case of log-phase cells) connected with pyrimidine dimers, and (3) that there is a mechanism of DNA repair in plateau-phase cells which is absent or weakly functioning in log-phase cells (Kantor, 1986).

We found that irradiation with HRRP at 271 nm induced replicative DNA synthesis in a fraction (subpopulation) of the plateau-phase HeLa cultures (Figs. 5, 12). The cells irradiated at low fluences not only retained the capacity for further growth but their proliferative capacity was stimulated (Figs. 4, 5).

Although most studies indicate that DNA-damaging agents inhibit DNA replication, a few papers suggest that some agents, UVR included, may also induce or stimulate DNA replication. A marked induction of DNA replication was observed in confluent diploid fibroblast cultures treated with low fluences of UVR (2–20 J/m²) or with carcinogenic agents N-methyl-N-nitrosourea and N-acetoxy-2-acetylaminofluorene (Cohn et al., 1984). Another carcinogenic agent, N-methyl-N-nitrosoquanidine, stimulated DNA replication in a post-confluent culture of Syrian hamster embryo cells (Mironescu et al., 1980). Gamma irradiation induced replicative synthesis of DNA in plateau-phase HeLa cultures (Synsyns and Saenko, 1986).

The authors of these investigations were interested in changes in DNA replication occurring after treatment with carcinogens, such changes considered to be necessary at the initial stages of carcinogenesis. The most interesting conclusion from these experiments (Mironescu *et al.,* 1980; Cohn *et al.,* 1984; Synsyns and Saenko, 1986) is that all DNA-damaging agents, each of them producing a different spectrum of damage (and different mutations, respectively), induced replicative DNA synthesis in a fraction of cells in plateau-phase cultures.

Recently, Peak *et al.* (1991) have demonstrated enhanced (almost 2-fold) gene expression for protein kinase C following brief exposure of cultured human epithelioid P3 cells to sunlight. They demonstrated that solar radiation induced a cellular transcription response similar to that found after administration of tumor-promoting agents and ionizing radiation, which suggests that solar radiation may function as a tumor promotor. It has been shown that DNA damage induces complex mechanisms in mammalian cells allowing the cells to handle or accommodate DNA lesions. At the molecular level, these mechanisms include enhanced expression of a number of genes (Ronai *et al.,* 1990). UVR is known to activate viral DNA sequences in transformed cells and to induce deletion of growth suppression genes (Ananthaswamy and Pierceall, 1990).

A new interest into UVR-induced stimulative processes has arisen recently in connection with the findings that 254-nm UVR or sunlight can induce a human immunodeficiency virus (HIV) promoter and stimulate growth of the complete virus in human cells (Valerie *et al.,* 1988; Zmudzka and Beer, 1990). UVR-sensitive sites in RNA transcripts of hepatitis delta virus and potato spindle fiber viroid have been found (Branch *et al.,* 1989). In these cases, the DNA damage was found to be a prerequisite for UVR-induced activation of the genes.

Last but not least, one should remember that a number of reports appear from time to time indicating that UVR stimulates cell division (Alpatov and Nastjukova, 1933, Carlson *et al.,* 1961; Walicka and Beer, 1973; Samoilova, 1979; Abbaszade, 1986; Belenkina *et al.,* 1990). Most of this work has been done with microorganisms, except that of Walicka and Beer (1973). Two strains of murine leukemic lymphoblasts, L 5178Y-S and L 5178Y-R, were exposed to various fluences of 254-nm UVR, ranging from 1.1 to 53.5 J/m^2, and cultured for up to 70 days after irradiation. UV-exposure stimulated growth in about 70% of the cultures; growth started immediately after post-irradiation growth disturbances and lasted several tens of generations. Proliferative activity was more greatly enhanced for L 5178Y-R cells (a less x-ray-sensitive culture) than for the L 5178-S strain: maximal shortening of the mean doubling time for the L 5178Y-R strain was 43%, whereas it was 20% for L 5178-S cells (Walicka and Beer, 1973). The most important result of this investigation was that growth stimulation also occurred at later stages of postexposure development, demonstrating UVR-induced heritable changes.

Taken together this research suggests that the induction of DNA replication by low doses of damaging agents is not a unique phenomenon. The various damaging processes have some common qualitative and quantitative characteristics. First, the variation of replicative DNA synthesis in plateau-phase cellular cultures with fluence of UVR, dose of γ-radiation, or concentration of carcinogenic chemicals is qualitatively similar: a bell-shaped curve with a rather sharp maximum, the effect rapidly decreasing to the control level at higher doses of the agents (Mironescu et al., 1980; Karu et al., 1982, 1983a; Cohn et al., 1984, Saenko and Synsyns, 1986; Figs. 6, 7, 10). Similar bell-shaped fluence dependencies were also established for HIV-1 promoter induction in HeLa cells by UVR (Zmudzka and Beer, 1990). As established by Cohn et al. (1984), the bell-shaped fluence dependence of replicative DNA synthesis was quite different from the curve for reparative DNA synthesis. Reparative DNA synthesis continued to increase up to a fluence of 20 J/m^2 and reached a plateau between 20 and 40 J/m^2.

Second, optimal UVR fluences (3 J/m^2, Cohn et al., 1984; 0.5 J/m^2, Karu et al., 1983a; Figs. 6b, 7) and magnitudes of effects (1.6, Cohn et al., 1984; 1.4–1.6, Karu, 1982, 1983a; Fig. 7) have been found to be very similar.

Third, replicative synthesis of DNA is induced only in a relatively small subpopulation of plateau-phase cells. Cohn et al. (1984) established this fraction to be $\approx 6\%$ in plateau-phase cultures and $\approx 18\%$ in serum-arrested cultures. We did not perform detailed measurements, but an analysis of autoradiographic data and clone size distributions allows us to estimate this fraction to be 10–15%. Cohn et al. (1984) found that the number of morphological transformants correlated with the percentage of cells in the carcinogen-responsive subpopulation which incorporated [³H]thymidine.

Under certain experimental conditions, the stimulation of [³H]thymidine incorporation into plateau-phase cells was a specific cellular response to HRRP; the CW UVR administered at practically the same wavelength, the same fluence, and during the same postirradiation period had no effect (Fig. 6b) or was inhibitive (Fig. 9c). In these experiments, the measurements were made during the first 5 h of postirradiation. With HRRP, the first point with increased [³H]thymidine incorporation was 2 h after exposure (Fig. 9a), which is a very rapid response of plateau-phase cells compared to the responses of these cells to UVR at 254 nm and to γ-radiation. With UVR at 254 nm, replicative DNA synthesis started to increase only after 12 h and was at a maximum 24 h after irradiation (Cohn et al., 1984). The incorporation of [³H]thymidine was inhibited a few hours after irradiation of plateau-phase HeLa cells with γ-radiation, and stimulated 24 h after exposure (Synsyns and Saenko, 1986). It is quite possible that after the initial inhibition which occurs during the first 2 hours after irradiation with CW UVR at 270 nm Fig. 9b), there will be an increase in [³H]thymidine incorporation. Measurements to determine this were not performed in our work, principally because there are no reasons to believe that the induction of DNA replicative synthesis by HRRP at 271 nm is a specific cellular response as compared to CW UVR at 270 nm. More probably, this response is specific for particular experimental conditions, and occurs very soon after exposure.

Upon reaching the plateau-phase, populations of normal cells accumulate in the early G_1 phase of the cell cycle, while most cells of transformed lines, HeLa included, accumulate in a late G_1 phase (Hittelman and Potu, 1978). Similar differences in the morphology of the G_1 phase between cells from normal tissues and solid tumors are reported (Grdina et al., 1977).

It has been shown in cell fusion experiments that the rate of initiation of replicative DNA synthesis in the nuclei of HeLa cells in the G_1 phase can be very rapid (≈ 1.5 h). UV irradiation of the quiescent cells before fusion induced decondensation of chromatin in a fluence-dependent manner in the nuclei of these cells (Rao and Smith, 1981). Decondensation of chromatin must occur before reparative (Schor et al., 1975) and replicative (Baserga, 1979) DNA synthesis can take place. It is quite possible that a subpopulation of the plateau-phase cells arrested at a late G_1 phase is responsible for the effects of HRRP described in this chapter. One can speculate that in this particular subpopulation, the damage caused by HRRP induces a chromatin template configuration promoting the beginning of replicative DNA synthesis.

3. The Effect of Powerful Ultrashort Pulses on Transport Processes in the Membranes of Plateau-Phase Cells

3.1. Does Two-Quantum Excitation Occur in Living Cells?

As explained in the introduction, PPP can excite absorbing molecules by a two-quantum mechanism (see Figs. 1c,d). The two-quantum mechanism of excitation occurs in DNA bases in solutions, in viruses, and in bacteria (Nikogosyan and Letokhov, 1983; Letokhov, 1983; Nikogosyan, 1990). It would be of interest to estimate the probability of two-quantum excitation of DNA molecules when mammalian cells are irradiated by ultrashort pulses with $\lambda = 266$ nm. If the target is a DNA molecule, cells with a higher quantity of DNA have a greater chance of being damaged. In other words, as the organism becomes more complex, the number of DNA bases increases and the fluence needed to cause damage to at least one base in the DNA chain decreases. This relationship, which is shown in Fig. 13, is also

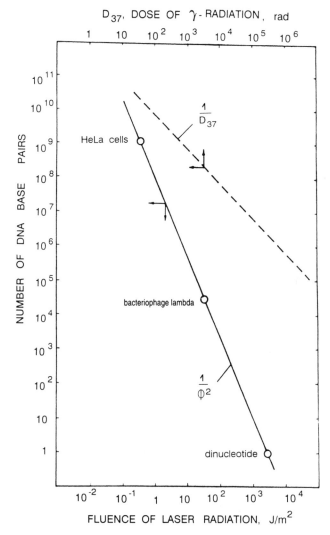

Figure 13. DNA-damaging fluence (ϕ) of picosecond UV pulses (λ = 266 nm) (solid line) as a function of the number of DNA base pairs in dinucleotides, viruses, and mammalian cells. The dashed line denotes the same type of dependence for γ-irradiation.

observed for γ-radiation (Kaplan and Moses, 1964). If the ordinate represents the quantity of DNA (number of base pairs, K_p) and the abscissa, the dose of radiation, D_{37} (a dose at which 63% of DNA molecules are damaged), one observes an inverse dependence, $K_p \sim 1/D_{37}$, between damage and dose (Fig. 13).

Let us estimate the minimal damaging fluence for HeLa cells irradiated with UV PPP at 266 nm. Figure 14a presents the changes in DNA synthesis rate after exposing exponentially growing HeLa cells to PPP at $\lambda = 266$ nm. [³H]thymidine incorporation was found to increase with fluences in the range 0.1–10 J/m², with a maximum at 1 J/m². Irradiation with PPP at 1,064 nm

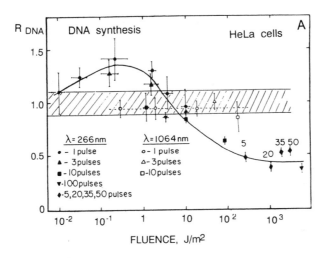

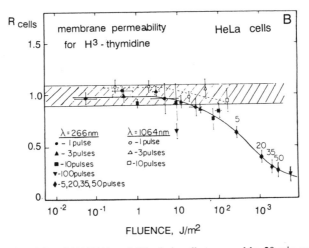

Figure 14. Radioactivity of (A) DNA and (B) whole cells measured by 20-min pulse labelings 2.5 h (starting point of the pulse label) after the exposure of exponentially growing HeLa cells to powerful picosecond UV ($\lambda = 266$ nm) and IR (infrared) ($\lambda = 1,064$ nm, dashed line) pulses. R is defined as the ratio of radioactivity values in irradiated and control samples.

produced no change in [³H]thymidine incorporation (Fig. 14a). Because this wavelength (λ = 1064 nm) is not absorbed by DNA, one can suggest that the effects at λ = 266 nm are connected with the absorption by DNA. Membrane permeability for [³H]thymidine did not differ from the control level with fluences varying from 0.1 to 10 J/m² (Fig. 14b). The trypan blue exclusive test showed that the cells were viable after irradiation.

There is every reason to believe that the stimulation of [³H]thymidine incorporation in the fluence range between 0.1 and 1 J/m² (Figure 14a) is a response to DNA damage. HeLa cells have 2.9×10^9 base pairs. The bacteriophage lambda contains 4.65×10^4 base pairs, and its damaging fluence (on the basis of survival curves) lies between 10 and 10^2 J/m² (Nikogosyan and Letokhov, 1983). Experiments with PPP on DNA fragments have shown that the lethal fluence for nucleotides is 30 J/m² (Nikogosyan and Letokhov, 1983). Since the damage yield at this fluence is about 1% (Letokhov, 1983), the fluence needed for guaranteed damage of a dinucleotide lies in the range 10^3-10^4 J/m².

Figure 13 shows the dependence of the damaging fluence of PPP on the number of base pairs. This dependence is quadratic, $K_p \sim 1/\phi_2$, and thus suggests a two-quantum damage mechanism in dinucleotides, bacteriophage lambda, and HeLa cells irradiated with PPP at 266 nm.

3.2. Irradiation with Two Ultrashort Pulses Increases the Permeability of Membranes of Plateau-Phase Cells

Nucleosides (precursors of DNA and RNA synthesis) in concentrations less than 2 μM are taken up into mammalian cells by means of a passively mediated transport mechanism, also known as *facilitated diffusion* (Hopwood et al., 1975). The nucleosides that have entered the cell are phosphorylated at once and thus lose their ability to diffuse out (Schlotissek, 1968). Since [³H]thymidine incorporation into the DNA of cells irradiated with PPP does not differ from the control level (Fig. 15, curve 4), this suggests that an increase in [³H]thymidine-labeled cells is due to an alteration of diffusion and/or transport properties of the cellular membrane.

When plateau-phase HeLa cells were irradiated with two UV (266 nm) PPP, no increase in [³H]thymidine uptake occurred when the interval between the two pulses was 1–2 s (Fig. 15, curves 1 and 2). Uptake increased as the interval between the two pulses increased, and a maximal increase occurred at a pulse interval of 4 to 6 s. Further increase in the pulse interval from 8 to 100 s caused the effect to be reduced to the control level. [³H]thymidine uptake did not depend on wavelength; the same effect occurred at 266, 532 (Fig. 15, curves 1 and 3) and 1,064 nm (Karu, 1986). Increase in the intensity of the pulses, however, decreased the effect (curves 1 and 2 in Fig. 15).

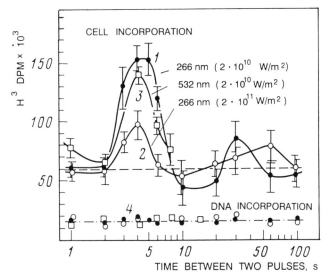

Figure 15. Changes in the permeability of cellular membranes to [³H]thymidine measured 1.5 h after exposure of plateau-phase HeLa monolayers to two PPP at 266 or 532 nm, as a function of the time between the two pulses. DPM = decays per minute.

These effects are specific for plateau-phase HeLa cells. The membrane permeability of cells in the exponential phase of growth did not vary from the control level (Karu *et al.*, 1984a).

The results obtained from irradiating cells with two PPP (Fig. 15) enable us to assume that the first pulse is responsible for some change in the cell transport system for thymidine which may be realized within several seconds with another pulse but which itself cannot cause a change in membrane permeability (Karu *et al.*, 1984a). If the second pulse comes at the right moment, a change in the transport system occurs which accelerates the passage of nucleosides through the membrane. If the second pulse arrives too late, when the preparatory reactions caused by the first pulse have decayed, no change in cell membrane permeability is observed.

Using a powerful, single, nanosecond pulse (third harmonic of Nd³⁺:YAG laser, λ = 335 nm), it is possible to perforate the cell membrane so that some allogenic material penetrates into the cell (Tsukakoshi *et al.*, 1984). The threshold energy used was about 1 mJ, and the hole in the membrane closed after a second. It should be emphasized that the fluence and pulse energies used in our experiments were essentially less than those used in the work of Tsukakoshi *et al.* (1984).

It is hardly probable that the increased permeability of the cellular membrane found in our experiments is due to purely mechanical destruction of the membrane since (1) there is a complex nonlinear dependence of the effect on the time between the two pulses (Fig. 15, curves 1–3), (2) the effect is reduced at increased light intensities (Fig. 15, curves 1 and 2), and (3) the increased permeability lasts for hours (at least 2.5 h; Karu, 1986).

One possible explanation for increased permeability is that irradiation results in some conformational changes in the structure of the thymidine transport system. Even slight changes of this kind may be important because some components of the transport systems of nucleotides are allosteric proteins (Eilam and Cabantchik, 1977; Pardee and Palmer, 1973).

What mechanisms could explain our results? A photochemical mechanism seems unlikely because there is no dependence of permeability on wavelength (Fig. 15). It is possible that the modulated transport functions are connected with local heating. The permeability of membranes of ascite tumor cells for fluoresceine and trypan blue was shown to increase with increasing temperature (Strom et al., 1973). In this experiment, tumor cells were incubated at different temperatures for hours. This kind of total heating is impossible in our experiments because pulse duration is only 3×10^{-11} s. The total increase of temperature will not exceed 10^{-3}–10^{-2}°C (Karu, 1986).

Due to the nonhomogeneity of the cellular membrane, irradiation with PPP can cause local heating. The possible local increase of temperature caused by one PPP was estimated to be ≈ 36°C (Karu, 1986). This short-interval increase is sufficient for causing some conformational changes in the molecules of the cellular membrane transport system.

4. Concluding Remarks

Two characteristic responses of plateau-phase HeLa cells to pulsed UVR are described: an increase in cellular membrane permeability to nucleosides after irradiation of the cells with two ultrashort UV pulses, and the induction of replicative DNA synthesis in a cellular subpopulation after irradiation with high-repetition-rate UV pulses. The first effect is actually a specific response to pulsed UVR for a very simple reason: the parameters afforded by pulsed UVR are not possible with ordinary CW UVR sources. It would be more correct to say that this effect is laser specific.

Whether induction of replicative DNA synthesis by HRRP at 271 nm is a specific cellular response to pulsed UVR is not perfectly understood. It is certainly a specific effect under particular experimental conditions, and occurs very soon after exposure. CW UVR at 270 nm might have a similar effect under other experimental conditions (e.g., with longer postexposure times).

It is a common misconception that UVR is photochemically much more active than visible light (Smith, 1980). It is also believed that UVR has mainly negative effects (inactivation of cellular reactions) and that laser radiation has special properties.

The experiments discussed in this chapter suggest that UVR has not only negative but positive effects: stimulation of colony-forming ability (Fig. 4a), enhancement of proliferation (Figs. 4b, 5), and modulation of transport processes in cellular membranes (Fig. 15). As to the photochemical activity of visible light, it is possible to stimulate DNA synthesis with this kind of radiation as well (Fig. 16b). In this case, DNA synthesis is the last step in a long chain of photosignal transduction and amplification reactions triggered by the absorption of visible light in mitochondria (Karu 1987, 1988, 1989). This scheme forms the basis for low-power laser therapy ("laser biostimulation"). Effecting DNA synthesis in this way is completely different from the mechanism discussed in this chapter, as one can see from a comparison of Figs. 16a and b.

Many people believe that properties of laser radiation such as coherence, high intensity, and monochromaticity make laser radiation unusual. An example is the so-called laser biostimulation, a purely photobiological phenomenon, the effects of which depend on the wavelength and fluence of the light used for irradiation (Fig. 16b), but not on its coherence (Karu, 1987). In this case, lasers just have the benefit of being a convenient, monochromatic light source.

Lasers can provide light with parameters that vary greatly from those of light from ordinary sources, e.g., the intense, ultrashort pulses used in the experiments discussed in this chapter (Section 3). But even here, where the primary excitation mechanism (two-quantum excitation, Fig. 1c,d) is entirely different from the classical one-quantum excitation (Fig. 1a), the laws of classical photobiology are valid. As an example, Fig. 17 illustrates the lethal action of PPP at 266 and 532 nm on plateau-phase and log-phase HeLa cells. It is known that visible light is ineffective in causing lethal and mutagenic effects. PPP at 532 nm kills both plateau-phase and log-phase cells, but with a very low efficiency not differing from the lethal action of CW green light (Karu *et al.*, 1988). This finding confirms the rule known in classical photobiology: the damage inflicted by visible radiation is not responsible for the death of cells, or in other words, even intense laser pulses cannot be effective when the wavelength is wrong. Plateau-phase cells have a higher resistance than log-phase cells to PPP at 266 nm (Fig. 17). The same is true for plateau-phase and log-phase cells exposed to CW low-intensity 254-nm UVR, as discussed in Section 2.4. These examples only illustrate some misconceptions about light. Readers may find better examples in their surroundings.

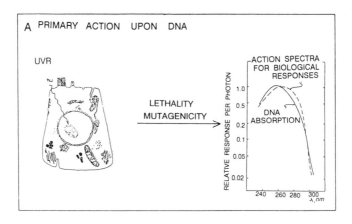

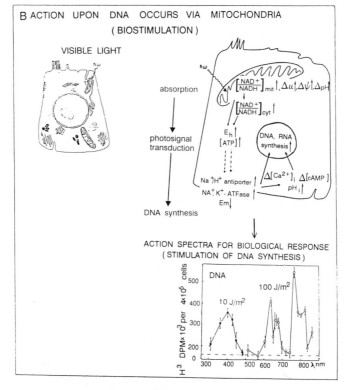

Figure 16. Two different mechanisms for exerting effects on DNA synthesis: (A) primary action after absorption of UVR by DNA forms the basis for lethal and mutagenic effects (from Coohill, 198), and (B) a mechanism whereby visible light is absorbed by mitochondrial pigments, and stimulation of DNA synthesis occurs as a final step after many dark reactions in a cell (photosignal transduction and amplification). This scheme forms a basis for low-power laser therapy (Karu, 1987, 1988, 1989).

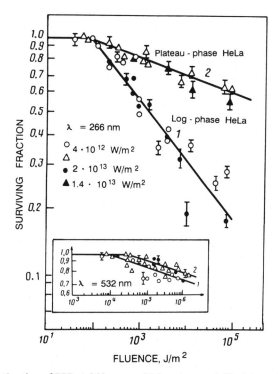

Figure 17. Lethal action of PPP at 266 nm on (1) log-phase and (2) plateau-phase HeLa cells measured by trypan blue exclusive test. The insert presents the same data for 532-nm PPP.

References

Abbaszade, I. G., 1986, Action of low doses of UVR on *E. coli,* in: *Molecular Biophysics of Cells and Cellular Processes,* Azerbaijan State University Publ., Baku, pp. 17–23 (in Russian).

Alpatov, W. W., and Nastjukova, O. K., 1933, The influence of different quantities of ultra-violet radiation on the division rate in *Paramecium, Protoplasma* 18:281–285.

Ananthaswamy, H. N., and Pierceall, W. E., 1990, Molecular mechanisms of ultraviolet radiation carcinogenesis, *Photochem. Photobiol.* 52:1119–1136.

Baserga, R. (ed.), 1971, *The Cell Cycle and Cancer,* Marcel Dekker, New York.

Baserga, R., 1979, *Multiplication and Division of Mammalian Cells,* Marcel Dekker, New York and Basel.

Belenkina, N. S., Strakhovskaya, M. G., and Fraikin, G. Ya., 1990, Activation effect of UV laser radiation on yeast growth, *Biofizika* 35:618–620.

Branch, A. D., Benenfield, B. J., Barondy, B. M., Wells, F. V., Gerin, J. L., and Robertson, H. D., 1989, An ultraviolet-sensitive RNA structural element in a viroid-like domain of the hepatitis delta virus, *Science* 243:649–652.

Cadet, J., and Vigny, P., 1990, The photochemistry of nucleic acids, in: *Bioorganic Photochemistry* (H. Morrison, ed.), Vol. 1, Wiley Interscience, New York, pp. 1–272.

Carlson, J. G., Gaulden, M. E., and Jagger, J., 1961, Mitotic effects of monochromatic ultraviolet irradiation of the nucleolus, in: *Progress in Photobiology* (B. C. Christiansen and B. Buchman, eds.), Elsevier, Amsterdam, pp. 251–253.

Chan, G. L., and Little, J. B., 1979, Responses of plateau-phase mouse embryo fibroblasts to UV light, *Int. J. Radiat. Biol.* 35:101–110.

Cohn, S. M., Krawicz, B. R., Dresler, S. L., and Lieberman, M. W., 1984, Induction of replicative DNA synthesis in quiescent human fibroblasts by DNA damaging agents, *Proc. Natl. Acad. Sci. U.S.A.* 81:4828–4832.

Coohill, T. P., 1986, Virus-cell interaction as probes for vacuum-ultraviolet radiation damage and repair, *Photochem. Photobiol.* 44:359–363.

Eilam, G., and Cabantchik, Z. I., 1977, Nucleoside transport in mammalian cell membranes: A specific inhibitory mechanism of high affinity probes, *J. Cell Physiol.* 92:185–202.

Epifanova, O. N., 1977, Metabolism of proliferating and resting cells, *Cytologia* 21:1379–1396 (in Russian).

Epifanova, O. I., Terskikh, V. V., and Polunovskii, V. A., 1983, *Resting Cells,* Nauka, Moscow (in Russian).

Giese, A. C., 1964, Studies on ultraviolet radiation action upon animal cells, in: *Photophysiology* (A. C. Giese, ed.), Vol. 2, Academic Press, New York, pp. 203–245.

Gill, R. F., and Coohill, T. P., 1987, A comparison of mammalian cell sensitivity to either 254 nm or artificial "solar" stimulated radiation, *Photochem. Photobiol.* 45:264–271.

Grdina, D. W., Hittelman, R. A., White, R. N., and Meistrich, M. L., 1977, Relevance of density, size and DNA content of tumour cells to the lung colony assay, *Br. J. Cancer* 36:659–669.

Hahn, G. M., 1975, Radiation and chemically induced potentially lethal lesions in non-cycling mammalian cells: Recovery analysis in terms of x-ray and UV-like synthesis, *Radiat. Res.* 64:533–545.

Hahn, G. M., and Little, J. B., 1972, Plateau-phase cultures of mammalian cells, *Current Topics in Radiat. Res.* 8:39–83.

Haushka, P. V., 1973, Analysis of nucleotide pools in animal cells, in: *Methods in Cell Biology* (D. M. Prescott, ed.), Academic Press, New York, pp. 361–462.

Hittelman, W. N., and Potu, N. R., 1978, Mapping G_1 phase by the structural morphology of the prematurely condensed chromosomes, *J. Cell. Physiol.* 95:333–342.

Hopwood, L. E., Dewey, W. C., and Hejny, W., 1975, Transport of thymidine during cell cycle in mitotically synchronized CHO cell, *Exp. Cell Res.* 96:425–429.

Kantor, G. J., 1986, Effects of UV, sunlight and x-ray radiation on quiescent human cells in culture, *Photochem. Photobiol.* 44:371–378.

Kantor, G. J., and Ritter, C., 1983, Sunlight-induced killing of nondividing human cells in culture, *Photochem. Photobiol.* 37:533–538.

Kantor, G. J., Sutherland, J. C., and Setlow, R. B., 1980, Action spectra for killing nondividing normal human and xeroderma pigmentosum cells. *Photochem. Photobiol.* 31:459–464.

Kaplan, H. S., and Moses, L. E., 1964, Biological complexity and radiosensitivity, *Science* 145: 21–25.

Karu, T. I., 1986, Action of pulsed UV radiation on membrane permeability of nonproliferating tumor cells HeLa, *Radiobiologiya* 26:793–797 (in Russian).

Karu, T. I., 1987, Photobiological fundamentals of low-power laser therapy, *IEEE J. Quantum Electron.* 23:1707–1717.

Karu, T. I., 1988, Molecular mechanisms of the therapeutic effect of low-intensity laser radiation, *Lasers in the Life Sciences* 2:53–74.

Karu, T. I., 1989, *Photobiology of Low-Power Laser Therapy,* Harwood Academic, London.

Karu, T. I., and Kalendo, G. S., 1987, Response of plateau-phase HeLa cells to low-intensity pulsed and continuous-wave UV light near 270 nm, Abstr. of 2nd Congress of ESP, Padova, 6–10.09.1987, Abstr. No C-98, p. 111.

Karu, T. I., Kalendo, G. S., Letokhov, V. S., Matveets, Yu. A., Semchishen, V. A., 1981, Action of ultrashort UV laser pulses upon tumor cells HeLa, *Kvantovaya Electroni.* 8:2540–2545 (Engl. translation: *Sov. J. Quantum Electron.* 11:1550–1553, 1981).

Karu, T. I., Kalendo, G. S., Letokhov, V. S., and Lobko, V. V., 1982, Response of proliferating and resting tumor HeLa cells to high repetition rate laser UV pulses, *Dokl. Akad. Nauk SSSR* 262:1498–1501 (Engl. translation: Dokl. Biophys. Febr. 1982, pp. 30–33).

Karu, T. I., Kalendo, G. S., Letokhov, V. S., and Lobko, V. V., 1983a, Different responses of proliferating and resting tumour HeLa cells to pulsed high repetition rate low-intensity laser light at 271 nm, *Laser Chemistry* 1(3):153–161.

Karu, T. I., Fedoseeva, G. E., Yudakhina, Ye. V., Kalendo, G. S., and Lobko, V. V., 1983b, Action of low-intensity high repetition rate UV laser pulses on nucleic acids synthesis rate in proliferating and resting cells, *Tsitologiya,* 25:1207–1212 (in Russian).

Karu, T. I., Kalendo, G. S., and Letokhov, V. S., 1984a, Comparison of action of ultrashort powerful UV pulses on replicative and transcriptive function of DNA in proliferating and nonproliferating HeLa cells, *Radiobiologiya,* 24:17–20 (in Russian).

Karu, T. I., Kalendo, G. S., Letokhov, V. S., and Lobko, V. V., 1984b, Action of pulsed UVR on proliferating and resting tumour cells HeLa, *Radiobiologiya* 24:273–276 (in Russian).

Karu, T. I., Kalendo, G. S., Letokhov, V. S., and Lobko, V. V., 1984c, Biostimulation of HeLa cells by low intensity visible light. III. Stimulation of nucleic acid synthesis in plateau phase cells, *Il Nuovo Cimento D* 3:319–325.

Karu, T., Pyatibrat, L., and Kalendo, G., 1987, Biostimulation of HeLa cells by low-intensity visible light. V. Stimulation of cell proliferation *in vivo* by He-Ne laser radiation. *Il Nuovo Cimento D* 9:1485–1494.

Karu, T. I., Pyatibrat, L. V., Tiphlova, O. A., and Nikogosyan, D. N., 1988, Investigation into specificity of lethal and mutagenic action of picosecond laser pulses at 532 nm, *Radiobiologiya* 28:499–502 (in Russian).

Karu, T., Kalendo, G., and Pyatibrat, L., 1990, On the role of cells attachment and spreading in biostimulation effects, *Lasers in the Life Sci.* 3:229–232.

Letokhov, V. S., 1983, *Nonlinear Laser Chemistry,* Springer-Verlag, Berlin, Chap. 8.

Little, J. B., 1969, Repair of sub-lethal and potentially lethal radiation damages in plateau phase cultures of human cells, *Nature* 224:804–806.

Mironescu, G. D., Epstein, S. M., and DiPaolo, J. A., 1980, Relationship between morphological transformation and H³-thymidine incorporation stimulated by a chemical carcinogen in postconfluent cultures of hamster embryo cell, *Cancer Res.* 40:2411–2416.

Moan, J., and Peak, M., 1989, Effect of UV radiation on cells, *Photochem. Photobiol.* 4:21–34.

Nikogosyan, D. N., 1990, Two-quantum UV photochemistry of nucleic acids: Comparison with conventional low-intensity UV photochemistry and radiation chemistry. *Int. J. Radiat. Biol.* 57:233–299.

Nikogosyan, D. N., and Letokhov, V. S., 1983, Nonlinear laser photophysics, photochemistry and photobiology of nucleic acids, *Riv. Nuovo Cimento* 6:1–74.

Painter, R. B., 1970, The action of ultraviolet light on mammalian cells, in: *Photophysiology* (A. C. Giese, ed.), Vol. 5, Academic Press, New York, pp. 169–189.

Pardee, A. B., and Palmer, L. M., 1973, Regulation of transport systems by means of controlling metabolic rates. *Proc. Soc. Exp. Biol.* 27:133–144.

Paribok, V. P., and Semyonova, F. G., 1970, Unscheduled synthesis of DNA and reparation of sublethal damages in HeLa J-63 cells, *Tsitologiya* 12:1423–1431 (in Russian).

Peak, M. J., and Peak, J. G., 1989, Solar-ultraviolet-induced damage to DNA, *Photodermatology* 6:1–15.

Peak, J. G., Woloschak, G. E., and Peak, M. J., 1991, Enhanced expression of protein kinase C gene caused by solar radiation, *Photochem. Photobiol.* 53:395–397.

Rao, P. N., and Smith, M. L., 1981, Differential response of cycling and noncycling cells to inducers of DNA synthesis and mitosis, *J. Cell Biol.* 88:649–653.

Rauth, A. M., 1970, Effects of ultraviolet light on mammalian cells in culture, in: *Current Topics in Radiation Research* (M. Ebert and A. Howard, eds.), Vol. 6, North Holland, Amsterdam and London, pp. 197–247.

Ronai, Z. A., Lambert, M. E., and Weinstein, I. B., 1990, Inducible cellular responses to ultraviolet light irradiation and other mediators of DNA damage in mammalian cells, *Cell Biol. Toxicol.* 6:105–126.

Rothman, R. H., and Setlow, R. B., 1979, An action spectrum for cell killing and pyrimidine dimer formation in Chinese hamster V-79 cells, *Photochem. Photobiol.* 29:57–61.

Samoilova, K. A., 1979, Light action on cells: Morphological, cytogenetic, physiological and biochemical aspects, in: *Photobiology of Living Cells,* Nauka, Leningrad, pp. 167–185 (in Russian).

Schlotissek, Ch., 1968, Studies on the uptake of nucleic acid precursors into cells in tissue culture, *Biochim. Biophys. Acta* 158:435–447.

Schor, S. L., Johnson, R. T., and Waldren, C. A., 1975, Changes in the organization of chromosomes during the cell cycle: Response to ultraviolet light, *J. Cell Sci.* 17:539–565.

Schulte-Frohlinde, D., Simic, M. G., and Görner, H., 1990, Laser-induced strand break formation in DNA and polynucleotids, *Photochem. Photobiol.* 52:1137–1151.

Smith, K. C., 1980, Common misconceptions about light, in: *Lasers in Photomedicine and Photobiology* (R. Pratesi and C. A. Sacchi, eds.), Springer-Verlag, Berlin, pp. 23–25.

Strom, R., Santoro, A. S., Criefo, C., Bozzi, A., Mondovi, B., and Fanelli, A. R., 1973, The biochemical mechanism of selective heat sensitivity of cancer cells. IV. Inhibition of RNA synthesis, *Eur. J. Cancer* 9:103–112.

Synsyns, B. I., and Saenko, A. S., 1986, DNA synthesis in stationary-growing HeLa culture after γ-irradiation, *Radiobiologiya* 26:800–802.

Tsukakoshi, M., Kurata, S., Nomiya, G., Ikawa, G., and Kasuya, T., 1984, A novel method of DNA transfection by laser microbeam cell surgery, *Appl. Phys. B* 35:135–140.

Valerie, K., Delers, A., Bruck, C., Thiriart, C., Rosenberg, H., Debouck, C., and Rosenberg, M., 1988, Activation of human immunodeficiency virus type 1 by DNA damage in human cells. *Nature* 333:78–81.

Valeriote, F., and van Putten, L., 1975, Proliferation dependent cytotoxicity of anticancer agents: A review, *Cancer Res.* 35:2619–2630.

Walicka, M., and Beer, J. Z., 1973, UV-light as a stimulating factor for growth of murine lymphoma L 5178Y cells *in vivo, Stud. Biophys.* 36/37:165–173.

Wechelbaum, R. R., Nove, J., and Little, J. B., 1980, Radiation response of human tumor cells *in vitro,* in: *Radiation Biology in Cancer Research* (E. Meyn and H. R. Witherom, eds.), Raven Press, New York, pp. 345–351.

Zmudska, B., and Beer, J. Z., 1990, Activation of human immunodeficiency virus by ultraviolet radiation, *Photochem. Photobiol.* 52:1153–1162.

Zubrod, G. G., 1978, Selective toxicity of anticancer drugs: Presidental address, *Cancer Res.* 38: 4374–4377.

9

Ultraviolet Radiation and Skin Cancer: Epidemiological Data from Australia

Adèle Green and Gail Williams

1. Skin Cancer: A Public Health Problem in Australia

1.1. Historical Background

The genesis of the problem of skin cancer in Australia did not occur until the first European settlement in the late eighteenth century, since Aboriginal Australians are rarely affected by skin cancer. From first settlement until the midtwentieth century, inhabitants of Australia were almost entirely of British descent, with high proportions of Irish and Scots. Their migration and settlement signified a shift from northern latitudes of 51° to 59° to much lower southern latitudes, e.g., Sydney at 34° and Brisbane at 27°, and to a land which received up to double the annual hours of sunshine of the British Isles (Scott, 1972).

Until the First World War, Australia was a primary producer, and the majority of the population worked outdoors, which entailed long periods of high sun exposure. A suntanned and weather-beaten complexion was the hallmark of the working Australian and was synonymous with toughness and manliness. By contrast, the early colonial upper class, oriented to indoor life,

Adèle Green • Epidemiology Department, Queensland Institute of Medical Research, Brisbane, 4029, Australia.   **Gail Williams** • Master of Public Health Program, University of Queensland, Brisbane, 4029, Australia.

Environmental UV Photobiology, edited by Antony R. Young *et al.* Plenum Press, New York, 1993.

valued the fair complexion of English society, and this fashion became more widespread as the middle class in urban Australia grew. However, after Australians served next to their paler European counterparts during the First World War, the legend of the "bronzed Australian" gradually became incorporated into the national image (Scott, 1972). This legend was augmented by a relaxation of standards of modesty in Western society, and by the European cult of the suntan, whereby a tanned skin now became synonymous with beauty, affluence, and leisure in sunny places, despite the universal availability of sunshine to Australians. Holidays at the beach are a national tradition year-round in this country, and holidays on the tropical islands off the eastern coast remain sought after. It is only recently, with the rapidly rising toll of skin cancer, that the consequence of the sun worship promoted by tourist brochures has become apparent.

1.2. Incidence and Mortality Patterns

The well-known gradient of skin cancer inversely associated with latitude was first described by Lancaster (1956) in relation to melanoma mortality within Australia, as well as worldwide. It is still observable in Australia overall for both incidence and mortality from melanoma (Giles et al., 1987). Incidence rates of other skin cancers also increase with decreasing latitude (Giles et al., 1989), and the magnitude of this increase closely resembles that seen in Scandinavia, particularly in the Norwegian population (see Chapter 10, this volume). Although Queensland has the highest officially reported rates of skin cancer in the world, the Caucasian residents of the Northern Territory of Australia (Saint-Yves, 1988), whose capital city, Darwin, is situated at a very low latitude, 12°S, may well have the highest rates of all. In the Northern Territory, Aborigines account for about 25% of the total population, and registration of cancer is voluntary, and thus the true incidence rates in the Caucasian population are unknown.

1.2.1. Melanoma

Based on compilation of statistics produced by the individual state and territory registries, the national incidence of cutaneous melanoma in 1982 was 18.0 per 100,000 in men and 17.6 per 100,000 in women standardized to the World Standard Population (Giles et al., 1987). The national incidence of melanoma between 1978 and 1987 has increased annually by an average of 6.5% in males and 4.4% in females (Giles et al., 1989). With respect to mortality, between 1985 and 1987 the age-standardized rates were 4.8 and 2.6 per 100,000 in males and females, respectively, and since 1930, there has been an average annual increase in melanoma mortality of 4.4% in males and 3.4% in females in Australia.

More recent age-standardized incidence rates of melanoma are available for individual Australian states for 1987, allowing direct comparison (Table 1). The highest incidence rates of invasive melanoma ever recorded were seen in the state of Queensland in 1987, namely, 48.9 per 100,000 in men and 39.7 in women (MacLennan *et al.*, 1992). Comparison with rates in the Queensland population in 1979–80 showed that incidence of invasive disease increased in the interim 7.5-year period by 116% in males, mostly those over 50 years, and by 53% in females. Preinvasive melanoma also increased by around 50% in both sexes in the 7.5-year period to 1987, suggesting that increased awareness and earlier diagnosis had accompanied the increase in incidence (MacLennan *et al.*, 1992).

Analysis of Queensland incidence data for 1987 according to detailed anatomic site, and taking relative surface areas into account, confirmed the extraordinarily high incidence of melanoma on chronically sun-exposed sites, namely, the face and ears of men and the face of women, followed by rates on the neck, shoulders, and back in men (Green *et al.*, 1993). Moreover, invasive disease increased significantly on all these sites in the 7.5 years prior to 1987. In contrast, on sites which receive least sun exposure—the buttocks of both sexes and the scalp in women—incidence was negligible.

1.2.2. Basal Cell Carcinoma (BCC)

The overall incidence of treated BCC in Australia was most recently estimated during a national polling survey carried out in 1985 in which a

**Table 1. Age-Standardized Incidence
of Melanoma in Australian States in 1987[a]**

State	Male	Female
Queensland	48.9	39.7
New South Wales[b]	27.4	23.8
Victoria[c]	19.5	19.7
South Australia	21.2	22.1
Tasmania	15.2	25.2
Western Australia	28.4	22.8
Northern Territory	13.5	4.4

Note. Rates per 100,000 (World Population).
[a] Personal communication from G. Giles, Chairman,
  Australasian Association of Cancer Registries.
[b] From McCredie *et al.* (1991).
[c] From Giles *et al.* (1991).

representative sample of 30,976 people aged 14 years or more were interviewed about skin cancers treated by a doctor in the preceding year; medical confirmation of the reported cancer was also sought (Giles *et al.*, 1988). The standardized annual incidence rates of treated BCC in Australia in 1985 were thereby estimated to be 510 and 377 per 100,000 in men and women, respectively (Table 2). The relative site distribution of BCCs (the calculated density of BCCs treated on various anatomic sites) showed a 7-fold excess on the head and neck, and a deficit on all other major sites, compared to the density on the whole body.

In addition to the national survey, field surveys of skin cancer have been conducted in several diverse Australian communities (Table 2). In southeast Queensland, a survey of skin cancer in adults aged between 20 and 69 years was conducted in 1986 in Nambour, an urban center situated at 26°S; the estimated prevalence of BCC was 4.2% (Green *et al.*, 1988). In a follow-up postal survey, information regarding all treated skin cancers occurring in survey participants in the preceding 2 years was obtained, and verified against doctors' records. Taking baseline prevalence into account, the estimated minimum annual incidence of treated BCC in this Queensland population was 1,772 per 100,000 in men and 1,610 in women, aged 20 to 69 years. In the age group 20–29 years, the incidence was 329 and 315 per 100,000 in men and women, respectively, increasing to 4,362 and 4,248 in men and women aged 60–69 (Green and Battistutta, 1990). A similar population-based survey was conducted in the Western Australian town of Geraldton, situated at 29°S, and gave estimates of incidence of BCC of 1,681 per 100,000 in men and 914 in women, in the narrower age range 40–64 years (Kricker *et al.*, 1990). In both the Queensland and Western Australian surveys, a considerable proportion of prevalent BCCs were found on the trunk, but these lesions appeared to be treated far less often than those on the head, neck, and arms. In a study between 1982 and 1986 based in the Victorian town of Maryborough, at 37°S, the average annual incidence of BCC on the head, neck, forearms, and backs of hands among persons 40–49 years was estimated to be 763 per 100,000, rising to 1,513 in the age group 60–69 years (Marks *et al.*, 1989).

The most southern state of Australia, Tasmania (40–44°S), is the only population for which cancer registry statistics on trends in the incidence of BCC are available. Over a 9-year period of observation, 1978–1986, the average increase in age-standardized incidence of routinely notified BCC was 7%: from 84 to 143 per 100,000 in men and 54 to 113 per 100,000 in women (Tasmanian Cancer Registry, 1990).

1.2.3. Squamous Cell Carcinoma (SCC)

Data for SCC (Table 2) have usually been gathered as for BCC above, and in the national survey (Giles *et al.*, 1988) the standardized rate ratio for

Table 2. Estimated Age-Standardized Incidence of Basal and Squamous Cell Carcinoma in Various Populations in Australia

Population	Latitude	Age (yrs)	BCC		SCC	
			Men	Women	Men	Women
Queensland[a]	26°S	20–69	1,772	1,610	600	298
Western Australia[b]	29°S	40–64	1,681	914	1,285	577
Victoria[c]	37°S	≥40	2,244[d]	1,069[d]	735	256
Tasmania 1986[e]	40–44°S	All	143	113	67	22
Australia 1985[f]		All	510	377	147	78

Note. Rates per 100,000 (World Population).

[a] Green and Battistutta (1990).

[b] Kricker *et al.* (1990).

[c] Marks *et al.* (1989).

[d] Sites restricted to head and neck, forearms, and dorsae of hands.

[e] Tasmanian Cancer Registry (1990).

[f] Giles *et al.* (1988).

BCC to SCC was around 4:1. The minimum annual incidence rates of treated SCC in Australia overall were estimated to be 147 and 78 per 100,000 in men and women, respectively (Giles *et al.,* 1988). Calculated relative density of SCC treated on various sites showed the largest excess on the head and neck, a small excess on the arm, and a deficit elsewhere (compared to the body as a whole).

In Nambour, Queensland, the estimated annual incidence of SCC, 1985–1987, was 600 and 298 per 100,000 among men and women, respectively, aged 20–69 years. While the overall rate ratio of BCC to SCC was 4.5:1, among men and women aged 60–69, the ratio was 1.4:1 and 3.3:1 respectively (Green and Battistutta, 1990). The estimates of incidence of medically confirmed SCC in Geraldton, Western Australia, were 1,285 per 100,000 in men and 577 in women 40–64 years, with overall rate ratios of BCC to SCC of 1.5:1 in men and 2.8:1 in women (Kricker *et al.,* 1990). The majority (84%) of all treated SCCs were on the head and neck. These rates were much higher by comparison with rates in more southern populations: in Maryborough, Victoria, the average annual incidence of SCC on the head, neck, forearms, and backs of hands among persons 60–69 years was estimated to be 815 per 100,000 (Marks *et al.,* 1989), and in the state of Tasmania the incidence of SCC in 1986 was 67 per 100,000 in men and 22 per 100,000 in women. In Tasmania the rates of notified SCC had increased by 4–5% annually since 1978 (Tasmanian Cancer Registry, 1990).

Since 1930, annual mortality in Australia from BCC and SCC combined (the majority due to SCC) has decreased by 10% in males and by 17% in females, reflecting improved diagnosis and treatment. Between 1985 and 1987, standardized mortality rates were 1.6 and 0.4 per 100,000 in men and women, respectively (Giles *et al.,* 1989), representing about 200 deaths (mostly elderly men in poor social circumstances) currently each year in Australia.

1.3. Limitations of Skin Cancer Surveillance

In general, to monitor the occurrence of skin cancer and to evaluate preventive programs requires a surveillance system. However, as well as documenting incidence rates, such a system should monitor changes in the factors influencing diagnosis of cancer if the estimated rates are to reflect the true incidence rates in the population.

Limitations in accuracy of epidemiological evaluation include selective factors such as heightened medical and public interest in skin cancer, which influences promptness of presentation, and the stage of disease when medically diagnosed (Green and MacLennan, 1989). For example, in Queensland, where public awareness of melanoma and abnormally pigmented lesions is high, melanomas are diagnosed relatively early. In 1987, 80% of incident melanomas in the Queensland population had a measured thickness of less than 1.5 mm (MacLennan *et al.*, 1992) in contrast to low-risk white populations such as in Scotland, where from 1985 to 1989, the corresponding proportion was 53% (MacKie *et al.*, 1992). Even in a brief period, features of "incident" melanomas can change dramatically. In 1977, preinvasive lentigo maligna and lentigo maligna melanoma constituted 15% of cases in Queensland, and 2.5 years later in 1979–80, this percentage rose to 23% of new cases annually (Green, 1982). By 1987, the rates of preinvasive lentigo maligna melanoma had dropped by about 50%, but the rates of lentigo maligna only had risen by over 300% (MacLennan *et al.*, 1992).

Similar changes in seeking medical treatment for other skin cancers are apparent in Australia. Patients present increasingly early with skin cancers which are small and difficult to identify, and solar keratoses are commonly brought to medical attention. In a survey of private dermatologic practice in Australia for 2 weeks in 1973 and 1974, 13% of new patients presented with solar keratoses as their main complaint (Nurse, 1975). Furthermore, because of the high prevalence of BCC and SCC in the Australia and the resultant clinical workload, many cases are not histologically diagnosed; nor are keratoses and small BCCs and SCCs distinguished when treated destructively in doctors' offices.

Variability in the diagnosis of skin cancer by clinicians also limits the validity of monitored rates, and similarly, variability in pathological diagnosis, which may be substantial for melanoma (Larsen *et al.*, 1980; Heenan *et al.*, 1984), will also affect the interpretation of incidence rates, especially if trends in definition and histologic diagnosis occur, e.g., as for "dysplastic" versus neoplastic melanocytic lesions (Swerdlow and Green, 1987).

For those skin cancers that are recorded by cancer registries, lack of a uniform registration policy makes comparisons of incidence rates difficult, nationally and internationally. Sources of systematic variation in the reporting of melanoma in Australia include registration of noninvasive (*in situ*) malignancy, coding of anatomic subsites, and reporting of multiple primary tumors (Green and MacLennan, 1989). The lack of routine histologic verification or cancer registration of BCC and SCC makes surveillance of these cancers even more difficult. In Tasmania, where the cancer registry data have been analyzed, reported trends in BCC and SCC may have been influenced by fluctuations in biopsy rates, and how representative registered cancers may be of all incident BCCs and SCCs in the state is not known.

Finally, changes in people's behavior which affect the likelihood that a medical diagnosis of skin cancer is made need to be taken into account when interpreting changes in incidence of skin cancer. For example, the underlying beliefs of the population relevant to reporting skin cancer to doctors could be monitored by national polling surveys, whether based on face-to-face or telephone interviews. These are a most effective means of canvassing opinions from representative samples of a population, and if employed to obtain attitudes about skin cancer (as distinct from obtaining disease statistics), they are not limited by accuracy of recall, or accuracy of medical records (Green and MacLennan, 1989).

2. Risk Factors for Cancer of the Skin in Australia

2.1. Melanoma

2.1.1. Genetic Factors

It is likely that the increase in risk of cutaneous melanoma due to UV exposure depends on the interaction with genetic factors which determine level of susceptibility. For a small proportion of the Australian population who have a family history of melanoma, around 10%, genetic factors may be particularly important. Linkage analyses are currently being carried out among several large Australian melanoma pedigrees in an attempt to identify the location of genes that predispose these individuals to melanoma, and early findings indicate that familial melanoma may be genetically heterogeneous. In a recent Australian study, localization of the familial melanoma locus to a region on the distal short arm of chromosome 1p, as reported previously (Bale *et al.,* 1989), was excluded during linkage studies in seven kindreds (Nancarrow *et al.,* 1992).

The epidemiological evidence of the risk factors for cutaneous melanoma in Australia comes from two population-based case–control studies carried out in Western Australia (Holman *et al.,* 1986) and in Queensland (Green *et al.,* 1986) in the early 1980s. Pigmentary characteristics were significant determinants of melanoma risk in both studies. The strongest association between melanoma and pigment phenotype was with benign melanocytic nevi (moles) on the arms: with a relative risk (RR) of melanoma of 11.3 for 10+ raised nevi on the arms in Western Australia (Holman *et al.,* 1986), and for any nevi on the arm, flat or raised, the estimated RR was 22.8 in the Queensland study (Green *et al.,* 1986), compared to persons who had no nevi counted. Melanoma was crudely associated with poor tanning ability, a tendency to sunburn, fair or red hair color, blue or green eyes, and pale innate skin color (Green *et al.,* 1986; Holman *et al.,* 1986), but in a logistic regression model where these factors were included simultaneously, only skin susceptibility to tanning and burning and hair color had independent effects (Green *et al.,* 1986). While it is believed that cases' reported sun sensitivity or phenotype may bias RR estimates upwards (Weinstock *et al.*, 1991), the strength and consistency of these associations across studies in Northern Hemisphere populations, as well as their plausibility, support their validity.

2.1.2. Environmental Factors

Sun exposure, or UV "dose," i.e., high sun exposure at the level of the dermis or melanocyte (Green *et al.*, 1986), was the principal risk factor for melanoma investigated in both the Queensland (Green *et al.*, 1986) and the Western Australian (Holman *et al.*, 1986) populations. The classification of the level of an individual's UV exposure was mostly a function of that individual's recalled sun exposure at various ages. Given the substantial but unavoidable misclassification error inherent in this method of assessing past sun exposure, in addition to contention over the approach to analysis (Armstrong, 1988), it is not surprising that findings often varied about the relation between melanoma and "sun exposure." Furthermore, the data in both studies were analyzed according to pathological subtype, despite the lack of evidence of an etiologic correlation, which may have reduced the power to identify causal associations with reported sun exposure variables.

To summarize the studies' findings, the salient indices of high sun exposure, based on recall, which were strongly associated with risk of melanoma were migration to Australia (Holman and Armstrong, 1984; Green *et al.*, 1986); experience of multiple painful sunburns (Green *et al.*, 1985); high total hours of sun exposure calculated over a lifetime (Green, 1984), or according to places of residence over a lifetime (Holman and Armstrong, 1984); and history of other skin cancer (Holman and Armstrong, 1984).

More importantly, some *objective* indicators of high-UV dose, namely, presence of solar keratoses or other skin cancers on the face (Green and O'Rourke, 1985) and presence of dermal elastosis inferred from the surface pattern of exposed skin on the back of the hand (Holman and Armstrong, 1984), were strongly and significantly associated with melanoma in the Queensland and Western Australian populations.

A range of risk factors other than solar UV were investigated in the Australian studies, but, overall, few positive associations with melanoma were found, perhaps reflecting the relative lack of power to detect true associations which may exist, e.g., with some chemical exposures (Green *et al.*, 1986). Other major factors assessed and for which no relation with melanoma was found included hormones, diet, and smoking (Green *et al.*, 1986; Holman *et al.*, 1986).

A conceptual schema that would be consistent with a dose-dependent model of carcinogenesis has been proposed (Green *et al.*, 1986). Given that UV dose received at the target cell level does not equate with UV exposure received at the skin surface, but is conditional upon the degree of epidermal transmission, a high-UV dosage to the target cell would occur with very high ambient UV incident on tanned skin, or moderate ambient exposure on very pale skin. A high cumulative UV dose to the target cell may occur at different dose-rates, e.g., continuously or intermittently. Either method will result in a high level of accumulated UV to the pigment cell, which has been shown using objective methods to be significantly related to melanoma irrespective of dose-rate. Melanocytes, normal or atypical (including genetically), may first undergo initiation mediated by susceptibility to mutation (initiation could occur spontaneously or as a result of a high dose of UV or some other agent, e.g., a chemical carcinogen), and promotion to neoplasia could occur by further action of such agents (Green et al., 1986). However, this schema may be useful only at the most general level: the complexity of the relationship between melanoma and sun exposure is still not understood, any current model notwithstanding. A new heuristic model of UV carcinogenesis for melanoma is required which incorporates not only the modulating effects of genetic susceptibility but also site-specificity of exposure and disease.

2.2. Basal Cell Carcinoma

In the Queensland population, persons with fair or red hair (Green and Battistutta, 1990), and, in Western Australia, those with freckling on the arm or with an inability to tan (Kricker et al., 1991) had a significant and at least 3-fold increase in risk of BCC compared to those who did not have these traits. In Queensland, of the reported sun exposure variables, neither number of painful sunburns nor predominantly outdoor occupational or recreational exposure were associated with BCC. However in both study populations, other objective indicators of sun-induced skin damage, namely, facial telangiectasia, elastosis of the neck, and high prevalence of solar keratoses, were strongly and significantly associated with BCC (Green and Battistutta, 1990; Kricker et al., 1991). The weight of objective evidence therefore supports a causal relationship between excessive sun exposure and the development of BCC. A possible explanation for the failure to find positive relations with variables like fair skin and reported high levels of sun exposure may lie in the choice of comparison groups, none of whom were affected by incident BCCs in the 12-month periods covered by each study, but an unknown proportion of whom would have had BCCs prior to this, given the high incidence and recurrence rates of BCC (Emmett, 1990). Moreover, those who were recently diagnosed with BCC may have systematically under- or overestimated their previous sun exposure.

The detailed site distribution of BCC remains ill understood in relation to sun exposure. For example, of a series of 1,814 incident BCCs treated by a Queensland plastic surgeon, 16% occurred on the trunk and only 11% on the limbs (Emmett, 1990); the reasons for the low frequency of BCC on the exposed upper limbs, in particular the backs of the hands and forearms, are not clear in relation to UV carcinogenesis. Furthermore, of the total 1,814 lesions in this personal series, 16% occurred on or around the nose, 14% around the eye (mostly on the inner canthus and lower lid), and 14% on the cheek, compared to 8% which occurred on the forehead and 5% on the temple. Diffey et al. (1979) have shown that the correlation between tumor density on facial subsites and UV dose is relatively poor, and it is uncertain why the characteristic clustering of BCCs occurs around the nose and eyes.

2.3. Squamous Cell Carcinoma

SCC was found to be associated with pale skin and hair color, inability to tan and propensity to freckle, and more strongly, with cutaneous signs of chronic sun exposure, including elastosis of the neck, solar keratoses (Green and Battistutta, 1990; Kricker et al., 1991), and with a reported permanent color difference between the neck and adjacent protected areas (Kricker et al., 1991). Regarding history of sun exposure, risk of SCC was raised 3-fold though not significantly in persons who reported multiple painful sunburns, while persons who had worked mainly in outdoor occupations all their lives had a significant 5-fold increase in risk of SCC (Green and Battistutta, 1990). The site distribution of SCC, with high incidence on the heavily exposed face and hands and forearms, is consistent with a direct association between SCC and excessive solar UV exposure.

3. Ultraviolet Radiation in Australia

3.1. Ultraviolet-B Distribution

Paltridge and Barton (1978) have mapped the UV-B erythemal field over Australia. The calculations were based on a clear sky parameterized model and included climatological estimates of monthly average total ozone, and total cloud cover data from 43 mainland meteorological stations. The maps confirm the high daily mean of ambient UV-B levels to which Australians are exposed (Figs. 1–3) compared to the majority of white populations (the clear sky daily totals and daily maxima, functions of maximum solar elevation, are also given). Average contour plots of erythemal UV distribution for various months (Figs. 2 and 3) show the deviations from parallelism with latitude

ANNUAL MEAN

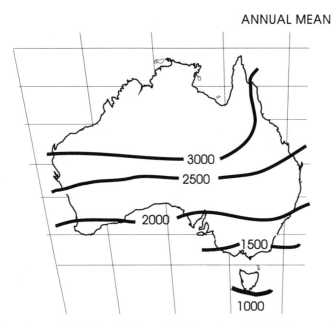

Figure 1. Annual mean distribution of daily total global erythemal UV radiation over Australia when cloud cover is taken into account. From G. W. Paltridge and I. J. Barton, 1978, *Erythemal Ultraviolet Radiation Distribution over Australia-the Calculations, Detailed Results and Input Data including Frequency Analysis of Observed Australian Cloud Cover,* Division of Atmospheric Physics Technical Paper 33, CSIRO, Melbourne, Australia. Reprinted with permission.

circles due to cloud variability, which are greatest in the Southern Hemisphere summer months, e.g., January (Fig. 2).

3.2. Monitoring UV Radiation and Ozone

At the Australian Radiation Laboratory near Melbourne, Victoria (at 38°S), solar UV radiation has been measured since 1982, and using broadband monitors since 1986 (Roy *et al.,* 1989). With expansion of the measurement program, broadband measurements are now made in Queensland (Brisbane, 27°S, and Mackay, 21°S), and in the Australian Antarctic bases of Casey, Davis, and Mawson. When comparing the annual measured UV-B profiles in the Queensland and Victorian centers, two features are notable. First, the maximum summer irradiances of tropical Mackay (2.7 W m^{-2}) and temperate Melbourne (2.3 W m^{-2}) are very similar; second, the winter irradiance in Mackay is about 50% of summer values, while the corresponding figure in Melbourne is 14% (Roy *et al.,* 1989). These features are clearly seen in the

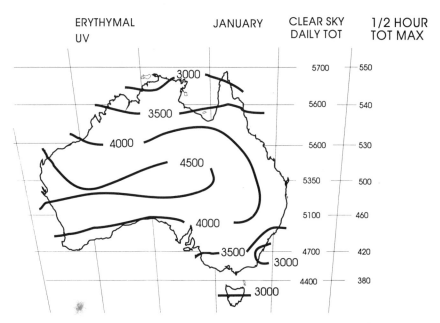

Figure 2. Distribution of daily total global erythemal UV radiation over Australia when cloud cover is taken into account, for the month of January. From G. W. Paltridge and I. J. Barton, 1978, *Erythemal Ultraviolet Radiation Distribution over Australia-the Calculations, Detailed Results and Input Data including Frequency Analysis of Observed Australian Cloud Cover,* Division of Atmospheric Physics Technical Paper 33, CSIRO, Melbourne, Australia. Reprinted with permission.

maps of calculated erythemal UV, for summer and winter months (Figs. 2 and 3). The importance of local variations in erythmal radiation flux density due to climatic conditions which can overide the latitude dependence of erythemal UV is also apparent when centers in New Zealand and tropical Queensland are compared, with the summer dose sometimes even higher in the New Zealand centers (Smith, 1990).

With respect to monitoring the effects of stratospheric ozone depletion, it would not be possible to detect the UV radiation change resulting from ozone depletion of a few percent per decade with existing equipment (Roy *et al.,* 1989). For example, the total solar UV-B level in Melbourne in December 1987, following the breakup of the ozone hole over Antarctica, was not significantly different to that in December 1988, taking cloud cover into account, despite the average ozone concentration over southern Australia in December 1987 being the lowest recorded for a decade (Roy *et al.,* 1989). Indeed, given the established international monitoring of atmospheric ozone, a case has

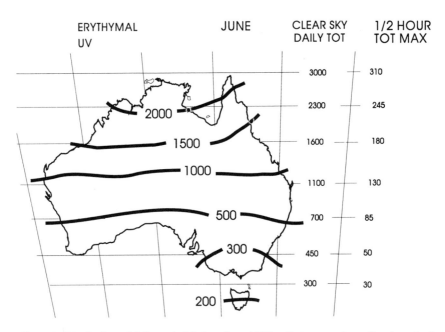

Figure 3. Distribution of daily total global erythemal UV radiation over Australia when cloud cover is taken into account, for the month of June. From G. W. Paltridge and I. J. Barton, 1978, *Erythemal Ultraviolet Radiation Distribution over Australia-the Calculations, Detailed Results and Input Data including Frequency Analysis of Observed Australian Cloud Cover,* Division of Atmospheric Physics Technical Paper 33, Melbourne, CSIRO, Australia. Reprinted with permission.

been made for a comparable global monitoring program of solar UV radiation (Basher, 1989) that meets the emerging needs of Europe and North America (Frederick *et al.,* 1991; Stolarski *et al.,* 1992) as well as the established needs of Southern Hemisphere white populations.

4. Estimation of Future Trends

4.1. Predicting Melanoma Incidence Rates

The prediction of the effect of ozone depletion upon melanoma incidence in future years is a complex problem. The components of any basic model include the rate at which the ozone layer is changing, the relationship between the thinning of the ozone column and the amount of UV radiation reaching the surface of the earth, the relationship between the amount of UV radiation at the earth's surface and the effective part of the spectrum (UV-B), the enhancement of melanoma risk associated with a given increase of UV-B, and the latency period or time lag for the increased UV radiation to exert its effect. In addition, predictions for any particular population will be affected by factors such as seasonal variations in UV-B, phenotypic distributions in the population, and sun exposure behavior. Complicating most of these issues are measurement difficulties, particularly in relation to changes in the thickness of the ozone layer and the amount of UV-B radiation being received.

4.1.1. Changes in Stratospheric Ozone

The level of stratospheric ozone at a particular position and time is a result of its rate of production by the action of solar UV radiation on molecular oxygen and its rate of destruction by reactions with substances such as nitrogen, hydrogen, chlorine, and bromine oxides. A decline in the ozone layer was first predicted in the early 1970s, specifically based on concern that supersonic air traffic would result in the increasing production of chlorofluorocarbons and halogens. The total thickness of the ozone layer was first measured from a ground station at Payerne in Switzerland in 1926; this monitoring has continued, and a network of stations, mostly in the Northern Hemisphere, has developed. Satellite measurement of total ozone began in 1978; altitude-specific measures of stratospheric ozone began in 1979 (Stolarski et al., 1992). While extraction of long-term trends from these measurements is rendered difficult because of variation with latitude, seasonality, and cloud cover as well as calibration problems, analyses of data collected since 1950 appear to leave little doubt that total stratospheric ozone is generally decreasing (Bojkov et al., 1990; Stolarski et al., 1992), although not all investigators agree (Scotto et al., 1988). The importance of this with respect to skin cancer lies in ozone's role in reducing the amount of UV-B reaching the earth. Not only is this waveband the most sensitive to fluctuations in ozone concentrations (UV-A and UV-C wavebands being virtually unaffected), but it is the most biologically effective for producing damage such as single- and double-strand breaks in DNA, production of mutations, and production of skin erythema and skin cancer (Urbach, 1989).

Of particular relevance to Australia is the repeated springtime appearance of a "hole" in the ozone layer over Antarctica and the observation based on satellite data that the maximum global decrease in ozone concentration is presently at 20°S in the upper stratosphere (an estimated decrease of 5–8% per decade) and at 25°S in the lower stratosphere (2–5% per decade) (Stolarski et al., 1992). This latitude band lies across the center of Australia.

4.1.2. UV Radiation Change

As described earlier (Section 3.2), detection of UV radiation change of the order of a few percent per decade is highly problematic. A further complication is that changes in stratospheric ozone are expected to produce disproportionately greater changes in biologically effective UV radiation; it is estimated that a 1% decrease in total ozone would increase the biologically effective UV-B radiation by 1.25–3% (Urbach, 1989). Calculations based on measured ozone levels from 1970 to 1988 indicate a statistically significant upward trend in high latitudes of Europe and North America of 2.1–3.1% per decade, depending on the action spectrum chosen.

4.1.3. Predicting Melanoma Incidence from UV-B Radiation

Just as the biological processes which underlie the relationship between melanoma and sun exposure are not yet fully understood, so the degree of enhancement of skin cancer risk associated with increased UV-B irradiation is presently unclear. This problem has been approached empirically by ecological models which calculate the gradient of skin cancer rates with latitude and, in turn, UV dosage (Elwood et al., 1974, Fears et al., 1977). More recently, regression models which include population-based covariable information on current phenotypic factors and sun exposure behavior, as well as current measured UV-B levels, have been fitted to 1978–1981 North American melanoma incidence rates (Scotto and Fears, 1987). These models show a biological amplification factor (percentage change in melanoma risk associated with a 10% increase in UV-B level) of 4% to 10%, depending on site, sex, and covariables included.

However, it is well known that ecological models must be interpreted with caution, because they are not based on within-person risk factors and outcomes. A specific additional difficulty in interpreting results of the models such as those mentioned above lies in the present uncertainty as to importance of the timing of the relevant exposure and therefore the ages at which sun exposure behavior is relevant. It has been suggested that risk depends upon the dose level of sun exposure, the timing of the dose, with childhood exposures thought to be more important, and the regularity of the dose, with some data suggesting that intermittent high exposures carry greatest risk (Elwood, 1989).

4.2. Predicting Skin Cancer Rates in Australia from Ozone Depletion

4.2.1. The Model

A carcinogenesis model has been proposed (Green et al., 1986, Section 2.1.2 above) which, while accommodating variable dose levels and skin types, focuses on accumulated UV dose to target cells. An alternative approach to predicting future trends based on this accumulated dose model is to predict the contraction in time to skin cancer onset associated with an increase in dose density.

Suppose UV-B dose/year is a constant D_0. Then after N years, an individual's accumulated UV-B exposure is ND_0. However, if UV-B exposure increases at the rate μ per year, starting from a base value of D_0, then an individual's accumulated exposure, over a period of N_1 years is

$$D_0 \frac{e^{\mu N_1} - 1}{\mu} \tag{1}$$

which means that the number of years required to produce an accumulated dose equivalent to that produced by a constant dose D_0 over N years is

$$N_1 = \frac{\log(1 + \mu N)}{\mu} \tag{2}$$

and it follows that the age-specific disease incidence curve will effectively be brought forward by

$$\Delta(age) = N - \frac{\log(1 + \mu N)}{\mu} \qquad \text{years} \tag{3}$$

Moreover, the relative increase in mean UV-B exposure over a period of N_1 years is

$$\frac{\Delta(age)}{N_1} \tag{4}$$

For $\mu = 0.01$ (a decrease of 0.5% per year in the ozone layer and assuming an optical amplification factor of 2 for UV-B at the earth's surface) and $N = 20$, the above calculations indicate that, over 20 years, the age-specific disease incidence curve will be brought forward by 1.78 years of age and the population would be subjected to an increase of 9.8% in mean UV-B exposure. For $\mu = 0.015$ [the upper limit that has been suggested for Australian latitudes (Stolarski *et al.*, 1992)], the time contraction would be 2.51 years of age and the increase in mean UV-B exposure would be 14.4%.

This formulation was used to predict the relative increase in skin cancer rates by fitting a regression model to present age-specific rates, followed by projections based on the appropriate age shift. In the calculations presented below, a power model, $\log(rate) = \alpha + \beta \log(age)$, was used for fitting age-specific rates, as this provided the best fit to available data sets. If b is the estimate of β obtained from the fitted regression, then the proportional increase in skin cancer rates over a period of N_1 years is

$$\left(1 + \frac{\Delta(age)}{age}\right)^b \tag{5}$$

or approximately

$$1 + \frac{b\Delta(age)}{age} \tag{6}$$

It follows that the relative increase in disease rate associated with a dose increasing at the rate of μ per year over N_1 years is approximately

$$\frac{b\Delta(age)}{age} \tag{7}$$

The biological amplification factor (BAF) has generally been defined as the relative increase in disease risk associated with a 10% increase in UV-B: as seen above this percentage increase occurs over an approximate 20-year period under the generally accepted assumption of a 5% decrease in the ozone layer per decade. The above formula thus provides a theoretical estimate of the BAF under the cumulative exposure model. It will be noted under this model that BAF is a function of age; younger age groups are expected to experience greater excess relative risks. Mathematically this arises because of the use of the power model, rather than a log-linear model, in describing the age-dependence of incidence rates. The power model assumes a constant relative increase in incidence for a given relative increase in age.

4.2.2. Model Fitting

Various Australian available data sets were fitted using a weighted least-squares regression (weights equal to the age-specific number of cases of skin cancer, age taken at the midpoint of available age intervals) for projections of 20 years and $\mu = 0.01$ and 0.015. For comparison, a log-rate model, $\log(rate) = \alpha + \beta age$, was also fitted, using weighted regression (weights equal to the number of persons in the age group). This model did not fit as well as the power model but yielded non-age-dependent relative increases in disease rates and provides a convenient way of obtaining a measure averaged over age. Results of all model fits are presented in Table 3.

Table 3. Relative Increase in Risk of Skin Cancer Associated with Depletion of the Ozone Layer over 20 Years for the Power and Log-Rate Models

Data source		Relative increase in risk after 20 years (%)							
		$\mu = 0.01$ (5% decrease per decade in ozone)				$\mu = 0.015$ (8% decrease per decade in ozone)			
		Power[a]			Log	Power			Log
		A	B	C		A	B	C	
Queensland 1987[b]									
Invasive melanoma	M	7	4	16	8	10	6	24	11
	F	4	2	9	5	5	3	13	7
Queensland 1985[c]									
Melanoma	M	6	4	15	8	7	5	22	11
	F	3	2	6	3	3	2	8	5
Victoria 1987[d]									
Melanoma	M	6	3	13	6	8	5	19	9
	F	4	2	9	5	5	3	13	7
Australia 1982[e]									
Melanoma	M	6	4	15	8	9	6	22	11
	F	3	2	7	4	4	3	10	6
Western Australia[f]									
BCC+SCC	M	14	12	18	16	20	17	26	23
Queensland[g]									
BCC	M+F	9	7	19		12	10	28	

Note. M, males; F, females.
[a] Entries in the body of the table for the power model are relative increases in risk (%) for: (A) 50–54, (B) 85+, and (C) 20–24 year age groups for melanoma data; relative increases in risk (%) for (A) 50–54, (B) 60–64, and (C) 40–44 year age groups for BCC+SCC data; and relative increases in risk (%) for (A) 50–59, (B) 60–69, and (C) 20–29 year age groups for BCC data.
[b] MacLennan *et al.* (1992).
[c] Queensland Cancer Registry (1990).
[d] Giles *et al.* (1991).
[e] Giles *et al.* (1987).
[f] Kricker *et al.* (1990).
[g] Green and Battistutta (1990).

Results predict that males will experience double the relative increase in risk associated with increased UV-B exposure due to ozone layer depletion. Figures for $\mu = 0.01$ over 20 years, equivalent to a 10% increase in average UV-B exposure, are consistent with the BAFs (4–10% for males, 4–11% for females) empirically obtained by Scotto and Fears (1987), although their analyses did not show a sex differential. The present findings are consistent with the approximately 100% and 50% observed increases in melanoma incidence between 1964 and 1985 for males and females, respectively, from southern to northern Australia associated with an approximately 100% increase in the magnitude of mean relative increase in annual erythymal UV (Fig. 1) (Khlat *et al.*, 1992).

The present model predicts that the influence of ozone depletion on BCC incidence is expected to be greater than that for melanoma. This arises because of the strong age-dependence of BCC incidence rates. This prediction should be treated with caution, however, as it is derived from population-based surveys with comparatively small numbers of incident cases.

In placing the predicted increase in an Australian context, it should be recalled (Section 1.2.1) that melanoma incidence between 1978 and 1987 increased by about 6.5% (males) and 4.4% (females) per year. The increase in risk attributed to the depletion of the ozone layer amounts to an annual increase of 0.5–0.6%.

References

Armstrong, B. K., 1988, Epidemiology of malignant melanoma: Intermittent or total accumulated exposure to the sun, *J. Dermatol. Surg. Oncol.* 14:835–849.

Bale, S. J., Dracopoli, N. C., Tucker, M. A., Clark, W. H., Fraser, M. C., Stanger, B. Z., Green, P., Donis-Keller, H., Housman, D. E., and Greene, M. H., 1989, Mapping the gene for hereditary cutaneous malignant melanoma-dysplastic nevus to chromosome 1p, *N. Engl. J. Med.* 320:1367–1372.

Basher, R. E., 1989, Perspectives on monitoring ozone and solar ultraviolet radiation, in: *Transactions of the Menzies Foundation* (Sir William Refshauge, ed.), Vol. 15, The Menzies Foundation, Melbourne, pp. 81–88.

Bojkov, R. D., Bishop, L., Hill, W. J., Reinsel, G. C., and Tiao, G. C., 1990, A statistical trend analysis of revised Dobson total ozone data over the Northern Hemisphere, *J. Geophys. Res.* 95:9785–9807.

Diffey, B. L., Tate, T. J., and Davis, A., 1979, Solar dosimetry of the face: The relationship of natural ultraviolet radiation exposure to basal cell carcinoma localisation, *Phys. Med. Biol.* 24:931–939.

Elwood, J. M., 1989, The epidemiology of melanoma: Its relationship to ultraviolet radiation and ozone depletion, in: *Transactions of the Menzies Foundation* (Sir William Refshauge, ed.), Vol. 15, The Menzies Foundation, Melbourne, pp. 95–107.

Elwood, J. M., Lee, J. A. H., Walter, S. D., Mo, T., and Green, A. E. S., 1974, Relationship of melanoma and other skin cancer mortality to latitude and ultraviolet radiation in the United States and Canada, *Int. J. Epidemiol.* 3:325–332.

Emmett, A. J. J., 1990, Surgical analysis and biological behaviour of 2277 basal cell carcinomas, *Aust. N. Z. J. Surg.* 60:855–863.

Fears, T. R., Scotto, J., and Schneiderman, M. A. H., 1977, Mathematical models of age and ultraviolet effects on the incidence of skin cancer among whites in the United States, *Amer. J. Epidemiol.* 105:420–427.

Frederick, J. E., Weatherhead, E. C., and Haywood, E. K., 1991, Long-term variations in ultraviolet sunlight reaching the biosphere: Calculations for the past three decades, *Photochem. Photobiol.* 54:781–788.

Giles, G. G., Armstrong, B. K., and Smith, L., (eds.), 1987, *Cancer in Australia 1982,* Anti-Cancer Council of Victoria, Melbourne.

Giles, G. G., Marks, R., and Foley, P., 1988, Incidence of non-melanocytic skin cancer treated in Australia, *Br. Med. J.* 296:13–17.

Giles, G. G., Dwyer, T., and Coates, M., 1989, Trends in skin cancer in Australia: An overview of the available data, in: *Transactions of the Menzies Foundation* (Sir William Refshauge, ed.), Vol. 15, The Menzies Foundation, Melbourne, pp. 143–147.

Giles, G. G., Farrugia, H., and Silver, B., 1991, *Victorian Cancer Registry 1987 Statistical Report,* Anti-Cancer Council of Victoria, Melbourne.

Green, A., 1982, Incidence and reporting of cutaneous melanoma in Queensland, *Aust. J. Dermatol.* 23:105–109.

Green, A., 1984, Sun exposure and the risk of melanoma, *Aust. J. Dermatol.* 25:99–102.

Green, A., and Battistutta, D., 1990, Incidence and determinants of skin cancer in a high risk Australian population, *Int. J. Cancer* 46:356–361.

Green, A., and MacLennan, R., 1989, Monitoring and surveillance of skin cancer, in: *Transactions of the Menzies Foundation* (Sir William Refshauge, ed.), Vol. 15, The Menzies Foundation, Melbourne, pp. 193–199.

Green, A., and O'Rourke, M., 1985, Cutaneous melanoma in association with other skin cancer. *J. Natl. Cancer Inst.* 74:977–80.

Green, A., Siskind, V., Bain, C., Alexander, J., 1985, Sunburn and malignant melanoma, *Br. J. Cancer* 51:393–397.

Green, A., Bain, C., MacLennan, R., and Siskind, V., 1986, Risk factors for cutaneous melanoma in Queensland, in: *Recent results in cancer research* (R. P. Gallagher, ed.), Vol. 102, Springer-Verlag, Berlin, pp. 76–97.

Green, A., Beardmore, G., Hart, V., Leslie, D., Marks, R., and Staines, D., 1988, Skin cancer in a Queensland population, *J. Am. Acad. Dermatol.* 19:1045–52.

Green, A., MacLennan, R., Youl, P., and Martin, N., 1993, Site distribution of cutaneous melanoma in Queensland, *Int. J. Cancer* 53:232–236.

Heenan, P. J., Matz, L. R., Blackwell, J. B., Kelsall, G. R. H., Singh, A., Ten Seldam, R. E. J., and Holman, C. D. J., 1984, Inter-observer variation between pathologists in the classification of cutaneous malignant melanoma in Western Australia, *Histopathology* 8:717–729.

Holman, C. D. J., and Armstrong, B. K., 1984, Relationship of solar keratoses and history of skin cancer to objective measures of actinic skin damage, *Br. J. Dermatol.* 110:129–138.

Holman, C. D. J., Armstrong, B. K., Heenan, P. J., Blackwell, J. B., Cumming, F. J., English, D. R., Holland, S., Kelsall, G. R. H., Matz, L. R., Rouse, I. L., Singh, A., Ten Seldam, R. E. J., Watt, J. D., and Xu, Z., 1986, The causes of malignant melanoma: Results from the West Australian Lions melanoma research project, in: *Recent Results in Cancer Research* (R. P. Gallagher ed.), Vol. 102, Springer-Verlag, Berlin, pp. 18–37.

Khlat, M., Vail, A., Parkin, M., and Green, A., 1992, Mortality from melanoma in migrants to Australia: Variation by age at arrival and duration of stay, *Am. J. Epidemiol.* 135:1103–13.

Kricker, A., English, D. R., Randell, P. L., Heenan, P. J., Clay, C. D., Delaney, T. A., and Armstrong, B. K., 1990, Skin cancer incidence in Geraldton, Western Australia: A survey of incidence and prevalence, *Med. J. Aust.* 152:399–407.

Kricker, A., Armstrong, B. K., English, D. R., and Heenan, P. J., 1991, Pigmentary and cutaneous risk factors for non-melanocytic skin cancer—A case-control study, *Int. J. Cancer* 48:650–662.

Lancaster, H. O., 1956, Some geographical aspects of the mortality from melanoma in Europeans, *Med. J. Aust.* 1:1082–1087.

Larsen, T. E., Little, J. H., Orell, S. R., and Prade, M., 1980, International pathologists congruence survey on quantitation of malignant melanoma, *Pathology* 12:245–253.

MacKie, R., Hunter, J. A. A., Aitchison, T. C., Hole, D., McLaren, K., Rankin, R., Blessing, K., Evans, A. T., Hutcheon, A. W., Jones, D. H., Soutar, D. S., Watson, A. C. H., Cornbleet, M. A., and Smyth, J. F., 1992, Cutaneous melanoma in Scotland, 1979–89, *Lancet* 339: 971–975.

MacLennan, R., Green, A., Martin, N., McLeod, R., 1992, Increasing incidence of cutaneous melanoma in Queensland, *J. Natl. Cancer Inst.* 84:1427–1432.

Marks, R., Jolley, D., Dorevic, A. P., and Selwood, T. S., 1989, The incidence of non-melanocytic skin cancers in an Australian population: Results of a five-year prospective study, *Med. J. Aust.* 150:475–478.

McCredie, M., Coates, M., Churches, T., and Taylor, R., 1991, Cancer incidence in New South Wales, Australia, *Eur. J. Cancer* 27:928–931.

Nancarrow, D. J., Palmer, J. M., Walters, M. K., Kerr, B. M., Hafner, G. J., Garske, L., Mcleod, G. R., and Hayward, N. K., 1992, Exclusion of the familial melanoma locus (MLM) from the PND/D1S47 and LMYC regions of chromosome arm 1p in 7 Australian pedigrees, *Genomics* 12:18–25.

Nurse, D. S., 1975, Dermatology in Australia: A survey of private practice, *Aust. J. Dermatol.* 16:127.

Paltridge, G. W., and Barton, I. J., 1978, *Erythemal Ultraviolet Radiation Distribution over Australia-the Calculations, Detailed Results and Input Data including Frequency Analysis of Observed Australian Cloud Cover,* Division of Atmospheric Physics Technical Paper 33, CSIRO, Melbourne, Australia.

Queensland Cancer Registry, 1990, *Cancer in Queensland: Incidence and Mortality 1985.* Queensland Department of Health, Brisbane.

Roy, C. R., Gies, P., and Elliott, G., 1989, The A.R.L. solar ultraviolet radiation measurement programme, in: *Transactions of the Menzies Foundation* (Sir William Refshauge, ed.), Vol. 15, The Menzies Foundation, Melbourne, pp. 71–76.

Saint-Yves, I. F. M., and Honari, M., 1988, Skin cancers in Australia's Northern Territory 1981–85, *J. Royal Soc. Health* 2:69–74.

Scott, G., 1972, Some sociologic observations on skin cancer in Australia, in: *Melanoma and Skin Cancer,* (W. H. McCarthy, ed.), pp. 1–22, Government Printer, Sydney.

Scotto, J., and Fears, T. R., 1987, The association of solar ultraviolet and skin melanoma incidence in Caucasians in the United States, *Cancer Investig.* 5:275–283.

Scotto, J., Cotton, G., Urbach, F., Berger, D., and Fears, T. R., 1988, Biologically effective ultraviolet radiation: Surface measurements in the United States, 1974 to 1985, *Science* 239:762–764.

Smith, G. J., 1990, A solar erythemal radiation monitoring programme in New Zealand and Queensland, Australia, *N. Z. Med. J.* 103:5–6.

Stolarski, R., Bojkov, R., Bishop, L., Zerefos, C., Staehelin, J., and Zawodny, J., 1992, Measured trends in stratospheric ozone, *Science* 256:342–349.

Swerdlow, A. J., and Green, A., 1987, Melanocytic naevi and melanoma: An epidemiologic perspective, *Br. J. Dermatol.* 117:137–146.

Tasmanian Cancer Registry, 1990, *Cancer in Tasmania: Incidence and Mortality 1986.* Artemis Publishing Consultants, Hobart.

Urbach, F., 1989, Potential effects of altered ultraviolet radiation on human skin cancer, *Photochem. Photobiol.* 50:507–513.

Weinstock, M. A., Colditz, G. A., Willett, W. C., Stampfer, M. J., Rosner, B., and Speizer, F. E., 1991, Recall (report) bias and reliability in the retrospective assessment of melanoma risk, *Am. J. Epidemiol.* 133:240–245.

10

Ultraviolet Radiation and Skin Cancer: Epidemiological Data from Scandinavia

Johan Moan and Arne Dahlback

1. Introduction

Skin cancer is a serious problem in the Scandinavian countries, as it is in most countries with white populations (Magnus, 1989, 1991; Muir and Nectoux, 1982). Scandinavia is located at high latitudes (>54°N) and therefore receives relatively small annual exposures of ultraviolet (UV) radiation from the sun. In spite of this, skin cancer (cutaneous malignant melanoma, squamous cell carcinoma, and basal cell carcinoma) is more frequent than any other cancer form.

Worldwide, there is a clear north-to-south gradient in the incidence rates of cutaneous malignant melanoma (CMM) and squamous cell carcinoma (SCC) (Fig. 1). As can be seen, the data for the Scandinavian countries fit well into this rough picture: the incidence rates for SCC lie at or slightly below the regression line, while those for CMM lie at or above the regression line. As will be discussed, CMM is unexpectedly frequent in Norway and Sweden.

Johan Moan • Department of Biophysics, Institute for Cancer Research, The Norwegian Radium Hospital, Montebello, N-0310 Oslo, Norway. Arne Dahlback • Norwegian Institute for Air Research, 2001 Lillestrøm, Oslo, Norway.

Environmental UV Photobiology, edited by Antony R. Young *et al.* Plenum Press, New York, 1993.

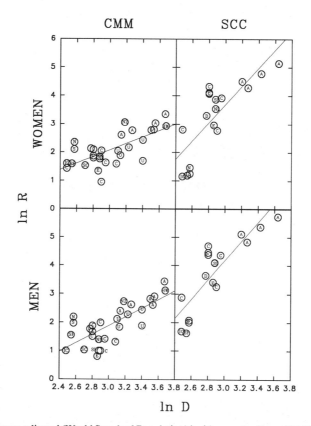

Figure 1. The age-adjusted (World Standard Population) incidence rates R, per 100,000, of CMM and SCC between 1978 and 1982 in different countries, as given by Muir *et al.* (1987), plotted as functions of ln$D$, where D is the annual exposure to carcinogenic solar radiation calculated for the CIE spectrum and a cylinder geometry as described in the text. The white populations of the following countries are included: Australia (A), Hawaii (HW), New Zealand (NZ), USA (U), Canada (C), England (E), The Netherlands (NL), Scotland (SC), Southern Ireland (SI), Iceland (IC), Denmark (D), Finland (SF), Sweden (S), and Norway (N).

In all Scandinavian countries the incidence rates of CMM are increasing rapidly with time (Fig. 2). Typical doubling times are 11–16 years. Could this increase be due to improved and earlier diagnosis? Obviously not, since the mortality rates are also increasing. In Norway the mortality rates are doubled in 17 and 20 years for women and men, respectively. The corresponding doubling times for incidence rates are 11.2 and 12.4 years. Whether early diagnosis or improved treatment contributes most to making doubling times for mortality rates longer than those for incidence rates will not be discussed here.

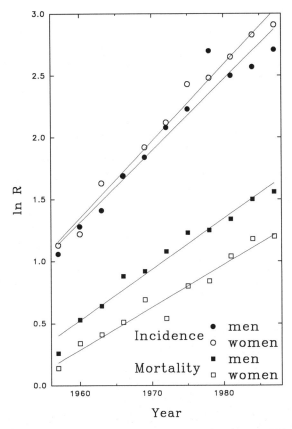

Figure 2. Trends in incidence and mortality rates (*R*) (age-adjusted to the European standard population) of CMM in Norway.

The populations of the Scandinavian countries are relatively homogeneous with respect to skin type. Lapps and immigrants with darkly pigmented skin constitute very small minorities. All CMMs and SCCs are carefully registered by national population-based cancer registries. Some epidemiological data also exist for basal cell carcinoma (BCC). The Scandinavian countries cover a long north-to-south distance; the annual exposure to carcinogenic radiation from the sun is 70% greater in the south than in the north. In view of this, Scandinavian populations are ideally suited for an evaluation of the relationship between UV exposure from the sun and skin cancer incidence, as well as for an evaluation of the impact of ozone depletion. With respect to UV fluence (and hence skin cancer incidence of a population), ozone depletion is equivalent to moving southward. We will try to evaluate the ozone-related amplification factor A for the different forms of skin cancer. A is defined as the percentage increase in skin cancer incidence resulting from the increase in UV exposure caused by a 1% ozone depletion.

It should be kept in mind that skin cancer induction is a slow process. Thus, if the ozone layer is suddenly and permanently reduced by n percent, the skin cancer incidence rate does not immediately increase by An percent but will asymptotically approach this value over several years, or perhaps decades, provided habits of sun exposure do not change.

Sun exposure behavior and styles of dress change with time and are clearly important factors related to skin cancer. Information about these trends can be obtained from cohort analyses and from the time trends of the age-specific incidence rates of skin cancer at different body sites. A rough indication of the significance of charter tours to southern countries at lower latitudes can be obtained by comparing the skin cancer incidence rates in populations for which the average time per person spent on charter tours vary.

2. Choice of Action Spectra for Calculations of UV Exposures

An action spectrum shows the wavelength dependence of the efficiency of UV radiation and light in producing a given effect, such as skin cancer. The action spectrum ϕ_c for induction of skin cancer in humans is not known. ϕ_c may even be different for the different types of skin cancer: different chromophores may be involved, and the process may start at different depths of the skin. BCC and SCC are thought to arise in the dividing cells in or at the basal layer, while melanomas originate from transformation of melanocytes located closer to the skin surface.

In the past, the DNA absorption spectrum, corrected for the transmission of light through the epidermis, was the action spectrum used (Setlow, 1974), and there still are arguments for this choice. UV-induced cyclobutyl pyrimidine dimers and (6-4)-photoadducts may result in carcinogenesis (Hart et al., 1977, Mitchell and Nairn, 1988). The action spectra for both these photoproducts are similar to the absorption spectrum of DNA (Matsunaga et al., 1991, Chan et al., 1986). Therefore, one of the action spectra used in this chapter is the same as the absorption spectrum of DNA (from Sutherland and Griffin, 1981), corrected for the transmission $T(\lambda)$ of UV radiation through the epidermis (Bruls and Van der Leun, 1982). This spectrum agrees well with the action spectrum for the formation of pyrimidine dimers in human skin (Freeman et al., 1989). Unfortunately, in the wavelength region above 310–320 nm there may be significant experimental error in these spectra, as well as in the spectra described below. The DNA spectrum used by Setlow (1974) decreases faster with increasing wavelength than does the spectrum of Sutherland and Griffin (1981). This spectral uncertainty is unfortunate, since the fluence rate of solar radiation per nanometer increases by four orders of magnitude from 290 to 320 nm, and by two orders of magnitude from 300 to 320 nm. Thus, the contribution of radiation with wavelengths larger than 310 nm to the carcinogenic exposure is significant as seen from the effectiveness spectra $E(\lambda, t) \cdot \phi_c(\lambda)$ (Moan et al., 1989b).

The second action spectrum used in our work is the CIE (Commission Internationale de l'Eclairage) erythema spectrum (Fig. 3) (McKinlay and Diffey, 1987). This spectrum is similar to an average mutation spectrum, corrected for $T(\lambda)$, as well as to the mouse carcinogenesis spectrum of Steerenborg (1987), as shown previously (Moan et al., 1989a). The mouse carcinogenesis

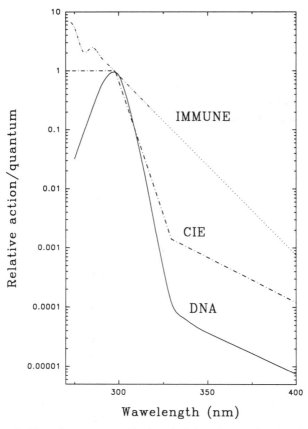

Figure 3. The action spectra used in the calculations. See text for discussion.

spectrum of Steerenborg has been slightly modified by Slaper (1987) and Kelfkens *et al.* (1990). This mouse carcinogenesis action spectrum is similar to the mouse edema action spectrum of Cole *et al.* (1986). All of these spectra weight the wavelength region above 310 nm more strongly than does the DNA spectrum.

The long wavelengths are weighted even more strongly in the third spectrum chosen: the action spectrum for immune suppression as determined by De Fabo and Noonan (1983) (Fig. 3). Beyond a doubt, immunologic processes play a role in skin carcinogenesis (Kripke and Fischer, 1976; Kripke, 1990; Romerdal *et al.*, 1988). However, the chromophore related to immunologic responses has not been agreed upon. Some investigations indicate that it is urocanic acid, which has an absorption spectrum similar to the spectrum of De Fabo and Noonan (De Fabo and Noonan, 1983; Reeve *et al.*, 1989; Reilly and De Fabo, 1991). In addition to the evidence presented in these papers, an argument for this spectrum is the fact that UV-B-blocking sunscreens, in some cases, seem not to protect against immunosuppression (Reeve *et al.*, 1991). Other investigations are in favor of an action spectrum resembling the absorption spectrum of DNA because of the fact that, at least in some systems, UV-induced immunosuppression can be reversed by photoreactivation (Applegate *et al.*, 1989). Since the three chosen action spectra differ significantly in the UV-A region, they will serve to evaluate how critically dependent the amplification factor is on the choice of action spectrum.

3. Ozone Data and Calculations of Doses of Carcinogenic UV Radiation

3.1. Ozone Measurements

The amount of ozone in the atmosphere over Oslo has been measured with a Dobson instrument at the Institute of Physics, University of Oslo. This instrument has a double monochromator and measures the radiance at a selected wavelength pair consisting of a short and a long wavelength, with strong and weak radiative absorption by ozone, respectively. The Dobson instrument in Oslo was modified and calibrated in 1977 in Boulder, Colorado, at the National Oceanic and Atmospheric Administration (NOAA) laboratory and was in the intercomparison campaign there.

The basic measurement is of the direct sun. In cloudy weather one has to measure the scattered light from the zenith sky, and empirically determined corrections have to be made for the effect of clouds. When the sun is too low, as it is during the period from 20 November to 20 January, the sun's UV radiation is too weak to be measured. During this period, light from the moon and from the zenith sky is used.

The total amount of ozone in a vertical column is usually expressed as the thickness in millicentimeters of an ozone layer converted to normal pressure and temperature. This is the Dobson unit (DU). Normally, the ozone amount over Oslo varies from about 300 DU in September and October to about 430 DU in March and April. The Dobson method makes use of a number of assumptions:

1. The major part of the ozone is assumed to be in the lower stratosphere, between 15 and 30 km above the earth's surface.
2. For the selected wavelength pairs, the relative radiance, or intensity ratio, at the top of the atmosphere is assumed not to change with time.
3. Atmospheric absorption of UV is assumed to be caused only by ozone.

Any cloud cover will introduce uncertainty, and even a clear sky observation will give uncertain values because the evaluation is sensitive to a change in the ozone's vertical distribution. Cloud cover will have an influence on the transport processes and may magnify real fluctuations of the ozone amount during the winter season. The maximum error of a single observation is about 10% to 15%.

The use of a mean value for the period of December through March, and its deviation from a long-term mean, is a reasonable method with which to follow the real change in ozone amount from year to year. This method has also been used by Bojkov in his work on ozone in the northern polar region (Bojkov, 1988). Our results are in agreement with Bojkov's data, as shown elsewhere (Larsen and Henriksen, 1990).

3.2. UV Calculations

The fluence rate of carcinogenically effective solar radiation is defined by the expression $E_c(t) = \int E(\lambda, t)\phi_c(\lambda)d\lambda$, the integration being performed over the wavelength region of the solar spectrum. $E(\lambda, t)$ is solar irradiance at the earth's surface, $\phi_c(\lambda)$ is the action spectrum for carcinogenesis, and t is time.

$E(\lambda, t)$ was determined by using a discrete-ordinates algorithm to calculate the propagation of light in vertically inhomogeneous, plane, parallel media (Stamnes et al., 1988). The model atmosphere used was the U.S. Standard Atmosphere 1976, which was divided into 39 homogeneous layers, each with a thickness of 2 km. We used the extraterrestrial solar radiation spectra, as well as all orders of scattered light (Rayleigh scattering) from the atmosphere.

The ground albedo (i.e., the ratio of the upward light flux to the downward light flux; Chandrasekhar, 1960) was set equal to 0.2, which is close to the climatological mean value for continental vegetation (Kondratyev, 1969). This value may be too high for UV radiation. However, the calculations of the relative UV exposures are insensitive to the choice of albedo. The absorption spectrum of ozone was taken from the publication *Atmospheric Ozone 1985* (World Meteorological Organization, 1985). The annual exposure to carcinogenic radiation from the sun is $D = \int E_c(t)dt$, the integral being taken over 1 year. In our calculations, the integrals were approximated by the sums

$$D = \Sigma\Sigma E(\lambda t)\phi_c(\lambda)\Delta\lambda\Delta t$$

with $\Delta t = 1$ h and $\Delta\lambda = 1$ nm. The seasonal average ozone levels at different latitudes were used.

In most calculations, the geometric shape of the human body was approximated by a cylinder with its axis oriented vertically, excluding its top and bottom (Dahlback and Moan, 1990). Calculations were also made for a horizontally oriented plane and for a sphere. Mean summer and winter values for the ozone level from measurements at 12 stations north of 59°N for the period 1957–1986 (Bojkov, 1988) were used in the calculations. For the period 1987–1991, the ozone data were obtained from the University of Oslo. No corrections were made for different sky covers in different regions. Such corrections are not likely to change the results drastically (Moan *et al.*, 1989a). Values for annual UV exposures, measured by means of a Robertson–Berger sunburn meter, and global irradiance and ozone in Norrköping, Sweden, were obtained from Dr. W. Josefsson at the Swedish Meteorological and Hydrological Institute (SMHI).

The effectiveness spectra as well as the annual UV doses at different latitudes, corresponding to the three chosen action spectra, are shown in Fig. 4.

4. Epidemiological Data

Data for incidence rates of skin cancers in Norway were provided by the Norwegian Cancer Registry. All incidence rates were age-adjusted to the European standard population (Hill, 1971).

Norway was divided into six regions: North (mean latitude 69.5°), Central (mean latitude 64°), West (mean latitude 61°), South/East (mean latitude 59.5°), Oslo (mean latitude 60°), and South (mean latitude 59°). Oslo was treated separately, since people living in Oslo spend significantly more time

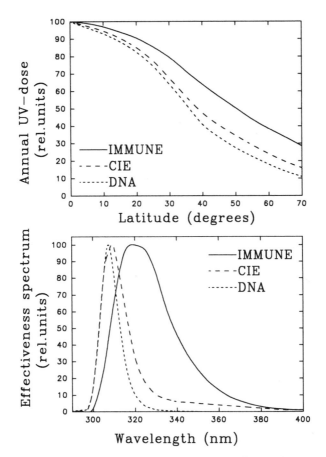

Figure 4. Above: The annual doses of carcinogenic radiation at different latitudes calculated by use of the action spectra shown in Fig. 3. A cylindrical representation of the exposed surface was used. The annual variation of the ozone layer at the different latitudes was taken into account. Below: Effectiveness spectra calculated for the three action spectra shown in Fig. 3. An ozone level of 350 DU and a solar zenith angle of 50° was used.

on vacations in Mediterranean and other southern countries than people in the other regions (see below) and since Oslo is the most urbanized region of Norway. All other cities in Norway are less densely populated than Oslo, and none of them dominate in population size in their regions.

All cases of SCC, CMM, and LMM are reported by the pathological laboratories to the Norwegian Cancer Registry, where coding and classification take place. Further details about the registration procedures can be found elsewhere (Jensen *et al.*, 1988).

Since not all BCC lesions are treated in hospitals, there may have been underreporting of this cancer form. To our knowledge there are no regional differences in the reporting rate of BCC, although this cannot be ruled out.

Practically all inhabitants in Norway are Caucasian, and we have no reason to believe that there is any difference between regions with respect to the distribution of persons with different skin types. In the present work we use the following definition of relative tumor density:

$$RTD_s = \frac{\text{annual age-adjusted incidence rate at body site } s}{\text{fraction } f \text{ of the skin surface occupied by } s} = \frac{R_s}{f}$$

Others have defined RTD slightly differently (Pearl and Scott, 1986). Their RTD values should be multiplied by a constant (R_t = the total incidence rate) to obtain our values. The fraction f of the human body occupied by different sites (head, neck, upper extremities, lower extremities, trunk, etc.) is from the work of Lund and Browder (1944).

5. Time Trends

5.1. Annual Exposures to Carcinogenic UV Radiation

Figure 5 shows the annual exposure D to carcinogenic radiation calculated for southern Norway, using known ozone values, the CIE action spectrum,

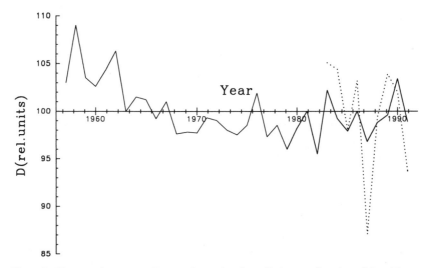

Figure 5. The annual exposures D to carcinogenic solar radiation as a function of time. Known ozone values for southern Norway, which agree with the values given by Bojkov (1988), were used. The dotted curve shows the UV exposures measured by a Robertson–Berger meter in Norrköping by W. Josefsson (see Josefsson, 1989).

and a cylinder representation for the exposed surface. The exposures measured by a Robertson–Berger meter in Norrköping are shown for comparison. The annual fluctuations around the mean value are typically around 5% for the calculated values, and somewhat larger, 10–15%, for the measured values. The variation in cloud cover during the summer months is the main cause of the latter variation. This is clearly indicated by Fig. 6, which shows that the ratio of daily doses of global radiation to the Robertson–Berger doses gives a curve which closely parallels the curve for the ozone variations. Thus, whenever there are data for global radiation and ozone level, it is possible to estimate the doses of carcinogenic UV radiation from the sun with significant accuracy.

As clearly shown by Fig. 5 there is no significant increasing trend of the annual doses of carcinogenic radiation over a long period of time. Therefore, in agreement with our earlier work (Moan *et al.*, 1990), we conclude that the increasing trend of skin cancer incidence rates (see below) has explanations other than a decreasing ozone level.

5.2. Trends in Skin Cancer Incidence Rates

The incidence rates of the most common cancers in Norway show an increasing trend, as shown by Fig. 7. BCC is the most common cancer. BCC, SCC, and CMM constitute 30–40% of all cancers. CMM has the steepest increasing trend of all cancers, with a doubling time of 11 to 12 years (Figs. 2 and 7). Similar trends for CMM are found in the other Nordic countries (Fig. 8). If these trends continue, CMM in Norwegian men will be more common than cancer of the prostate by the year 2020, and CMM in Norwegian women will be more common than breast cancer by the year 2010.

The increasing trend of skin cancers is often attributed to changing sun exposure behavior and clothing styles that lead to increased total exposures. In view of the increasing trends of several other cancers, which are certainly not induced by sun exposure, one should be careful when making such a general statement. Even though sun exposure is the dominating risk factor for skin cancers, as we will see below, other factors may contribute significantly to increased risk. The increasing trend might be related to a changing pattern of sun exposure toward more or less brief and intense exposure periods. A change from outdoor to indoor occupations, as well as an increase in leisure time to enjoy recreational sunbathing at home or abroad would certainly change the exposure pattern drastically.

The significance of changes in clothing styles to skin cancer incidence is demonstrated by the data shown in Fig. 9. The rates of increase of CMM in women in Norway vary between body sites. The slowest increase is found for CMM occurring on the scalp, face, and neck, followed by that for trunk and the lower extremities. The fastest increase is found for the female breast. This

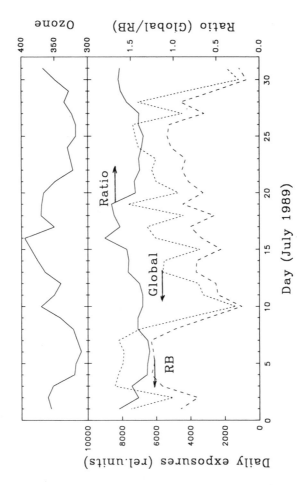

Figure 6. Daily R–B doses, doses of global radiation (below) and ozone values (above) for Norrköping, July 1989, provided by Dr. W. Josefsson (personal communication). The ratio of global to R–B doses is shown in the lower panel.

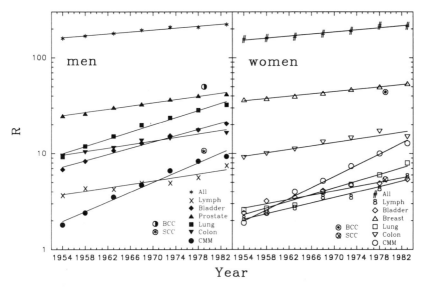

Figure 7. Trends in incidence rates (age-adjusted to the European standard population) of the major cancer types in Norway. The average rates for BCC and SCC for 1976 to 1985 are also given. Data from The Norwegian Cancer Registry.

increase is almost certainly related to topless swimming suits, which made their appearance in the sixties. The data shown in Fig. 9 strongly indicate that sun exposure is the main inducing agent for CMM: The skin of the female breast is supposedly not significantly different from the skin of the rest of the trunk. Still, prior to 1980, CMM arose 10 times less frequently per unit skin area of the breast than per unit skin area of the rest of the trunk.

There is a significant increase in the incidence rates of CMM from north to south in Norway (Fig. 10). However, the time trend of incidence rates is similar for all regions of the country, with a possible exception of the county furthest to the north (Finnmark). This similarity in time trend is an argument for the homogeneity of the population with respect to sun exposure behavior and clothing styles, as well as to exposure to promoting agents.

The incidence rates of BCC and SCC are frequently assumed to be related to accumulated sun exposure during the lifetime (Elwood *et al.,* 1989). One should then expect the time trends of the incidence rates of these two cancers to be similar. The data from Finland conflict with this (Fig. 11): The incidence rate of SCC is increasing much more slowly than that of BCC. This may, of course, be due to incomplete and variable registration of SCC, as well as to false early diagnosis of BCC (L. Teppo, personal communication), but it may also indicate that time trends are due to varying exposure to promoting factors,

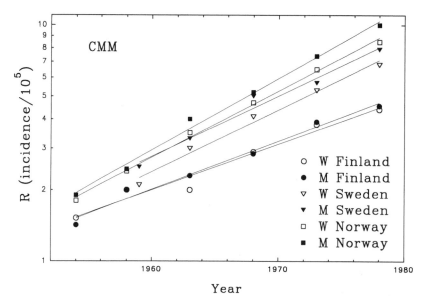

Figure 8. Trends in the incidence rates of CMM for women (W) and men (M) in Norway, Sweden, and Finland. Data from Hakulinen *et al.* (1986).

and that there are different promoting factors for BCC and SCC. In this context it is relevant to note that the Nordic profile of skin cancer is characterized by a high and variable ratio of BCC to SCC incidence. Magnus (1991) reported this ratio to be 6 for Denmark, 8 for Finland, and 9 for Norway, while a ratio of about 4 has been reported for some other countries (Scotto *et al.,* 1983; Giles *et al.,* 1988). This may indicate that BCC, being the most common and least feared skin cancer form, is relatively completely reported in the Nordic countries. It may, however, indicate that different associated factors to sun exposure are involved in the induction of BCC and SCC and that the level of such factors vary between populations. For instance, it is known that burn scars and skin tuberculosis, as well as arsenic exposure, are risk factors for SCC.

The Norwegian BCC to SCC ratio, which is not dependent on latitude (Moan *et al.,* 1989a), is 4.4 ± 0.7 for men and 7.3 ± 1.5 for women (means, weighting the different regions equally). The rate of increase of SCC incidence in Norway is similar for all age groups (Fig. 12). (The aberrant trend of the age group around 35 years should be noted but is probably not significant since the number of cases is small.) The time trends for incidence rates are similar for SCC arising at various places on the body (Fig. 13). Only data for

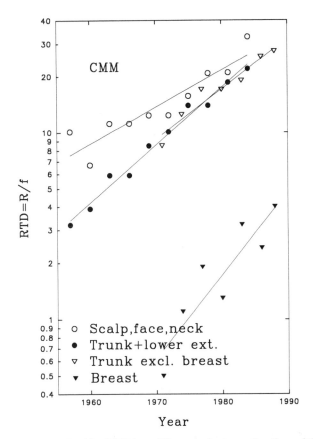

Figure 9. Relative tumor densities (RTDs) on different body sites as functions of time. RTD = R/f, where R = annual age-adjusted rate (European standard population) per 100 000, and f = fraction of the body surface occupied by the actual sites (Lund and Browder, 1944). The fraction occupied by the female breasts was set at 0.05. Data from The Norwegian Cancer Registry.

women are shown, since those for men are similar. Thus, the pattern of the sun exposure inducing SCC seems not to have changed drastically. Since most SCCs are found on the face and neck, this is reasonable.

Some investigators have reported that a wave of high incidence of CMM tends to follow some years after each sunspot maximum (Wigle, 1978, Houghton *et al.*, 1978). Even though the fluence rate of ozone-producing UV radiation of wavelengths shorter than 230 nm is 3–4% larger at sunspot maximum than at sunspot minimum (Lean, 1987), the ozone level fluctuates by hardly more than 1–2% (Bojkov, 1988). This will cause a 1–2% fluctuation

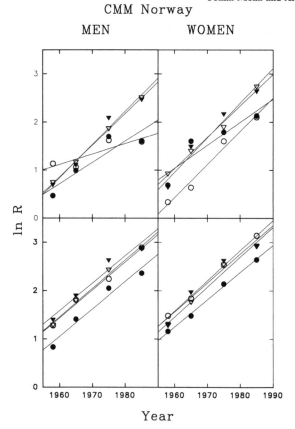

Figure 10. Trends in the incidence rates of CMM in different regions of Norway for men and women. Upper panel: ○ Finnmark, ● Nordland, ▽ Møre, Trøndelag, ▼ western Norway. Lower panel: ▼ Oslo, ▽ southeastern Norway, ● eastern Norway, ○ southern Norway.

of UV fluence, which is much less than annual fluctuations caused by meteorological events (see Fig. 5). Thus, we are forced to conclude that the reported relationship between sunspot activity and incidence of CMM is a matter of arbitrary coincidence.

6. Relative Tumor Densities at Various Body Sites

Relative tumor density, RTD, is here defined as the annual age-adjusted (European standard population) incidence rate (per 100,000) of skin cancer at a given body site divided by the fraction of the total skin area occupied by that site.

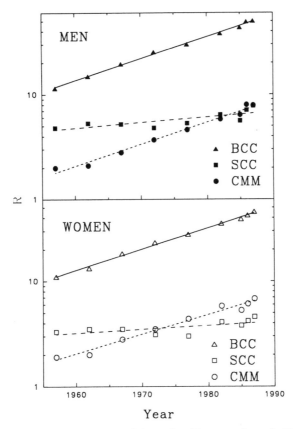

Figure 11. Trends in the incidence rates of the major skin cancer forms in Finland (Cancer Society of Finland, 1991).

If one assumes that incidence of SCC as well as BCC is related to accumulated UV exposure, one should expect to find RTD(SCC) = constant · RTD(BCC) for different body sites receiving different doses of UV. As can be seen from Fig. 14, this is not generally true. RTD(BCC) at the trunk is unexpectedly large as compared with RTD(SCC). On the other hand, RTD(BCC) at the upper extremities is smaller than expected in relation to the value of RTD(SCC) at the same site. The sex differences are more or less as expected for RTD of both SCC and BCC at the lower extremities and at the trunk. At the upper extremities, RTD(SCC) is surprisingly high for men. The large variance between the RTD values at the ears for men and women is, presumably, due to differences in hair fashion. However, one should note that BCC is more frequent than SCC at the ears of women, while the opposite

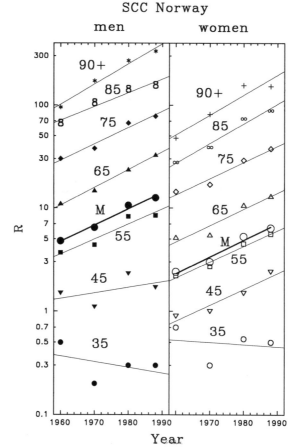

Figure 12. Time trends in the age-specific and age-adjusted (large symbols, M) incidence rates of SCC in Norway. Data from The Norwegian Cancer Registry.

is true for men (Fig. 14). Similar considerations are valid for the RTDs at the upper extremities.

On the body as a whole, BCC is almost as frequent in women as in men, while SCC is twice as frequent in men as in women. Thus, it is quite clear that factors related to sex (i.e., factors other than hair fashion and clothing styles) are important in skin carcinogenesis. Also, it is not generally true that the incidence rates of both BCC and SCC are related to total sun exposure. A close inspection of Fig. 14 suggests that the "total dose" hypothesis may be correct for SCC but not for BCC, notably since the upper extremities obviously get higher exposures than the trunk.

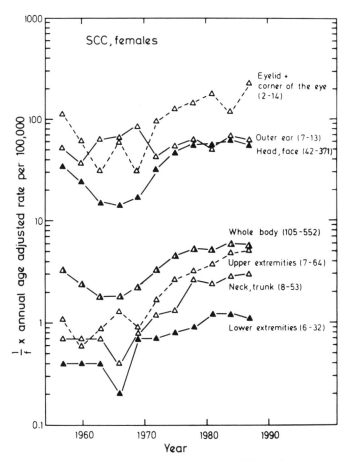

Figure 13. Time trends in the age-adjusted incidence rates of SCC in females at various body sites occupying a fraction *f* of the skin. The values of *f* are taken from Lund and Browder (1944). Data from The Norwegian Cancer Registry.

The RTD of both BCC and SCC is higher at the face than at the lower extremities or the trunk. This is also the case for CMM for people born before 1910 (Fig. 15). For men born later, CMM RTD is in some cases greater on the back than on the face and neck, as shown by a ratio larger than unity in Fig. 15. For women born after 1910, CMM RTD is in most cases greater on the lower extremities than on the face and neck. Combined with the data shown in Fig. 14, these data present a rather strong indication that the incidence of CMM is not directly related to total sun exposure.

Lentigo maligna melanoma has an RTD pattern more similar to non-melanoma skin cancers than to CMM.

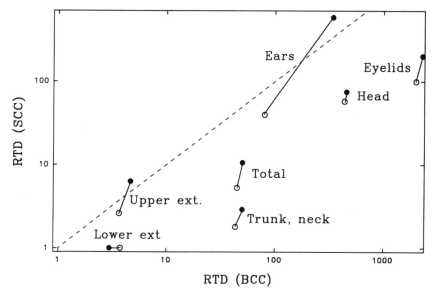

Figure 14. Relative tumor densities (RTDs) of SCC at different body sites as a function of the RTDs of BCC at the same body sites. Data for 1960–1990 from The Norwegian Cancer Registry. Data for men (●) and women (○).

7. Age-Specific Incidence Rates

When age-specific incidence rates are considered, the data for CMM differ qualitatively from those for BCC and SCC. The incidence rates of the latter skin cancers increase more or less uniformly with age (Magnus, 1991). This is true for different body sites such as head and trunk, on which RTDs vary widely (data not shown). For the head and neck, the age-specific incidence rate of CMM increases with age (Fig. 16), in agreement with the data for SCC and BCC. For the trunk, however, the age-specific incidence rate is highest at age 50–60 years (somewhat earlier for women than for men; data not shown), after which it falls drastically (Fig. 16). The fact that this trend holds for all regions of Norway indicates that the population behaves homogeneously with respect to changing habits of sun exposure. Similar data have been published for Denmark (Østerlind et al., 1988). This pattern agrees with the strong birth cohort effect observed (Magnus, 1981) and indicates that exposure to melanoma-inducing UV radiation has taken place early in life, and that the exposure has increased for successive young generations (Magnus, 1981, Østerlind et al., 1988).

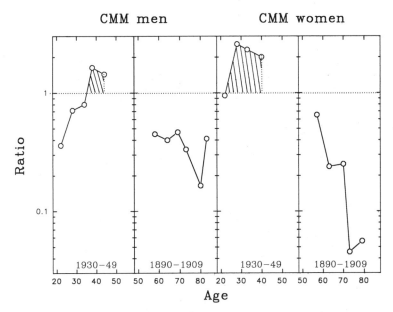

Figure 15. The age dependence of the ratio of RTD(CMM) on the trunk of men (on the lower extremities of women) to RTD(CMM) on the head/neck. Age-specific rates for cohorts born between 1890 and 1909 and 1930 and 1949 (Magnus, 1981) were used in the calculations.

8. North–South Gradients

In Norway there is a clear north–south gradient for the incidence of the three major skin cancer forms (Fig. 17). For BCC and SCC our data are in excellent agreement with those from Australia. The biological amplification factor, which is identical with the slope of the curves shown in Fig. 17, is very similar for the two countries, in both the case of BCC and SCC.

In the case of CMM, however, the north–south gradient is much steeper for Norway than for Australia. Furthermore, the incidence rate of CMM is higher in Norway than what one might expect in view of the Australian data. The incidence rates of CMM for the other Nordic countries are also unexpectedly high, although lower than those for Norway (Fig. 1).

From these data we may conclude that the relationship between sun exposure and skin cancer incidence is not the same for CMM as it is for BCC and SCC. The exposure pattern that leads to CMM is obviously different in the Nordic countries from that in Australia and several other countries. A tentative explanation might be that intermittent, intense sun exposure, for instance during recreation and vacation, is much more common in Scandi-

1976−85

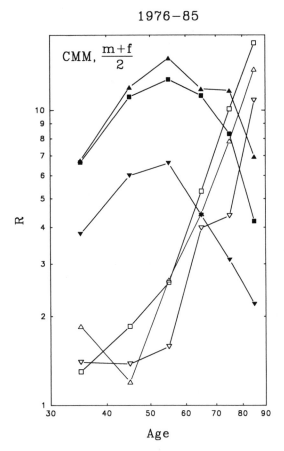

Figure 16. Age-specific incidence rates of CMM (averaged for men and women) at the trunk (closed symbols) and head/neck (open symbols) as a function of age. Data for 1976 to 1985 for different regions of Norway: (▲, △) Oslo; (■, □) South/East; (▼, ▽) North. From The Norwegian Cancer Registry.

navian countries than in Australia. If so, our data support the intermittent exposure hypothesis for CMM.

As can be seen from Fig. 18, the amplification factors for BCC and SCC are within error limits the same for heavily sun-exposed skin areas, such as those on the head, as for less exposed areas with RTD values more than 50 times lower. This indicates that a power relationship (RTD = const. $\cdot D^{A_b}$, where D = exposure, A_b = biological amplification factor) is a better description of our data than an exponential relationship (RTD = const. $\cdot e^{KD}$, where D = exposure, K = constant). The latter relationship would give a biological am-

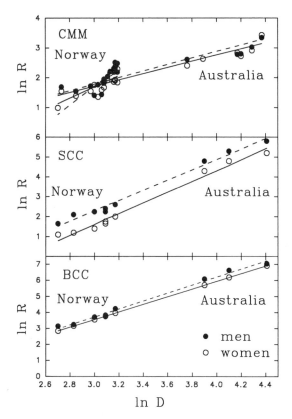

Figure 17. The dependence of the age-adjusted incidence rates (R) of CMM, SCC, and BCC for Norway (1976–1985) and Australia (1985, SCC, BCC; 1978–1982, CMM) on the calculated annual exposure to carcinogenic UV from the sun (see Fig. 4). The data for Australia are from Giles *et al.* (1988) and Muir *et al.* (1987).

plification factor that varied with D, while our data indicate that A_b is constant. Using the power law, the A_b values, and the RTD values from Fig. 18, it can be estimated that (within the Norwegian population) people expose the skin areas of their head to a UV dose that is more than 5 times larger than the average dose received by body sites other than the head, neck, and trunk. By using personal UV dosimeters of polysulfonate we will test the validity of these estimations.

Data for Sweden and Finland are in agreement with the data for Norway. Figure 19, which compares the three countries, indicates that for the period 1970–1979, A_b values for CMM were slightly larger for women than for men. Similar data have been published for the USA (Scotto and Fears, 1987). A

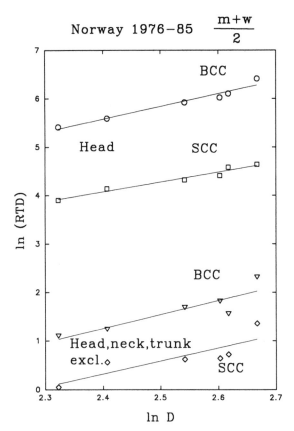

Figure 18. The dependence of BCC and SCC RTD (head, skin areas excluding head, neck, and trunk) on annual exposure D (Fig. 4). Data for 1976 through 1985 for men and women averaged. Incidence data from The Norwegian Cancer Registry.

possible explanation might be that a warmer and slightly sunnier climate in the south encourages vacation-related intermittent sun exposure, and that this activity may be indulged in more by women than by men (Moan and Dahlback, 1992). However, after 1979, there is no difference between the A_b values for men and women in Norway (data not shown).

The differences between urban and rural populations are often a focus in skin cancer epidemiology. In the case of SCC, there is no significant urban/rural difference in Norway, neither of the A_b values nor of the absolute incidence rates. CMM, however, is 30–40% more prevalent in urban than in rural regions, as can be estimated from the regression lines in Fig. 20.

A_b values are not significantly different for urban and rural populations (Fig. 20). Furthermore, these values are similar for different time periods,

CMM

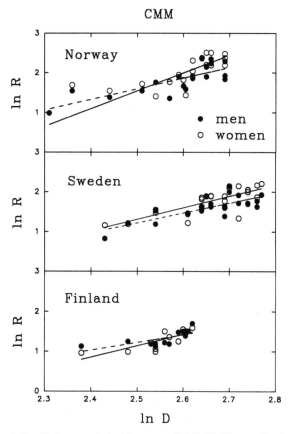

Figure 19. The relationship between the incidence rate of CMM (in Norway, Sweden, and Finland) and annual UV exposure (see Fig. 4). Data from Jensen *et al.* (1988).

although there seems to be a slight (insignificant) increasing trend of the A_b values over time, notably for men (Figure 20), a trend, that contradicts the Finnish data. Teppo *et al.* (1978) found a north–south gradient for rural regions but not always for urban areas. The authors concluded that when the overall rates were adjusted for the urban/rural ratio of the population, the north–south gradient almost disappeared for the periods 1953–1959 and 1961–1970.

A closer inspection of the work of Teppo *et al.* (1978) indicates that the scatter of the data makes the error limits too large to permit any conclusion. In any case, the north–south gradient of CMM incidence in Norway cannot be explained by more urbanization in the south than in the north. The male/

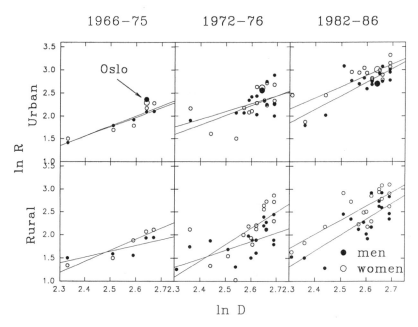

Figure 20. The relationship between the incidence rate of CMM in urban and rural districts of Norway and the exposure D to carcinogenic radiation from the sun (see Fig. 4). Incidence data for three time intervals were provided by The Norwegian Cancer Registry.

female ratio of CMM incidence is slightly higher for urban than for rural regions, but it is not dependent on the latitude (analysis not shown).

In the southern part of Norway, the urban/rural ratio of CMM incidence on head and neck is close to or lower than unity, while the ratios for other and less heavily sun-exposed sites are significantly higher. This is in agreement with the work of Eklund and Malec (1978) for Sweden, and supports the intermittent exposure hypothesis if we assume that vacation-related sun exposure is more common among the urban than the rural population.

For CMM in women in Norway, the A_b values are within the error limits similar for head/neck, trunk, and lower limbs (excluding feet) (Fig. 21). For CMM in men, however, the A_b value for the trunk is significantly larger than that for the head and neck. A_b values for CMM incidence on the feet, which certainly receive very small UV exposures, are small for both sexes (0.5 ± 0.6 and 0.9 ± 0.9 for men and women, respectively), indicating that at this site inducing factors other than UV radiation are significant.

A_b values for CMM seem to be almost similar for the different age groups (data not shown). However, after 1960 the older age groups (>60 years) have slightly lower A_b values than the younger age groups. Thus, the north–south

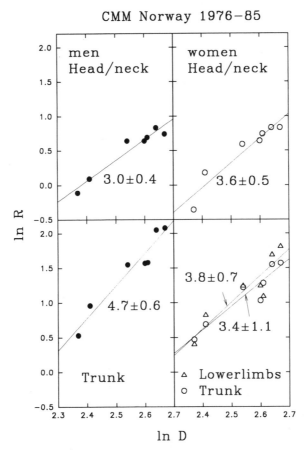

Figure 21. The relationship between the incidence rate of CMM at different body sites and the exposure D (see Fig. 4).

gradient increases slightly with decreasing age, which is in agreement with the data for the USA (Lee, 1989), and which is as one might expect, since the younger generations probably enjoy vacation-related sun exposure more than older generations, and the warmer climate in the south may stimulate such an activity.

The fact that the male/female ratio for CMM incidence in the USA decreases with decreasing latitude has been used to cast doubt on the role of solar UV radiation in the etiology of CMM (Baker-Blocker, 1980). In Norway this ratio is not dependent on the latitude, so this argument is not supported.

9. Vacations to Southern Latitudes

Klepp and Magnus (1979) found that the proportion of melanoma patients that had been in southern Europe for sunbathing during a 5-year period was 19%, while the corresponding figure among matched controls was 9%. The relative risk of CMM for those going on such vacations was 2.5, which was found to be on the border of statistical significance at the 5% level.

However, for the population of Norway as a whole it seems that charter tours to southern countries do not play a major role in the etiology of CMM. Thus, according to a report from The Norwegian Transport Economical Institute (1988), people in Oslo spent on average more than 50% more time on vacations to southern latitudes than people in the surrounding areas (Moan and Dahlback, 1992). Figure 20 shows that the incidence of CMM in Oslo is not alarmingly high, especially considering the fact that Oslo is the most urbanized region of Norway.

10. Amplification Factors

10.1. The Radiation Amplification Factor A_r

A_r is defined as the ratio of the increment in annual fluence of carcinogenic UV radiation to the decrement in ozone. The dependence of A_r on latitude, action spectrum, and geometric shape of the exposed area is shown in Fig. 22. Neither the latitude dependence nor the dependence on geometric shape of the exposed area is large. However, A_r varies widely for the three different action spectra, ranging from about 0.4 for the "immune" spectrum to about 1.7 for the DNA spectrum. This is mainly due to differences between the spectra in the UV-A region.

10.2. The Biological Amplification Factor A_b

Since in all our data there seems to be a linear dependence of $\ln R$ on $\ln D$, A_b can be determined as the slope of curves like those shown in Fig. 23. In this figure we have used data for RTD for BCC on the head, since this RTD is large and since significant variations in clothing fashion are not expected for this body site. The CIE action spectrum was used. In the calculations, we have assumed that all skin cancer cases are related to UV radiation from the sun. This is certainly not true. However, the RTDs for body sites that receive small exposures are small, while those for the head are large. Thus, our A_b values are too small, but the errors are supposedly not significant since

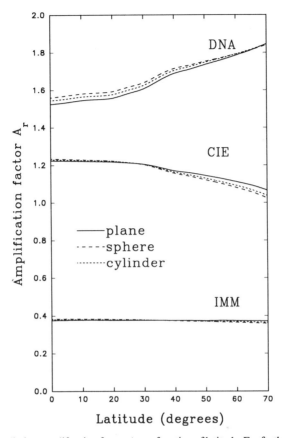

Figure 22. The radiation amplification factor A_r as a function of latitude. For further explanation, see the text.

we have used RTDs for head and neck, which are supposedly much larger than "background, non-UV-related" RTDs.

10.3. The Total Amplification Factor $A = A_r \cdot A_b$

A comparison of Figs. 22 and 23 shows that the action spectrum that gives the largest value of A_r gives the smallest value of A_b. This fact has led several investigators to state that the choice of action spectrum is rather unimportant to the magnitude of the total amplification factor A, since $A = A_r \cdot A_b$; a change of the action spectrum that makes A_r go up, simply makes A_b go down! However, Fig. 24 clearly shows that the choice of action spectrum

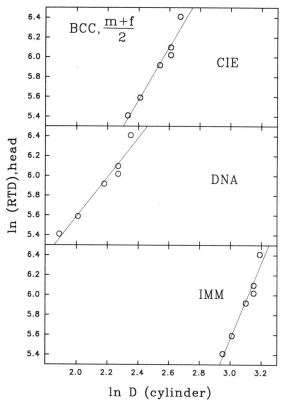

Figure 23. Plots to determine the biological amplification factor for BCC by using different action spectra. Data are for incidence rates of BCC (1976–1985) at the head, averaged for men and women in Norway.

is important. For all geometric representations of the exposed skin area, of which the vertically oriented cylinder is the most relevant, A is 2.6–2.7 times larger when the DNA action spectrum is used than when the immune spectrum is used.

If one should be forced to choose between the action spectra, one would probably choose the CIE spectrum since it is very similar to the skin carcinogenesis spectrum for mice. The cylinder representation makes A slightly larger than does the plane representation (Fig. 23). The value for the CIE spectrum plane representation (2.4) is close to the value published earlier (Moan *et al.,* 1989a). Thus, with this representation, a 1% ozone depletion gives about a 2.4% increase in the incidence of BCC. In view of Fig. 18, and in agreement with our earlier work, A is 20–30% lower for SCC than for BCC, i.e., about 1.8 for plane geometry and about 2.0 for a cylinder geometry.

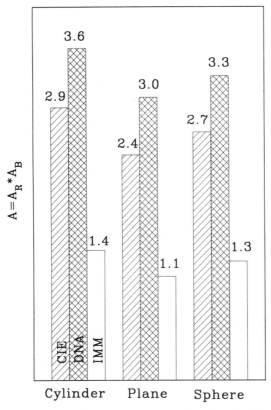

Figure 24. The total amplification factors ($A = A_r \cdot A_b$) obtained by using different action spectra (Fig. 3) and different geometric representations for the exposed human skin surface.

The value of A for immunological responses ranges from 1.1 to 1.4, while that for DNA damage in the basal layer ranges from 3.0 to 3.6 (Fig. 24). Thus, an ozone depletion would be more dangerous for the DNA in the basal layer of the skin than for the immune system. The situation for skin cancer induction may be somewhere in between. This fits excellently with our knowledge that both DNA damage and immunological effects play roles in skin cancer induction. Future research will show whether these two factors should be weighted differently for the different skin cancer types. It has been proposed that the effect of occasional sun exposure on CMM development is mediated through stimulation of the melanocyte system and perturbation of the immune system rather than through DNA damage (Holman *et al.*, 1983). If this is correct, the action spectrum for melanoma induction may be different from all the spectra considered in the present work. It may also be possible that absorption of radiation by the melanin in pigmented nevi plays a role in CMM induction. Melanin has, indeed, a very strong absorption in the UV-A and visible spectral regions.

10.4. Amplification Factors for Melanoma Induction

Several authors have expressed serious doubts about the dependence of CMM incidence on UV exposure (Baker-Blocker, 1980; Rampen and Fleuren, 1987; Cascinelli and Marchesini, 1989). They present the following evidence:

1. The rates of CMM are low in persons with outdoor occupations.
2. CMM is relatively common on skin sites which receive low exposures of sunlight (trunk, lower extremities, etc.).
3. Little solar elastosis is found in the vicinity of CMM, in contrast to what is found for BCC and SCC.
4. Some authors have reported reduced incidence rates of CMM with increasing UV exposure (Graham *et al.*, 1985; see references in Elwood *et al.*, 1989, and in Holman *et al.*, 1986).
5. CMM cannot be induced by UV radiation alone in mice.

Most of these arguments can be met with other arguments such as:

1. The dose relationship of CMM may be as proposed by Holman *et al.* (1986); up to a certain limiting exposure, which varys with skin type, the incidence rate may increase, while at higher exposures it may level off or even decrease.
2. The rates of CMM may be related to intermittent exposures but not to the total exposure.
3. UV exposure is also known to stimulate melanocytes in unexposed skin areas (Stierner *et al.*, 1989).
4. UV exposure can induce melanomas in other animals than mice.
5. While CMM tends to occur densely on thick skin (cheeks, ear lobes), BCC and SCC occur more densely on thin skin (e.g., bridge of the nose) (Scott and Straf, 1977).
6. Generally, CMM RTD *is* greater for heavily sun-exposed skin sites than for shielded sites (Figs. 9 and 15; Pearl and Scott, 1986).

Most epidemiological investigations support the intermittent exposure hypothesis (Elwood *et al.*, 1989), as does the present one. For instance, why would the RTD of CMM be so much lower for the female breast than for the rest of the trunk (Fig. 9) if CMM were unrelated to sun exposure? Why would the north–south gradient be so convincing throughout?

If the intermittent exposure hypothesis is correct, is it then appropriate to define an amplification factor related to CMM? Empirically, for Nordic populations the incidence rates of CMM are related to the annual UV exposures in the same manner as are the incidence rates of nonmelanomas, i.e., according to a power relationship (Figs. 17, 19, 20, and 21). Therefore, if we assume that the UV-exposure habits of á population remain constant, then it is possible to assign an amplification factor to CMM. For the Norwegian and Swedish populations, biological amplification factors in the range 2.5–4.7 have been ascertained (Figs. 19, 20, 21). However, populations with different or nonuniform habits of sun exposure cannot be used to evaluate A_b for CMM. This is clearly indicated by Fig. 17.

11. Applications of Amplification Factors

The main application of amplification factors is, of course, to estimate the biological implications of a future ozone depletion. However, they may also have other applications, as described below.

The incidence rate of SCC in Norway has doubled during the 20 years from 1960 to 1980 (Fig. 12). Since $R(SCC) = \text{const.}\ D^{A_b}$, where $A_b \sim 2.0$ for SCC (Fig. 18), we find that in order to double $R(SCC)$, the accumulated dose of carcinogenic UV radiation must be increased by 40%. It is not unlikely that the population received an accumulated average sun exposure in 1980 that was 40% larger than the average exposure received by the population in 1960. However, increased UV exposure may be only one of the reasons for the increasing trend of skin cancers (cf. Fig. 7). Thus, the calculated 40% is a maximal value. Figure 12 shows that between 1985 and 1990, people 55 years of age had about the same age-specific incidence rate as people 65 years of age had in 1960. To accumulate the same dose at these two ages, the average exposure rate must have been increased by 18% for the 55-year-olds as compared to that for the 65-year-olds. This figure is in agreement with the considerations addressed above.

All the data presented in this work show that $R(CMM)$ is dependent upon the annual UV exposure according to a relationship identical to that for SCC and BCC, i.e., $R(CMM) = \text{const.}\ D^{A_b}$, where $A_b \sim 2.2$ for head and neck and $A_b \sim 3.4$ for the trunk (data not shown). Thus, under the assumption that the population does not change its habits of sun exposure, we can use similar considerations as for SCC and BCC to estimate the impact of an ozone depletion on the incidence of CMM. In the period from 1960 to 1980 the incidence rate of CMM on the head and neck increased by a factor of 2.5, while that of CMM on the trunk increased by a factor of 4.2 (Fig. 9). Both of these factors are consistent with a 50% increase of the "CMM-generating" exposure from 1960 to 1980 ($1.5^{2.2} \approx 2.5$; $1.5^{3.4} \approx 4.2$). Thus, the CMM-generating exposure has increased more rapidly with time than the SCC-generating exposure. This indicates a change in the exposure pattern toward a more intermittent, CMM-generating one. Thus, these data also support the intermittent exposure hypothesis. The change in exposure pattern with the time of birth for different cohorts was clearly pointed out by Magnus (1987). His data show that Norwegians born in 1940 had the same age-specific rate of CMM on the head and neck (1.5 per 100,000) at an age of about 36 years as Norwegians born in 1900 had at an age of 60 years. The former birth cohort had the same age-specific rate of CMM on the trunk and lower extremities (2.5 per 100,000) at an age of about 23 years as the latter cohort had at an age of about 62 years. Thus, the younger birth cohorts have shifted their sun exposure habits toward a pattern that gives relatively more CMM on the trunk and the lower extremities.

The difference between a CMM generating and an SCC-generating exposure pattern can be illustrated as follows. Assume that SCC incidence on a given body site is dependent on the accumulated dose "D." Since lnRTD = A_b ln"D" + const., we can estimate relative values for "D." As shown by Fig. 25, lnRTD (CMM) is not a function increasing linearly with increasing values of "D." Notably, the trunk of men is exposed in an extremely CMM-generating pattern. It is tempting to focus on intense exposures (sunburns?) occurring when people take off their shirts and enjoy sunbathing in the spring and early summer. This observation should also be seen in relation to the hypothesis of Holman et al. (1986), which says that CMM incidence increases with carcinogenic UV exposures up to a limiting value, but decreases with exposures above this value.

12. Conclusions

Epidemiological data from Scandinavia are excellently suited for an evaluation of the relationship between sun exposure and incidence rates of different skin cancers. The cancer registries are well organized, the Scandinavian populations are relatively homogeneous with respect to skin type and habits of sun exposure, and the Finnish, Swedish, and Norwegian populations are spread over a long north to south distance creating a large gradient in annual doses of carcinogenic radiation from the sun. Annual fluctuations of UV exposure from the sun caused by ozone fluctuations are around 5%, while fluctuations caused by the cloud cover during the summer are larger, i.e., around 10–15%.

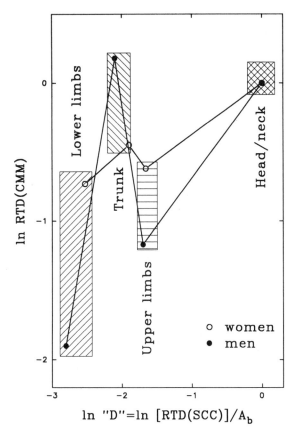

Figure 25. The relationship between RTD of CMM at different body sites and carcinogenic exposures "D" received by the same sites, determined by use of the RTD values for SCC at the same sites. It is assumed that for SCC, $\ln RTD = A_b \ln "D" + const$. Incidence data for 1976 to 1985 from The Norwegian Cancer Registry.

Our data (i.e., the latitude gradients and the magnitude of RTDs at different body sites) strongly indicate that sun exposure is the main cause of all three major skin cancers (BCC, SCC, and CMM).

The rapid increase in incidence rates of all major skin cancers with time is not due to ozone depletion. Part of the increase may be related to similar factors as those causing the increased incidence of several "non-UV-related" cancers. Our data are consistent with an assumption that the populations expose themselves to ever greater UV exposure from the sun. This increasing exposure seems to have taken place mainly at home. Thus, for the Norwegian population as a whole, vacations to southern countries have not yet occurred at an extent that makes them major determinants for skin cancer induction. There seems to be a change in the sun exposure pattern toward a CMM-generating one. Our data for different body sites are in agreement with the intermittent exposure hypothesis for CMM, which states that an intermittent exposure pattern during vacations and holidays is notably CMM-generating.

The choice of action spectrum for carcinogenesis is of great importance for the magnitude of amplification factors. The total ozone-related amplification factor found when using an action spectrum similar to the CIE spectrum is about 2 for nonmelanomas and somewhat larger for melanomas (≥ 2.2 for head and neck and ~ 3.4 for the trunk). When the action spectrum similar to the absorption spectrum of urocanic acid is used (i.e., a spectrum for perturbations of the immune system), the total amplification factor is smaller by about a factor of 2.

ACKNOWLEDGMENTS. This chapter was supported by the Norwegian Research Council for Science and the Humanities. The authors want to thank the Norwegian, Swedish, and Finnish cancer registries for valuable help. We want to express a special thanks to Dr. Lyly Teppo at the Finnish Cancer Registry and to Steinar Hansen at the Norwegian Cancer Registry. The advice of Dr. Knut Magnus, the nestor in skin cancer epidemiology in Norway, was extremely valuable, as were the data sent to us from the Swedish Meteorological Institute by Dr. Weine Josefsson in Norrköping. Vladimir Iani was of great help to us in preparing most of the figures, and Mette Jebsen was very patient in typing and correcting errors in the manuscript.

References

Applegate, L. A., Ley, R. D., Alcalay, J., and Kripke, M., 1989, Identification of the molecular target for the suppression of contact hypersensitivity by ultraviolet radiation, *J. Exp. Med.* 170:1117–1131.

Baker-Blocker, A., 1980, Ultraviolet radiation and melanoma mortality in the United States, *Environ. Res.* 23:24–28.

Bojkov, R. D., 1988, Ozone variations in the northern polar region, *Meteorol. Atmos. Phys.* 38: 117–130.

Bruls, V. A., and Van der Leun, J. C., 1982, The use of diffusers in the measurement of transmission of human epidermal layers, *Photochem. Photobiol.* 36:709–713.

Cancer Society of Finland, 1991, *Cancer Incidence in Finland, 1987*, Publ. No. 49, Helsinki.

Cascinelli, N., and Marchesini, R., 1989, Increasing incidence of cutaneous melanoma: Ultraviolet radiation and the clinician, *Photochem. Photobiol.* 50:497–505.

Chan, G. L., Peak, M. J., Peak, J. G., and Haseltine, W. A., 1986, Action spectrum for the formation of endonuclease-sensitive sites and (6-4) photoproducts in a DNA fragment by ultraviolet radiation, *Int. J. Radiat. Biol.* 50:641–648.

Chandrasekhar, S., 1960, *Radiative Transfer,* Academic Press, New York, p 147.

Cole, A. C., Forbes, P. D., and Davies, R. E., 1986, An action spectrum for photocarcinogenesis, *Photochem. Photobiol.* 43:275–284.

Dahlback, A., and Moan, J., 1990, Annual exposures to carcinogenic radiation from the sun at different latitudes and amplification factors related to ozone depletion. The use of different geometrical representations of the skin surface receiving the ultraviolet radiation, *Photochem. Photobiol.* 52:1025–1028.

De Fabo, E. C., and Noonan, F. P., 1983, Mechanism of immune suppression by ultraviolet irradiation in vivo, *J. Exp. Med.* 157:84–98.

Eklund, G., and Malec, E., 1978, Sunlight and incidence of cutaneous malignant melanoma in Sweden, *Scand. J. Plast. Reconstr. Surg.* 12:231–241.

Elwood, J. M., Whitehead, S. M., and Gallagher, R. P., 1989, Epidemiology of human malignant skin tumors with special reference to natural and artificial ultraviolet radiation exposures, in: *Skin Tumors. Experimental and Clinical Aspects* (C. J. Conti, T. J. Slaga, and A. J. P. Klein-Szanto, eds.), Raven Press, New York, pp. 55–84.

Freeman, S. E., Hacham, H., Gange, R. W., Maytum, D. J., Sutherland, J. C., and Sutherland, B. M., 1989, Wavelength dependence of pyrimidine dimer formation in DNA of human skin irradiated in situ with ultraviolet light, *Proc. Natl. Acad. Sci. U.S.A.* 86:5605–5609.

Giles, G., Marks, R., and Foley, P., 1988, Incidence of non-melanocytic skin cancer treated in Australia. *Br. Med. J.* 296:13–17.

Graham, S., Marshale, J., Haughey, B., Stoll, H., Zielezny, M., Brasure, J., and West, D., 1985, An inquiry into the epidemiology of melanoma. *Am. J. Epidemiol.* 122:606–619.

Hakulinen, T., Andersen, A., Malker, B., Pukkala, E., Schou, G., and Tulinius, H., 1986, *Trends in Cancer Incidence in the Nordic Countries,* A Collaborative Study of the Five Nordic Cancer Registries, Helsinki.

Hart, R. W., Setlow, R. B., and Woodhead, A. D., 1977, Evidence that pyrimidine dimers in DNA can give rise to tumors, *Proc. Natl. Acad. Sci. U.S.A.* 74:5574–5578.

Hill, A. B., 1971, *Principles of Medical Statistics,* The Lancet, London, pp. 204–210.

Holman, C. D. A. J., Armstrong, B. K., and Heenan, P. J., 1983, A theory of the etiology and pathogenesis of human cutaneous malignant melanoma, *J. Natl. Cancer Inst.* 71:651–656.

Holman, C. D. A. J., Armstrong, B. K., and Heenan, P. J., 1986, Relationship of cutaneous malignant melanoma to individual sunlight-exposure habits. *J. Natl. Cancer Inst.* 76:403–414.

Houghton, A., Munster, E. W., and Viola, M. V., 1978, Increased incidence of malignant melanoma after peaks of sunspot activity. *Lancet* 1, 759–760.

Jensen, O. M., Carstensen, B., Glattre, E., Malker, B., Pukkala, E., and Tulinius, H., 1988, *Atlas of Cancer Incidence in the Nordic Countries, Nordic Cancer Union,* A Collaborative Study of the Five Nordic Cancer Registries, Helsinki, pp. 135–140.

Josefsson, W., 1989, Testing of the MED-meter and a proposal of a solar UV-network in Sweden. *Report from the Swedish Meteorological and Hydrological Institute,* 60176 Norrköping, Sweden, p. 29.

Kelfkens, G., De Gruijl, F., and Van der Leun, J. C., 1990, Ozone depletion and increase in annual carcinogenic ultraviolet dose, *Photochem. Photobiol.* 52:819–823.

Klepp, O., and Magnus, K., 1979, Some environmental and bodily characteristics of melanoma patients. A case control study, *Int. J. Cancer* 23:282–286.

Kondratyev, K. Y., 1969, *Radiation in the Atmosphere,* Dover, New York.

Kripke, M. L., 1990, Effects of UV radiation on tumor immunity, *J. Natl. Cancer Inst.* 17:1392–1395.

Kripke, M. L., and Fisher, M. S., 1976, Immunologic parameters of ultraviolet carcinogenesis, *J. Natl. Cancer Inst.* 57:211–215.

Larsen, S. H. H., and Henriksen, T., 1990, Persistent arctic ozone layer, *Nature* 343:124.

Lean, J., 1987, Solar ultraviolet irradiance variations: A review, *J. Geophys. Res.* 92:839–868.

Lee, J. A., 1989, The relationship between malignant melanoma of skin and exposure to sunlight, *Photochem. Photobiol.* 50:493–496.

Lund, C. C., and Browder, N. C., 1944, The estimation of area of burns, *Surg. Gynecol. Obstet.* 79:352–361.

Magnus, K., 1981, Habits of sun exposure and risk of malignant melanoma: An analysis of incidence rates in Norway 1955–1977 by cohort, sex, age and primary tumor site, *Cancer* 48:2329–2335.

Magnus, K., 1987, Epidemiology of malignant melanoma of the skin, in: *Cutaneous Melanoma* (U. Veronesi, N. Cascinelle, and M. Santinami, eds.), Academic Press, London, pp. 1–13.

Magnus, K., 1989, Sunlight and skin cancer: Epidemiological studies, in: *Radiation and Cancer Risk,* (T. Brustad, F. Langmark, and J. B. Reitan, eds.), Hemisphere, New York, pp. 89–100.

Magnus, K., 1991, The Nordic profile of skin cancer incidence. A comparative epidemiological study of the three main types of skin cancer, *Int. J. Cancer* 47:12–19.

Matsunaga, T., Hieda, K., and Nikaido, O., 1991, Wavelength dependent formation of thymine dimers and (6-4) photoproducts in DNA by monochromatic ultraviolet light ranging from 150 to 365 nm, *Photochem. Photobiol.* 54:403–410.

McKinlay, A. F., and Diffey, B. L., 1987, A reference action spectrum for ultraviolet-induced erythema in human skin, in: *Human Exposure to Ultraviolet radiation: Risk and Regulations* (W. F. Passchier and B. F. Bosnajakovic, eds.), Elsevier, Amsterdam, pp. 83–87.

Mitchell, D. L., and Nairn, R. S., 1988, The (6-4) photoproduct and human skin cancer, *Photodermatology* 5:61–64.

Moan, J., and Dahlback, A., 1992, The relationship between skin cancers, solar radiation and ozone depletion, *Br. J. Cancer* 65:916–921.

Moan, J., Dahlback, A., Henriksen, T., and Magnus, K., 1989a, Biological amplification factors for sunlight-induced nonmelanoma skin cancer at high latitudes, *Cancer Res.* 49:5207–5212.

Moan, J., Dahlback, A., Larsen, S., Henriksen, T., and Stamnes, K., 1989b, Ozone depletion and its consequences for the fluence of carcinogenic sunlight, *Cancer Res.* 49:4247–4250.

Moan, J., Dahlback, A., Larsen, S., and Henriksen, T., 1990, Ozone depletion and some of its biological consequences, *Trends in Photochemistry and Photobiology* 1:95–106.

Muir, C. S., and Nectoux, J., 1982, Time trends: Malignant melanoma of skin, in: *Trends in Cancer Incidence: Causes and Practical Implications* (K. Magnus, ed.), Hemisphere, Washington, DC, pp. 365–385.

Muir, C., Waterhouse, J., Mack, T., Powell, J., and Whelan, S. (eds.), 1987, *Cancer Incidence in Five Continents,* Vol. V, IARC Scientific Publications No. 88, International Agency for Research on Cancer, Lyon.

Pearl, D. K., and Scott, E., 1986, The anatomical distribution of skin cancers, *Int. J. Epidemiol.* 15:502–506.

Rampen, F. H. J., and Fleuren, E., 1987, Melanoma of the skin is not caused by ultraviolet radiation but by a chemical xenobiotic, *Med. Hypothesis* 22:341–346.

Reeve, V. E., Greenoak, G. E., Canfield, P. J., Boem-Vilcox, C., and Gallagher, C. H., 1989, Topical urocanic acid enhances UV-induced tumor yield and malignancy in the hairless mouse, *Photochem. Photobiol.* 49:459–464.

Reeve, V. E., Bosnic, M., Bochan-Wilcox, C., and Ley, R. D., 1991, Differential protection by two sunscreens from UV radiation-induced immunosuppression, *J. Invest. Dermatol.* 97:624–628.

Reilly, S. K., and De Fabo, E. C., 1991, Dietary histidine increases mouse skin urocanic acid levels and enhances UVB-induced immune suppression of contact hypersensitivity, *Photochem. Photobiol.* 53:431–438.

Report from The Norwegian Transport Economical Institute, 1988, Norway.

Romerdal, C. A., Donawho, C., Fidler, I. J., and Kripke, M. L., 1988, Effect of ultraviolet-B radiation on the in vitro growth of murine melanoma cells, *Cancer. Res.* 48:4007–4010.

Scott, E. L., and Straf, M. L., 1977, Ultraviolet radiation as a cause of cancer, in: *Origins of Human Cancer* (A. Book, H. Hiatt, J. D. Watso, and J. A. Winsten, eds.), Plenum, London, pp. 529–635.

Scotto, J., and Fears, T. R., 1987, The association of solar ultraviolet and skin melanoma incidence among caucasians in the United States, *Cancer Invest.* 5:275–283.

Scotto, J., Fears, T. R., and Fraumeni, J. F., Jr., 1983, *Incidence of nonmelanoma skin cancer in the United States,* DHEW Publ. (NIH) National Cancer Inst., Bethesda, 83:2433.

Setlow, R. B., 1974, The wavelength in sunlight effective in producing skin cancer. A theoretical analysis, *Proc. Natl. Acad. Sci. U.S.A.* 71:3363–3366.

Slaper, H., 1987, *Skin Cancer and UV Exposure: Investigations on the Estimations of Risks,* Ph.D. thesis, University of Utrecht, The Netherlands, pp. 147–154.

Stamnes, K., Tsay, S., Wiscombe, W., and Jayaweera, K., 1988, Numerically stable algorithm for discrete-ordinate-method radiative transfer in multiple scattering and emitting layered media, *Appl. Opt.* 32:2502–2509.

Steerenborg, H. J. C. M., 1987, *Investigations on the Action Spectrum of Tumorgenesis by Ultraviolet Radiation,* Ph.D. thesis, University of Utrecht, The Netherlands.

Stierner, U., Rosdal, I., Augustson, A., and Kågedal, B., 1989, UVB irradiation induces melanocyte increase in both exposed and shielded human skin, *J. Invest. Dermatol.* 92:561–564.

Sutherland, J. C., and Griffin, K. P., 1981, Absorption spectrum of DNA for wavelengths greater than 300 nm, *Radiat. Res.* 86:399–409.

Teppo, L., Pakkanen, M., and Hakulinen, T., 1978, Sunlight as a risk factor of malignant melanoma of the skin, *Cancer,* 41:2018–2027.

Wigle, D. T., 1978, Malignant melanoma and sunspot activity, *Lancet* 1:38.

World Meteorological Organization, 1985, Global ozone research and monitoring project, Ref. No. 16, *Atmospheric Ozone 1985,* 1:355–358.

Østerlind, A., Hou-Jensen, K., and Jensen, O. M., 1988, Incidence of cutaneous malignant melanoma in Denmark 1978–1982. Anatomical distribution, histological types, and comparison with non-melanoma skin cancer, *Br. J. Cancer* 58:385–391.

Ultraviolet Radiation and Skin Cancer: Epidemiological Data from the United States and Canada

Martin A. Weinstock

The three major forms of skin cancer (melanoma, squamous cell carcinoma, and basal cell carcinoma) are all epidemic in the Caucasian population of North America, and each has therefore been the subject of numerous investigations. This chapter reviews the descriptive and analytic epidemiology of melanoma and nonmelanoma skin cancer. The data reviewed are based primarily on populations in the United States and Canada; the (largely parallel) results that characterize the Scandinavian and Australasian literature are reviewed in Chapters 9 and 10.

1. Melanoma

1.1. Incidence

1.1.1. Summary Numbers and Sources of Data

Melanoma is included in all major cancer registries throughout the United States and Canada; these are the major sources of information about the

Martin A. Weinstock • Dermatoepidemiology Unit, Departments of Medicine, VA Medical Center, Roger Williams Medical Center, and Brown University, Providence, Rhode Island 02908.

Environmental UV Photobiology, edited by Antony R. Young *et al.* Plenum Press, New York, 1993.

incidence of this tumor. The 1987–1988 estimate for the United States was 10.2×10^{-5} new cases/year, adjusted for age to the 1970 U.S. standard (Ries *et al.,* 1991).

There is evidence that melanoma may be underreported in the United States. The cancer registries that serve as the basis for the incidence estimates are the participants in the Surveillance, Epidemiology, and End Results (SEER) program of the National Cancer Institute, which covers approximately 10% of the population. Reports are based primarily on hospital discharges, death certification, and pathology laboratories (Young *et al.,* 1981). Since melanoma is usually cured and typically does not require hospitalization, adequacy of registration is critically dependent on the completeness of ascertainment from pathology laboratories. There is evidence that some of the SEER registries have had inadequate coverage of pathology laboratories in their areas, particularly those laboratories that are located in a doctor's office or not associated with a hospital. This defect in the registration network has been exacerbated by an increase in the number of specimens that are mailed to pathology laboratories outside of the region covered by the registry. The proportion of cases not registered by these SEER registries is unknown, but one published estimate suggested that from 1974 to 1984, the proportion increased from 2% to 21% (Karagas, 1991). Out-of-state laboratories diagnosed approximately 16% of melanomas among residents of Connecticut, another SEER area (Bolognia *et al.,* 1992). Non-SEER registries in the United States often have no coverage of any nonhospital pathology laboratories, and so have an even more severe underregistration problem (Koh *et al.,* 1991), but they are not typically part of the basis for nationwide incidence statistics.

Health Maintanence Organizations (HMOs) are a potential source of additional data for the United States, since they provide comprehensive health care for their members. A study from the northwest coast of the United States noted that melanoma incidence during 1980 to 1986 was 20×10^{-5}/year for men and 17×10^{-5}/year for women, considerably higher than in a nearby SEER registry area (Glass and Hoover, 1989). HMO members, however, may be unrepresentative of the general population.

A final source of incidence data is the record-linkage system of Rochester, the small southeastern Minnesota community with a major tertiary care medical center. Melanoma incidence in this community during the 1970s and early 1980s was similar to the incidence reported by SEER registries at similar latitudes (Popescu *et al.,* 1990).

In Canada, melanoma registration is more uniform nationwide, so underregistration does not seem to be as severe a problem, particularly in those provinces that have made skin cancer registration a priority.

1.1.2. Race

Dark-skinned racial groups have a much lower incidence of melanoma. Among U. S. blacks, the 1987–1988 incidence rate was 0.9×10^{-5}/year, compared to 10.9×10^{-5}/year among whites (Ries et al., 1991). Melanomas in blacks tend to occur on acral sites (the palms, soles, fingernails, and toenails) and on mucous membranes; these are proportionately uncommon sites in Caucasians (Fleming et al., 1975; Krementz et al., 1982; Crowley et al., 1991). However, the overall incidence of acral melanomas does not appear to differ between blacks and whites (Stevens et al., 1990). Furthermore, there appears to be relatively minor differences, if any, in the incidence of nonocular noncutaneous melanomas between whites and blacks (Weinstock, 1992b).

In the United States, Hispanic populations are intermediate between non-Hispanic Caucasians and blacks in pigmentation and patterns of melanoma occurrence, and their incidence of melanoma is substantially lower than similarly located Anglo populations (Bergfelt et al., 1989; Black et al., 1987; Vázquez-Botet et al., 1990). In Los Angeles county, large numbers of non-Hispanic whites, Hispanics, and blacks live in close geographic proximity; the melanoma incidence in these groups ($\times 10^{-5}$/year) was, respectively, 12.1, 3.3, and 1.1 among men and 10.0, 3.6, and 1.0 among women during 1972 to 1982 (Mack and Floderus, 1991). Native Americans, native Hawaiians, and Americans of East Asian ancestry also have low melanoma rates, more similar to those of Hispanics and blacks than non-Hispanic whites (Horm et al., 1984; Black and Wiggins, 1985).

1.1.3. Latitude Gradient and Migrant Studies

Within North America, melanoma incidence among whites is generally higher in more southerly latitudes (Ries et al., 1991; Scotto and Fears, 1987). For example, the incidence ($\times 10^{-5}$/year) among non-Hispanic Caucasian men and women in Los Angeles, California, is 12.4 and 10.9; among Caucasians in Connecticut, 8.4 and 7.7; and among Canadians, 5.3 and 6.0 (Whelan et al., 1990). In contrast, ocular melanoma shows no latitude gradient (Schwartz and Weiss, 1988; Scotto et al., 1976), and noncutaneous, nonocular melanoma is strikingly more frequent in more northerly latitudes, in sharp contrast to cutaneous melanoma (Weinstock, 1992b).

Two studies of migrants within the United States have suggested that southerly latitude of residence early in life is an important risk factor for melanoma independent of latitude of residence at the time of diagnosis (Mack and Floderus, 1991; Weinstock et al., 1989a). The only discrepant note concerns migrants to Hawaii, where ethnicity may be primarily responsible for lower risk among the native-born (Hinds and Kolonel, 1980; Lee, 1982).

1.1.4. Gender and Anatomic Site

Incidence rates are generally higher among men than among women (Ries *et al.*, 1991; Bureau of Cancer Epidemiology, 1989; Parkin *et al.*, 1990; Byers *et al.*, 1991), although the opposite is found in some areas (Gallagher *et al.*, 1990; Popescu *et al.*, 1990). The most striking and consistent differences between men and women are the patterns of anatomic distribution: women have a consistent excess of melanomas on the legs, whereas among men, the back is usually the more common site, even after adjusting for the relative area of skin at each site (Elwood and Gallagher, 1983). One analysis has suggested that melanoma incidence at "usually exposed" sites (head, neck, and upper extremities) is more strongly associated than other sites (trunk and lower extremities) with ambient ultraviolet-B flux at the geographic area of residence at diagnosis (Scotto and Fears, 1987).

1.1.5. Secular Trends

Melanoma incidence has been increasing at a rapid rate in both the United States and Canada. In the United States, data from the Connecticut Tumor Registry document an age-adjusted incidence rate of 1.0×10^{-5}/year during 1935 to 1939, the first years of the registry. That has increased steadily, and in 1989 the corresponding rate was 12.4×10^{-5}/year (J. Flannery, unpublished data, 1992). Other registries have been in operation for shorter periods of time, but document similar trends. In British Columbia, incidence more than doubled (2.3-fold increase among men, 2.0-fold increase among women) in the 12 years from 1973–75 to 1985–87 (Gallagher *et al.*, 1990). In Rochester, Minnesota, there was an approximately 3-fold increase (2.8 among men, 2.7 among women) in incidence from 1950–58 to 1977–85 (Popescu *et al.*, 1990). In New York State (excluding New York City), the increase from 1950 to 1985 was 5-fold in men and more than 3-fold among women (Bureau of Cancer Epidemiology, 1989). In Hawaii, the incidence tripled among Caucasians from 1960 to 1977 (Hinds and Kolonel, 1980). In a northwestern U.S. Health Maintenance Organization, incidence among men and women increased 3.5- and 4.6-fold from the 1960s to the 1980s (Glass and Hoover, 1989). In the SEER registry areas, incidence among whites increased 74% from 1973 to 1988 (Ries *et al.*, 1991). Indeed, between 1973 and 1988, melanoma has increased faster in reported incidence than cancer of any major noncutaneous site (Frey and Hartman, 1991).

In the SEER registries, the incidence among whites in 1988 was 9% less than the incidence in 1987; however, it is not clear that this decrease represents a real reversal in the long-term trend. Underregistration of melanoma has become an increasingly important concern in the SEER system of registries in recent years as outpatient diagnosis and interstate mailing of pathological specimens have become more commonplace (Karagas *et al.,* 1991). Changes in diagnostic criteria could also be operative; as dysplastic nevi are more commonly diagnosed, moderate to severely atypical melanocytic lesions may be less frequently classified as melanoma. Alternatively, it certainly can be argued that the American lifestyle with respect to sun exposure became maximally melanomagenic many years ago, so incidence in the mid- to late 1980s represents the maximum that could be achieved by a population of that ethnic composition living with the present level of solar ultraviolet flux. Sufficient data have not yet been analyzed to document or refute this argument.

The converse has also been argued, i.e., that the long-term increase in melanoma incidence is due to factors other than a true rise in disease frequency. For example, over time there may have been a decreased misclassification of melanoma as a malignancy arising from another site, changes in diagnostic criteria for melanoma, or progressively more complete registration of melanomas diagnosed. These factors are unlikely to account for the observed dramatic increase, however. Misclassification of melanoma deaths is not frequent (Percy *et al.,* 1981), and melanoma mortality has been increasing along with incidence, although not as rapidly (see Section 1.3). Melanoma incidence follows a birth cohort pattern, yet changes in criteria, ascertainment, and misclassification would be expected to follow a period pattern (see Section 1.1.6). Finally, efforts to adjust observed trends for indices of completeness of ascertainment have not suggested a substantial artifactual component of observed incidence (Roush *et al.,* 1988). Hence, the observed increase is generally viewed as real.

The most detailed analyses of secular trends in incidence by anatomic site include Connecticut data regarding 2,966 melanomas diagnosed from 1935 through 1974. These document the most rapid increases in trunk melanomas among men and leg melanomas among women. The scalp and neck and the face exhibited the slowest rates of increase for both gender groups (Houghton *et al.,* 1980; Stevens and Moolgavkar, 1984). Similar observations were reported regarding 3,777 melanomas diagnosed from 1973 through 1987 in British Columbia (Gallagher *et al.,* 1990). The smaller study in the northwestern United States noted a particularly sharp increase in melanoma of the trunk (Glass and Hoover, 1989). Among Hawaiian Caucasian women, however, no increase was noted in lower-leg melanomas from 1960 to 1977, although the number of cases was relatively small (Hinds and Kolonel, 1980).

No change in incidence of cutaneous melanoma was noted among non-Caucasians in Hawaii during 1960 to 1977 (Hinds and Kolonel, 1980), nor among blacks in the SEER registry areas during 1973 to 1981 (Horm *et al.,* 1984).

1.1.6. Age and Cohort Effects

Time can be represented by three interrelated variables in descriptive epidemiological analyses: age of individuals, their year of birth (cohort), and the calander year (period). Most cancers increase in incidence as a power function of age (see Section 2.1.6). When incidence is changing, it may change primarily as a function of the calander year (period)—e.g., if the entire population were exposed to a new carcinogen—or it may change primarily as a function of year of birth e.g., if childhood exposures are the primary determinant of later risk. The pattern of rising incidence of melanoma is most consistent with the latter, a "cohort" pattern (Roush et al., 1987). When analyzed by site (face, scalp and neck, trunk, upper extremities, lower extremities), a similar pattern describes each site, i.e., increase in incidence as a power function of age and as a function of birth cohort, but not period (Stevens and Moolgavkar, 1984). However, the sites differ in the rapidity of the increase in incidence with successive birth cohorts (see Section 1.1.5). The overall melanoma incidence rates in recent cohorts are similar to rates of colon cancer (presently the third most common malignancy) (Roush et al., 1985).

1.2. Case Fatality

Overall 5-year relative survival (United States, 1981–1987) was 82% among whites, but only 70% among blacks, although both appear to be improving. Blacks more commonly than whites present with advanced disease and lesions in poor prognosis locations. Among whites, the survival was greater for women (87%) than men (77%); this may be due to differences in anatomic distribution of lesions (Ries et al., 1991).

The prognosis of melanoma is primarily dependent on the stage of disease: at least 75% to 80% of patients with localized melanoma are long-term survivors, whereas only 15% to 20% of those with regional metastases appear to be curable, and the prognosis for those with distant metastases is dismal (median survival is approximately 6 months). If metastases are absent, survival depends primarily on the "Breslow" thickness of the primary lesion: long-term survival is greater than 90% if the primary is less than 3/4 mm, but less than 50% if it is greater than 4 mm. A variety of other factors influence prognosis, including (for metastatic disease) the pattern of metastasis, and, for localized disease, ulceration of the primary lesion, mitotic rate, anatomic site, and gender (women fare better) (Weinstock et al., 1993; Vollmer, 1989; Clark et al., 1989; Soong, 1992; Balch et al., 1992a,b).

1.3. Mortality

1.3.1. Person and Place

Mortality data from the United States and Canada is based on universal mandatory death certification. Certification of melanoma deaths appears to be accurate (Percy *et al.,* 1981). Overall melanoma mortality ($\times 10^{-5}$/year) among whites in the United States from 1987 to 1988 was 2.4 (3.4 among men and 1.7 among women). Among blacks, the corresponding rates were 0.4 for both men and women (Ries *et al.,* 1991). When mortality among Caucasians is examined by state or province of residence, a strong negative correlation is found with latitude (Elwood *et al.,* 1974).

1.3.2. Time

Melanoma mortality rates have been increasing by 2% to 3% per year in the United States and Canada for several decades (Lee and Carter, 1970; Lee *et al.,* 1979). As with incidence, these mortality rate increases have followed a "cohort" pattern (Venzon and Moolgavkar, 1984). However, in the United States, the rates have been stabilizing in cohorts born after 1930, and actual declines in mortality rates have been noted among men born since the 1950s and women born since the 1930s (Scotto *et al.,* 1991; Roush *et al.,* 1992b). These declines are manifest as decreasing mortality among those in the youngest age groups, despite increasing mortality among the elderly. Since most deaths occur among the older age groups, the overall melanoma mortality continues to rise, but is projected to stabilize by the year 2020 (Scotto *et al.,* 1991). This projection assumes that other factors relevant to melanoma mortality remain constant.

There are many factors which may affect melanoma mortality, however, including melanoma incidence, which may be affected by stratospheric ozone levels, sunscreen use, and lifestyle changes, and melanoma case fatality, which may be affected by public education campaigns, screening efforts, and therapeutic advances. Indeed, the factors that have led to the observed mortality changes in successive birth cohorts are somewhat unclear because of the ambiguities in incidence rates due to uncertainty regarding the magnitude of underregistration of new cases. It is clear, however, that decreased case fatality has played a major role. It can be further hypothesized that additional factors have led to the appearance of a cohort mortality rate decline, such as the relative difficulty of early melanoma diagnosis among the elderly, who have many benign cutaneous neoplasms such as seborrheic keratoses and angiomata.

1.4. Risk Factors

Race, age, latitude, gender, and birth cohort are important risk factors for melanoma, as described above. Melanocytic lesions, sun-related factors, and family history are also major risk factors. These and other risk factors are documented in case–control and cohort studies. Those including North American populations will be reviewed here, although the findings are generally consistant with similar studies conducted elsewhere.

1.4.1. Melanocytic Lesions

1.4.1.1. Definitions. It may be that all melanocytic proliferations are precursors and risk factors for melanoma, and that atypical proliferations present the greatest risk, but this has not been demonstrated. Attention has focused on dysplastic nevi, congenital nevi, lentigo maligna, and counts of nevi or of large nevi. Unfortunately, all of these entities are plagued by ambiguities or disagreements regarding their definition.

Dysplastic nevi have proven particularly troublesome to define. Their clinical appearance differs from common acquired nevi in the presence of one or more features often associated with melanoma, such as asymmetry, variation in color, and large size. Histologic findings include melanocytic atypia, typically associated with certain architectural features. Little about dysplastic nevi is completely free from dispute, including their name (they are also referred to as "atypical nevi" and "Clark's nevi") and their very existence. The reader is referred to the reference section for further discussion (Barnhill and Roush, 1991; Clemente *et al.*, 1991; Friedman *et al.*, 1985; Greene, 1991; Kelly *et al.*, 1986; Piepkorn *et al.*, 1989; Rhodes *et al.*, 1989; Roth *et al.*, 1991).

It is generally agreed that congential nevi are nevomelanocytic lesions that were present at birth ("birth moles"). These nevi have characteristic clinical and histologic features, but none of these features appear to be entirely specific to congenital nevi, nor are they always present in these nevi (Rhodes *et al.*, 1985). Hence, the distinction between congenital and acquired nevi in studies of melanoma may be subject to a degree of error.

Lentigo maligna may be considered a precursor of lentigo maligna melanoma that is defined by its typical clinical and histologic characteristics (Clark and Mihm, 1969; Jackson *et al.*, 1966). However, small foci of invasive melanoma can often be found in cases of lentigo maligna, and some consider lentigo maligna to be lentigo maligna melanoma *in situ* (Pennys, 1987). The precise clinical or histologic features required for the diagnosis of lentigo maligna may also differ from center to center, despite agreement on more typical cases.

Nevi in general can be defined by their histologic features, but for epidemiological purposes one must rely primarily on clinical definitions, and these become very difficult for small lesions (e.g., less than 2 or 3 mm), where the distinction from lentigenes (freckles) and seborrheic keratoses may be particularly troublesome, even for clinical experts. Substantial difficulty may also arise among larger lesions in the diagnosis of amelanotic nevi, and some studies have excluded these nonpigmented nevi from their counts of total nevi.

1.4.1.2. Studies of the Link with Melanoma. We do not yet know the risk of melanoma in melanocytic precursor lesions or among individuals who have these lesions, whether or not the lesions are routinely excised. It is clear, however, that in some circumstances the risk is quite high. A group of 14 families with melanoma in two or more members were followed prospectively with close dermatologic surveillance including full-body photography and excision of suspicious lesions. Family members with dysplastic nevi had an approximately 300-fold increased risk of developing melanoma (Greene *et al.*, 1985a,b; Kraemer and Greene, 1985).

Cohort study of dysplastic nevus patients referred to a single practitioner suggests that the increased risk of melanoma also applies to those without a family history of melanoma, but few melanomas have been found so far (Tiersten *et al.*, 1991; Rigel *et al.*, 1989).

Several case–control studies suggest that dysplastic nevi, many nevi, or large nevi are important melanoma risk factors under more general circumstances. A clinic-based sample of melanoma patients and controls at the University of California were examined for nevi greater than 2 mm in diameter by a dermatologist and a dermatology fellow (Holly *et al.*, 1987; Kelly *et al.*, 1989). Counts of dysplastic nevi and total nevi were independently associated with melanoma risk: the relative risks associated with 6 or more dysplastic nevi and 100 or more nondysplastic nevi were 11 and 17 (crude) and 6 and 10 (adjusted for other risk factors). A clinic-based sample of melanoma patients and a workplace-based sample of controls in Philadelphia were examined by dermatologists or other physicians skilled in the diagnosis of pigmented lesions (Halpern *et al.*, 1991). Any dysplastic nevi and 25 or more nondysplastic nevi greater than 2 mm in diameter were each associated with melanoma (relative risks 9 and 7). When the association of dysplastic nevi with melanoma was adjusted for nondysplastic nevi and other risk factors, the relative risk was 7; the reverse adjustment was not reported. A Boston-based study noted a relative risk of 24 (unadjusted) for dysplastic nevi (Rhodes *et al.*, 1980). Other studies involving counts (including self-reported counts) of nevi or large nevi have also shown substantial relative risks (Reynolds and Austin, 1984; Dubin *et al.*, 1986, 1989, 1990; Bain *et al.*, 1988; Weinstock *et al.*, 1989b).

An intriguing observation is the apparent lack of a direct site-specific association between mole counts at various anatomic sites and the site of melanoma (Weinstock *et al.*, 1989b). In this case–control study nested in a cohort study of female nurses, leg mole counts were more closely associated with melanoma risk than arm mole counts, but arm mole counts were more closely associated with leg melanomas, and leg mole counts with arm melanomas, than arm mole counts with arm melanomas. Similarly, sites of dysplastic nevi did not correlate with sites of melanoma (Roush *et al.*, 1992a). These findings suggest that even though nevus counts and dysplastic nevi are closely associated with melanoma, the factors that cause these lesions to occur at specific anatomic sites may differ from the factors that cause melanoma to occur at a specific anatomic site. Hence, nevi and dysplastic nevi may be primarily indications of systemic melanoma risk.

The evidence that links small congenital nevi with melanoma is more indirect because of the difficulty in accurate retrospective diagnosis of a congenital nevus. The link is based on the prevalence of small congenital nevi and the frequency with which melanomas are noted to be histologically contiguous with them or historically associated with them, although the magnitude of the association remains unclear (Elder, 1985; Rhodes and Melski, 1982).

"Giant" or "bathing trunk" congenital nevi are said to give rise to melanoma with particularly high frequency, even in early childhood. However, data regarding the magnitude of this risk are fragmentary. One study followed 47 patients with congenital nevi at least 20 cm in diameter for an average of 4.5 years, and noted only one melanoma (Gari *et al.*, 1988). That melanoma arose in the central nervous system of a 2-month-old girl.

Lentigo maligna is the least common of these melanocytic precursors of melanoma. Its typically large size and the frequent development of melanoma within these lesions has led to general acceptance of this association. The precise frequency with which melanoma develops within lentigo maligna is quite uncertain, however. An indirect estimate of this frequency was obtained from national prevalence and incidence data, and this estimate indicates that a small proportion of clinically defined lentigo maligna lesions develop invasive melanoma, but more direct measures would be desirable (Weinstock and Sober, 1987).

1.4.2. Family History of Melanoma

Some families are prone to melanoma. A prospective investigation of 14 families in which at least two members had been diagnosed with melanoma found that the other family members were at high risk. A particularly high incidence of melanoma ($1,430 \times 10^{-5}$/year) was noted among members who had dysplastic nevi; their cumulative risk to 60 years of age was over 50% (Greene *et al.*, 1985a).

Case–control investigations usually rely on the participant's report for family history of melanoma. This source of information is somewhat inaccurate because subjects may be unaware of their relatives' diagnoses, particularly those that do not require hospitalization, and subjects may be given inaccurate information regarding a family member's diagnosis. The report of a Buffalo, New York, case–control study is particularly notable because cancer registry data was used to find and confirm melanoma diagnoses among first-degree relatives of cases and controls living in New York State (Duggleby *et al.*, 1981). Five melanoma diagnoses were confirmed among case relatives versus none among control relatives, and 0.625 expected based on the age and gender of the case relatives and the melanoma rates in New York State. A case–control study conducted in a (medically aware) cohort of nurses found a relative risk of 2.3 associated with a report of melanoma in a sibling or parent (Weinstock *et al.*, 1989a).

It is also noteworthy that melanomas occur more frequently than expected among parents of retinoblastoma patients and among retinoblastoma survivors (Sanders *et al.*, 1989; Traboulsi *et al.*, 1988).

Investigators in several centers in North America and elsewhere have studied large families with multiple cases of melanoma to clarify potential patterns of inheritance and potential genetic links between melanoma and nevi or dysplastic nevi; consensus has not yet been reached (Bale *et al.*, 1989; Goldgar *et al.*, 1991; Kefford *et al.*, 1991; van Haeringen *et al.*, 1989).

1.4.3. Sun-Related Risk Factors

Increased risk of melanoma among sun-sensitive individuals is a uniform finding among all studies of these questions. Various characteristics have been used to measure sun sensitivity by questionnaire, including hair and eye color, susceptibility to sunburn, and ability to tan. There is evidence that indices based on combinations of these characteristics may be better indicators of overall sun sensitivity and more closely associated with melanoma risk than any single characteristic (Weinstock, 1992a; Weinstock *et al.*, 1991b). The risk in albinos has not been determined, but appears to be consistent with the risk in fair-skinned Caucasians (Weinstock 1992c).

Several studies from the United States and Canada speak to the issue of sun exposure. In one cohort study of deaths over a 35-year period among 50,000 male alumni of Harvard University and the University of Pennsylvania, outdoor employment before college was a significant risk factor for melanoma (Paffenbarger *et al.*, 1978). This study may be criticized because of its lack of measures of potential confounding variables, such as sun sensitivity, but it has the key feature that exposure was ascertained far in advance of diagnosis, therefore eliminating the possibility of biased recall.

A measure frequently used in case–control studies is number of sunburns, severe sunburns, or blistering sunburns, which is consistently associated with melanoma in North American investigations (Lew *et al.*, 1983; Holly *et al.*, 1987; Elwood *et al.*, 1984, 1985a; Weinstock *et al.*, 1989a). However, sunburn counts indicate both exposure and sensitivity to the sun. One attempt to separate these is to control for measures of sun sensitivity, which typically and predicably attenuates the effect of sunburns, yet typically does not eliminate that effect. Another approach compared blistering sunburns in the teenage years to blistering burns after age 30, and found that the former was a significantly better predictor of melanoma risk (Weinstock *et al.*, 1989a). Although this last finding is consistent with data implicating early-life exposures, alternate explanations can be considered. For example, variation in sun exposures among teenagers may be less relative to variation in sun sensitivity than variation in sun exposures among adults older than 30 years, hence the former may more closely correlate with sun sensitivity.

Several studies have inquired directly about exposures to the sun. Assessment of these exposures is complex both logistically and conceptually, since people are exposed to solar radiation for varying durations at various ages and with intensities that vary with clothing worn, season of the year, time of the day, and other factors. The major North American studies that have studied lifetime cumulative exposure to the sun have noted no association with melanoma (Elwood *et al.*, 1985b; Dubin *et al.*, 1986; Weinstock *et al.*, 1991b), a direct association (Gellin *et al.*, 1969), or an inverse association (Graham *et al.*, 1985). The largest of these studies was the Western Canada Melanoma Study, which included 595 patients with melanoma diagnosed during a 2-year period, and 595 age- and gender-matched controls (Elwood *et al.*, 1984, 1985a,b; Gallagher *et al.*, 1986). Both cases and controls were population based, thereby minimizing selection bias and maximizing generalizability to the population from which they were drawn (the provinces of British Columbia, Alberta, Saskatchewan, and Manitoba). The investigators distinguished between more chronic, occupational sun exposures and more intermittent, recreational and vacation sun exposures. Increasing recreational and vacation exposures were associated with melanoma risk, although the trends with occupational exposure was more complex: those with virtually no summer occupational exposure and those with moderate and heavy summer occupational exposure were at lower melanoma risk than those with approximately 1–8 h per week of summer occupational exposure (Elwood *et al.*, 1985b). Intriguing observations were also provided from a nested study in the Nurses' Health Study cohort, in which both cases and controls were derived from the same group of female nurses, thereby minimizing selection bias. These analyses suggested that intense sun exposures at ages 15 to 20 years, represented by frequency of summer outdoor activities while wearing a swim suit, was associated with melanoma risk among sun-sensitive women, but not among sun-resistant women, suggesting that a tan may be protective (Weinstock et al., 1991b). Both these studies suggest that intense, intermittent exposures are important in the etiology of melanoma.

Two Canadian studies have examined the relation between sunlamp and sunbed use and melanoma risk; one found an association and one did not (Walter et al., 1990; Gallagher et al., 1986). Melanoma has been reported in patients who have received psoralen photochemotherapy, but it is not clear that the incidence rate in this group is higher than in the general population (Gupta et al., 1988). Melanoma has been found to be more frequent in workers with indoor occupations (Lee and Strickland, 1980), and is also more common in census tracts with higher mean income and among individuals with higher educational attainment (Kirkpatrick et al., 1990; Lee et al., 1992).

1.4.4. Other Factors

Immunosuppression has been linked to melanoma in studies of Hodgkin's disease (Tucker et al., 1985), other lymphoproliferative malignancies (Greene and Wilson, 1985), and organ transplant recipients (Greene et al., 1981; Gupta et al., 1986; Witherspoon et al., 1989).

Other factors have been suggested as possible causes of melanoma, including exogenous hormone use (Holly, 1986), chemical exposures (Dubrow, 1986; Gallagher et al., 1986; Marsh et al., 1991; Rampen and Fleuren, 1987; Wright et al., 1983), fluorescent light exposure (Elwood, 1986; Walter et al., 1992), and diet (Stryker et al., 1990), but their link with melanoma has not been established to date.

2. Basal and Squamous Cell Carcinoma of the Skin

2.1. Incidence

2.1.1. Sources of Data

Basal and squamous cell cancers are often not tabulated by cancer registries. Those registries which do include these cancers rely on sources other than hospitals and death certification for case ascertainment, such as pathology laboratories, radiation therapy facilities, and direct reports from physician offices; nevertheless, some cases may be missed. The special skin cancer surveys conducted by the National Cancer Institute (1971–1972 and 1977–1978), which utilized a broad array of sources, is the major source of data on basal and squamous cell carcinoma incidence in the United States (Scotto *et al.,* 1983; Scotto *et al.,* 1974). Some Canadian provinces are able to routinely register these tumors because virtually all health care involves the provincial health plans (Gallagher *et al.,* 1990; Muir *et al.,* 1987).

An alternative to cancer registration is self-report. A study of U.S. nurses relied on self-report, and in this study, examination of medical records on a subsample confirmed the self-report in 27 of 28 evaluable cases (Hunter *et al.,* 1990).

A third approach is to study incidence among members of a prepaid health care plan (Glass and Hoover, 1989). Since virtually all health care is provided by one organization, more complete ascertainment may be possible for members of a prepaid plan than for a population which receives its health care from many sources. However, plan members may differ in important respects from the general population, so the incidence in a prepaid plan may be a biased reflection of the incidence in the community.

With all of these methods, the completeness of ascertainment and accuracy of diagnosis may be imperfect and may differ between basal cell carcinoma (BCC) and squamous cell carcinoma (SCC). The approaches reviewed above may also differ in accuracy and completeness of the investigators' access to all diagnoses in the target populations. Similarities among estimates derived by different methods are reassuring in this regard. Although the reader must be cautious regarding possible bias, these data provide important guideposts to the understanding of BCC and SCC.

Two final precautions pertain to the calculation of age-adjusted incidence rates. Some studies count persons with cancer; others count cancers, so one person with multiple malignancies may be counted several times. Also, age-adjustment of incidence is always calculated with respect to a standard population. For conditions most common among the elderly, such as BCC and SCC, the choice of the standard population may have a substantial impact. For example, use of the World Standard Population will lead to a lower age-adjusted incidence rate than use of the 1970 United States standard population.

2.1.2. Latitude

Data on the incidence of BCC among whites has been published for certain areas in the United States and Canada (Table 1). The most detailed description of incidence derives from the two special surveys of selected areas in the United States conducted by the National Cancer Institute in the 1970s (Scotto *et al.,* 1983; Scotto *et al.,* 1974; Fears and Scotto, 1982). The overall rates for 1971–72 and 1977–78, respectively, were 202×10^{-5}/year and 247×10^{-5}/year for men, and 116×10^{-5}/year and 150×10^{-5}/year for women; 31,514 cases were registered. The incidence in more southerly locations (30–35°N latitude) was approximately double the incidence in northern areas (40–50°N latitude).

The geographic distribution of cutaneous SCC is quite similar to that of BCC. Among Caucasians, the incidence is greatest in the more equatorial latitudes within North America (Table 1). The gradient with latitude is slightly greater with SCC than BCC, particularly in the United States, but this finding is not consistent internationally and the magnitude is small. The BCC to SCC ratio among Caucasians is high, generally 3:1 to 5:1 among men, and 4:1 to 8:1 among women.

2.1.3. Gender

Both these keratinocyte-derived malignancies are more common in men than women. For BCC, the male-to-female incidence rate ratio was between 1.3 and 1.9 in major studies (Table 1). For SCC, the male-to-female ratio was generally 2.0 to 3.4, considerably higher than the ratio for BCC.

2.1.4. Secular Trends

BCC appears to be increasing in incidence, although data are sparse. In the British Columbia registry, increases of 61% and 48% were noted among men and women over the 12-year period 1973–75 to 1985–87 (Gallagher et al., 1990). The two National Cancer Institute surveys documented an 18% increase in incidence over 6 years in California and Minnesota during the mid-1970s (Fears and Scotto, 1982). The rates of increase in men and women were similar. Incidence studies from Minnesota in 1963, and from eight cities in the United States in 1947, found BCC to be considerably less frequent than would be expected on the basis of the incidence rates at similar latitudes in the 1970s and 1980s (Haenszel, 1963; Lynch et al., 1970). This is consistent with the trends in nonmelanoma skin cancer (predominantly BCC) from the Texas registry (MacDonald and Heinze, 1978).

SCC is increasing in reported incidence over time. Among members of a large HMO in the northwestern United States, the incidence increased 2.6-fold in men and 3.1-fold in women from 1960–69 to 1980–86 (Glass and Hoover, 1989). In British Columbia, increases of 59% and 67% were noted from 1973–75 to 1985–87 (Gallegher et al., 1990). In California and Minnesota, small increases were noted over 4- to 6-year periods. The 1963 Minnesota survey found a much lower incidence than surveys in the 1970s, but this may be due to different methods employed and the restriction of the latter surveys to the Minneapolis–St. Paul area.

Table 1. Basal and Squamous Cell Carcinoma Incidence among Caucasians in the United States and Canada

	Latitude	Year	Basal cell				Squamous cell				BCC:SCC ratio		Standard population	Reference
			N of cases	Incidence ($\times 10^{-5}$/year) Men	Women	Male:Female ratio	N of cases	Incidence ($\times 10^{-5}$/year) Men	Women	Male:Female ratio	Men	Women		
1970s and 1980s														
Kauai, Hawaii	22°N	1983	80	656		—	19	156		—	4.2		Crude	Stone et al. 1986
New Orleans, Louisiana	35°N	1977–8	2114	410	215	1.9	653	153	49	3.1	2.7	4.4	1970 U.S.	Scotto et al. 1983
Dallas–Ft. Worth, Texas	33°N	1971–2	2442	394	205	1.9	776	145	54	2.7	2.7	3.8	1970 U.S.	Scotto et al. 1983
Atlanta, Georgia	34°N	1977–8	3214	423	229	1.8	836	131	53	2.5	3.2	4.3	1970 U.S.	Scotto et al. 1983
New Mexico	30–37°N	1977–8	2549	346	205	1.7	638	98	42	2.3	3.5	4.9	1970 U.S.	Scotto et al. 1983
Anglo	37°N	1977–8	2376	495	279	1.8	600	143	55	2.6	3.5	5.1	1970 U.S.	Scotto et al. 1983
Hispanic	37°N	1977–8	173	64	48	1.3	38	13	12	1.1	5.0	4.0	1970 U.S.	Scotto et al. 1983
San Francisco–Oakland, California	38°N	1977–8	5355	239	145	1.6	1010	56	18	3.1	4.3	8.1	1970 U.S.	Scotto et al. 1983
Utah	37–42°N	1971–2	2103	198	117	1.7	427	52	16	3.3	3.8	7.3	1970 U.S.	Scotto et al. 1983
Iowa	41–44°N	1977–8	2610	327	198	1.7	817	123	46	2.7	2.7	4.3	1970 U.S.	Scotto et al. 1983
		1971–2	1489	123	69	1.8	507	51	14	3.6	2.4	4.9	1970 U.S.	Scotto et al. 1983
Detroit, Michigan	42°N	1977–8	3871	142	97	1.5	634	30	11	2.7	4.7	8.8	1970 U.S.	Scotto et al. 1983
Vermont and New Hampshire	43–45°N	1979–80	2022	159	87	1.8	285	32	8	4.0	5.0	10.9	1970 U.S.	Serrano et al. 1991

Location	Latitude	Period											Reference	Citation
Rochester, Minnesota	45°N	1976–84	657	175	124	1.4	169	63	23	2.8	2.8	5.5	1980 U.S.	Chuang et al., 1990a,b
Minneapolis–St. Paul, Minnesota	45°N	1977–8	2939	213	144	1.5	382	37	12	3.1	5.8	12.0	1970 U.S.	Scotto et al., 1983
Portland, Oregon	46°N	1971–2	1018	165	102	1.6	175	37	12	3.1	4.5	8.5	1970 U.S.	Scotto et al., 1983
		1960–86	—	—	—	—	1874	81	24	3.4	—	—	1970 U.S.	Glass and Hoover, 1989
Seattle, Washington	48°N	1977–8	1810	210	125	1.7	325	47	16	2.9	4.5	7.8	1970 U.S.	Scotto et al., 1983
British Columbia	49–60°N	1973	1684	71	62	1.1	344	17	9	1.8	4.3	6.5	1971 Can.	Gallagher et al., 1990
		1980	2736	96	81	1.2	622	26	13	2.0	3.7	6.2	1971 Can.	Gallagher et al., 1990
		1987	4152	120	92	1.3	963	31	17	1.8	3.9	5.5	1971 Can.	Gallagher et al., 1990
Pre-1970														
Honolulu, Hawaii	21°N	1955	135	68			124	62		—		1.1	Crude	Allison and Wong, 1957
Southern Arizona	32°N	1969	1331	317			435	104		—		3.0	Crude	Schreiber et al., 1971
United States, 4 southern cities	30–34°N	1947	—	61	41	1.5	—	32	14	2.3	1.9	2.3	1950 U.S.	Auerbach, 1961; Haenszel, 1963
United States, 4 northern cities	40–42°N	1947	—	16	12	1.3	—	9	5	1.7	1.8	2.8	1950 U.S.	Auerbach, 1961; Haenszel, 1963
Upstate New York		1958–60	—	27	19	1.4	—	7	3	2.5	3.7	6.7	1950 U.S.	Haenszel, 1963
Minnesota (entire state)	44–49°N	1963	1897[a]	55	40	1.4	384	15	5	3.0	3.7	8.0	1950 U.S.	Lynch et al., 1970

[a] Approximate.

2.1.5. Ethnicity

A direct comparison of BCC rates between Hispanic and Anglo populations is available from the 1977–1978 survey in New Mexico. Hispanics had one seventh the Anglo incidence of BCC, and the male to female ratio was 1.3, which is similar to the ratio noted in some Anglo populations. Data from Texas on nonmelanoma skin cancer suggests that incidence more than doubled (from 14×10^{-5}/year to 30×10^{-5}/year) among Hispanics between 1949 and 1966 (Macdonald and Heinze, 1978). The Hispanic rate was one fifth of the Anglo rate, and the male to female ratio was 1.1 (Table 2). These data were not stratified by histology, although contemporaneous studies from that geographic area indicate that most nonmelanoma skin cancers among Hispanics were BCC (Macdonald and Bubendorf, 1964). Data from Puerto Rico suggest that overall skin cancer incidence rates (predominantly BCC) increased from 42.5×10^{-5}/year in 1974 to 51.5×10^{-5}/year in 1981 (Quintero et al., 1985).

The incidence of SCC in Hispanics was also compiled for New Mexico in 1977–78. SCC affected Hispanic males at one eleventh the rate within the Anglo population, and Hispanic women at one fifth the Anglo rate. The male-to-female ratio among Hispanics was 1.1, which is dramatically lower than the ratios seen in Anglo populations (Table 1).

BCC incidence in nonwhite populations is not extensively documented. A 1955–1956 survey of Honolulu dermatologists regarding BCC in Japanese-Americans noted seven cases, an incidence of 1×10^{-5}/year (Allison and Wong, 1957). However, in Kauai, Hawaii, the (crude) incidence in Japanese-Americans was 74×10^{-5}/year in a 1983–1985 survey (23 total cases), which is one ninth the corresponding Caucasian rate (Leong et al., 1987; Stone et al., 1986). The high rate on Kauai may be due, in part, to a single family with xeroderma pigmentosum.

The special surveys in Hawaii have also contributed estimates of the incidence of SCC. In 1983 to 1985 in Kauai, 15 cases were found in patients

Table 2. Incidence of Nonmelanoma Skin Cancer
in Texas, 1949–1966[a]

	Males	Females	Male:Female ratio
Anglo	133	74	1.8
Hispanic	22	21	1.1
Nonwhite	5	4	1.2

Note. Cases $\times 10^{-5}$/year (age-adjusted to the 1970 U.S. population).
[a] From Macdonald and Heinze (1978).

of Japanese origin (49 × 10^{-5}/year), although as mentioned above, many of these cases may represent a single family with xeroderma pigmentosum. In 1955–56 in Honolulu, 10 cases were found among the non-Caucasian, predominantly Asian-American population (2 × 10^{-5}/year). This is more consistent with the overall incidence of nonmelanoma skin cancer in East Asian populations (Muir *et al.,* 1987).

Blacks also have a much lower incidence of BCC and SCC. In the United States in the 1970s, the incidence of nonmelanoma skin cancer among blacks was 3.4 × 10^{-5}/year (based on 68 cases), which is one sixty-eighth of the Caucasian rate in the same survey (Scotto *et al.,* 1983). Among all nonwhites in the areas covered by the Texas registry during 1949 to 1966, the incidence of nonmelanoma skin cancer was 4.3 × 10^{-5}/year, based on 169 cases (Table 2). The rate for the final 5-year period (1962–1966) was 3.9 × 10^{-5}/year, which suggests the absence of any substantial change in incidence over the 18 years of reported data (Macdonald and Heinze, 1978).

2.1.6. Age

The incidence of both BCC and SCC rises sharply with age in all populations studied. The age-specific incidence rates noted in the 1977–1978 skin cancer survey of eight areas in the United States is representative (Table 3).

A variety of mathematical models have been formulated to describe the age-specific incidence of cancer. For many cancer sites, the data are consistent

Table 3. Age- and Gender-Specific Incidence of BCC and SCC among Caucasians in Eight Regions of the United States, 1977–1978[a]

| Age | BCC | | | SCC | | | BCC:SCC ratio | |
	Males	Females	Male:Female ratio	Males	Females	Male:Female ratio	Males	Females
0–14	0.6	0.6	1.0	0.1	0.1	1.0	6.0	6.0
15–24	3.6	6.5	0.6	0.7	0.3	2.3	5.1	21.7
25–34	33.9	34.9	1.0	4.0	2.5	1.6	8.5	14.0
35–44	138.7	120.8	1.1	18.7	9.4	2.0	7.4	12.9
45–54	370.8	259.8	1.4	77.0	26.7	2.9	4.8	9.7
55–64	671.5	397.4	1.7	170.6	56.8	3.0	3.9	7.0
65–74	1,084.2	586.7	1.8	300.4	102.6	2.9	3.6	5.7
75–84	1,475.4	765.6	1.9	517.4	183.8	2.8	2.9	4.2
85+	1,528.2	780.7	2.0	577.1	278.1	2.1	2.6	2.8

Note. Rates are ×10^{-5}/year.
[a] From Scotto *et al.* (1983).

with a power curve, i.e., incidence $=$ constant $\times$ (age)^k, where k varies with the specific site and is usually between 3 and 6 (Cook et al., 1969). For cancer sites that follow this relation, the graph of the logarithm of incidence versus the logarithm of age is linear, with a slope of k. This relation is consistent with certain two-stage models of carcinogenesis as well as a multistage model of carcinogenesis with k stages (Gaffrey and Altshuler, 1988).

The value of the parameter k can be estimated from the published incidence data for whites (Table 4). This value is greater for SCC than for BCC; hence the ratio of SCC to BCC increases with age (Table 3). Similarly, the value of k is greater for men than for women for both BCC and SCC.

2.1.7. Anatomic Site

Both BCC and SCC are concentrated on the chronically sun-exposed skin of the face. For BCCs, 70% to 90% are on the head and neck in most population-based series (Table 5) (Haenszel, 1963; Lynch et al., 1970; Pearl and Scott, 1986). The site distribution among men is generally similar to the site distribution among women. However, women have a substantial and consistent excess of lower extremity BCCs compared to men, even though only a small proportion occur at this site. The distribution of BCCs among

Table 4. Estimates of the Exponent of Age Using the Power Function Model of BCC and SCC Incidence

Country	Exponent $\pm$ standard error		Source of data
	Men	Women	
BCC			
U.S.	3.54 $\pm$ .24	2.89 $\pm$ .22	Scotto et al., 1983
U.S. (nurses)	—	2.82 $\pm$ .37	Hunter et al., 1990
U.S. (Minnesota)	3.49 $\pm$ .11	2.93 $\pm$ .13	Chuang et al., 1990a
Switzerland	4.03 $\pm$ .33	3.16 $\pm$ .17	Levi et al., 1988
Netherlands	4.21 $\pm$ .15	3.70 $\pm$ .13	Coebergh et al., 1991
Wales	3.38 $\pm$ .47	3.44 $\pm$ .54	Roberts, 1990
SCC			
U.S.	4.44 $\pm$ .25	4.19 $\pm$ .08	Scotto et al., 1983
U.S. (prepaid health plan)	5.56 $\pm$ .27	5.44 $\pm$ .27	Glass and Hoover, 1989
Switzerland	6.47 $\pm$ .22	4.91 $\pm$ .40	Levi et al., 1988
Netherlands	6.32 $\pm$ .55	5.26 $\pm$ .60	Coebergh et al., 1991
England	5.64 $\pm$ .53	4.47 $\pm$ .38	Whitaker et al., 1979

Note. Calculated from the published data by linear regression of age-specific incidence vs. ln(age), weighted by the number of cases. Only population-based age-specific incidence rates from studies involving 100 or more cases were included in this table.

Table 5. Anatomic Distribution of BCC and SCC among Caucasians
in the United States, 1977–1978[a]

Site	Basal cell carcinoma		Squamous cell carcinoma	
	Male (%)	Female (%)	Male (%)	Female (%)
Face, head, neck	81.2	84.1	74.8	60.1
Scalp or forehead	18.3	18.7	17.9	13.3
Eyelids	5.1	5.8	2.0	2.7
Ears	8.0	1.6	14.4	1.9
Nose	18.3	25.4	8.3	14.8
Lips	3.2	6.4	11.1	5.5
Cheek, chin, or jaw	20.0	19.4	14.2	17.0
Neck	7.4	6.1	6.1	4.2
Head, neck, NOS[b]	0.8	0.7	0.8	0.8
Trunk	12.0	8.9	4.5	5.3
Front	4.6	3.4	2.6	2.8
Back	7.0	5.3	1.7	2.3
NOS	0.3	0.3	0.2	0.2
Upper extremities	4.9	3.4	18.1	25.8
Arms	4.0	2.7	5.7	8.8
Hands	0.8	0.6	11.9	16.4
Arms/hands, NOS	0.1	0.1	0.5	0.6
Lower extremities	1.3	2.9	1.3	5.7
Legs	1.2	2.7	0.9	5.1
Feet	0.1	0.1	0.3	0.5
Legs/feet, NOS	0.0	0.0	0.0	0.1
Other sites, NOS	0.5	0.7	1.4	3.2
Genitals	0.1	0.3	0.8	2.8
Skin, NOS	0.4	0.4	0.6	0.4

[a] From Scotto et al. (1983).
[b] NOS, not otherwise specified.

subsites on the head and neck is also similar between men and women with one notable exception, i.e., the substantially higher frequency of BCC on the ear among men. These differences between men and women in incidence of BCC on the ear and leg parallels differences at these sites in melanoma, and is presumably due to sun protection by hair among women and by clothing among men (Elwood and Gallagher, 1983). BCCs are located in areas of solar elastosis, although the correlation is not absolute (Brodkin et al., 1969; Zaynoun et al., 1985). In Minneapolis–St. Paul, Minnesota, and San Francisco–Oakland, California, there were significant increases in the incidence of BCC both on the head and neck and on trunk and extremities between 1971–72 and 1977–78 (Scotto et al., 1983). In the Texas registry, despite the rise in overall incidence on nonmelanoma skin cancer, there were no consistent increases in incidence on nonexposed surfaces between 1949 and 1966 in either Anglos or Hispanics (Macdonald and Heinze, 1978). Hence, the proportion of nonmelanoma skin cancer in nonexposed areas declined between 1949 and 1953 and between 1962 and 1966: from 13% to 7% in Anglo men and from 15% to 7% in Anglo women, and from 18% to 11% in Hispanic men and from 22% to 7% in Hispanic women.

For SCC, the majority of tumors also occur on the head and neck (Scotto *et al.,* 1983; Glass and Hoover, 1989; Haenszel, 1963; Lynch *et al.,* 1970; Pearl and Scott, 1986). However, the proportion which occurs in this area is greater for men (75% to 80%) than for women (60% to 70%) in most population-based series. The proportion of SCCs which occur on the trunk and upper extremities is greater in women than men, but the most dramatic female excess is in the proportion occurring on the lower extremities. As with BCC, the proportion occurring on the ear is much greater among men than women. SCC is substantially more common than BCC on the hands for both sexes. The best available data on secular trends was obtained from a HMO based in Portland, Oregon, which suggests similar rates of increase in incidence for head and neck, trunk, and extremity SCCs between 1960 and 1986 (Glass and Hoover, 1989).

Urbach (1969) used mannequin heads and a chemical UV dosimeter to document the distribution of ultraviolet flux on the human face, scalp, and neck under various conditions. The areas of high flux were the nose, cheekbones, lower lip, and in males, the back of the neck and the ear. Areas of the face which received little UVR were the orbital area, nasolabial fold, upper lip, center of chin, anterior neck, and lower retroauricular area. Urbach then plotted the anatomic locations of BCCs and SCCs from two case series (his own [Urbach, 1969] and a Scandinavian series [Magnusson, 1935]) and noted that most head and neck SCCs and BCCs occurred on the areas of high flux. He also noted that BCCs in the series were more likely than SCCs to occur in areas receiving less than 20% of the maximum UV dose, and that the anatomic distribution of SCCs on the ear was more closely correlated with UV dose than the distribution of BCCs. A more quantitative study of the distribution of UVR on the head confirmed Urbach's findings with respect to BCC (Diffey *et al.,* 1979). Furthermore, a series of BCCs and SCCs of the ear with precise anatomic subsite data also confirmed this pattern (Huriez *et al.,* 1962). Most of the population-based data is not sufficiently precise in anatomic description to test this hypothesis. However, the 1977–78 U.S. skin cancer survey data (Scotto *et al.,* 1983) indicate that 5% of BCCs and 2% of SCCs occur on the eyelids, which supports Urbach's suggestion. At the other protected facial site for which data are available, the ear in women, both histologic types occur in equal proportions. Contrary to the hypothesis, 20% of BCCs occurred on the nose, compared to 10% of SCCs. Similar trends were noted in the 1971–72 U.S. skin cancer survey for BCCs versus SCCs: 6% versus 2% located on the eyelids, 1.7% versus 2.1% among women on the ears, and 21% versus 8% on the nose. Observations from both the earlier and subsequent Minnesota studies are consistent with these data (Chuang *et al.,* 1990a,b; Lynch *et al.,* 1970).

Among people of African and Asian ethnicity, the vast majority of BCCs also occur on the head, particularly the (sun-exposed) face (Oettle, 1963; Atkinson *et al.,* 1963; Davies *et al.,* 1968; Oluwasanmi *et al.,* 1969; Budhraja *et al.,* 1972; Pringgoutomo and Pringgoutomo, 1963; Tada and Miki, 1984; Fleming *et al.,* 1975; Mora and Burris, 1981; Ademiluyi and Ijaduola, 1987). However, different patterns, including greater concentrations on the lower extremities, are noted in SCC, particularly in populations with the darkest skins (Table 6). One study tabulated BCCs and SCCs of the scalp among natives of Bombay. In the men, among whom this site is relatively more sun-exposed than among women, there were 11 BCCs and 24 SCCs in this series. Yet among women, only 1 BCC and 11 SCCs were noted (Paymaster *et al.,* 1971). This may be attributed to the predominance of solar factors in the etiology of BCC, but not SCC, among darker-skinned ethnic groups.

2.1.8. Additional Data on Nonmelanoma Skin Cancer (NMSC)

The above discussion has focused primarily on analyses which distinguished among the major histologic types of skin cancer, i.e., BCC, SCC, and melanoma. Some noteworthy early surveys did not make any of these distinctions (Dorn, 1944; Segi *et al.,* 1981), and there is a substantial body of information on NMSC which excludes melanoma, but does not distinguish BCC from SCC. Most of the provincial cancer registries in Canada track the incidence of NMSC; summary incidence data are displayed in Table 7 (Muir *et al.,* 1987).

BCC and SCC share many features, yet the differences noted above in their incidence patterns imply some differences in etiology, which justifies the focus on histology-specific data.

2.2. Mortality Rates

Mortality from NMSC has been difficult to ascertain due to substantial misclassification (Weinstock *et al.,* 1992). The origin of this misclassification was investigated in Rhode Island, where it was noted that many deaths due to mucosal SCC were listed on the death certificate as due to cancer or squamous cancer of the neck or of the head and neck, and coded with cutaneous carcinomas arising from the scalp and neck. Fortunately, scalp and neck carcinomas comprise a small portion of true cutaneous carcinomas, so elim-

Table 6. Anatomic Distribution of SCC among Darker-Skinned Ethnic Groups

Location	Year	N (site known)	Distribution by site (%)				Reference
			Head and neck	Trunk	Upper extremity	Lower extremity	
Blacks							
Transvaal (Banta)	1949–53	121	21	7	7	64	Oettle, 1963
Soweto	1966–75	88	34	3	14	49	Isaacson, 1979
New Guinea	1959–61	89	7	11	2	80	Atkinson et al., 1963
Nigeria	1960–67	157	39	8	3	50	Oluwasanmi et al., 1969
Uganda	1947–60	483	9	1	5	85	Davies et al., 1968
New Orleans	1948–79	163	48	20	11	28	More and Perniciaro, 1981
Memphis	1961–73	38	34	8	8	50	Fleming et al., 1975
Washington, DC	Before 1950	61	79	8	0	13	Hazen, 1950
Other dark-skinned ethnic groups							
Bangkok	1949–60	407	49	18	10	22	Tansurat, 1963
Taiwan	1946–61	187	43	12	8	37	Yeh, 1963
Manila	1953–59	180[a]	67	18	5	9	Pantangco et al., 1963
Singapore (Chinese)	1955–61	127	42	24	8	27	Shanmugaratnam and La'Brooy, 1963
(Others)	1955–61	47	32	30	13	26	Shanmugaratnam and La'Brooy, 1963
Pondicherry	1966–71	45	31	7	9	53	Budhraja et al., 1972
Bombay	1941–65	632	34	30	11	26	Paymaster et al., 1971
Calcutta	1963–65	71	24	23	14	39	vorty and DuttaChoudhuri, 1968

Note. Genital SCC is excluded.
[a] Presumably includes genital SCC.

Table 7. Incidence of Nonmelanoma Skin Cancer
in Canada, 1978–1982[a]

Province[b]	Males	Females
Alberta	80.1	58.7
British Columbia	109.1	75.5
Manitoba	84.0	56.2
New Brunswick	79.2	50.5
Nova Scotia	39.0	22.8
Prince Edward Island	99.3	65.8
Newfoundland	57.6	33.8
Northwest Territories & Yukon	17.8	16.5
Quebec	25.7	15.9
Saskatchewan	86.1	60.8

Note. Cases $\times 10^{-5}$/year, age-adjusted to the world standard.
[a] From Muir *et al.* (1987).
[b] No data available for Ontario.

ination of these allows for reasonable estimation of true NMSC mortality. Care must also be taken to avoid deaths due to AIDS-associated Kaposi's sarcoma, which was also coded with cutaneous carcinomas through 1986 (Weinstock *et al.,* 1992).

Application of this type of correction to U.S. mortality data from 1969 to 1988 leads to several observations. Mortality from NMSC rises sharply with age, particularly among whites. NMSC mortality declined during the 20-year period among both whites and blacks, and among both women and men (see Table 8). The 1987–88 mortality rate ratio for whites versus blacks was 1.8 among men and 1.2 among women, more than an order of magnitude less than the population-based incidence rate ratios reviewed above. Among whites, the male-to-female mortality rate ratio was 2.2, similar to or slightly

Table 8. Nonmelanoma Skin Cancer Mortality
in the United States[a]

	Rate 1987–88 ($\times 10^{-5}$/year)	Decline from 1969–1970 (%)
White men	0.67	21
White women	0.30	28
Black men	0.37	17
Black women	0.24	38

[a] Weinstock (1993).

less than the corresponding incidence rate ratio for SCC. Overall, approximately 1,200 deaths per year are attributable to NMSC in the United States (Weinstock, in press).

More detailed evaluation of true NMSC deaths in Rhode Island noted that most were due to SCC, and that a great number of these were due to lesions arising on the ear. When BCC led to death, it almost always occurred in the elderly (mean age 85 years), was often associated with refusal of treatment, and typically was due to direct extension of the tumor (Weinstock *et al.*, 1991a).

Although international comparisons must be viewed cautiously because of possible systematic misclassification, analyses of NMSC mortality rates among the states and provinces of the United States and Canada for 1950–1967 demonstrated a strong association of mortality with southerly latitude (Elwood *et al.*, 1974).

2.3. Measures of Frequency Other Than Incidence and Mortality

Prevalence of skin cancer is an important measure of the burden of this disease in the community. It reflects not only the incidence but also the rate at which existing cancers are cured and the rate at which people with skin cancer die.

The prevalence of BCC has been estimated in a large probability sample of the U.S. population during 1971 to 1974 (Johnson and Roberts, 1978). The estimate, based on examination by third-year dermatology residents, was 0.47% and 0.35% among men and women less than 75 years of age. The prevalence was directly associated with age, and reached a maximum of 2.0% in the group 65–74 years of age. A smaller study of a sample of Caucasian residents of Tipton County, Tennessee, during 1969 to 1971 noted a much higher prevalence of BCC (3.4%) among adults, based on 33 cases verified clinically by a dermatologist (Zagula-Mally *et al.*, 1974). The prevalence in the group 65–74 years of age was 8% (11/140).

Among participants in the American Academy of Dermatology (AAD) skin cancer screenings during 1985 to 1991, the prevalence of suspected BCC was 7.3% (Weinstock, 1992d). The validity of these diagnoses is unknown. However, in the 1986–87 Massachusetts screenings, the predictive value of a positive screening examination for BCC was 39% (57% among those with follow-up), and the (crude) prevalence of confirmed BCCs (i.e., excluding cases without follow-up data) was 3.2% (Koh *et al.*, 1990). In the 1988 Connecticut screening, the corresponding predictive values were 30% and 43%, and the crude prevalence of confirmed BCCs 2.4% (Bolognia *et al.*, 1990). It is likely that the true prevalence among individuals screened is considerably greater than the prevalence in the population from which they are derived, since most who present for screening have some warning sign of cutaneous malignancy (Weinstock, 1990).

The prevalence of SCC was 0.03% in the U.S. survey and 1.0% (10 cases) in the Tennessee survey. The prevalence of suspected SCCs in the AAD skin cancer screenings was 0.9%, and the crude prevalences of confirmed SCCs among all participants in the Massachusetts and Connecticut screenings were 0.35% and 0.40%.

2.4. Risk Factors for BCC and SCC

The leading cause of BCC and SCC in Caucasian populations is solar radiation. The importance of the sun in the etiology of BCC and SCC is established by demonstrating a strong and consistent association in diverse settings, a gradient of response (i.e., dose-response relation), and coherence with other epidemiological and laboratory data. The associations with proximity to the equator, race, ethnicity, anatomic areas of high sun exposure, and gender differences by anatomic site have already been discussed. The trends in incidence over time are also consistent with the prominent role of sunlight. Additional sun- and ultraviolet-related risk factors, exclusive of precursor lesions, are discussed below. The induction of cutaneous keratinocyte-derived malignancy by ultraviolet light in laboratory animals is well established and its parameters defined (Blum, 1959; Epstein, 1983; Findlay, 1928; Forbes, 1981; Fry and Ley, 1989; Winklmann, 1963).

2.4.1. Sun Sensitivity

As with melanoma, case–control studies have linked BCC risk with traits indicative of sun sensitivity, including red or blonde hair color, blue, green, or grey eye color, fair complexion, ethnic heritage (high risk is consistently associated with Irish, Welsh, and Scottish ancestry), susceptibility to sunburn and ease and depth of tanning (Finn, 1958; Gellin *et al.,* 1965; Hogan *et al.,* 1989; Kopf, 1979; Urbach *et al.,* 1972; Vitaliano and Urbach, 1980). Although many of these studies have methodological problems, such as poor response rates, poorly defined control groups, failure to match or adjust adequately for age, or inclusion of non-Caucasians (without stratification of subsequent analyses), the results consistently link sun sensitivity to BCC risk. Other studies of NMSC in which BCC was the predominant histologic type support the above analyses of the relation of sun sensitivity to BCC (Hall, 1950; Molesworth, 1927; Scotto *et al.,* 1982; Ward, 1952).

Several studies have used alternate methods to confirm the associations with sun sensitivity. A prospective study of patients receiving psoralen photochemotherapy (PUVA) confirmed the association of skin type with BCC

(Stern *et al.*, 1988). Finally, a cohort study of nurses in the United States linked BCC with hair color (relative risk 2.5 for red hair versus 0.7 for black hair), tendency to sunburn (relative risk 4.0), and inability to tan (relative risk 2.5) (Hunter *et al.*, 1990).

Experimental measures of sun sensitivity have also been studied in relation to BCC. The minimal erythema dose (MED) of UVR was not associated with BCC risk in several studies (Alcalay *et al.*, 1989; Aubin *et al.*, 1989). However, BCC patients have been noted to have more intense and prolonged erythema and less pigmentary response than controls after an ultraviolet irradiation (Aubin *et al.*, 1989; Jung *et al.*, 1980; Tanenbaum *et al.*, 1976).

SCC is also associated with red or blonde hair, blue, green, or grey eyes, fair complexion, ethnic heritage, and sun-sensitive skin types in case–control studies (Urbach *et al.*, 1972; Vitaliano and Urbach, 1980; Aubry and MacGibbon, 1985; Hogan *et al.*, 1990). Follow-up studies of PUVA patients have also noted an association between SCC and sun-sensitive skin types (Stern *et al.*, 1988; Foreman *et al.*, 1989).

2.4.2. Albinism

Among African blacks, the incidence of SCC in albinos is very high (approximately 500×10^{-5}/year) (Oettle, 1963; Okoro, 1975). Albino African blacks also have an increased incidence of BCC, and the BCC to SCC ratio is typically 0.4 or less (Cohen *et al.*, 1952; Oluwasanmi *et al.*, 1969; Shapiro *et al.*, 1953; Mora and Burris, 1981; Ademiluyi and Ijaduola, 1987; Mora and Perniciaro, 1981; Urbach, 1981). Asian Indian albinos follow similar patterns (Paymaster *et al.*, 1971). No increase in BCC or SCC incidence has been demonstrated among patients with vitiligo, although reports of SCC arising in vitiligenous skin have been published (Paymaster *et al.*, 1971).

2.4.3. Sun Exposure

Actual lifetime sun exposure is more difficult to ascertain than sun sensitivity, and published studies use various methods for assessment. Kopf and colleagues asked participants the average number of hours per day they spent outdoors during their lifetime (Kopf, 1979). For BCC, the relative risks were 4.9 for 6 or more hours, and 3.6 for 3 to 5 hours, compared to the 0- to 2-hour per day reference group. Hogan and colleagues observed a crude relative risk of 1.4 both for those who spent 6 or more hours per day outdoors in the summer and for those who spent 3 or more hours per day outdoors in the winter (Hogan *et al.*, 1989). Urbach measured cumulative sun exposure by a combination of vocational, military, sunbathing, and sports participation history and found a strong association with BCC (relative risk of 9.9 for the group with greatest exposure) (Urbach *et al.*, 1972).

Similar observations have been made for SCC. Urbach found that cumulative sun exposure was more closely related to SCC risk (relative risk 21.3 for the group with greatest exposure) than to BCC risk, although this comparison was confounded by age (Urbach *et al.*, 1972). Aubry and colleagues also noted a strong association of SCC and sun exposure (Allison and Wong, 1957). A study from Saskatchewan noted that farmers were at increased risk (Hogan *et al.*, 1990). Finally, in a study of Maryland watermen, a past or current history of SCC but not of BCC was associated with average annual intensity of sun exposure (Strickland *et al.*, 1989; Vitasa *et al.*, 1990).

2.4.4. Sun-Induced Damage

Both BCC and SCC have been linked to evidence of sun-induced damage such as sunburn history (Hunter *et al.*, 1990; Hogan *et al.*, 1989, 1990; Urbach *et al.*, 1972) and physical signs of dermatoheliosis (Engel *et al.*, 1988).

2.4.5. Xeroderma Pigmentosum

Xeroderma pigmentosum is a genetic disorder characterized by defective DNA repair and marked photosensitivity (Robbins, 1988). During the first two decades of life, patients with this disorder have a several thousand-fold increased risk of melanoma, nonmelanoma skin cancer, and cancers of certain other sites (Kraemer *et al.*, 1984). The nonmelanoma skin cancers include both BCC and SCC, although a majority are SCC, and 97% occur on the head and neck (Kraemer *et al.*, 1987).

2.4.6. Ultraviolet Phototherapy

Psoralen photochemotherapy (PUVA) has been associated with the development of SCC. One large American cohort of PUVA-treated patients were found to have an approximately 10-fold increased incidence of SCC 5 to 10 years after initiation of PUVA. Those who received the highest doses had the greatest (more than 50-fold increased) risks of SCC (Stern *et al.*, 1988). The same group noted the incidence of BCCs was more than double the rate expected in the general population, and the relative risk for BCC in the highest PUVA-dose group was 7 (Stern *et al.*, 1988). A smaller increase in SCC risk was noted in another U.S. study (Foreman *et al.*, 1989). Studies in European cohorts (Eskelinen *et al.*, 1985; Henseler *et al.*, 1987) have not confirmed these findings, although several trends noted in these studies parallel the findings from the United States. The nature and magnitude of the discrepancies are debated, and may be due to various factors, including dosing schedule, background rate of SCC, and exposure to other carcinogens.

The literature regarding an association between nonmelanoma skin cancer and ultraviolet-B (UV-B) phototherapy is scanty. One case–control study of nonmelanoma skin cancer among psoriatics noted a relative risk of 4.7 for high exposure to tar and UV-B (Stern *et al.,* 1980). Another suggested a link between SCC and sunlamp use (Aubry and MacGibbon, 1985). Others have failed to find an excess risk with these exposures (Maughan *et al.,* 1980; Pittelkow *et al.,* 1981). One study noted an excess of BCC among psoriatics, which was apparently not attributable to sunlamp use, tar exposure, or ionizing radiation, although unassessed sun exposure may explain this observation (Stern *et al.,* 1985).

2.4.7. Other Types of Radiation

Shortly after their discovery, X-rays were noted to cause SCC in exposed areas. Subsequent studies have shown that conventional X-ray therapy for tinea capitis, acne, and other benign dermatoses is associated with BCC more commonly than SCC, particularly for treatments involving the face or scalp. The latency period is typically in excess of 20 years for SCC, and may be greater for BCC (up to 64 years has been reported). Exposure to other carcinogens, particularly sunlight, appears to further increase the risk of malignancy among those exposed to X-rays (Cade, 1957; Davis *et al.,* 1989; Martin *et al.,* 1970; Shore *et al.,* 1984; Traenkle, 1963; Ron *et al.,* 1991; Hildreth *et al.,* 1985).

Grenz rays (ultrasoft X-rays) have also been associated with an increased risk of SCC (Dabski and Stoll, 1986; Lindelof and Eklund, 1986; Mortensen and Kjeldsen, 1987). Many of these patients had been exposed to other carcinogens. The magnitude of the increase appears modest: a large series with excellent follow-up noted a 45% increase in overall incidence of skin cancers other than melanoma and BCC (Lindelof and Eklund, 1986). However, the magnitude of the increase in incidence is higher if only the cutaneous sites exposed to the Grenz rays are considered. Carcinomas are usually diagnosed after a latency period of 10 or more years (Lindelof and Eklund, 1986). A link between BCC and Grenz-ray exposure has not been established or refuted.

A 42-year follow-up of individuals exposed to radiation from the atomic bomb explosion at Nagasaki has also revealed a marked increase in cutaneous malignancies. The greatest increases occurred among those closest to the epicenter, and the skin cancers were most likely to occur more than 30 years after exposure (Sadamori *et al.,* 1989). Gold rings contaminated with radon have also been linked to cutaneous carcinoma after 30- to 40-year latent periods (Callary, 1989; Gerwig and Winer, 1968). Uranium workers may also have increased risk of BCC (Šecová *et al.*, 1978).

2.4.8. Arsenic Exposure

The ingestion of arsenic in drinking water was noted since at least 1809 to produce cutaneous malignancy. Arsenical contamination of drinking water has been responsible for several notable epidemics. "Reichenstein disease" produced a high incidence of malignancies for many years in the town near Klodzko, Poland, for which the disease was named. The arsenical fumes produced by the smelting of the gold-containing ores arsenopyrite and loellingite were precipitated by rain and entered the drinking water. Reichenstein disease disappeared after changes were made in the method of smelting, and a safe water supply was provided in 1928 (Neubauer, 1947).

Other areas of chronic exposure to arsenic include Córdoba, Argentina (Bergoglio, 1964), Antofagasta, Chile (Zaldívar, 1974), Region Lagunera, Mexico (Cebrián *et al.*, 1983), and elsewhere, although not all of these were linked to cutaneous malignancies (Morton *et al.*, 1976). An episode in Taiwan is particularly noteworthy for the epidemiological investigations of "Blackfoot disease" and the associated cutaneous malignancies, keratoses, and pigmentary changes, which documented their association with the concentration of arsenic in artesian well water (Tseng, 1977; Tseng *et al.*, 1968; Chen and Wang, 1990). Most of the cutaneous malignancies were invasive or *in situ* SCCs (Yeh *et al.*, 1968).

Occupational arsenic exposures have primarily involved arsenical insecticides and arsenical powders used to treat mange in sheep. The predominant cause of arsenic-related malignancy, arsenic-containing medicinal preparations, such as Fowler's solution, Donovan's solution, and Asiatic pills, were linked to cutaneous malignancy over 100 years ago (Eggers, 1932; Hutchinson, 1887), yet cases continue to be reported (Jackson and Grainge, 1975), and the author is aware of at least one patient who was given an arsenical medication during his European travels in the 1980s. The arsenicals were most often used for psoriasis, which may explain reported associations of nonmelanoma skin cancer with psoriasis (Neubauer, 1947; Currie, 1947).

2.4.9. Soot, Shale Oil, and Tar and Petroleum Products

The study of chemical carcinogenesis in humans may be said to have begun in 1775 with Percival Pott, who observed that scrotal cancer "seems to derive its origin from a lodgement of soot in the rugae of the scrotum" among men employed as chimney sweeps (Levin, 1963; Pott, 1963; Potter, 1963). Coal, pitch, tar, shale and cutting oils, and other petroleum products have since been implicated, particularly in occupational settings, such as spinners exposed to shale oil in the cotton industry (Gordon and Silverstone, 1976; Hueper, 1963). Workplace conditions and sanitary standards in developed countries are now vastly superior to those which were prevalent in eighteenth-century England, and the frequency of these chemically induced cancers correspondingly low (Eggers, 1932; Waterhouse, 1971). A recent study of scrotal cancer in New York State noted both its rarity (1 case per 2 million men per year) and the apparent absence of plausible occupational exposures in most cases (A. L. Weinstein *et al.*, 1989). Carcinogenic exposures from psoriasis therapy—including tar, arsenic, and PUVA—and human papillomavirus infection may be more common etiologic factors of scrotal carcinoma (Andrews *et al.*, 1991; Stern, 1990).

2.4.10. Immunosuppression

Among transplant recipients, the incidence of SCC is particularly high. BCC risk may be elevated as well, but to a lesser degree (Kinlen *et al.*, 1983). The magnitude of the elevation is associated with more equatorial latitudes of residence, greater sun sensitivity and sun exposure prior to the transplant, although sun exposure after transplantation does not appear to be a predominant factor (Boyle *et al.*, 1984; Kelly *et al.*, 1987). A large Canadian series noted an 18-fold increase in risk of SCC after renal transplantation (Gupta *et al.*, 1986). Other studies have found similar trends and have also noted an increased risk of metastasis among SCCs in transplant recipients (Gupta *et al.*, 1986; Disler *et al.*, 1981; Hoxtell *et al.*, 1977; Koranda *et al.*, 1974; Penn, 1987). Immunosuppression with cyclosporin appears to increase the risk of cutaneous malignancy, particularly SCC (Oxholm *et al.*, 1989; Penn, 1988). Patients with mycosis fungoides also have an increased incidence of nonmelanoma skin cancer, particularly SCC (Abel *et al.*, 1986; Kravitz and McDonald, 1978; Vonderheid *et al.*, 1989). In addition to immunosuppression from the lymphoma, they are typically exposed to carcinogenic treatments, including nitrogen mustard, electron beam, and PUVA. Finally, associations of both Hodgkin's disease and chronic lymphatic leukemia with nonmelanoma skin cancer have been noted (Eastcott, 1963; Kaldor *et al.*, 1987). Immunosuppression due to UVR may be one mechanism by which sunlight induces skin cancers (Kripke, 1988; Morison, 1989).

2.4.11. Diet and Smoking

Synthetic retinoids have been used successfully to reduce the incidence of nonmelanoma skin cancer in patients with xeroderma pigmentosum and basal cell nevus syndrome (Goldberg *et al.*, 1989; Hodak *et al.*, 1987; Kraemer *et al.,* 1988), and may be useful in certain other patients at very high risk (Peck *et al.,* 1982). Nevertheless, no consistent association has been demonstrated between either BCC or SCC and diet, including the use of selenium, vitamin A, carotene, and other nutrients (Hunter *et al.,* 1990; Clark *et al.,* 1984, 1987). One case–control study with a poor response rate suggested an association between smoking and SCC (Aubry and MacGibbon, 1985), although another did not (Hogan *et al.,* 1990). A subsequent cohort study found no association between smoking and BCC (Hunter *et al.,* 1990).

2.4.12. Human Papillomavirus

Human papillomaviruses (HPVs) are small DNA viruses which most commonly give rise to warts of little malignant potential (Cobb, 1990; Howley and Schlegel, 1985). However, certain HPV types (particularly type 16) have been linked to SCC in several settings (Howley, 1987; Syrjanen, 1987; Pierceall *et al.,* 1991). The evidence is particularly striking in the setting of genital carcinomas, including cervical, penile, and anal SCCs (Boon *et al.,* 1989; Holly *et al.,* 1989; Zur, 1989). Among patients with epidermodysplasia verruciformis, SCC has been associated with a variety of HPV types, of which type 5 is the most frequent (Ostrow *et al.,* 1982; Yutsudo *et al.,* 1985). These patients usually develop the SCC on sun-exposed sites, and SCC appears to be uncommon among blacks with epidermodysplasia verruciformis (Jacyk and Subbuswamy, 1979); hence, sunlight appears to be an important cofactor in the etiology of HPV-associated SCC in this patient group. A similar relationship may pertain to some SCCs in renal transplant recipients (Boyle *et al.,* 1984; Lutzner *et al.,* 1983; Dyall-Smith *et al.,* 1991). Recently, several reports of HPV-associated nongenital cutaneous SCC, particularly SCC of the finger, have been published (Grimmel *et al.,* 1988; Kawashima *et al.,* 1986; Moy *et al.,* 1989; Ostrow *et al.,* 1989; Eliezri *et al.,* 1990; Kettler *et al.,* 1990; McDonnell *et al.,* 1989; Ashinoff *et al.,* 1991). In the more typical SCCs on sun-exposed skin of otherwise healthy individuals, the role of HPV (if any) remains to be determined (Kawashima *et al.,* 1990; Quan and Moy, 1991).

The epidemiology of nonmelanoma skin cancer of the genitalia follows quite different patterns than that of other anatomic sites and is not reviewed here.

2.4.13. Burns, Scars, Chronic Inflammation, and Other Dermatoses

Other etiologies of BCC and SCC have been identified, largely on the basis of case series. Marjolin's ulcer is a carcinoma arising in a burn scar, although in the older literature its definition was broader. Burn scar cancers are predominantly SCCs and typically have a latent period of decades (Macdonald and Bubendorf, 1964; Ikegawa *et al.*, 1989; Mosberg *et al.*, 1988; Novick *et al.*, 1977; Treves and Pack, 1930). Other types of thermal injury, such as erythema ab igne and frostbite, have been linked to SCC (DiPirro and Conway, 1966; Peterkin, 1955). In certain areas, local customs may give rise to a high prevalence of thermal injury, hence a high proportion of SCCs arising from this source. For example, in Kashmir a pot tied to the abdomen and containing burning leaves or charcoal has been used for warmth, and has been linked to thousands of cases of SCC ("Kangri cancer") (Mulay, 1963). Similarly, in Lanchow (northwest China), sleeping on heated brick beds has been linked to "Kang" cancers of the skin over the trochanter (Laycock, 1948). Scars of other etiologies (including vaccination, nonthermal trauma, lupus vulgaris, discoid lupus, and recessive dystrophic epidermolysis bullosa) have been associated with BCC or SCC in many cases (Arons *et al.*, 1965; Belisario, 1959; Dix, 1960; Gardiner, 1959; Marmelzat, 1968; Noodleman and Pollack, 1986). Other chronic inflammatory processes, such as sinus tracts due to osteomyelitis and hidradenitis superativa, lichen planus, chromoblastomycosis, lobomycosis, chronic filariasis, necrobiosis lipoidica diabeticorum, herpetic infections, acne conglobata, Hansen's disease, and other chronic ulcers and infections, have been thought to give rise to SCC and, perhaps occasionally, to BCC, in individual published cases or case series (Chakravorty and Dutta-Choudhuri, 1968; Sedlin and Fleming, 1963; Wyburn-Mason, 1955, 1957). "Dhoti" cancer in men and "sari" cancer in women are SCCs, typically located on the lateral waist, due to chronic irritation from the garments worn by many South Asians, after which these cancers are named (Chakravorty and Dutta-Choudhuri, 1968; Paymaster *et al.*, 1971). Nevus sebaceous may give rise to BCC and, rarely, SCC (Jones and Heyl, 1970). Certain other dermatoses and syndromes, such as porokeratosis of Mibelli (and, in isolated cases, other forms of porokeratosis), severe eruptions due to quinacrine, and keratitis, ichthyosis, and deafness (KID) syndrome, have been reported to give rise to SCC (Bauer, 1981; Cort and Abdel-Aziz, 1972; Grob *et al.*, 1987; Guss *et al.*, 1971; Hazen *et al.*, 1989; Shrum *et al.*, 1982). SCC has been reported in skin tubes used for antethoracic esophageal reconstruction and in epidermal cysts (Nakayama *et al.*, 1971; Shah *et al.*, 1989). BCC has been reported in Rombo's syndrome (Michaelsson *et al.*, 1981). Basal cell nevus syndrome is an autosomal-dominant disorder involving multiple organ systems including the skin (Gorlin, 1987). Affected individuals have an extraordinarily high frequency and multiplicity of BCCs.

3. Summary

The epidemiological characteristics of malignant melanoma, basal cell carcinoma, and squamous cell carcinoma are dominated by those related to sun exposure, although other factors play a role in the etiology of each. Primary prevention efforts for these tumors will therefore naturally overlap. Strategies geared toward early detection, however, may differ substantially due to dissimilarities in high-risk groups and mortality patterns.

References

Abel, E. A., Sendagorta, E., and Hoppe, R. T., 1986, Cutaneous malignancies and metastatic squamous cell carcinoma following topical therapies for mycosis fungoides, *J. Am. Acad. Dermatol.* 14:1029–1038.

Ademiluyi, S. A., and Ijaduola, G. T. A., 1987, Occurrence and recurrence of basal cell carcinoma of the head and neck in negroid and albinoid africans, *J. Laryngol. Otol.* 101:1324–1328.

Alcalay, J., Goldeberg, L. H., Kripke, M. L., and Wolf, J. E., 1989, The sensitivity of Langerhans cells to simulated solar radiation in basal cell carcinoma patients, *J. Invest. Dermatol.* 93: 746–750.

Allison, S. D., and Wong, K. L., 1957, Skin cancer: Some ethnic differences, *Arch. Dermatol.* 76: 737–739.

Andrews, P. E., and Farrow, G. M., Oesterling, J. E., 1991, Squamous cell carcinoma of the scrotum: Long-term followup of 14 patients, *J. Urol.* 146:1299–1304.

Arons, M. S., Lynch, J. B., Lewis, S. R., and Blocker, T. G., 1965, Scar tissue carcinoma: I. A clinical study with special reference to burn scar carcinoma, *Ann. Surg.* 161:170–188.

Ashinoff, R., Li, J. J., Jacobson, M., Friedman-Klein, A. E., and Geronemus, R. G., 1991, Detection of human papillomavirus DNA in squamous cell carcinoma of the nail bed and finger determined by polymerase chain reaction, *Arch. Dermatol.* 127:1813–1818.

Atkinson, L., Farago, C., Forbes, B. R. V., and ten Saldam, R. E. J., 1963, Skin cancer in New Guinea native peoples, *Natl. Cancer Inst. Monogr.* 10:167–179.

Aubin, F., Zultak, M., Blanc, D., Terrasse, F., Quencez, E., and Agache, P., 1989, Reaction to UV-induced erythema in young patients with basal cell carcinoma, *Photodermatology* 6: 118–123.

Aubry, F., and MacGibbon, B., 1985, Risk factors of squamous cell carcinoma of the skin, *Cancer* 55:907–911.

Auerbach, M., 1961, Geographic variation in incidence of skin cancer in the United States, *Public Health Rep.* 76:345–348.

Bain, C., Colditz, G. A., Willett, W. C., Stampfer, M. J., Green, A., Bronstein, B. R., Mihm, M. C., Rosner, B., Hennekens, C. H., and Speizer, F. E., 1988, Self-reports of mole counts and cutaneous malignant melanoma in women: Methodological issues and risk of disease, *Am. J. Epidemiol.* 127:703–712.

Balch, C. M., Soong, S., Shaw, H. M., Urist, M. M., and McCarthy, W. H., 1992a, An analysis of prognostic factors in 8500 patients with cutaneous melanoma, in: *Cutaneous Melanoma,* C. M. Balch, A. N. Houghton, G. W. Milton, A. J. Sober, and S. Soong, (eds.), 2nd ed., J.B. Lippincott, Philadelphia, pp. 165–187.

Balch, C. M., Cascinelli, N., Drzewiecki, K. T., Eldh, J., Mackie, R. M., McCarthy, W. M., McLeod, R., Morton, D. L., Seigler, H. F., Shaw, H. M., Sim, F. H., Sober, A. J., Soong, S.-J., Takematsu, H., Tonak, J., and Wong, J., 1992b, A comparison of prognostic factors worldwide, in: *Cutaneous Melanoma,* C. M. Balch, A. N. Houghton, G. W. Milton, A. J. Sober, and S. Soong, (eds.), 2nd ed., J.B. Lippincott, Philadelphia, pp. 188–199.

Bale, S. J., Dracopoli, N. C., Tucker, M. A., Clark, W. H., Fraser, M. C., Starger, B. Z., Green, P., Donis-Keller, H., Housman, D. E., and Green, M. H., 1989, Mapping the gene for hereditary cutaneous malignant melanoma-dysplastic nevus to chromosome 1p. *N. Engl. J. Med.* 320:1367–1372.

Barnhill, R. L., and Roush, G. C., 1991, Correlation of clinical and histopathologic features in clinically atypical melanocytic nevi, *Cancer* 67:3157–3164.

Bauer, F., 1981, Quinacrine hydrochloride drug eruption (tropical lichenoid dermatitis): Its early and late sequelae and its malignant potential: A review, *J. Am. Acad. Dermatol.* 4:239–248.

Belisario, J. C., 1959, *Cancer of the Skin.* Butterworth, London.

Bergfelt, L., Newell, G. R., Sider, J. G., and Kripke, M. L., 1989, Incidence and anatomic distribution of cutaneous melanoma among United States hispanics, *J. Surg. Oncol.* 40:222–226.

Bergoglio, R. M., 1964, Mortalidad por cáncer en zonas de aguas arsenicales del la Provincia de Córdoba, República Argentina, *Prensa. Med. Argent.* 51:994–998.

Black, W. C., and Wiggins, C., 1985, Melanoma among southwestern American Indians, *Cancer* 55:2899–2902.

Black, W. C., Goldhahn, R. T., and Wiggins, C., 1987, Melanoma within a southwestern hispanic population, *Arch. Dermatol.* 123:1331–1334.

Blum, H. F., 1959, *Carcinogenesis by Ultraviolet Light,* Princeton University Press, Princeton, New Jersey.

Bolognia, J. L., Berwick, M., and Fine, J. A., 1990, Complete follow-up and evaluation of skin cancer screening in Connecticut, *J. Am. Acad. Dermatol.* 23:1098–1106.

Bolognia, J. L., Headley, A., Fine, J., and Berwick, M., 1992. Histologic evaluation of pigmented lesions in Connecticut and its influence on the reporting of melanoma, *J. Am. Acad. Dermatol.* 26:198–202.

Boon, M. E., Susanti, I., Tasche, M. J. A., and Kok, L. P., 1989, Human papillomavirus (HPV)-associated male and female genital carcinomas in a Hindu population: The male as vector and victim, *Cancer* 64:559–565.

Boyle, J., Mackie, R. M., Briggs, J. D., Junor, B. J. R., and Aitchison, T. C., 1984, Cancer, warts, and sunshine in renal transplant patients: A case control study. *Lancet* 1:702–705.

Brodkin, R. H., Kopf, A. W., and Andrade, R., 1969, Basal cell epithelioma and elastosis: A companson of distribution, in: *The Biologic Effect of Ultraviolet Radiation* (F. Urbach ed.), Pergamon Press, New York.

Budhraja, S. N., Pillai, V. C. V., Periyanayagam, W. J., Kaushik, S. P., and Bedi, B. M. S., 1972, Malignant neoplasms of the skin in Pondicherry, *Indian J. Cancer* 9:284–295.

Bureau of Cancer Epidemiology, 1989, Changing trends in malignant melanoma in New York State, *N. Y. State J. Med.* 89:239–241.

Byers, R. L., Lantz, P. M., Remington, P. L., and Phillips, J. L., 1991, Malignant melanoma: Trends in Wisconsin, 1980–1989, *Wis. Med. J.* 90:305–307.

Cade, S., 1957, Radiation induced cancer in man, *Br. J. Radiol.* 30:393–402.

Callary, E. M., 1989, Cancer caused by radioactive gold rings, *Can. Med. Assoc. J.* 141:507.

Cebrián, M. E., Albores, A., Aguilar, M., and Blakely, E., 1983, Chronic arsenic poisoning in the north of Mexico, *Human Toxicol.* 2:121–133.

Chakravorty, R. C., and Dutta-Choudhuri, R., 1968, Malignant neoplasms of the skin in eastern India: An analysis of cases seen at its Chittaranjan Cancer Hospital, Calcutta, during the years 1963–65 inclusive, *Indian J. Cancer* 5:133–144.

Chen, C. -J., and Wang, C. -J., 1990, Ecologic correlation between arsenic level in well water and age-adjusted mortality from malignant neoplasms, *Cancer Res.* 50:5470–5474.

Chuang, T. -Y., Popesca, A., Su, W. P. D., and Chute, C. G., 1990a, Basal cell carcinoma: A population-based incidence study in Rochester, Minnesota, *J. Am. Acad. Dermatol.* 22:413–417.

Chuang, T. -Y., Popescu, A., Su, W. P. D., and Chute, C. G., 1990b, Squamous cell carcinoma: A population-based incidence study in Rochester, Minn., *Arch. Dermatol.* 126:185–188.

Clark, L. C., Graham, G. F., Crounse, K. G., Grimson, R., Hulka, B., and Shy, C. M., 1984, Plasma selenium and skin neoplasms: A case-control study, *Nutr. Cancer* 6:13–21.

Clark, L. C., Turnbull, B. W., Graham, G., Hulka, B. S., Bray, J., and Shy, C. M., 1987, Non-melanoma skin cancer and plasma selenium: A prospective cohort study, in: *International Symposium on Selenium in Biology and Medicine,* G. F. Coombs, J. E. Spallholz, O. A. Levander, and J. E. Oldfield, (eds.), AVI Publishing, New York, pp. 1122–1134.

Clark, W. H., and Mihm, M. C., 1969, Lentigo maligna and lentigo maligna melanoma, *Am. J. Pathol.* 55:39–67.

Clark, W. H., Elder, D. E., Guerry, D., Braitman, L. E., Trock, B. J., Schultz, D., Synnestvedt, M., and Halpern, A. C., 1989, A model predicting survival in stage I melanoma based upon tumor progression, *J. Natl. Cancer Inst.* 81:1893–1904.

Clemente, C., Cochran, A. J., Elder, D. E., Levene, A., Mackie, R. M., Mihm, M. C., Rilke, F., Cascinelli, N., Fitzpatrick, T. B., and Sober, A. J., 1991, Histopathologic diagnosis of dysplastic nevi: Concordance among pathologists convened by the World Health Organization Melanoma Programme, *Hum. Pathol.* 22:313–319.

Cobb, M. W., 1990, Human papillomavirus infection, *J. Am. Acad. Dermatol.* 22:547–566.

Coebergh, J. W. W., Neumann, H. A. M., Vrints, L. W., van der Heijden, L., Meijer, W. J., and Verhagen-Teulings, M. T., 1991, Trends in the incidence of non-melanoma skin cancer in the SE Netherlands 1975–1988: A registry-based study, *Br. J. Dermatol.* 125:353–359.

Cohen, L., Shapiro, M. P., and Kern, P., Henning, A. J. H., 1952, Malignant disease in the Transvaal: I. Cancer of the skin, *S. Afr. Med. J.* 26:932–944.

Cook, P. J., Doll, R., and Fellingham, S. A., 1969, A mathematical model for the age distribution of cancer in man, *Int. J. Cancer* 4:93–112.

Cort, D. F., and Abdel-Aziz, A. H. M., 1972, Epithelioma arising in porokeratosis of Mibelli, *Br. J. Plastic Surg.* 25:318–328.

Crowley, N. J., Dodge, R., Vollmer, R. T., and Seigler, H. F., 1991, Malignant melanoma in black Americans: A trend toward improved survival, *Arch. Surg.* 126:1359–1365.

Currie, A. N., 1947, The role of arsenic in carcinogenesis, *Br. Med. Bull.* 4:402–405.

Dabski, K., and Stoll, H. L., 1986, Skin cancer caused by Grenz rays, *J. Surg. Oncol.* 31:87–93.

Davies, J. N. P., Tank, R., Meyer, R., and Thurston, P., 1968, Cancer of the integumentary tissues in Uganda Africans: The bases for prevention, *J. Natl. Cancer Inst.* 41:31–51.

Davis, M. M., Hanke, C. W., Zollinger, T. W., Montebello, J. F., Hornback, N. B., and Norins, A. L., 1989, Skin cancer in patients with chronic radiation dermatitis, *J. Am. Acad. Dermatol.* 20:608–616.

DiPirro, E., and Conway, H., 1966, Carcinoma after frostbite: A case report, *Plast. Reconstr. Surg.* 38:541–543.

Disler, P. B., MacPhail, A. P., Meyers, A. M., and Myburgh, J. A., 1981, Neoplasia after successful renal transplantation, *Nephron* 29:119–123.

Dix, C. R., 1960, Occupational trauma and skin cancer, *Plast. Reconstr. Surg.* 26:546–554.

Dorn, H. F., 1944, Illness from cancer in the United States, *Public Health Rep.* 59:33–48, 65–77, 97–115.

Dubin, N., Moseson, M., and Pasternack, B. S., 1986, Epidemiology of malignant melanoma: Pigmentary traits, ultraviolet radiation, and the identification of high risk populations, *Recent Results Cancer Res.* 102:56–75.

Dubin, N., Moseson, M., and Pasternack, B. S., 1989, Sun exposure and malignant melanoma among susceptible individuals *Environ. Health Perspectives* 81:139–151.

Dubin, N., Pasternack, B. S., and Moseson, M., 1990, Simultaneous assessment of risk factors for malignant melanoma and non-melanoma skin lesions, with emphasis on sun exposure and related variables, *Int. J. Epidemiol.* 19:811–819.

Dubrow, R., 1986, Malignant melanoma in the printing industry, *Am. J. Indust. Med.* 10:119–126.

Diffey, B. L., Tate, T. J., and Davis, A., 1979, Solar dosimetry of the face: The relationship of natural ultraviolet radiation exposure to basal cell carcinoma localisation, *Phys. Med. Biol.* 24:931–939.

Duggleby, W. F., Stoll, H., Priore, R. L., Greenwald, P., and Graham, S., 1981, A genetic analysis of melanoma—polygenic inheritance as a threshold trait, *Am. J. Epidemiol.* 114:63–72.

Dyall-Smith, D., Trowell, H., Mark, A., and Dyall-Smith, M., 1991, Cautaneous squamous cell carcinomas and papillomaviruses in renal transplant recipients: A clinical and molecular biological study, *J. Dermatol. Sci.* 2:139–146.

Eastcott, D. F., 1963, Epidemiology of skin cancer in New Zealand, *Natl. Cancer Inst. Mongr.* 10:141–151.

Eggers, H. E., 1932, The etiology of cancer: II. Irritation, *Arch. Pathol.* 13:112–150.

Elder, D. E., 1985, The blind men and the elephant: Different views of small congenital nevi, *Arch. Dermatol.* 121:1263–1265.

Eliezri, Y. D., Silverstein, S. J., and Nuovo, G. J., 1990, Occurrence of human papillomavirus type 16 DNA in cutaneous squamous and basal cell neoplasms, 1990, *J. Am. Acad. Dermatol.* 23:836–842.

Elwood, J. M., 1986, Could melanoma be caused by fluorescent light? A review of relevant epidemiology, *Recent Results Cancer Res.* 102:127–136.

Elwood, J. M., and Gallagher, R. P., 1983, Site distribution of malignant melanoma. *Can. Med. Assoc. J.* 128:1400–1404.

Elwood, J. M., Lee, J. A. H., Walter, S. D., Mo, T., and Green, A. E. S., 1974, Relationship of melanoma and other skin cancer mortality to latitude and ultraviolet radiation in the United States and Canada, *Int. J. Epidemiol.* 3:325–332.

Elwood, J. M., Gallagher, R. P., Hill, G. B., Spinelli, J. J., Pearson, J. C. G., and Threlfall, W., 1984, Pigmentation and skin reaction to sun as risk factors for cutaneous melanoma: Western Canada Melanoma Study, *Br. Med. J.* 288:99–102.

Elwood, J. M., Gallagher, R. P., Davison, J., Hill, G. B., 1985a, Sunburn, suntan and the risk of cutaneous malignant melanoma—the Western Canada Melanoma Study, *Br. J. Cancer* 51:543–549.

Elwood, J. M., Gallagher, R. P., Hill, G. B., Pearson, J. C. G., 1985b, Cutaneous melanoma in relation to intermittent and constant sun exposure—the Western Canada Melanoma Study, *Int. J. Cancer* 35:427–33.

Engel, A., Johnson, M. -L., and Haynes, S. G., 1988, Health effects of sunlight exposure in the United States: Results from the first National Health and Nutrition Examination Survey, 1971–1974, *Arch. Dermatol.* 124:72–79.

Epstein, J. H., 1983, Photocarcinogenesis, skin cancer and aging, *J. Am. Acad. Dermatol.* 9:487–502.

Eskelinen, A., Halme, K., Lassus, A., and Idanpaan-Heikkila, J., 1985, Risk of cutaneous carcinoma in psoriatic patients treated with PUVA, *Photodermatol.* 2:10–14.

Fears, T. R., and Scotto, J., 1982, Changes in skin cancer morbidity between 1971–72 and 1977–78, *J. Natl. Cancer Inst.* 69:365–370.

Findlay, G. M., 1928, Ultraviolet light and skin cancer, *Lancet* 2:1070–1073.

Finn, O. A., 1958, Skin colouring in rodent ulcer with special reference to jute workers, *Br. J. Dermatol.* 70:218–219.

Fleming, I. D., Barnawell, J. R., Burlison, P. E., and Rankin, J. S., 1975, Skin cancer in black patients, *Cancer* 35:600–605.

Forbes, P. D., 1981, Experimental photocarcinogenesis, an overview, *J. Invest. Dermatol.* 77: 139–143.

Foreman, A. B., Roenigk, H. H., Caro, W. A., and Magid, M. L., 1989, Long-term follow-up of skin cancer in the PUVA-48 cooperative study, *Arch. Dermatol.* 125:515–519.

Frey, C., and Hartman, A., 1991, U.S. trends in melanoma incidence and mortality, *J. Natl. Cancer Inst.* 83:1705.

Friedman, R. J., Heilman, E. R., Rigel, D. S., and Kopf, A. W., 1985, The dysplastic nevus: Clinical and pathologic features, *Dermatol. Clin.* 3:239–249.

Fry, R. J. M., and Ley, R. D., 1989, Ultraviolet radiation induced skin cancer, *Carcinog. Compr. Surv.* 11:321–337.

Gaffrey, M., and Altshuler, B., 1988, Examination of the role of cigarette smoke in lung carcinogenesis using multistage models, *J. Natl. Cancer Inst.* 80:925–931.

Gallagher, R. P., Elwood, J. M., Threlfall, W. J., Band, P. R., and Spinelli, J. J., 1986, Occupation and risk of cutaneous melanoma, *Am. J. Indust. Med.* 9:289–294.

Gallagher, R. P., Ma, B., McLean, D. I., Yang, C. P., Mo, V., Carruthers, J. A., and Warshawski, L. M., 1990, Trends in basal cell carcinoma, squamous cell carcinoma, and melanoma of the skin from 1973 through 1987, *J. Am. Acad. Dermatol.* 23:413–421.

Gardiner, A. W., 1959, Trauma and squamous skin cancer, *Lancet* 1:760–761.

Gari, L. M., Rivers, J. K., and Kopf, A. W., 1988, Melanomas arising in large congenital nevocytic nevi: A prospective study, *Pediatr. Dermatol.* 5:151–158.

Gellin, G. A., Kopf, A. W., and Garfinkel, L., 1965, Basal cell epethelioma: A controlled study of associated factors, *Arch. Dermatol.* 91:38–45.

Gellin, G. A., Kopf, A. W., and Garfinkel, L., 1969, Malignant melanoma: A controlled study of possibly associated factors, *Arch. Dermatol.* 99:43–48.

Gerwig, T., and Winer, M. N., 1968, Radioactive jewelry is a cause of cutaneous tumor, *J. Am. Med. Assoc.* 205:595–596.

Glass, A. G., and Hoover, R. N., 1989, The emerging epidemic of melanoma and squamous cell skin cancer, *J. Am. Med. Assoc.* 262:2097–2100.

Goldberg, L. H., Hsu, S. H., and Alcalay, J., 1989, Effectiveness of isotretinoin in preventing the appearance of basal cell carcinomas in basal cell nevus syndrome, *J. Am. Acad. Dermatol.* 21:144–145.

Goldgar, D. E., Cannon-Albright, L. A., Meyer, L. J., Piepkorn, M. W., Zone, J. J., and Skolnick, M. H., 1991, Inheritance of nevus number and size in melanoma and dysplastic nevus syndrome kindreds, *J. Natl. Cancer Inst.* 83:1726–1733.

Gordon, D., and Silverstone, H., 1976, Worldwide epidemiology of premalignant and malignant cutaneous lesions, in: *Cancer of the Skin: Biology, Diagnosis, Management* (R. Andrade *et al.,* eds.), Saunders, Philadelphia, pp. 405–434.

Gorlin, R. J., 1987, Nevoid basal-cell carcinoma syndrome. *Medicine* 66:98–113.

Graham, S., Marshall, J., Haughey, B., Stoll, H., Zielezny, M., Brasure, J., and West, D., 1985, An inquiry into the epidemiology of melanoma, *Am. J. Epidemiol.* 122:606–619.

Greene, M. H., 1991, Rashomon and the procrustean bed: A tale of dysplastic nevi, *J. Natl. Cancer Inst.* 83:1720–1724.

Greene, M. H., and Wilson, J., 1985, Second cancer following lymphatic and hematopoietic cancers in Connecticut, 1935–82, *Natl. Cancer Inst. Monogr.* 68:191–217.

Greene, M. H., and Wilson, J., 1985, Second cancer following lymphatic and hematopoietic cancers in Connecticut, 1935–82, *Natl. Cancer Inst. Monogr.* 68:191–217.

Greene, M. H., Clark, W. H., Tucker, M. A., Kraemer, K. H., Elder, D. E., and Fraser, M. C., 1985a, High risk of malignant melanoma in melanoma-prone families with dysplastic nevi, *Ann. Intern. Med.* 102:458-465.

Greene, M. H., Clark, W. H., Tucker, M. A., Elder, D. E., Kraemer, K. H., Guerry, D., Witmer, W. K., Thompson, J., Matozzo, I., and Fraser, M. C., 1985b, Acquired precursors of cutaneous malignant melanoma: The familial dysplastic nevus syndrome, *N. Engl. J. Med.* 312:91-97.

Grimmel, M., DeVilliers, E. -M., Neumann, C., Pawlita, M., and Zur Hausen, M., 1988, Characterization of a new human papillomavirus (HPV 41) from disseminated warts and detection of its DNA in some skin carcinomas, *Int. J. Cancer.* 41:5-9.

Grob, J. J., Breton, A., Bonafe, J. L., Sauvan-Ferdani, M., and Bonerandi, J. J., 1987, Keratitis, icthyosis and deafness (KID) syndrome: Vertical transmission and death from multiple squamous cell carcinomas, *Arch. Dermatol.* 123:777-782.

Gupta, A. K., Cardella, C. J., and Haberman, H. F., 1986, Cutaneous malignant neoplasms in patients with renal transplants, *Arch. Dermatol.* 122:1288-1293.

Gupta, A. K., Stern, R. S., Swanson, N. A., and Anderson, T. F., 1988, PUVA Follow-up Study. Cutaneous melanomas in patients treated with psoralens plus ultraviolet A: A case report and the experience of the PUVA Follow-up Study, *J. Am. Acad. Dermatol.* 19:67-76.

Guss, S. B., Osbourn, R. A., and Lutzner, M. A., 1971, Porokeratosis plantaris, palmaris, et disseminata: A third type of porokeratosis, *Arch. Dermatol.* 104:366-373.

Haenszel, W., 1963, Variations in skin cancer incidence within the United States, *Natl. Cancer Inst. Monogr.* 10:225-243.

Hall, A. F., 1950, Relationships of sunlight, complexion and heredity to skin carcino-genesis, *Arch. Dermatol. Syphilol.* 61:589-610.

Halpern, A. C., Guerry, D., Elder, D. E., Clark, W. H., Synnestvedt, M., Norman, S., and Ayerle, R., 1991, Dysplastic nevi as risk markers of sporadic (nonfamilial) melanoma: A case-control study, *Arch. Dermatol.* 127:995-999.

Hazen, H. H., and Freeman, C. W., 1950, Skin cancer in the American negro, *Arch. Dermatol. Syphilol.* 62:622-623.

Hazen, P. G., Carney, P., and Lynch, W. S., 1989, Keratitis, ichthyosis, and deafness syndrome with development of multiple cutaneous neoplasms, *Int. J. Dermatol.* 28:190-191.

Henseler, T., Christophers, E., Honigsman, H., Wolff, K., *et al.*, 1987, Skin tumors in the European PUVA study: Eight-year follow-up of 1,643 patients treated with PUVA for psoriasis, *J. Am. Acad. Dermatol.* 16:108-116.

Hildreth, N. G., Shore, R. E., Hempelmann, L. H., and Rosenstein, M., 1985, Risk of extrathyroid tumors following radiation treatment in infancy for thymic enlargement, *Radiat. Res.* 102: 378-391.

Hinds, M. W., and Kolonel, L. N., 1980, Malignant melanoma of the skin in Hawaii, 1960-1977, *Cancer* 45:811-817.

Hodak, E., Ginzburg, A., David, M., and Sandbank, M., 1987, Etretinate treatment of the nevoid basal cell carcinoma syndrome: Therapeutic and chemopreventive effect, *Int. J. Dermatol.* 26:606-609.

Hogan, D. J., To, T., Gran, L., Wong, D., and Lane, P. R., 1989, Risk factors for basal cell carcinoma, *Int. J. Dermatol.* 28:591-594.

Hogan, D. J., Lane, P. R., Gran, L., and Wong, D., 1990, Risk factors for squamous cell carcinoma of the skin in Saskatchewan, Canada, *J. Dermatol. Sci.* 1:97-102.

Holly, E. A., 1986, Cutaneous melanoma and oral contraceptives: A review of case-control and cohort studies, *Recent Results Cancer Res.* 108-117.

Holly, E. A., Kelly, J. W., Shpall, S. N., and Chiu, S-H., 1987, Number of melanocytic nevi as a major risk factor for malignant melanoma, *J. Am. Acad. Dermatol.* 17:459-468.

Holly, E. A., Whittemore, A. S., Aston, D. A., Ahn, D. K., Nickoloff, B. J., and Kristiansen, J. J., 1989, Anal cancer incidence: Genital warts, anal fissure or fistula, hemorrhoids, and smoking, *J. Natl. Cancer Inst.* 81:1726–1731.

Horm, J. W., Asire, A. J., Young, J. L., Pollack, E. S., eds., 1984, *SEER Program: Cancer Incidence and Mortality in the United States 1973–81,* National Cancer Institute (NIH Pub. No. 85-1837), Bethesda, Maryland.

Houghton, A., Flannery, J., and Viola, M. V., 1980, Malignant melanoma in Connecticut and Denmark, *Int. J. Cancer* 25:95–104.

Howley, P. M., 1987, The role of papillomaviruses in human cancer, *Important Adv. Oncol.* 55–73.

Howley, P. M., and Schlegel, R., 1985, The human papillomaviruses: An overview, *Am. J. Med.* Suppl 2A:155–158.

Hoxtell, E. O., Mandel, J. S., Murray, S. S., Schuman, L. M., and Goltz, R. W., 1977, Incidence of skin carcinoma after renal transplantation, *Arch. Dermatol.* 113:436–438.

Hueper, W. C., 1963, Chemically induced skin cancers in man, *Natl. Cancer Inst. Monogr.* 10:377–391.

Hunter, D. J., Colditz, G. A., Stampfer, M. J., Rosner, B., Willett, W. C., and Speizer, F. E., 1990, Risk factors for basal cell carcinoma in a prospective cohort of women, *Ann. Epidemiol.* 1:13–23.

Huriez, C., LeBeurre, R., and LePerre, B., 1962, Etude de 126 tumers auriculaires malignes observees en 9 ans a la clinique dermatologique, Universitaire de Lille, *Bull. Soc. Fr. Dermatol. Syphiligr.* 69:886–892.

Hutchinson, J., 1887, Arsenic cancer, *Br. Med. J.* 2:1280–1281.

Ikegawa, S., Saida, T., Takizawa, Y., Tokuda, Y., Ito, T., Fujioka, F., Sakaki, T., Uchida, N., Arase, S., and Takeda, K., 1989, Vimentin-positive squamous cell carcinoma arising in a burn scar, *Arch. Dermatol.* 125:1672–1676.

Isaacson, C., 1979, Cancer of the skin in urban blacks of South Africa, *Br. J. Dermatol.* 100:347–350.

Jackson, R., and Grainge, J. W., 1975, Arsenic and cancer, 1975, *Can. Med. Assoc. J.* 113:396–399.

Jackson, R., Williamson, G. S., and Beattie, W. G., 1966, Lentigo maligna and malignant melanoma, *Can. Med. Assoc. J.* 95:846–851.

Jacyk, W. K., and Subbuswamy, S. G., 1979, Epidermodysplasia verruciformis in Nigerians, *Dermatologica* 159:256–265.

Johnson, M-L. T., and Roberts, J., 1978, *Skin conditions and related need for medical care among persons 1–74 years, United States, 1971–74* (DHEW Publication No. PHS 79-1660). Hyattsville, Maryland, National Center for Health Statistics.

Jones, E. W., and Heyl, T., 1970, Nevus sebaceous: A report of 140 cases with special regard to the development of secondary malignant tumors, *Br. J. Dermatol.* 82:99–117.

Jung, E. G., Furtwangler, M., Klostermann, G., and Bohnert, E., 1980, Light-induced skin cancer and prolonged UV erythema, *Arch. Dermatol. Res.* 267:33–36.

Kaldor, J. M., Day, N. E., Band, P., Choi, N. W., Clarke, E. A., Coleman, M. P., Hakama, M., Koch, M., Langmark, F., Neal, F. E., Petterson, F., Pompe-Kim, V., Prior, B., and Storm, H. H., 1987, Second malignancies following testicular cancer, ovarian cancer and Hodgkin's disease: An international collaborative study among cancer registries, *Int. J. Cancer* 39:571–585.

Karagas, M. R., Thomas, D. B., Roth, G. J., Johnson, L. K., and Weiss, N. S., 1991, The effects of changes in health care delivery on the reported incidence of cutaneous melanoma in western Washington State, *Am. J. Epidemiol.* 133:58–62.

Kawashima, M., Jablonska, S., Favre, M., Obalek, S., Croissant, O., and Orth, G., 1986, Characterization of a new type of human papillomavirus found in a lesion of Bowen's disease of the skin, *J. Virol.* 57:688–692.

Kawashima, M., Favre, M., Obalek, S., Jablonska, S., and Orth, G., 1990, Premalignant lesions and cancers of the skin in the general population: Evaluation of the role of human papillomaviruses, *J. Invest. Dermatol.* 95:537–542.

Kefford, R. F., Salmon, J., Shaw, H. M., Donald, J. A., McCarthy, W. H., 1991, Hereditary melanoma in Australia: Variable association with dysplastic nevi and absence of genetic linkage to chromosome 1p, *Cancer Genet. Cytogenet.* 51:45–55.

Kelly, G. E., Mahony, J. F., Sheil, A. G. R., Meikle, W. D., Tiller, D. S., and Horvath, J., 1987, Risk factors for skin carcinogenesis in immunosuppressed kidney transplant recipients, *Clin. Transplantation* 1:271–277.

Kelly, J. W., Crutcher, W. A., and Sagebiel, R. W., 1986, Clinical diagnosis of dysplastic nevi: A clinicopathologic correlation, *J. Am. Acad. Dermatol.* 14:1044–1052.

Kelly, J. W., Holly, E. A., Shpall, S. N., and Ahn, D. K., 1989, The distribution of melanocytic naevi in melanoma patients and control subjects, *Australas. J. Dermatol.* 30:1–8.

Kettler, A. H., Putledge, M., Tschen, J. A., and Buffone, G., 1990, Detection of human papillomavirus in nongenital Bowen's disease by in situ DNA hybridization, *Arch. Dermatol.* 126: 777–781.

Kinlen, L., Doll, R., and Peto, J., 1983, The incidence of tumors in human transplant recipients, *Transplantation Proc.* 15:1039–1042.

Kirkpatrick, C. S., Lee, J. A. H., and White, E., 1990, Melanoma risk by age and socioeconomic status, *Int. J. Cancer* 46:1–4.

Koh, H. K., Caruso, A., Gage, I., Geller, A. C., Prout, M. N., White, H., O'Connor, K., Balash, E. M., Blumenthal, G., Rex, I. M., Wax, F. D., Rosenfeld, T. L., Gladstone, G. C., Shama, S. K., Koumars, J. A., Baler, G. R., and Lew, R. A., 1990, Evaluation of melanoma/skin cancer screening in Massachusetts: Preliminary results, *Cancer* 65:375–379.

Koh, H. K., Clapp, R. W., Barnett, J. M., Barnett, J. M., Nannery, W. M., Tahan, S. R., Geller, A. C., Bhawan, J., Harrist, T. J., Kwan, T., Stadecker, M., Okun, M. R., Dong, J. A., Beattie, M., Prout, M. N., Murphy, G. F., and Lew, R. A., 1991, Systematic underreporting of cutaneous malignant melanoma in Massachusetts: Possible implications for national incidence figures, *J. Am. Acad. Dermatol.* 24:545–550.

Kopf, A. W., 1979, Computer analysis of 3531 basal cell carcinomas of the skin, *J. Dermatol.* 6: 267–281.

Koranda, F. C., Dehmel, E. M., Kahn, G., and Penn, I., 1974, Cutaneous complications in immunosuppressed renal homograft recipients, *JAMA* 229:419–424.

Kraemer, K. H., and Greene, M. H., 1985, Dysplastic nevus syndrome: Familial and sporadic precursors of cutaneous melanoma, *Dermatol. Clin.* 3:225–237.

Kraemer, K. H., Lee, M. M., and Scotto, J., 1984, DNA repair protects against cutaneous and internal neorplasia, *Carcinogenesis* 5:511–514.

Kraemer, K. H., Lee, M. M., and Scotto, J., 1987, Xeroderma pigmentosum: Cutaneous, ocular, and neurologic abnormalities in 830 published cases, *Arch. Dermatol.* 123:241–250.

Kraemer, K. H., DiGiovanna, J. J., Moshell, A. N., Tarone, R. E., and Peck, G. L., 1988, Prevention of skin cancer in xeroderma pigmentosum with the use of oral isotretinoin, *N. Engl. J. Med.* 318:1633–1637.

Kravitz, P. H., and McDonald, C. J., 1978, Topical nitrogen mustard induced carcinogenesis, *Acta Derm. Venereol.* 58:421–425.

Krementz, E. T., Reed, R. J., Coleman, W. P., Sutherland, C. M., Carter, R. D., and Campbell, M., 1982, Acral lentigenous melanoma: a clinicopathologic entity, *Ann. Surg.* 195:632–645.

Kripke, M. L., 1988, Impact of ozone depletion on skin cancers, *J. Dermatol. Surg. Oncol.* 14: 853–857.

Laycock, H. T., 1948, The "Kang Cancer" of north-west China, *Br. Med. J.* 1:982.

Lee, J. A. H., 1982, Melanoma and exposure to sunlight, *Epidemiol. Rev.* 4:110–136.

Lee, J. A. H., and Carter, A. P., 1970, Secular trends in mortality from malignant melanoma, *J. Natl. Cancer Inst.* 45:91–97.

Lee, J. A. H., Petersen, G. R., Stevens, R. G., and Vesanen, K., 1979, The influence of age, year of birth, and date on mortality from malignant melanoma in the populations of England and Wales, Canada, and the white population of the United States, *Am. J. Epidemiol.* 110: 734–739.

Lee, J. A. H., and Strickland, D., 1980, Malignant melanoma: Social status and outdoor work, *Br. J. Cancer.* 41:757–763.

Lee, P. Y., Silvarman, M. K., Rigel, D. S., Vossaert, K. A., Kopf, A. W., Bart, R. S., Garfinkel, L., and Levenstein, M. J., 1992, Level of education and the risk of malignant melanoma, *J. Am. Acad. Dermatol.* 26:59–63.

Leong, G. K. P., Stone, J. L., Farmer, E. R., Scotto, J., Reizner, G. T., Burnett, T. S., and Elpern, D. J., 1987, Nonmelanoma skin cancer in Japanese residents of Kauai, Hawaii, *J. Am. Acad. Dermatol.* 17:233–238.

Levi, F., LaVecchia, C., Te, V.-C., and Mezzanotte, G., 1988, Descriptive epidemiology of skin cancer in the Swiss Canton of Vaud, *Int. J. Cancer* 42:811–816.

Levin, M. L., 1963, Addendum. In: Urbach F. Conference on Biology of Cutaneous Cancer, *Natl. Cancer Inst. Monogr.* 10:653–656.

Lew, R. A., Sober, A. J., Cook, N., Marvell, A., and Fitzpatrick, T. B., 1983, Sun exposure habits in patients with cutaneous melanoma: A case-control study, *J. Dermatol. Surg. Oncol.* 9: 981–986.

Lindelof, B., and Eklund, G., 1986, Incidence of malignant skin tumors in 14,140 patients after Grenz-ray treatment for benign skin disorders, *Arch. Dermatol.* 122:1391–1395.

Lutzner, M. A., Orth, G., Dutronquay, V., Ducasse, M.-F., Kreis, M., and Crosnier, J., 1983, Detection of human papillomavirus type 5 DNA in skin cancers of an immunosuppressed renal allograft recipient, *Lancet* 2:422–424.

Lynch, F. W., Seidman, H., and Hammond, E. C., 1970, Incidence of cutaneous cancer in Minnesota, *Cancer* 25:83–91.

Macdonald, E. J., and Bubendorf, E., 1964, Some epidemiologic aspects of skin cancer, in: *Tumors of the Skin,* Year Book Medical Publishers, Chicago, pp. 23–65.

Macdonald, E. J., and Heinze, E. B., 1978, *Epidemiology of Cancer in Texas: Incidence Analyzed by Type, Ethnic Group, and Geographic Location,* Raven Press, New York.

Mack, T. M., and Floderus, B., 1991, Malignant melanoma risk by nativity, place of residence at diagnosis, and age at migration, 1991, *Cancer Causes Contr.* 2:401–411.

Magnusson, A. H. W., 1935, Skin cancer: A clinical study with special reference to radium treatment, *Acta. Radiol. Suppl.* 22:1–287.

Marmelzat, W. L., 1968, Malignant tumors in smallpox vaccination scars: A report of 24 cases, *Arch. Dermatol.* 97:400–406.

Marsh, G. M., Enterline, P. E., and McCraw, D., 1991, Mortality patterns among petroleum refinery and chemical plant workers, *Am. J. Indust. Med.* 19:29–42.

Martin, H., Strong, E., and Spiro, R. H., 1970, Radiation-induced skin cancer of the head and neck, *Cancer* 25:61–71.

Maughan, W. Z., Muller, S. A., Perry, H. O., Pittelkow, M. R., and O'Brien, P. C., 1980, Incidence of skin cancers in patients with atopic dermatitis treated with coal tar: A 25-year follow-up study, *J. Am. Acad. Dermatol.* 3:612–615.

McDonnell, J. M., McDonnell, P. J., Stout, W. C., and Martin, W. J., 1989, Human papillomavirus DNA in a recurrent squamous carcinoma of the eyelid., *Arch. Opthalmol.* 107:1631–1634.

Michaelsson, G., Olsson, E., and Westermark, P., 1981, The Rombo syndrome: A familial disorder with vermiculate atrophoderma, milia, hypotrichosis, trichoepitheliomas, basal cell carcinomas and peripheral vasodilitation with cyanosis, *Acta. Derm. Venereol.* (Stockh) 61:497–503.

Molesworth, E. H., 1927, Rodent ulcer, *Med. J. Aust.* 1:878–899.

Mora, R. G., and Burris, R., 1981, Cancer of the skin in blacks: A review of 128 patients with basal cell carcinoma, *Cancer* 47:1436–1438.

Mora, R. G., and Perniciaro, C., 1981, Cancer of the skin in blacks, I. A review of 163 black patients with cutaneous squamous cell carcinoma, *J. Am. Acad. Dermatol.* 5:535–543.

Morison, W. L., 1989, Effects of ultraviolet radiation on the immune system in humans, *Photochem. Photobiol.* 50:515–524.

Mortensen, A. C., and Kjeldsen, H., 1987, Carcinomas following Grenz ray treatment of benign dermatoses, *Acta. Derm. Venereol.* 67:523–525.

Morton, W., Starr, G., Pohl, D., Stoner, J., Wagner, S., and Weswig, P., 1976, Skin cancer and water arsenic in Lane County, Oregon, *Cancer* 37:2523–2532.

Mosberg, D. A., Crane, R. T., Tami, T. A., and Parker, G. S., 1988, Burn scar carcinoma of the head and neck, *Arch. Otolaryngol. Head Neck Surg.* 114:1038–1040.

Moy, R. L., Eliezri, Y. D., Nuovo, G. J., Zitelli, J. A., Bennett, R. G., and Silverstein, S., 1989, Human papillomavirus type 16 DNA in periungual squamous cell carcinomas, *J. Am. Med. Assoc.* 261:2669–2673.

Muir, C., Waterhouse, J., Mack, T., Powell, J., Whelan, S. (eds.), 1987, *Cancer Incidence in Five Continents:* Volume V. International Agency for Research on Cancer, Lyon.

Mulay, D. M., 1963, Skin cancer in India, *Natl. Cancer Inst. Monogr.* 10:215–223.

Nakayama, K., Yazawa, C., Sakakibara, N., Suzuki, H., Yano, M., and Ide, H., 1971, A report on three cases with carcinomata developing after antethoracic reconstructive surgery of the esophagus (by skin graft), *Surgery* 69:800–804.

Neubauer, O., 1947, Arsenical cancer: A review, *Br. J. Cancer* 1:192–251.

Noodleman, F. R., and Pollack, S. V., 1986, Trauma as a possible etiologic factor in basal cell carcinoma, *J. Dermatol. Surg. Oncol.* 12:841–846.

Novick, M., Gard, D. A., Hardy, S. B., and Spira, M., 1977, Burn scar carcinoma: A review and analysis of 46 cases, *J. Trauma* 17:809–817.

Oettle, A. G., 1963, Skin cancer in Africa, *Natl. Cancer Inst. Monogr.* 10:197–214.

Okoro, A. N., 1975, Albinism in Nigeria: A clinical and social study, *Br. J. Dermatol.* 92:485–492.

Oluwasanmi, J. O., Williams, A. O., and Alli, A. F., 1969, Superficial cancer in Nigeria, *Br. J. Cancer* 23:714–728.

Ostrow, R., Bender, M., Niimura, M., Seki, T., Kawashima, M., Pass, F., and Faras, A. J., 1982, Human papillomavirus DNA in cutaneous primary and metastasized squamous cell carcinomas from patients with epidermo dysplasia verruciformis, *Proc. Natl. Acad. Sci. U.S.A.* 79:1634–1638.

Ostrow, R. S., Shaver, M. K., Turnquist, S., Viksnins, A., Bender, M., Vance, C., Kays, V., and Faras, A. J., 1989, Human papillomavirus-16 DNA in a cutaneous invasive cancer, *Arch. Dermatol.* 125:666–669.

Oxholm, A., Thomsen, K., and Menne, T., 1989, Squamous cell carcinomas in relation to cyclosporin therapy of non-malignant skin disorders, *Acta. Derm. Venereol.* 69:89–90.

Paffenbarger, R. S., Wing, A. L., and Hyde, R. T., 1978, Characteristics in youth predictive of adult-onset malignant lymphomas, melanomas, and leukemias: Brief communication, *J. Natl. Cancer Inst.* 60:89–92.

Pantangco, E. E., Canlas, M., Basa, G., and Sin, R., 1963, Observations on the incidence, biology, and pathology of skin cancer among Filipinos, *Natl. Cancer Inst. Monogr.* 10:109–125.

Parkin, W. E., Petrone, M. E., Harlan, D. M., Kohler, B. A., and Lewis, H. C., 1990, Malignant melanoma of the skin—New Jersey, 1979–1985, *MMWR* 39:341–343.

Paymaster, J. C., Talwalkar, G. V., and Gangadharan, P., 1971, Carcinomas and malignant melanomas of the skin in western India, *J. R. Coll. Surg. Edinburgh* 16:166–173.

Pearl, D. K., and Scott, E. L., 1986, The anatomic distribution of skin cancers, *Int. J. Epidemiol.* 15:502–506.

Peck, G. L., Grose, E. G., Butkus, D., and DiGiovanna, J. J., 1982, Chemoprevention of basal cell carcinoma with isotretinoin, *J. Am. Acad. Dermatol.* 6:815–823.

Penn, I., 1987, Neoplastic consequences of transplantation and chemotherapy, *Cancer Detection Prevention Suppl.* 1:149–157.

Penn, I., 1988, Cancers after cyclosporine therapy, *Transplant. Proc.* 20 Suppl. 1:276–279.

Penneys, N. S., 1987, Microinvasive lentigo maligna melanoma, *J. Am. Acad. Dermatol.* 17:675–680.

Percy, G., Stanek, E., and Gloeckler, L., 1981, Accuracy of cancer death certificates and its effect on cancer mortality statistics, *Am. J. Public Health* 71:242–250.

Peterkin, G. A. G., 1955, Malignant change in erythema ab igne, *Br. Med. J.* 2:1599–1602.

Piepkorn, M., Meyer, L. J., Goldgar, D., Seuchter, S. A., Cannon-Albright, L. A., Skolnick, M. H., and Zone, J. J., 1989, The dysplastic melanocytic nevus: A prevalent lesion that correlates poorly with clinical phenotype, *J. Am. Acad. Dermatol.* 20:407–415.

Pierceall, W. E., Goldberg, L. H., and Ananthaswamy, H. N., 1991, Presence of human papilloma virus type 16 DNA sequences in human nonmelanoma skin cancers, *J. Invest. Dermatol.* 97:880–884.

Pittelkow, M. R., Perry, H. O., Muller, S. A., Maughan, W. Z., and O'Brien, P. C., 1981, Skin cancer in patients with psoriasis treated with coal tar: A 25-year follow-up study, *Arch. Dermatol.* 117:465–468.

Popescu, N. A., Beard, C. M., Treacy, P. J., Winkelmann, R. K., O'Brien, P. C., and Kurland, L. T., 1990, Cutaneous malignant melanoma in Rochester, Minnesota: Trends in incidence and survivorship, 1950 through 1985, *Mayo Clin. Proc.* 65:1293–1302.

Pott, P., 1963, Chirurgical observations relative to the cataract, the polypus of the nose, the cancer of the scrotum, the different kinds of ruptures and the mortification of the toes and feet, London, Haves, Clarke, and Collins 1775:63–68. Reproduced in: Urbach F., ed., Conference on Biology of Cutaneous Cancer, *Natl. Cancer Inst. Monogr.* 10:7–13.

Potter, M., 1963, Percivall Pott's contribution to cancer research, In: Urbach F., ed., Conference on Biology of Cutaneous Cancer, *Natl. Cancer Inst. Monogr.* 10:1–6.

Pringgoutomo, S., and Pringgoutomo, S., 1963, Skin cancer in Indonesia, *Natl. Cancer Inst. Monogr.* 10:191–195.

Quan, M. B., and Moy, R. L., 1991, The role of human papillomavirus in carcinoma, *J. Am. Acad. Dermatol.* 25:698–705.

Quintero, A. L., Torres, S. M., and Sanchez, J. L., 1985, Skin cancer in Puerto Rico, *Bol. Assoc. Med. P. R.* 77:502–503.

Rampen, F. H. J., and Fleuren, E., 1987, Melanoma of the skin is not caused by ultraviolet radiation but by a chemical xenobiotic, *Med. Hypotheses* 22:341–346.

Reynolds, P., and Austin, D. F., 1984, Epidemiologic-based screening strategies for malignant melanoma of the skin, in: P. F. Engstrom, P. N. Anderson, and L. E. Mortenson (eds.), *Advances in Cancer Control: Epidemiology and Research,* Alan R. Liss, New York, pp. 245–254.

Rhodes, A. R., and Melski, J. W., 1982, Small congenital nevocellular nevi and the risk of cutaneous melanoma, *Pediatrics* 100:219–224.

Rhodes, A. R., Sober, A. J., and Fitzpatrick, T. B., 1980, Possible risk factors for primary cutaneous malignant melanoma, *Clin. Res.* 28:252A (Abstract).

Rhodes, A. R., Silverman, R. A., Harrist, T. J., and Melski, J. W., 1985, A histologic comparison of congenital and acquired nevomelanocytic nevi, *Arch. Dermatol.* 121:1266–1273.

Rhodes, A. R., Mihm, M. C., and Weinstock, M. A., 1989, Dysplastic melanocytic nevi: A reproducible histologic definition emphasizing cellular morphology, *Modern Pathol.* 2:306–319.

Ries, L. A. G., Hankey, B. F., Miller, B. A., Hartman, A. M., Edwards, B. K., eds., 1991, *Cancer Statistics Review 1973–88,* National Cancer Institute, NIH Pub. No. 91-2789.

Rigel, D. S., Rivers, J. K., Kopf, A. W., Friedman, R. J., Vinokur, A. F., Heilman, E. R., and Levenstein, M., 1989, Dysplastic nevi: Markers for increased risk for melanoma, *Cancer* 63:386–389.

Robbins, J. H., 1988, Xeroderma pigmentosum: Defective DNA repair causes skin cancer and neurodegeneration, *J. Am. Med. Assoc.* 260:384–388.

Roberts, D. L., 1990, Incidence of non-melanoma skin cancer in West Glamorgan, South Wales, *Br. J. Dermatol.* 122:399–403.

Ron, E., Modan, B., Preston, D., Alfandary, E., Stovall, M., and Boice, J. D., 1991, Radiation-induced skin carcinomas of the head and neck, *Radiat. Res.* 125:318–325.

Roth, M. E., Grant-Kels, J. M., Ackerman, A. B., Elder, D. E., Friedman, R. J., Heilman, E. R., Maize, J. C., and Sagebiel, R. W., 1991, The histopathology of dysplastic nevi: Continued controversy, *Am. J. Dermatopathol.* 13:38–51.

Roush, G. C., Schymura, M. J., and Holford, T. R., 1985, Risk for cutaneous melanoma in recent Connecticut birth cohorts, *Am. J. Public Health* 75:679–682.

Roush, G. C., Holford, T. R., Schymura, M. J., and White, C., 1987, *Cancer risk and incidence trends: The Connecticut perspective,* Hemisphere, Washington.

Roush, G. C., Schymura, M. J., and Holford, T. R., 1988, Patterns of invasive melanoma in the Connecticut Tumor Registry: Is the long-term increase real? *Cancer* 61:2586–2595.

Roush, G. C., Berwick, M., Barnhill, R. L., 1992a, Association of dysplastic nevi and melanoma by anatomic site, *J. Invest. Dermatol.* 98:604.

Roush, G. C., McKay, L., Holford, T. R., 1992b, A reversal in the long-term increase in deaths attributable to malignant melanoma, *Cancer* 69:1714–1720.

Sadamori, N., Mine, M., and Hori, M., 1989, Skin cancer among atom bomb survivors, *Lancet* 1:1267.

Sanders, B. M., Jay, M., Draper, G. J., and Roberts, E. M., 1989, Non-ocular cancer in relatives of retinoblastoma patients. *Br. J. Cancer* 60:358–365.

Schreiber, M. M., Shapiro, S. I., Berry, C. Z., Dahlen, R. F., and Friedman, R. P., 1971, The incidence of skin cancer in southern Arizona (Tucson), *Arch. Dermatol.* 104:124–127.

Schwartz, S. M., and Weiss, N. J., 1988, Place of birth and incidence of ocular melanoma in the United States, *Int. J. Cancer* 41:174–177.

Scotto, J., Fears, T. R., and Fraumeni, J. F., 1983, *Incidence of nonmelanoma skin cancer in the United States* (NIH Publication No. 83-2433), Public Health Service, Washington, DC.

Scotto, J., and Fears, T. R., 1987, The association of solar ultraviolet and skin melanoma incidence among caucasians in the United States, *Cancer Invest.* 5:275–283.

Scotto, J., Kopf, A. W., and Urbach, F., 1974, Nonmelanoma skin cancer among caucasians in four areas of the United States, *Cancer* 34:1333–1338.

Scotto, J., Fraumeni, J. F., and Lee, J. A. H., 1976, Melanomas of the eye and other noncutaneous sites: Epidemiologic aspects, *J. Natl. Cancer Inst.* 56:489–491.

Scotto, J., Fears, T. R., and Fraumeni, J. F., 1982, Solar radiation, in: D. Schottenfeld, and J. F. Fraumeni (eds.), *Cancer Epidemiology and Prevention,* Saunders, Philadelphia. pp. 254–276.

Scotto, J., Pitcher, H., and Lee, J. A. H., 1991, Indications of future decreasing trends in skin-melanoma mortality among whites in the United States, *Int. J. Cancer* 49:490–497.

Šecová, M., Ševc, J., and Thomas, J., 1978, Alpha irradiation of the skin and the possibility of late effects, *Health Physics* 35:803–806.

Sedlin, E. D., and Fleming, J. L., 1963, Epidermoid carcinoma arising in chronic osteomyelitis foci, *J. Bone Jt. Surg.* 45A:827–838.

Segi, M., Tominaga, S., Aoki, K., and Fujimoto, I., 1981, *Cancer Mortality and Morbidity Statistics: Japan and the World,* Japan Scientific Societies Press, Tokyo.

Serrano, H., Scotto, J., Shornick, G., Fears, T. R., and Greenberg, E. R., 1991, Incidence of nonmelanoma skin cancer in New Hampshire and Vermont, *J. Am. Acad. Dermatol.* 24: 574–579.

Shah, L. K., Rane, S. S., and Holla, V. V., 1989, A case of squamous cell carcinoma arising in an epidermal cyst, *Indian J. Pathol. Microbiol.* 32:138–140.

Shanmugaratnam, K., and La'Brooy, E. B., 1963, Skin cancer in Singapore, *Natl. Cancer Inst. Monogr.* 10:127–140.

Shapiro, M. P., Keen, P., Cohen, L., and Murray, J. F., 1953, Skin cancer in the South African Bantu, *Br. J. Cancer* 7:45–57.

Shore, R. E., Albert, R. E., Reed, M., Harley, N., and Pasternack, B. S., 1984, Skin cancer incidence among children irradiated for ringworm of the scalp, *Radiat. Res.* 100:192–204.

Shrum, J. R., Cooper, P. H., Greer, K. E., and Landes, H. B., 1982, Squamous cell carcinoma in disseminated superficial actinic porokeratosis, *J. Am. Acad. Dermatol.* 6:58–62.

Soong, S., 1992, A computerized mathematical model and scoring system for predicting outcome in patients with localized melanoma, in: C. M. Balch, A. N. Houghton, G. W. Milton, A. J. Sober, and S. Soong (eds.), *Cutaneous Melanoma,* 2nd ed., Lippincott, pp. 200–212.

Stern, R. S., 1990, Photochemotherapy Follow-up Study, Genital tumors among men with psoriasis exposed to psoralens and ultraviolet A radiation (PUVA) and ultraviolet B radiation, *N. Engl. J. Med.* 322:1093–1097.

Stern, R. S., Zierler, S., and Parrish, J. A., 1980, Skin carcinoma in patients with psoriasis treated with topical tar and artificial ultraviolet radiation, *Lancet* 1:732–735.

Stern, R. S., Scotto, J., and Fears, T. R., 1985, Psoriasis and susceptibility to nonmelanoma skin cancer, *J. Am. Acad. Dermatol.* 12:67–73.

Stern, R. S., Lange, R., *et al.,* 1988, Non-melanoma skin cancer occurring in patients treated with PUVA five to ten years after first treatment, *J. Invest. Dermatol.* 91:120–124.

Stevens, N. G., Liff, J. M., and Weiss, N. S., 1990, Plantar melanoma: Is the incidence of melanoma of the sole of the foot really higher in blacks than whites? *Int. J. Cancer* 45:691–693.

Stevens, R. G., and Moolgavkar, S. H., 1984, Malignant melanoma: Dependence of site-specific risk on age, *Am. J. Epidemiol.* 119:890–895.

Stone, J. L., Reizner, G., Scotto, J., Elpern, D. J., Farmer, E. R., and Pabo, R., 1986, Incidence of non-melanoma skin cancer in Kauai during 1983, *Hawaii Med. J.* 45:281–286.

Strickland, P. T., Vitasa, B. C., West, S. K., Rosenthal, F. S., Emmott, E. A., and Taylor, H. R., 1989, Quantitative carcinogenesis in man: Solar ultraviolet B dose dependence of skin cancer in Maryland watermen, *J. Natl. Cancer Inst.* 81:1910–1913.

Stryker, W. S., Stampfer, M. J., Stein, E. A., Kaplan, L., Louis, T. A., Sober, A., and Willett, W. C., 1990, Diet, plasma levels of beta-carotene and alpha-tocopherol, and risk of malignant melanoma, *Am. J. Epidemiol.* 131:597–611.

Syrjanen, K. J., 1987, Human papillomavirus (HPV) infections and their associations with squamous cell neoplasia, *Arch. Geschwulstforsch.* 57:417–443.

Tada, M., and Miki, Y., 1984, Malignant skin tumors among dermatology patients in university hospitals of Japan, *J. Dermatol.* 11:313–321.

Tanenbaum, L., Parrish, J. A., Haynes, H. A., Fitzpatrick, T. B., and Pathak, M. A., 1976, Prolonged ultraviolet light-induced erythema and the cutaneous carcinoma phenotype, *J. Invest. Dermatol.* 67:513–517.

Tansurat, P., 1963, Regional incidence and pathology of skin cancer in Thailand, *Natl. Cancer Inst. Monogr.* 10:71–74.

Tiersten, A. D., Grin, C. M., Kopf, A. W., Gottlieb, G. J., Bart, R. S., Rigel, D. S., Friedman, R. J., and Levenstein, M. J., 1991, Prospective follow-up for malignant melanoma in patients with atypical-mole (dysplastic-nevus) syndrome, *J. Dermatol. Surg. Oncol.* 17:44–48.

Traboulsi, E. I., Zimmerman, L. E., and Manz, H. J., 1988, Cutaneous malignant melanoma in survivors of heritable retinoblastoma, *Arch. Opthalmol.* 106:1059–1061.

Traenkle, H. L., 1963, X-ray induced skin cancer in man, *Natl. Cancer Inst. Monogr.* 10:423–432.

Treves, N., and Pack, G. T., 1930, The development of cancer in burn scars, *Surg. Gynecol. Obstet.* 51:749–782.

Tseng, W.-P., Chu, H. M., How, S. W., Fong, J. M., Lin, C. S., and Yeh, S., 1968, Prevalence of skin cancer in an endemic area of chronic arsenicism in Taiwan, *J. Natl. Cancer Inst.* 40:453–463.

Tseng, W.-P., 1977, Effects and dose-response relationship of skin cancer and blackfoot disease with arsenic, *Environ. Health Perspect.* 19:109–119.

Tucker, M. A., Misfeldt, D., Coleman, N., Clark, W. H., and Rosenberg, S. A., 1985, Cutaneous malignant melanoma after Hodgkin's disease, *Ann. Intern. Med.* 102:37–41.

Urbach, F., 1969, Geographic pathology of skin cancer, in: F. Urbach (ed.), *Biological Effects of Ultraviolet Radiation,* Pergamon Press, New York, pp. 635–650.

Urbach, F., 1981, Skin cancer in man, in: O. D. Laerum and O. H. Iversen (ed.), *Biology of Skin Cancer (excluding Melanomas),* International Union Against Cancer, Geneva, pp. 58–86.

Urbach, F., Rose, D. B., and Bonnem, M., 1972, Genetic and environmental interactions in skin carcinogenesis, in: *Environment and cancer,* Williams and Wilkins, Baltimore, pp. 355–371.

van Haeringen, A., Bergman, W., Nelen, M. R., van der Kooij-Meijs, E., Hendrikse, I., Wijnen, J. T., Khan, P. M., Klasen, E. C., and Frants, R. R., 1989, Exclusion of the dysplastic nevus syndrome (DNS) locus from the short arm of chromosome 1 by linkage studies in Dutch families, *Genomics* 5:61–64.

Vázquez-Botet, M., Latoni, D., and Sánchez, J. L., 1990, Melanoma maligno en Puerto Rico, *Bol. Asoc. Med. P. R.* 82:454–457.

Venzon, D. J., and Moolgavkar, S. H., 1984, Cohort analysis of malignant melanoma in five countries, *Am. J. Epidemiol.* 119:62–70.

Vitaliano, P. P., and Urbach, F., 1980, The relative importance of risk factors in nonmelanoma carcinoma, *Arch. Dermatol.* 116:454–456.

Vitasa, B. C., Taylor, H. R., Strickland, P. T., Rosenthol, F. S., West, S., Abbey, H., Ng, S. K., Munoz, B., and Emmet, E. A., 1990, Association of nonmelanoma skin cancer and actinic keratosis with cumulative solar ultraviolet exposure in Maryland watermen, *Cancer* 65:2811–2817.

Vollmer, R. T., 1989, Malignant melanoma: a multivariate analysis of prognostic factors, *Pathol. Annu.* 24:383–407.

Vonderheid, E. C., Tan, E. T., Kantor, A. F., Shrager, L., Micaily, B., and Van Scott, E. J., 1989, Long-term efficacy, curative potential, and carcinogenicity of topical mechlorethamine chemotherapy in cutaneous T cell lymphoma, *J. Am. Acad. Dermatol.* 20:416–428.

Walter, S. D., Marrett, L. D., From, L., Hertzman, C., Shannon, H. S., and Roy, P., 1990, The association of cutaneous malignant melanoma with the use of sunbeds and sunlamps, *Am. J. Epidemiol.* 131:232–243.

Walter, S. D., Marrett, L. D., Shannon, H. S., From, L., and Hertzman, C., 1992, The association of cutaneous malignant melanoma and fluorescent light exposure, *Am. J. Epidemiol.* 135: 749–762.

Ward, W. H., 1952, The problem of skin cancer, *Aust. J. Dermatol.* 1:157–163.

Waterhouse, J. A. M., 1971, Cutting oils and cancer, *Ann. Occup. Hyg.* 14:161–170.

Weinstein, A. L., Howe, M. L., and Burnett, W. S., 1989, Sentinel health event surveillance: Skin cancer of the scrotum in New York State, *Am. J. Public Health* 79:1513–1515.

Weinstock, M. A., 1990, Prevalence of the early warning signs for melanoma among participants in the 1989 Rhode Island skin cancer screening, *J. Am. Acad. Dermatol.* 23:516–518.

Weinstock, M. A., 1992a, Assessment of sun sensitivity by questionnaire: Validity of items and formulation of a prediction rule, *J. Clin. Epidemiol.* 45:547–552.

Weinstock, M. A., 1992b, Incidence of malignant melanoma of sites never exposed to direct sunlight: Relation with race and latitude, *J. Invest. Dermatol.* 98:573 (Abstract).

Weinstock, M. A., 1992c, Human models of melanoma, *Clin. Dermatol.* 10:83–89.

Weinstock, M. A., 1992d, Statistics of interest to the dermatologist, in: A. J. Sober and T. B. Fitzpatrick (eds.), *1992: The Year Book of Dermatology,* Mosby-Year Book, St. Louis, pp. xxxiii–liv.

Weinstock, M. A., 1993, Nonmelanoma skin cancer mortality in the United States, 1969–1988, *Arch. Dermatol.,* in press.

Weinstock, M. A., and Sober, A. J., 1987, The risk of progression of lentigo maligna to lentigo maligna melanoma, *Br. J. Dermatol.* 116:303–310.

Weinstock, M. A., Colditz, G. A., Willett, W. C., Stampfer, M. J., Bronstein, B. R., Mihm, M. C., and Speizer, F. E., 1989a, Nonfamilial cutaneous melanoma incidence in women associated with sun exposure before 20 years of age, *Pediatrics* 84:199–204.

Weinstock, M. A., Colditz, G. A., Willett, W. C., Stampfer, M. J., Bronstein, B. R., Mihm, M. C., and Speizer, F. E., 1989b, Moles and site-specific risk of nonfamilial cutaneous malignant melanoma in women, *J. Natl. Cancer Inst.* 81:948–952.

Weinstock, M. A., Bogaars, H. A., Ashley, M., Litle, V., Bilodeau, E., and Kimmel, S., 1991a, Nonmelanoma skin cancer mortality: A population-based study, *Arch. Dermatol.* 127:1194–1197.

Weinstock, M. A., Colditz, G. A., Willett, W. C., Stampfer, M. J., Bronstein, B. R., Mihm, M. C., and Speizer, F. E., 1991b, Melanoma and the sun: The effect of swim suits and a "healthy" tan on the risk of nonfamilial malignant melanoma in women, *Am. J. Epidemiol.* 134:462–470.

Weinstock, M. A., Bogaars, H. A., Ashley, M., Litle, V., Bilodeau, E., and Kimmel, S., 1992, Inaccuracies in certification of nonmelanoma skin cancer deaths, *Am. J. Public Health* 82: 278–281.

Weinstock, M. A., Clark, J. W., and Calabresi, P., 1993, Melanoma, in: P. Calabresi and P. S. Schein (eds.), *Medical Oncology: Basic Principles and Clinical Management of Cancer,* 2nd ed., McGraw-Hill, New York.

Whelan, S. L., Parkin, D. M., and Masuyer, E., (eds.), 1990, *Patterns of Cancer in Five Continents* (IARC Scientific Publication No. 102.), International Agency for Research on Cancer, Lyon.

Whitaker, C. J., Lee, W. R., and Downes, J. E., 1979, Squamous cell skin cancer in the northwest of England, 1967–69, and its relation to occupation, *Br. J. Industr. Med.* 36:43–51.

Winklmann, R. K., 1963, Squamous carcinoma produced by ultraviolet light in hairless mice, *J. Invest. Dermatol.* 40:217–224.

Witherspoon, R. P., Fisher, L. D., Schoch, G., Martin, P., Sullivan, K. M., Sanders, J., Degg, J., Doney, K., Thomas, D., Storb, R., and Thomas, E. D., 1989, Secondary cancers after bone marrow transplantation for leukemia or aplastic anemia, *N. Engl. J. Med.* 321:784–789.

Wright, W. E., Peters, J. M., and Mack, T. M., 1983, Organic chemicals and malignant melanoma, *Am. J. Indust. Med.* 4:577–581.

Wyburn-Mason, R., 1955, Malignant change arising in tissues affected by herpes, *Br. Med. J.* 2: 1106–1109.

Wyburn-Mason, R., 1957, Malignant change following herpes simplex, *Br. Med. J.* 2:615–616.

Yeh, S., 1963, Relative incidence of skin cancer in Chinese in Taiwan with special reference to arsenical cancer, *Natl. Cancer Inst. Monogr.* 10:81–107.

Yeh, S., How, S. W., and Lin, C. S., 1968, Arsenical cancer of skin: Histologic study with special reference to Bowen's disease, *Cancer* 21:312–339.

Young, J. L., Percy, C. L., Asire, A. J., (eds.), 1981, Surveillance, epidemiology, and end results: Incidence and mortality data, 1973–77, *Natl. Cancer Inst. Monogr.* 57:1–1082.

Yutsudo, M., Shimakaga, T., Hakura, A., 1985, Human papillomavirus type 17 DNA in skin carcinoma tissue of a patient with epidermodysplasia verruciformis, *Virology* 144:295–298.

Zagula-Mally, Z. W., Roseberg, E. W., and Kashgarian, M., 1974, Frequency of skin cancer and solar keratoses in a rural southern county as determined by population sampling, *Cancer* 34:345–349.

Zaldívar, R., 1974, Arsenic contamination of drinking water and foodstuffs causing endemic chronic poisoning, *Beitr. Pathol. Bd.* 151:384–400.

Zaynoun, S., Ali, L. A., Shaib, J., and Kurban, A., 1985, The relationship of sun exposure and solar elastosis to basal cell carcinoma, *J. Am. Acad. Dermatol.* 12:522–525.

Zur Hausen, H., 1989, Papillomavirus in anogenital cancer: The dilemma of epidemiologic approaches, *J. Natl. Cancer Inst.* 81:1680–1682.

12

The Induction and Repair of DNA Photodamage in the Environment

David L. Mitchell and Deneb Karentz

1. Introduction

Since its discovery in bacteria 25 years ago, DNA repair, particularly in response to damage induced by UV radiation, has been a major focus of molecular biology. The roles played by DNA damage and repair in the aging and carcinogenic processes have been of particular interest. More recently, degradation of stratospheric ozone and the resultant increase in UV-B radiation at the earth's surface have focused our attention on the environmental effects of solar DNA damage. Although we have learned much about DNA damage tolerance mechanisms in prokaryotes and some lower eukaryotes, our understanding of the molecular events that determine biological effects of UVR in complex higher eukaryotes is far from complete.

Exposure to the ultraviolet portion of the solar spectrum is primarily responsible for skin cancer in the human population. The site of action of ultraviolet light in living cells is undoubtedly DNA, the primary chromophore, where damage accumulates and results in cell death or mutation. Two molecular mechanisms are currently considered important in the initiation of

David L. Mitchell • Department of Carcinogenesis, University of Texas System Cancer Center, Science Park/Research Division, Smithville, Texas 78957.   Deneb Karentz • Department of Biology, University of San Francisco, San Francisco, California 94117.

Environmental UV Photobiology, edited by Antony R. Young *et al.* Plenum Press, New York, 1993.

some cancers: (1) activation of protooncogenes, such as *ras,* and (2) inactivation of tumor-suppressor genes. Alterations in DNA sequence, such as point mutations or deletions, can cause either event.

The effects of a particular type of DNA damage are complex, determined by the interplay between its intrinsic mutagenicity and lethality. A lesion that inhibits genome replication or transcription of essential genes is cytotoxic; one that blocks DNA polymerization causes cell death and suppresses mutation induction. Hence, the more cytotoxic a lesion is (i.e., assuming its lethality results from termination of DNA synthesis), the less likely it will allow replicative bypass and mutation induction at that site (Brash *et al.,* 1987). Should the damage produce a mutation, several outcomes may result: (1) there may be no effect because the mutation occurs in a nontranscribed region of DNA or does not alter gene (protein) function, (2) the mutation may be harmful to the organism resulting in reduced fitness, or (3) the mutation may actually increase fitness.

Technical advances over the past few years have shown that DNA damage and repair in eukaryotic systems is diverse. The temporal and spatial distribution of photoproduct induction and repair in chromatin is complex, determined by DNA sequence and interactions between DNA and structural and regulatory proteins. First, sequencing analyses and other analytical techniques have shown extensive heterogeneity of UV photodamage. Although the cyclobutane pyrimidine dimer is the major damage induced, myriad minor lesions also occur which may be cytotoxic or elicit site-specific mutations. Second, the distribution of this damage is nonrandom, determined by DNA sequence and DNA–protein interactions within chromatin. Third, excision repair in mammalian cells is heterogeneous as well; cells repair different photoproducts at different rates in the overall genome and preferentially repair lesions in actively transcribing DNA.

2. Induction of Photodamage by Solar UV Radiation

2.1. DNA Photochemistry

Photon absorption rapidly converts a pyrimidine base to an excited state. This event promotes an electron in a filled bonding π orbital into a higher energy, empty π^* antibonding orbital, thus initiating photoproduct formation. Formation of an excited base occurs within 10^{-12} s after photon absorption. Various pathways are then available for resolution of this unstable electronic configuration (for review see Fisher and Johns, 1976; Patrick and Rahn, 1976). The major pathway involves rapid dissipation of the energy of the excited base to the ground state (10^{-9} s) by nonradiative transition or by fluorescence, yielding heat or light in the process. The excited base can then react with other molecules to form unstable intermediates (i.e., free radicals) or stable photoproducts. Finally, there is a low probability that intersystem crossing, a nonradiative pathway, can transfer a base from the excited singlet state to the excited triplet state. The lifetime of the triplet state is several orders of magnitude longer than the excited singlet state (10^{-3} s), increasing the chance of photoproduct formation. The cyclobutane pyrimidine dimer forms through the excited triplet state. Other photoproducts, such as the pyrimidine(6-4)pyrimidone photoproduct [or (6-4) photoproduct], form by some other mechanism (Umlas *et al.*, 1985).

Figure 1 illustrates the diverse spectrum of structural damage produced in DNA by UVR. Compared to pyrimidine dimers, monobasic damage occurs in UV-irradiated DNA at very low frequencies (≈ 1–2%) (Demple and Linn, 1980; Mitchell *et al.*, 1991a). Because of their rapid reversal and lability to DNA hydrolysis, the photobiology of cytosine photohydrates has been refractory to quantitative analysis. Recent data using sequencing analysis showed that thymine glycols do not occur in UV-irradiated DNA at a significant rate; hence, they do not appear to contribute significantly to the biological effects of UVR (Mitchell *et al.*, 1991a). Dimers containing purine, such as the 8,8-adenine dehydrodimer (Gasparro and Fresco, 1986) and a thymine–adenine dimer (Bose and Davies, 1984), have been isolated from heavily irradiated DNA (Fig. 1). These rare photoproducts occur at 1–10% the frequency of the cyclobutane dimer in DNA. In addition to the predominant cyclobutane dimer, sequencing and immunologic analyses have identified the (6-4) photoproduct and its photoisomer, the Dewar pyrimidinone (Figs. 1 and 2), as potentially important lethal and mutagenic determinants of solar UV (Mitchell and Nairn, 1989).

Recent evidence has accumulated which suggests that cytosine, more than thymine, may bear primary responsibility for the biological effects of dipyrimidine photoproducts (Mitchell and Cleaver, 1990). Although the thymine–thymine cyclobutane dimer (T$\langle$ $\rangle$T) occurs with greatest frequency, it is exceeded by the sum of dimers containing cytosine (C$\langle$ $\rangle$T + T$\langle$ $\rangle$C + C$\langle$ $\rangle$C). In addition, the (6-4) photoproduct and Dewar pyrimidinone are induced with much greater frequency between cytosine and adjacent pyrimidines (Lippke, *et al.*, 1981). The ability of cells to repair dimers containing cytosine much more efficiently than thymine–thymine cyclobutane dimers (Carrier *et al.*, 1982; Mitchell *et al.*, 1985) suggests that they may be more harmful. Cytosine dimers may, in fact, represent the primary lethal and mutagenic sites of action in UV-irradiated mammalian cells. Hence, the amount of cytosine in an organism's genome (AT:GC ratio) may be predictive of its sensitivity to UVR. Indeed, increased induction of dimeric damage containing

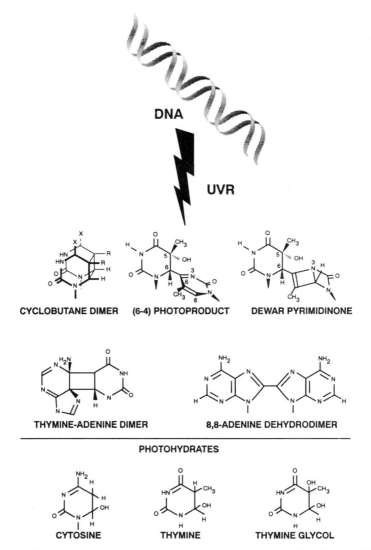

Figure 1. Structures of dimeric and monomeric photoproducts induced in DNA by UVR.

cytosine by the UV-B portion of the solar spectrum [for example, the Dewar photoisomer of the (6-4) photoproduct] may bear significant responsibility for sunlight-induced carcinogenesis in the human population, as well as contribute to the detrimental effects of UVR on the biosphere (Mitchell and Nairn, 1988; Taylor *et al.,* 1990).

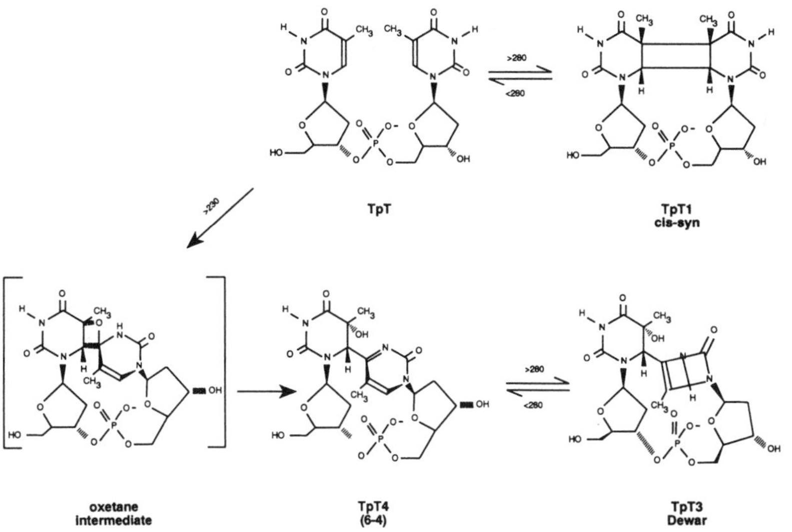

Figure 2. Reaction scheme for irradiation of the dinucleotide TpT (modified from Taylor and Cohrs, 1987). Photoproducts shown include the *cis–syn* cyclobutane dimer, (6-4) photoproduct, and Dewar pyrimidinone.

2.2. Methods for Detecting and Quantifying Photodamage in DNA

Methods for quantifying UV damage in DNA rely on (1) separation of modified bases by chromatography, (2) enzymatic or biochemical incision of DNA at sites of photoproducts, or (3) antibody binding to structural damage in DNA. Damage caused by UVR was first detected in DNA using chromatography (Setlow and Carrier, 1966). DNA is prelabeled with radioactive thymidine (^{3}H-TdR) during cell proliferation, extracted and reduced to individual bases by acid or enzymatic hydrolysis. Cyclobutane dimers are separated from unmodified bases by two-dimensional paper (Setlow and Carrier, 1966), thin-layer (Cook and Friedberg, 1976), or high-performance liquid chromatography (Breter *et al.*, 1974; Cadet *et al.*, 1980).

Photoproduct formation and repair also can be measured by incising DNA at sites of damage and quantifying the resultant single-strand breaks by alkaline-gradient centrifugation (Ganesan *et al.*, 1981) or gel electrophoresis. Analyses of photodamage as strand breaks require radioactive labeling to detect and quantify DNA on sucrose gradients or agarose gels. Although these techniques are more sensitive than chromatography, high-molecular-weight DNA is required for accurate quantitation. Cyclobutane dimers are specifically cleaved with a UV endonuclease from T4 phage (Haseltine *et al.*, 1980).

UvrABC exinuclease, a partial excision repair complex purified from *Escherichia coli,* is less specific, cleaving DNA on either side of damage produced by exposure to genotoxic chemicals or UVR (Sancar and Rupp, 1983). This broad-spectrum enzyme measures overall levels of photodamage in UV-irradiated DNA and non-cyclobutane dimer damage remaining after enzymatic photoreactivation. *E. coli* endonuclease III cleaves DNA at cytosine and thymine photohydrates as well as unidentified photoproducts occurring at sites of (6-4) photoproducts (Mitchell *et al.,* 1991a) and modified purines (Gallagher and Duker, 1986). Nonenzymatic procedures also incise DNA at sites of photodamage. The (6-4) photoproduct is converted to the alkali-labile Dewar pyrimidinone by extended secondary irradiation with UV-A light (>320 nm) (Fig. 2) (Mitchell *et al.,* 1990a). Prolonged incubation of the Dewar photoisomer in mild alkali produces a strand break at the site of the original damage. These photoinduced alkali-labile sites (or PALS) are quantified by the same techniques used to measure sites of enzyme digestion.

New techniques have been developed to explore the fine structure of UV damage induction and repair using strand scission. Photodamage at the sequence level was mapped as strand breaks in DNA fragments irradiated with high fluences of UV-C (240–280 nm) and UV-B (280–320 nm) light (Lippke *et al.,* 1981; Gallagher and Duker, 1986; Mitchell *et al.,* 1992; Mitchell *et al.,* 1992) and at the gene level using Southern blot hybridization of DNA fragments separated by agarose gel electrophoresis (Bohr *et al.,* 1985). Recently, the distribution of photoproducts at the sequence level in a gene promoter irradiated *in vivo* with low UV fluences was analyzed by ligation-mediated polymerase chain reaction (LMPCR) (Pfeiffer *et al.,* 1991).

Antisera raised against UV-irradiated DNA are potent and versatile reagents for the study of various photochemical and photobiological phenomena. Polyclonal and monoclonal antibodies recognize and bind a variety of photoproducts, including the cyclobutane dimer (Mori *et al.,* 1991), (6-4) photoproduct (Mitchell *et al.,* 1985), Dewar pyrimidinone (Mitchell and Rosenstein, 1987), and thymine glycol (Leadon and Hanawalt, 1983). Assays using antisera raised against UV-irradiated DNA have provided unique insights into the responses of several biological systems to these photoproducts. Prominent within the extensive arsenal of immunologic approaches adapted to the analysis of DNA damage and repair are quantitative immunofluorescence, immunoprecipitation, enzyme-linked immunoassays, and radioimmunoassays (RIAs). Each technique has its own unique attributes and applications. Unlike chromatographic techniques, immunologic assays do not require chemical or enzymatic degradation of DNA before analysis; unlike endonucleolytic assays, their sensitivities do not depend on the molecular weight (or purity) of sample DNA.

2.3. Wavelength Dependence of Photoproduct Formation

The composition of photodamage in DNA is wavelength dependent. The ratio of dimers containing cytosine to thymine homodimers (Cyt⟨ ⟩Thy + Thy⟨ ⟩Cyt + Cyt⟨ ⟩Cyt : Thy⟨ ⟩Thy) is smaller after UV-C than UV-B irradiation (Carrier et al., 1982). Analysis of the site specificity of photoproduct induction in a small fragment of DNA showed that the ratio of Thy⟨ ⟩Thy: Cyt⟨ ⟩Thy:Thy⟨ ⟩Cyt:Cyt⟨ ⟩Cyt cyclobutane dimers was 68:13:16:3 after UV-C irradiation and 52:19:21:7 after UV-B irradiation (Mitchell et al., 1992). Quantum yield measurements of cyclobutane pyrimidine dimers showed that at wavelengths below 300 nm saturation of cytosine-containing dimers occurs at a lower fluence and at a lower level than for thymine homodimers (Garces and Davila, 1982). The wavelength dependence for the induction of both types of photoproducts ran parallel to the absorption spectrum of thymine at 260 and 280 nm but not at 300 nm. Hence, below 300 nm cyclobutane dimer formation relies primarily on the excitation of thymine, whereas other mechanisms are involved in dimerization at wavelengths > 300 nm, in which cytosine plays a more prominent role.

Unlike cyclobutane dimers, the formation of (6-4) photoproducts does not proceed via an excited triplet state; hence it is less dependent on the photochemical behavior of thymine bases in DNA. Cyclobutane dimer and (6-4) photoproducts were measured in normal human skin fibroblasts exposed to 265–313-nm monochromatic UV light using RIA (Rosenstein and Mitchell, 1987). Action spectra for induction of these lesions were very similar for wavelengths between 265 and 302 nm; however, (6-4) photoproduct induction by 313-nm light was less than half that of cyclobutane dimers. An RIA was developed to detect Dewar pyrimidinones in DNA and, along with the (6-4) photoproduct RIA, used to generate action spectra for the loss and gain of antibody-binding sites associated with photoisomerization of (6-4) photoproducts (Mitchell and Rosenstein, 1987). Both activities were shown to parallel absorption by the (6-4) photoproduct, with maxima within the UV-B region of the solar spectrum (Fig. 3).

In contrast to the cyclobutane dimer, which forms primarily between adjacent thymines, the (6-4) photoproduct and Dewar pyrimidinone form preferentially between cytosines located 5' to thymine or cytosine (Lippke et al., 1981). As discussed above, the cytosine moieties of dipyrimidine photoproducts may determine their biological effects (Mitchell and Cleaver, 1990). Recent data strongly support this contention. Transition mutations occurring at sites of cyclobutane dimers and (6-4) photoproducts containing cytosine may be primarily responsible for human skin cancer (Brash et al., 1987, 1991). Hence, the production of cytosine-containing cyclobutane dimers, (6-4) photoproducts, and Dewar pyrimidinones by UV-B light may underlie the lethal, mutagenic, and carcinogenic effects of sunlight.

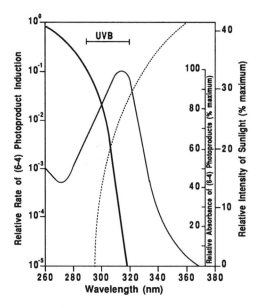

Figure 3. The relationship between action spectra for the induction (——) and absorption (photoisomerization) (——) of (6-4) photoproducts and the solar spectrum at the earth's surface (----) (Mitchell, 1988).

Unlike direct induction of DNA damage by UV-C and UV-B, UV-A (320–400 nm) produces damage indirectly through highly reactive chemical intermediates (Peak and Peak, 1990). Similar to ionizing radiation, UV-A generates oxygen and hydroxyl radicals, which react with DNA to form monomeric damage (photohydrates), strand breaks, and DNA–protein crosslinks. These photoproducts have very low induction rates; hence, their biological importance is problematic. However, data suggest that the UV-A component of sunlight may have significant biological effects. For example, early studies by Elkind and co-workers (1978) and Smith and Paterson (1982) suggested that dimeric damage was not primarily responsible for the lethal effects of sunlight. Tyrrell and Pidoux (1986, 1987) saw significant levels of cell killing and mutation induction in human epidermal cells after UV-A irradiation and found that endogenous photoprotectors can reduce the cytotoxic effects. More recently, Sterenborg and Van der Leun (1990) showed that UV-A induced significant levels of tumorigenesis in hairless mice.

2.4. Genomic Distribution of DNA Damage

The relative induction rate of (6-4) photoproducts and cyclobutane dimers in DNA has been a matter of some controversy (Mitchell and Nairn, 1989). Early chromatographic studies showed that the induction of (6-4) photoproducts relative to cyclobutane dimers varied widely, depending on base composition (AT:GC ratios); that is, 5.4% in Chinese hamster cell DNA, 30% in *E. coli* DNA irradiated *in vitro* or *in vivo,* and 276% in the conidia of *Streptomyces griseus* (Patrick and Rahn, 1976; Patrick, 1977). Using DNA-sequencing techniques, the incidence of (6-4) photoproducts and cyclobutane dimers in specific gene fragments was shown to be highly variable, depending on the sequence analyzed (Lippke *et al.,* 1981; Brash and Haseltine, 1982). Although cyclobutane dimers between two thymines were the major lesions detected, (6-4) photoproducts occurred with greater frequency at certain di-pyrimidine sites. In sequenced DNA fragments, (6-4) photoproducts formed most frequently between thymine, followed by a downstream cytosine (Lippke *et al.,* 1981; Sancar and Rupp, 1983). Recently developed biochemical and immunologic procedures yielded induction rates of 0.5–0.6 (6-4) photoproducts/10^8 daltons (300,000 bases) DNA/J/m^2 (23% the level of cyclobutane dimers) in purified DNA irradiated *in vitro* (Mitchell *et al.,* 1990a).

DNA secondary structure, determined by protein–DNA interactions, influences photoproduct distribution. Gale and co-workers (1987) have shown that chromatin structure modulates the sites of cyclobutane dimer induction in core nucleosome DNA (Fig. 4). They used the 3' → 5' exonuclease activity of T4 DNA polymerase to map UV photoproducts and found the distribution strongly modulated with a periodicity of 10.3 bases. Sites of maximum photoproduct induction on core DNA corresponded to sites where the phosphate backbone was farthest from the core histone surface. The limited flexibility of the DNA helix resulting from histone–DNA interactions inhibited cyclobutane dimer formation. Using RIA, we measured the relative induction of (6-4) photoproducts and cyclobutane dimers in nucleosomal (core) and internucleosomal (linker) DNA in UV-irradiated human chromatin (Mitchell *et al.,* 1991b). Cyclobutane dimers formed in equal amounts per nucleotide in core and linker DNA, and (6-4) photoproducts formed with 6-fold greater frequency per nucleotide in linker DNA. These data are consistent with those of Gale and Smerdon (1990), who showed a similar pattern. DNA–protein binding strongly influenced the formation of (6-4) photoproducts that occur more readily in DNA that is less tightly bound to nucleosomal proteins.

Distribution of (6-4) photoproducts in chromatin is also determined by base changes associated with metabolic activity. Sequencing analysis has shown that methylation of the 3' cytosine of a (6-4) photoproduct inhibits its formation in isolated DNA (Glickman *et al.,* 1986). Using LMPCR to map (6-4) photoproduct induction at single-nucleotide resolution, Pfeiffer and co-workers (1991) confirmed this finding in genomic DNA irradiated *in vivo.* Analysis of the human phosphoglycerate kinase gene promoter showed that the frequency of (6-4) photoproducts was different in DNA derived from active and inactive X chromosomes. Inactivation of the X chromosome by cytosine methylation may have inhibited (6-4) photoproduct formation.

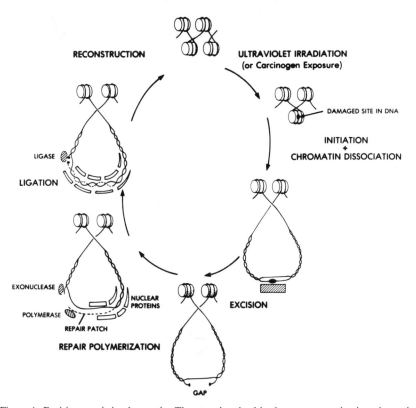

**NATIVE CHROMATIN
(Nucleosomal DNA)**

Figure 4. Excision repair in chromatin. The steps involved in damage processing in eukaryotic DNA are illustrated in a clockwise manner beginning with undamaged chromatin at the top. DNA-repair processes involving at least 8–10 damage recognition gene products are required to initiate chromatin dissociation before incision. The nucleosome structure is modified to allow excision repair of UVR damage. After excision, the chromatin is reassembled to its original conformation.

Recent studies show that (6-4) photoproducts occur with much greater frequency in actively transcribing genes than in nonexpressed genes in human fibroblasts (Cleaver *et al.,* 1990). Preferential induction of (6-4) photoproducts in functionally active regions of chromatin would serve to increase the lethal potential of this type of photodamage. Indeed, the predominance of these lesions in nonnucleosomal (metabolically active) regions of the genome may facilitate their removal (see below). In other words, DNA damage in more open chromatin conformations may be more accessible to excision repair enzymes than damage induced in more compact regions.

2.5. Mitigation of DNA Damage by Natural Mechanisms for Reducing UV Exposure

Nearly all organisms have behaviors or natural features that reduce exposure of DNA to solar UVR and reduce the amount of photodamage. Human behaviors include wearing clothes, hats, sunglasses, and sunscreens, and the general habit of living (and in many cases working) indoors. Many plants and animals have outer coverings that filter UVR as well. Biological components such as bark, cuticles, skin, fur, feathers, scales, and shells can strongly attenuate or eliminate the transmission of UVR to internal areas of cells or organisms. Habitat selection is also a significant factor. Setting up residence under a rock, deep in the water, on shaded sides of trees or landscapes, or inside another organism can reduce UVR exposure. Actively seeking shade to avoid direct sunlight is a common biological response. However, UVR can also act to impair motility, preventing an organism from moving to a "safer" location (Häder et al., 1990).

In some cases, protective behaviors (avoidance responses) and physical attributes may not have evolved in response to UVR exclusively but may have been triggered by visible light receptors. Since solar radiation includes UVR, there is, however, the additional benefit of UV protection. Some life strategies that lessen UV exposure may have evolved in response to other physical and environmental stresses, e.g., to deter predation. For example, shells of marine invertebrates provide a hard outer covering for internal soft tissues. Exoskeletons protect organisms from physical abrasion in the environment and serve to increase survival in the marine food chain by limiting predation. Physical and behavioral factors that affect UV exposure are important considerations in the evaluation of UVR effects on complex biosystems.

Morphology may determine the susceptibility of unicellular organisms to damage by UVR. Data using ionizing radiation indicate that small cells and organisms are more sensitive to physical damage than larger ones. Our studies show that nonionizing radiation (UVR) damage may also be mediated by morphological factors (Fig. 5). The amount of DNA damage in Antarctic phytoplankton correlates with the morphometric characteristics (ratio of cell surface to cell volume) of individual species (Karentz et al., 1991a). In smaller cells the distance between the cell surface and the nucleus (DNA) is shortened. The truncated light path reduces refraction and absorption by cytoplasmic components and increases the amount of UVR entering the nucleus. Similar principles can be applied to multicellular organisms. In humans, for example,

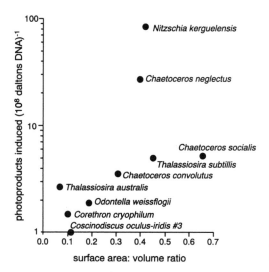

Figure 5. Relationship between surface area:volume ratios and total photoproducts in DNA induced by exposure to 2,500 J · m^{-2} UV-B light (Karentz *et al.*, 1991a).

the ratio of exposed surface area (skin and eyes) to internal tissues and organs is reduced compared to smaller animals.

At least several mechanisms for biochemical protection from UVR occur in nature. Pigmentation in the outer layers of plants and animals reduces the intensity and wavelength distribution of light reaching internal components. In some cases, light exposure induces pigment production. For example, the UVR-absorbing compounds melanin and anthocyanin are produced in human skin and in plants, respectively (Takahashi *et al.*, 1991). The increase of melanin and anthocyanin concentrations reduces UV penetration into pigment-containing cells and tissues and reduces DNA damage.

In addition to pigments, colorless UV-absorbing compounds have also been identified as possible UV-protective chemicals. Recent studies have focused on the relationship between UV exposure and the physiological processes involved in the synthesis of such compounds. Concentrations of flavonoids in terrestrial plants (Beggs *et al.*, 1986; Tevini and Teramura, 1989), mycosporine amino acids in fungi (Arpin and Bouillant, 1981) and mycosporine-like compounds in marine organisms (Shibata, 1969; Nakamura *et al.*, 1982; Dunlap *et al.*, 1986; Shick *et al.*, 1991) all correlate with UV exposure history. A most intriguing aspect of these compounds is the ubiquitous nature of their occurrence. For example, flavonoids are very common in terrestrial plants, and mycosporine-like compounds have been found in most marine organisms studied, regardless of the latitude in which they live (Karentz *et al.*, 1991b). Widespread geographic and phylogenetic occurrence of these molecules suggests their evolution may be associated with selective pressures in addition to UVR.

Researchers have identified over 25 different mycosporine-like compounds in marine algae and invertebrates. These molecules display different UV-B and UV-A absorption profiles (Table 1), with several (three or more) compounds occurring in a single species. This pattern of occurrence provides broadband UV coverage and optimizes protection from solar UVR (Dunkle and Shasha, 1989). Because of these attributes, these compounds are being investigated as possible human sunscreens.

3. DNA Damage Tolerance Mechanisms in Aquatic and Terrestrial Organisms

The amount of photodamage present in cellular DNA depends not only on the absorbed UV dose but on the ability of the organism to repair the damage. Strategies of UV damage tolerance vary, depending on the type of photoproduct encountered and the organism at risk, but are composed in whole or part of at least two well-studied mechanisms, photoenzymatic repair (PER) and nucleotide excision repair (NER) (Fig. 6). PER is an *in situ* repair process that directly and specifically reverses cyclobutane dimers in DNA by the combined action of an enzyme (photolyase) and visible light. Nucleotide excision repair is a much more complex mechanism involving damage recognition, incision of the DNA backbone at or near the site of the lesion, concomitant excision and resynthesis of the DNA around the damaged site, and ligation of the single-strand nick left after detachment of DNA polymerase.

Table 1. Maximum Wavelength of Absorption (λ_{max} in nm) of Some Common Mycosporine-like Amino Acids (MAAs)

MAA	λ_{max}	Reference
Mycosporine-glycine	310	Nakamura *et al.*, 1982
Palythine	320	Nakamura *et al.*, 1982
Asterna-330	330	Nakamura *et al.*, 1982
Palythinol	332	Nakamura *et al.*, 1982
Mycosporine-glycine:serine	332	Grant *et al.*, 1985
Porphyra-334	334	Nakamura *et al.*, 1982
Shinorine	334	Nakamura *et al.*, 1982
Mycosporine-glycine: valine	335	Karentz *et al.*, 1991b
Palythenic acid	337	Nakamura *et al.*, 1982
Palythene	360	Nakamura *et al.*, 1982

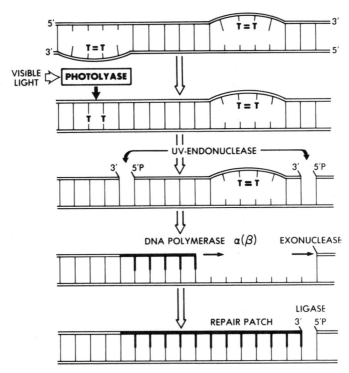

Figure 6. Photoenzymatic repair (PER) and nucleotide excision repair (NER). Two DNA-repair pathways involved in the removal of cyclobutane dimers (T = T) from UVR-irradiated DNA are illustrated. PER (left side) is the direct reversal of dimers by the combined action of an enzyme (photolyase) and visible light. NER (right side) is initiated by damage recognition followed by the removal of a segment of DNA (≈ 30 bases) around the damage by a repair complex. In eukaryotes, the process of damage recognition is unknown, and is made complex by the organization of DNA into chromatin (Fig. 4). Similarly, the incision step of the DNA repair process has not been characterized in mammalian cells. Illustrated is a UV endonuclease incision process characteristic of the *uvr*ABC enzyme complex of *E. coli* in which a single-strand nick is made on both sides of the damage. The nicked DNA is then recognized as a substrate for DNA polymerase–exonuclease. The basic processes of DNA damage tolerance in prokaryotes and eukaryotes is well summarized in Friedberg (1985).

Unlike PER, which repairs only cyclobutane dimers, the NER pathway is highly versatile, mending various classes of damage (bulky adducts) in addition to those induced by UV radiation.

3.1. DNA-Repair Measurements

The techniques described above for detecting DNA damage can be used to measure photoproduct removal by PER or NER in cells. Procedures are also available for determining other stages of the NER process, such as incision, DNA resynthesis, and ligation. The technique of alkaline elution measures transient incisions associated with the initial step of NER (Fornace *et al.*, 1976). Alkaline lysis of cells on filters results in denaturation (melting) of DNA and its elution at a rate dependent on its molecular weight; that is, the smaller fragments containing more nicks elute faster than the larger ones. The single-strand molecular weight is calculated from the elution profiles and used to determine incision (and ligation). To facilitate detection of the very rapid incision stage of NER, base analogues that block progression of DNA polymerase along the DNA helix are used to accumulate strand breaks for measurement by alkaline elution.

The sensitivity of incision measurements depends on the ability to discriminate between strand breaks associated with DNA repair and those occurring during the DNA replication process. This can be accomplished by blocking DNA replication with specific inhibitors (e.g., hydroxyurea) or using senescent (G_0) or terminally differentiated nonproliferating cell populations. Studies with mammalian cells show a rapid accumulation of strand breaks within the first few minutes (0–15) after UV irradiation followed by gradual restoration to the original molecular weight within an hour (Friedberg, 1985).

Protocols for assaying the resynthesis stage of excision repair also require inhibition of DNA synthesis. Unscheduled DNA synthesis (UDS) provided the first demonstration of repair synthesis (Rasmussen and Painter, 1964). Cells were grown for a short period in [^{3}H]thymidine and then examined by autoradiography as silver grains on a photographic plate. Cells that are undergoing DNA synthesis (S phase) show intense labeling of their nuclei. Nonreplicating cells undergoing repair (unscheduled) synthesis show much reduced labeling. Repair synthesis also can be determined using buoyant density centrifugation of DNA that has incorporated the base analog [^{3}H]5-bromodeoxyuridine in place of thymidine. Regions containing this base have a higher density than the surrounding DNA. DNA that has replicated semiconservatively (one entire new strand) migrates as a hybrid density band on cesium chloride gradients. Repair synthesis occurring in very small patches (≈ 30 bases) migrates to a position of normal density DNA (Friedberg, 1985). The amount of radioactivity incorporated into unreplicated DNA is then a measure of the total amount of repair synthesis.

3.2. Photoenzymatic Repair (Photoreactivation)

PER is a light-dependent process involving the enzyme-catalyzed reversal of cyclobutane dimers *in situ.* Reversal of UV damage by photoreactivation was observed first in plants by Hausser and von Oehmcke (1933). Photoreactivation of DNA damage was discovered simultaneously by Albert Kelner and Renato Dulbecco in 1949 (Kelner, 1949; Dulbecco, 1949) and has since been shown to be widespread throughout the plant and animal kingdoms (Table 2). Although PER appears to be a ubiquitous repair mechanism, there are several organisms in which it has not been found, including: *Haemophilus influenzae, Diplococcus pneumoniae, Bacillus subtilis, Micrococcus luteus,* and *Micrococcus radiodurans* in the Schizomycophyta (Friedberg, 1985); *Chaetoceros convolutus, Nitzschia kerguelensis, Thalassiosira australis* and *T. subtilis* among the diatoms (Table 3) (Karentz *et al.,* 1991a); *Phaseolus aureus* and *Haplopappus gracilis* of the angiosperms; and the nematode *Caenorhabditis elegans* (Hartman *et al.,* 1990). Although PER has been detected in many marsupial species (Wade and Trosko, 1983; Ley, 1984), its existence in placental mammals is a matter of some controversy (Ley *et al.,* 1978; Harm, 1976).

During PER a low-molecular-weight enzyme (photolyase) binds to a cyclobutane dimer, absorbs photons within the UV-A/visible range of light, cleaves the cyclobutyl ring of the dimer, then dissociates to search for additional damage. Although a simple enzymatic process, PER displays much diversity in various biological systems. For example, studies on two closely related marine fishes, the tautog and cunner, showed a 5-fold difference in the rate of PER (Regan *et al.,* 1982). Differences in action spectra, constitutive levels of photolyase, and cofactor concentrations may account for differences in the efficiency of PER measured in different organisms. The wavelength dependence for maximum PER efficiency is ≈ 380 nm in *E. coli* and *Euglena* spp., ≈ 400 nm in *Neurospora crassa,* and ≈ 440 nm in *Streptomyces griseus* (Harm, 1975). The lack of PER in many organisms and its evolutionary conservation in systems never exposed to sunlight (e.g., soil, enteric bacteria, and blind cave fish) suggests photolyase may function in NER as well as PER (Yamamoto *et al.,* 1984; Sancar and Smith, 1989). In support of this notion, it has been shown that *E. coli* photolyase stimulates excision repair *in vitro* (Sancar *et al.,* 1984).

Recent studies in Japan have further clarified the temporal and spatial distribution of PER in eukaryotic systems. The kinetics of NER and PER were recently examined in the *ras* oncogene of cultured goldfish cells (Komura *et al.,* 1991). Similar to mammalian cells, fish cells display preferential NER of cyclobutane dimers in actively transcribing sequences. In contrast, the efficiency of PER was the same in both transcribing and nontranscribing regions of the fish genome, suggesting that accessibility to chromatin does not affect PER. In another study, Yasuhira and co-workers (1991) showed that PER is inducible in goldfish cells. Preillumination of cells with visible light 8 h before UVR significantly increased PER of cyclobutane dimer removal and survival. This was the first clear demonstration of an inducible repair system in eu-

Table 2. Plants and Animals Capable of Removing Cyclobutane Dimers by Enzymatic Photoreactivation (PER)[a]

Group	Species	Common name
Cyanophyta	*Plectonema boreanum*	Blue-green alga
	Anacystis nidulans	Blue-green alga
Schizomycophyta	*Escherichia coli*	Colon bacterium
	Streptomyces griseus	Soil actinomycete
Eumycophyta	*Saccharomyces cerevisiae*	Baker's yeast
	Neurospora crassa	Bread mold
Euglenophyta	*Euglena gracilis*	
Angiospermae	*Phaseolus vulgaris*	Pinto bean
	Phaseolus lunatus	Lima bean
	Zea mays	Maize
	Nicotiana tabacum	Tobacco
Gymmospermae	*Gingko biloba*	Gingko tree
Protozoa	*Paramecium aurelia*	Paramecium
	Tetrahymena pyriformis	
Mollusca	*Physa* spp.	Pond snail
Echinodermata	*Arbacia punctulata*	Sea urchin
	Echinarachnius parma	Sand dollar
Arthopoda	*Anagasta kiihniella*	Flower moth
	Gecarcinus lateralis	Land crab
	Artemia salina	Brine shrimp
	Homarus americanus	Lobster
Insecta	*Trichoplusia ni*	Cabbage looper
	Drosophila melanogaster	Fruit fly
Teleosts	*Haemulon sciurus*	Bluestriped grunt
	Pimephales promelas	Fathead minnow
	Anoptichthys jordani	Blind cave fish
	Tautoga onitis	Tautog
	Tautogolabrus adsperus	Cunner
	Carassius auratus	Goldfish
Amphibia	*Bufo marinus*	Cowflop toad
	Xenopus laevis	African clawed toad
	Rana pipiens	Leopard frog
Reptilia	*Terrapene carolina*	Box turtle
	Iguana iguana	Iguana lizard
	Gekko gekko	Gecko lizard
Aves	*Gallus gallus*	Domestic chicken
Marsupialia	*Didelphis marsupialis*	American opossum
	Caluromys derbianus	Wooly opossum
	Potorous tridactylis	Rat kangaroo

[a] Modified from Rupert (1975) and Friedberg (1985).

Table 3. Photoproduct Induction and Repair in Antarctic Phytoplantkon [a]

Species	I [b]	PP$_{37}$ [c]	Photoproducts removed [d] (10^8 daltons DNA^{-1})		%E [e]
			y	w	
Cis–syn cyclobutane dimers					
Chaetoceros convolutus	1.9		1.019	1.290	21
Chaetoceros neglectus	15.1		0.000	4.615	100
Chaetoceros socialis	2.7		0.754	1.151	34
Corethron cryophilum	0.6	3.84	0.000	0.282	100
Coscinodiscus oculus-iridis	0.5	1.62	0.000	0.363	100
Nitzschia kerguelensis	53.7		0.000	0.000	0
Odontella weissflogii	1.4	2.19	0.000	0.083	100
Thalassiosira australis	0.9	0.25	0.428	0.512	16
Thalassiosira subtilis	3.1	3.42	1.076	0.780	0
Pyrimidine (6-4) pyrimidone photoproducts					
Chaetoceros convolutus	1.6		0.604	0.934	35
Chaetoceros neglectus	12.1		0.000	0.000	0
Chaetoceros socialis	2.7		1.313	1.568	16
Corethron cryophilum	0.9	5.39	0.831	0.813	0
Coscinodiscus oculus-iridis	0.4	1.35	0.000	0.283	100
Nitzschia kerguelensis	30.1		0.000	0.606	100
Odontella weissflogii	1.3	2.03	0.000	0.122	100
Thalassiosira australis	0.9	0.25	0.316	0.595	47
Thalassiosira subtilis	1.9	2.08	0.209	0.314	33

[a] From Karentz et al., (1991).
[b] I = number of photoproducts produced · (10^8 daltons DNA)$^{-1}$ by exposure to 2,500 J · m^{-2}.
[c] PP37 = number of photoproducts required to kill organism.
[d] Photoproducts removed in yellow (y) and white (w) light in 6 h.
[e] %E = percent enhancement of photoproduct removed when cells were incubated in white light.

karyotes. The authors suggested that the mechanism of enhanced photorepair may involve conversion of photolyase from an inactive to an active form.

PER occurs in photosynthetic organisms and has been reported in both terrestrial plants (Trosko and Mansour, 1969) and algae (Wu et al., 1967; Karentz et al., 1991a). Photosynthetically active radiation (PAR) ranges from 400 to 700 nm, overlapping with wavelengths in the lower visible region which promote PER. The dual role for blue and green wavelengths is an added complication for designing experiments to study PER in phototrophic organisms.

3.3. Nucleotide Excision Repair

Organisms that display efficient PER appear to have a reduced capacity for excision repair (Regan *et al.*, 1983). Studies on NER support this contention:

1. Several species of Antarctic diatoms, including *Chaetoceros neglectus, Corethron cryophilum, Coscinodiscus oculus-iridis,* and *Odontella weisflogii,* display PER without NER (Table 3; Fig. 7) (Karentz *et al.,* 1991a).
2. Very low levels of NER are seen in three species of cold-blooded vertebrates at 24 h after exposure to UVR ($\approx 5\%$ excision in the Amazon molly, $\approx 5\%$ in the Carolina box turtle, and negligible levels in the rainbow trout) (Woodhead *et al.,* 1980).
3. NER in two closely related marine fishes, the tautog and cunner, showed extremely low levels of cyclobutane dimer removal ($<10\%$ at 24 h), incision, and repair synthesis compared to human cells (Regan *et al.,* 1983).
4. Although frog cells remove $>90\%$ of the cyclobutane dimers within 1 h by PER, $<10\%$ are repaired in 24 h in the dark (Fig. 8) (Mitchell *et al.,* 1986).
5. Very low NER levels have been measured in cultured cells from the marsupial *Potorus tridactylis* (Wade and Trosko, 1983).

Conversely, many organisms that lack PER have a greater capacity to remove cyclobutane dimers by NER:

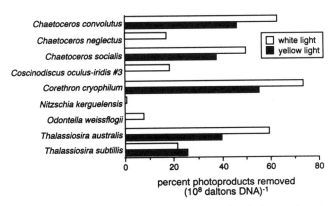

Figure 7. DNA repair in Antarctic diatoms. The percentage of total photoproducts [cyclobutane dimers and (6-4) photoproducts] removed from DNA after 6-h incubation in white or yellow light is shown. Photodamage was induced in clonal cultures of diatoms by 2,500 $J \cdot m^{-2}$ UV-B light (Karentz *et al.,* 1991a).

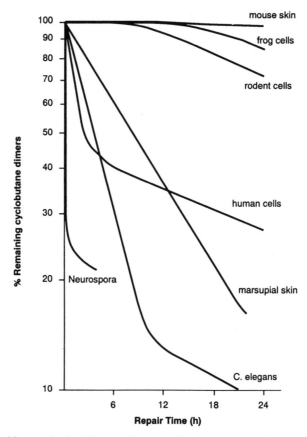

Figure 8. Excision repair of cyclobutane dimer in cells, tissues, and whole organisms. Rates of photoproduct removal were determined by RIA of DNA purified from samples irradiated with sublethal fluences of UV-C or UV-B light.

1. The closely related diatom species *Ch. convolutus, Ch. socialis, Thalassiosira australis,* and *Th. subtilis* showed efficient NER without PER (Table 3).

2. The nematode worm *C. elegans* excises 50% of the cyclobutane dimers by 6 h post-UVR (Hartman *et al.,* 1989), a rate very similar to that determined for human cells in culture (Fig. 8).

However, NER is a genetically complex and phenotypically diverse system (Figs. 8 and 9); its competency is dependent on the species and developmental state of the individual, as well as the type of photoproduct encountered and its genomic distribution. With some exceptions, organisms such as *Tetrahymena* spp., diatoms, frogs, fish, rodents, and humans, for example, excise (6-4) photoproducts much more rapidly than the cyclobutane dimer (Fig. 9; note time scale) (Mitchell *et al.,* 1985; Mitchell and Nairn, 1989). Very similar rates of excision of these two types of photodamage have been observed in *N. crassa* (Baker *et al.,* 1991), *C. elegans* (Hartman *et al.,* 1989), and the epidermis of the marsupial *Monodelphis domestica* (Mitchell *et al.,* 1990b).

Cyclobutane dimer and (6-4) photoproduct excision shows considerable variability in Antarctic diatoms (Table 3). Although *Ch. socialis* and *C. cryophilum* repair (6-4) photoproducts more rapidly than cyclobutane dimers, *Ch. convolutus, T. australis,* and *T. subtilis* show the opposite pattern. Comparative studies of cells derived from several mammalian species show similar diversity (Francis *et al.,* 1981). Even among closely related species the capacity for NER varies widely.

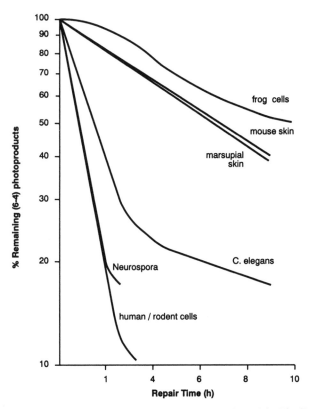

Figure 9. Excision repair of (6-4) photoproducts (see legend for Fig. 8).

Measures of excision repair in total cellular DNA can be misleading. Cyclobutane dimer excision in rodent cells is extremely low compared to human cells, yet both cell types display the same tolerance to UV radiation. Recent advances have shed some light on this interesting dilemma, termed the "rodent paradox." Preferential repair of cyclobutane dimers (Bohr *et al.,* 1985) and (6-4) photoproducts (Thomas *et al.,* 1989) in transcribing DNA suggests that damage in essential genes is responsible for cell killing. The inability of rodent cells to repair cyclobutane dimers in bulk chromatin is, therefore, inconsequential. Since rodent and human cells excise (6-4) photoproducts at the same rates (Fig. 9), the presence of these lesions in bulk chromatin also may contribute to UV cytotoxicity in mammalian cells (Mitchell and Nairn, 1989). Nonmammalian systems also show preferential repair of UV damage. Recent studies on goldfish cells have demonstrated preferential NER in specific DNA sequences in fish, and nonpreferential PER in these same genes (Komura *et al.,* 1989, 1991).

3.4. Photoenhanced Excision Repair

The development and application of methods for detecting (6-4) photoproducts in DNA have shown that removal of cyclobutane dimers by PER can increase excision of noncyclobutane dimer damage. This phenomenon, termed photoenhanced excision repair (PEER), was first seen in frog cells (Mitchell *et al.,* 1985) and has since been observed in human skin (Eggset *et al.,* 1983), marsupial skin (Mitchell *et al.,* 1990b), diatoms (Karentz *et al.,* 1991a), and fish cells (Mitani *et al.,* 1991). How PER influences (6-4) photoproduct excision in these systems is unknown. A putative pathway involves removal of a competing substrate (i.e., cyclobutane dimers) followed by commitment of a common repair system to the task of (6-4) photoproduct removal. Another, more intriguing possibility is that photoreactivating enzyme (PRE) might play an intrinsic role in the excision repair process in these cells. In support of this notion, *E. coli* DNA photolyase stimulates the broadly specific *uvr*ABC excinuclease *in vitro* (Sancar *et al.,* 1984), an enzyme system involved in (6-4) photoproduct repair (Franklin and Haseltine, 1985). Hence, the recovery of cells from the lethal effects of UV light by PER may not result from cyclobutane dimer repair alone, but may also involve enhanced excision of other damage, such as the (6-4) photoproduct.

In the future, analyses of DNA damage and repair in diverse organisms may reveal new, unique mechanisms of DNA repair. Data from our diatom studies show one such curiosity. *N. kerguelensis* cannot repair cyclobutane dimers or (6-4) photoproducts by NER nor remove cyclobutane dimers by PER. However, incubation of irradiated samples in white light, as opposed to yellow light, significantly reduces (6-4) photoproducts. These data suggest that light-dependent repair processes may have evolved in response to this type of damage, as they have for the cyclobutane dimer.

The apparently unique phenotype of *N. kerguelensis* must be viewed with some caution, however, since the same wavelengths responsible for PEER are also involved in photosynthesis. In obligate phototrophs, products of photosynthesis are the only substrates available for the ATP-generating pathways of respiration. The energy-requiring processes within cells, such as NER, rely exclusively on the net production of glucose from photosynthesis. Therefore, photoenhancement of repair could occur under light conditions that increase the efficiency of photosynthesis. For *N. kerguelensis,* exposure to white light represents a more complete spectrum of PAR than exposure to yellow light alone and may promote a higher synthetic rate.

UVR can also damage photosynthetic components and pathways important for survival (Sisson, 1986; Renger *et al.,* 1989; Teramura, 1990). The ability for a photosynthetic organism to survive the effects of UVR is thus contingent on the availability of excess energy for repair processes after basal metabolic requirements have been satisfied. Thus, conditions that inhibit photosynthetic production could significantly reduce the efficiency of DNA repair.

3.5. Developmental Regulation of DNA Repair

Collected data support the notion that DNA-repair efficiency depends on the developmental stage of a cell, tissue, or organism (Mitchell and Hartman, 1990). Both developmental and genetic attributes have made the nematode *C. elegans* a popular organism for probing many biological problems, including DNA repair (Hartman *et al.,* 1990). A striking phenotype of certain wild-type and mutant nematodes is the manifestation of stage-specific variation in UV hypersensitivity and excision repair capacity (Fig. 10). Young larvae (6 h) have significantly greater capacity to excise UV photoproducts compared to older (72 h) and estivating (dauer) nematodes. *Rad-3* mutant nematodes are only moderately hypersensitive when irradiated as embryos, but extremely hypersensitive when irradiated as adults. Consistent with this observation, although *Rad-3* embryos display a reduced capacity for excision repair, larvae assayed 24 h later are almost completely deficient. The correlation between UV hypersensitivity and reduced excision repair capacity in the nematode suggests that, in this organism, DNA damage is lethal and DNA repair is developmentally regulated.

DNA repair varies through development in higher eukaryotes. Excision repair of UV damage in cultured goldfish and medaka cells is very low, although significant NER is observed in primary embryonic cells from medaka (Kator and Egami, 1985). Similar studies on the fathead minnow (*Pimephales*

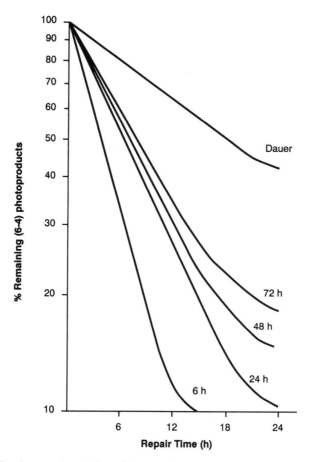

Figure 10. Developmental regulation of DNA repair in *C. elegans*. Kinetics of repair of (6-4) photoproducts in animals of different developmental stages are shown; 6, 24, 48, and 72 h refer to the times elapsed after populations were synchronized and correspond to embryos, stage-1 larvae, stage-2 larvae, and adults. Under harsh environmental conditions animals molt to become dauer larvae rather than stage-3 larvae, a senescent stage which may survive for several months with little effect on post-dauer lifespan.

promelas) indicate that, in contrast to repair-deficient adults, embryos are capable of removing 50% of the cyclobutane dimers after 9 h in the dark (Applegate and Ley, 1988). Exposure to photoreactivating light has no effect on cyclobutane dimer removal in adult mouse epidermal DNA or mouse fibroblasts in culture (Ley *et al.*, 1978), yet significantly reduces the number of these lesions in the epidermis of neonatal mice (Ananthaswamy and Fisher, 1981). A similar pattern is evident for excision repair. Unlike repair-deficient cells cultured from adult mice, those derived from early embryonic mice efficiently remove cyclobutane dimers by excision repair (Peleg *et al.,* 1976).

DNA damage resulting from UVR is more efficiently repaired in proliferating basal cells in rodent and human epidermis than in terminally differentiated keratinocytes (Mitchell and Hartman, 1989). A compelling generality one can derive from the developmental pattern of DNA repair in organisms and tissues is that more rapidly proliferating systems, such as stem cells in proliferating organs, embryos, fetuses, and larvae, have increased capacity for self-correction. Hence, those organisms most vulnerable to the lethal and mutagenic consequences of genetic change can tolerate it. DNA repair exhibits different developmental profiles in different organisms. It thus displays great variation in both ontogeny and phylogeny.

4. Lethal and Mutagenic Effects of Solar UVR

Support for the role of DNA damage in the biological effects of UVR on living systems comes from correlative studies of action spectra for various biological endpoints and direct effects caused by specific photoproducts in DNA. Wavelengths of light that are most efficiently absorbed by DNA coincide with those most effective in cell killing, mutation induction, and erythema (Coohill and Jacobson, 1981; Jones *et al.,* 1987; Peak and Peak, 1982; Tyrrell and Pidoux, 1987). Action spectra for a variety of biological endpoints correlate with induction of the major, dimeric photoproducts in DNA. However, comparative action spectrum studies on bacteria and eukaryotic microorganisms suggest that factors other than "gross DNA absorption" may modulate the effects of UV-B light (Calkins *et al.,* 1988). Factors that can cause deviations in action spectra of various biological endpoints from the DNA absorption spectrum are complex. For example, the wavelength dependence of UVR-absorbing compounds (photoprotection), penetrance through membranes and epidermal tissues, wavelength-dependent inducible NER, PER, and PEER systems, and differences in the intrinsic cytotoxicity and mutagenicity of individual photoproducts may affect action spectrum values. Although DNA is undoubtedly the primary chromophore, other non-DNA-mediated effects may also contribute to cell killing and mutation induction, especially in the UV-B and UV-A ranges of the solar spectrum.

More direct and compelling evidence for the importance of DNA damage comes from studies on DNA repair. Light-dependent removal of cyclobutane dimers by PER and removal of (6-4) photoproducts by PEER reverse the lethal, mutagenic, and tumorigenic effects of UVR in many biological systems. In addition to increasing survival (Table 2), PER reduces the frequency of

UVR-induced mutations in prokaryotes (Menningmann, 1972), lower eukaryotes, such as *Neurospora* (Kilbey and Serres, 1969) and the slime mold *Dictyostelium* (Okaichi *et al.*, 1989), *Drosophila* (Meyer, 1951), *Xenopus laevis* (van Zeeland *et al.*, 1980), and the marsupial *Potorus tridactylus* (Wade and Trosko, 1983). It was recently shown that PER reduces UVR-induced melanoma to background levels in a platyfish–swordtail hybrid (Setlow *et al.*, 1989).

The recessive disease xeroderma pigmentosum (XP) illustrates the importance of DNA as a chromophore for UVR. In this disease, major increases in squamous and basal cell carcinomas and melanomas correlate with a severe deficiency in NER (Cleaver and Kraemer, 1989; Cleaver and Mitchell, 1993). Studies of this disease and other sunlight-sensitive disorders have identified a large series of genetic loci that control the response of mammalian skin to photodamage. These loci are all characterized by significant increases in sensitivity to UVR and include the following groups: (1) Excision repair cross-complementing (ERCC) series identified by selection of UV-sensitive rodent cells in culture (≈ 10 complementation groups), (2) XP (≈ 9 complementation groups), (3) Cockayne syndrome (CS) (≈ 3 complementation groups), and (4) Trichothiodystrophy (TTD) (≈ 3 distinct phenotypic types).

These disorders show increased sensitivity to UV-B wavelengths due to recessive mutations and represent subsets within a large family of genes that regulate DNA repair. These subsets are not mutually exclusive; CS overlaps with XP groups B and H; TTD overlaps with XP group D. The *ERCC1, ERCC2, ERCC3,* and *ERCC6* genes have been cloned and exhibit DNA sequence similarity to bacterial and yeast genes: *ERCC1* shows sequence homology to *uvr* A and C of *E. coli* and *rad10* of yeast; *ERCC2* corresponds to XP group D and is similar to *rad3,* an essential gene of yeast; *ERCC3* corresponds to XP group B; *ERCC6* corresponds to CS group B. These data support the idea that DNA repair is a highly conserved enzymatic process essential to the maintenance of life and preservation of genetic diversity on earth.

5. Conclusions

Data regarding the effects of UVR on DNA are essential to a meaningful evaluation of the biological and environmental impact of stratospheric de-ozonation and increased UV-B at the earth's surface. The amount and composition of UVR damage induced in the DNA of a organism depends on a variety of factors, including: DNA sequence and chromatin structure; cellular anatomy, physiology, and morphology; and behavior and ecological niche.

Likewise, the number and types of photoproducts remaining in an organism's genome after or during the course of ambient solar UVR exposure depends on the repertoire and efficiency of its DNA-repair arsenal. Therefore, the ability to tolerate DNA photodamage is species dependent; each and every organism has its own unique approach to surviving the effects of solar UVR.

Not only do DNA damage tolerance mechanisms vary between species, but they also change as individuals grow and develop. An accurate assessment of the potential risk of an organism to increased UVR, therefore, requires quantitation of its sensitivity and molecular response to UVR damage throughout the life cycle. From these data, it should then be possible to rank species as to their relative vulnerability to UVR and predict the effects of deozonation on the biomass and species composition of complex ecosystems.

References

Ananthaswamy, H. N., and Fisher, M. S., 1981, Photoreactivation of ultraviolet radiation-induced pyrimidine dimers in neonatal BALB/c mouse skin, *Cancer Res.* 41:1829–1833.

Applegate, L. A., and Ley, R. D., 1988, Ultraviolet radiation-induced lethality and repair of pyrimdine dimers in fish embryos, *Mutat. Res.* 198:85–92.

Arpin, N., and Bouillant, M. L., 1981, Light and mycosporines, in: *The Fungal Spore: Morphogenic Controls, Proceedings of the 3rd International Fungal Spore Symposium* (G. Turian and H. R. Hohl, eds.), Academic Press, London, pp. 435–454.

Baker, T. I., Radloff, R. J., Cords, C. E., Engel, S. R., and Mitchell, D. L., 1991, The induction and repair of (6-4) photoproducts in *Neurospora crassa, Mutat. Res.* 255:211–218.

Beggs, C. J., Schneider-Ziebert, U., and Wellman, E., 1986, UV-B radiation and adaptive mechanisms in plants, in: *Stratospheric Ozone Reduction, Solar Ultraviolet Radiation and Plant Life* (R. C. Worrest and M. M. Caldwell, eds.), NATO NSI Series G: Ecological Sciences Vol. 8, Springer-Verlag, Berlin, pp. 235–250.

Bohr, V. A., Smith, C. A., Okumoto, D. S., and Hanawalt, P. C., 1985, DNA repair in an active gene: Removal of pyrimidine dimers from the DHFR gene of CHO cells is much more efficient than in the genome overall, *Cell* 40:359–369.

Bose, S. N., and Davies, R. J. H., 1984, The photoreactivity of T–A sequences in oligodeoxyribonucleotides and DNA, *Nucl. Acids Res.* 12:7903–7913.

Brash, D. E., and Haseltine, W. A., 1982, UV-induced hotspots occur at DNA damage hotspots, *Nature* 298:189–192.

Brash, D. E., Seetharam, S., Kraemer, K. H., Seidman, M. M., and Bredberg, A., 1987, Photoproduct frequency is not the major determinant of UV base substitution hot spots or cold spots in human cells, *Proc. Natl. Acad. Sci. U.S.A.* 84:3782–3786.

Brash, D. E., Rudolph, J. A., Simon, J. A., Lin, A., McKenna, G. J., Baden, H. P., Halperin, A. J., and Ponten, J., 1991, A role for sunlight in skin cancer: UV-induced p53 mutations in squamous cell carcinoma, *Proc. Natl. Acad. Sci. U.S.A.* 88:10124–10128.

Breter, H. J., Weinblum, D., and Zahn, R. K., 1974, A highly sensitive method for the estimation of pyrimidine dimers in DNA by high-pressure liquid cation-exchange chromatography. *Anal. Biochem.* 61:362–366.

Cadet, J., Voituriez, L., Hahn, B. S., and Wang, S. Y., 1980, Separation of cyclobutyl dimers of thymine and thymidine by high-performance liquid chromatography and thin-layer chromatography, *J. Chromatogr.* 195:139–145.

Calkins, J., Wheeler, J. S., Keller, C. I., Colley, E., and Hazle, J. D., 1988, Comparative ultraviolet action spectra (254–320 nm) of five "wild-type" eukaryotic microorganisms and *Escherichia coli, Radiat. Res.* 114:307–318.

Carrier, W. L., Snyder, R. D., and Regan, J. D., 1982, Ultraviolet-induced damage and its repair in human DNA, in: *The Science of Photomedicine* (J. D. Regan and D. D. Parrish, eds.), Plenum Press, New York, pp. 91–112.

Cleaver, J. E., and Kraemer, K. H., 1989, Xeroderma pigmentosum, in: *The Metabolic Basis of Inherited Disease* (C. R. Scriver, A. L. Beaudet, W. S. Sly, and D. Valle, eds.), Vol. II, 6th ed., McGraw-Hill, New York, pp. 2949–2971.

Cleaver, J. E., and Mitchell, D. L., 1993, Ultraviolet radiation carcinogenesis, in: *Cancer Medicine* (J. F. Holland, E. Frei III, R. C. Bast, Jr., D. W. Kufe, D. L. Morton, and R. R. Weichselbaum, eds.), Lea & Febiger, Philadelphia, Section III, Chapter 5 (in press).

Cleaver, J. E., Jen, J., Nguyen, T., Lutze, L. H., and Mitchell, D. L., 1990, Distribution and repair of pyrimidine(6-4)pyrimidone photoproducts in UV-irradiated human cells, *Proc. Am. J. Human Genetics* 49 (supplement), 2539.

Coohill, T. P., and Jacobson, E. D., 1981, Action spectra in mammalian cells exposed to ultraviolet radiation, *Photochem. Photobiol.* 33:941–945.

Cook, K. H., and Friedberg, E. C., 1976, Measurement of thymine dimers in DNA by thin-layer chromatography, *Anal. Biochem.* 73:419–422.

Demple, B., and Linn, S., 1980, DNA N-glycosylases and UV repair, *Nature* 287:203–208.

Dulbecco, R., 1949, Reactivation of ultraviolet inactivated bacteriophage by visible light, *Nature* 163:949.

Dunkle, R. L., and Shasha, B. S., 1989, Response of starch-encapsulated *Bacillus thuringiensis* containing ultraviolet screens to sunlight, *Entomolog. Soc. Amer.* 18:1035–1041.

Dunlap, W. C., Chalker, B. E., and Oliver, J. K., 1986, Bathymetric adaptations of reef-building corals at Davies Bay, Great Barrier Reef, Australia. III. UV-B absorbing compounds, *J. Exp. Mar. Biol. Ecol.* 104:239–248.

Eggset, G., Volden, G., and Krokan, H., 1983, U.v.-induced DNA damage and its repair in human skin *in vivo* studied by sensitive immunohistochemical methods, *Carcinogenesis* 4: 745–750.

Elkind, M. M., Han, A., and Chiang-Liu, C. -M., 1978, "Sunlight"-induced mammalian cell killing: A comparative study of ultraviolet and near-ultraviolet inactivation, *Photochem. Photobiol.* 27:709–715.

Fisher, G., and Johns, H., 1976, Pyrimidine photodimers, in: *Photochemistry and Photobiology of Nucleic Acids* (S. Y. Wang, ed.), Vol. I, Academic Press, New York, pp. 226–295.

Fornace, A. J., Jr., Kohn, K. W., and Kann, H. E., Jr., 1976, DNA single-strand breaks during repair of UV damage in human fibroblasts and abnormalities in xeroderma pigmentosum, *Proc. Natl. Acad. Sci. U.S.A.* 73:39.

Francis, A. A., Lee, W. H., and Regan, J. D., 1981, The relationship of DNA excision repair of ultraviolet-induced lesions to the maximum life span of mammals, *Mech. of Ageing Dev.* 16: 181–189.

Franklin, W. A., and Haseltine, W. A., 1985, Removal of UV light-induced pyrimidine-pyrimidone(6-4) products from *Escherichia coli* DNA requires the *uvr*A, *uvr*B, and *uvr*C gene products, *Proc. Natl. Acad. Sci. U.S.A.* 81:3821–3824.

Friedberg, E. C., 1985, *DNA Repair,* W. H. Freeman, New York.

Gale, J. M., and Smerdon, M. J., 1990, UV induced (6-4) photoproducts are distributed differently than cyclobutane dimers in nucleosomes, *Photochem. Photobiol.* 51:411–417.

Gale, J. M., Nissen, K. A., and Smerdon, M. J., 1987, UV-induced formation of pyrimidine dimers in nucleosome core DNA is strongly modulated with a period of 10.3 bases. *Proc. Natl. Acad. Sci. U.S.A.* 84:6644–6648.

Gallagher, P. E., and Duker, N. J., 1986, Detection of UV purine photoproducts in a defined sequence of human DNA, *Mol. Cell. Biol.* 6:707–709.

Ganesan, A. K., Smith, C. A., and van Zeeland, A. A., 1981, Measurement of the pyrimidine dimer content of DNA in permeabilized bacterial and mammalian cells with endonuclease V of bacteriophage T4, in: *DNA Repair: A Laboratory Manual of Research Procedures, Part A* (E. C. Friedberg and P. C. Hanawalt, eds.), Marcel Dekker, New York.

Garces, F., and Davila, C. A., 1982, Alterations in DNA irradiated with ultraviolet radiation-I. The formation process of cyclobutylpyrimidine dimers: Cross sections, action spectra and quantum yields, *Photochem. Photobiol.* 35:9–16.

Gasparro, F., and Fresco, J., 1986, Ultraviolet-induced 8,8-adenine dehydrodimers in oligo- and polynucleotides, *Nucl. Acids Res.* 14:4239–4251.

Glickman, B. W., Schaaper, R. M., Haseltine, W. A., Dunn, R. L., and Brash, D. L., 1986, The C-C (6-4) UV photoproduct is mutagenic in *Escherichia coli*, *Proc. Natl. Acad. Sci. U.S.A.* 83:6945–6949.

Grant, P. T., Middleton, C., Plack, P., and Thompson, R. H., 1985, The isolation of four aminocyclohexenimines (mycosporines) and a structurally related derivative of cyclohexan-1:3-dione (gadusol) from the brine shrimp, *Artemia, Comp. Biochem. Physiol.* 80B:755–759.

Häder, D.-P., Häder, M., Liu, S.-M., and Ullrich, W., 1990, Effects of solar radiation on photoorientation, motility and pigmentation in a freshwater Peridinium, *BioSystems* 23:335–343.

Harm, H., 1976, Repair of UV-irradiated biological systems: Photoreactivation, in: *Photochemistry and Photobiology of Nucleic Acids* (S. Y. Wang, ed.), Vol. II, Academic Press, New York, pp. 219–261.

Hartman, P. S., Hevelone, J., Dwarakanath, V., and Mitchell, D. L., 1989, Excision repair of UV radiation-induced DNA damage in *Caenorhabditis elegans*, *Genetics* 122:379–385.

Hartman, P. S., Mitchell, D. L., Swensen, B-A., and Reddy, J., 1990, DNA repair in the nematode *Caenorhabditis elegans,* in: *UCLA Symposium on Molec. and Cell Biol., New Series,* Vol. 123, (C. E. Finch and T. E. Johnson, eds.), Alan R. Liss, New York, pp. 67–80.

Haseltine, W. A., Gordon, L. K., Lindan, C. P., Grafstrom, R. H., Shaper, N. L., and Grossman, L., 1980, Cleavage of pyrimidine dimers in specific DNA sequences by a pyrimdine dimer DNA-glycolsylase of *M. luteus, Nature* 285:634–641.

Hausser, K. W., and von Oehmcke, H., 1933, *Strahlentherapie* 48:223.

Jones, C. A., Huberman, E., Cunningham, M. L., and Peak, M. J., 1987, Mutagenesis and cytotoxicity in human epithelial cells by far- and near-ultraviolet radiations: Action spectra, *Radiat. Res.* 110:244–254.

Karentz, D., Cleaver, J. E., and Mitchell, D. L., 1991a, Cell survival characteristics and molecular responses of Antarctic phytoplankton to ultraviolet-B radiation, *J. Phycol.* 27:326–341.

Karentz, D., McEuen, F. S., Land, M. C., and Dunlap, W. C., 1991b, Survey of mycosporine-like amino acid compounds in Antarctic marine organisms: Potential protection from ultraviolet exposure, *Mar. Biol.* 108:157–166.

Kator, K., and Egami, N., 1985, Repair of UV damage in cultured fish cells. IV. Excision repair in primary cultured embryonic cells of Medaka, *Oryzias latipes, J. Radiat. Res.* 26:44.

Kelner, A., 1949, Effect of visible light on the recovery of *Streptomyces griseus* conidia from ultraviolet irradiation injury, *Proc. Natl. Acad. Sci. U.S.A.* 35:73.

Kilbey, B. J., and Serres, F. J., 1969, Quantitative and qualitative aspects of photoreactivation of premutational ultraviolet damage at the *ad-3* loci of *Neurospora crassa, Mutat. Res.* 4:21–29.

Komura, J. -I., Mitani, H., and Shima, A., 1989, More efficient excision repair of pyrimidine dimers in the specific DNA sequence than in the genome overall in goldfish cells, *Photochem. Photobiol.* 49:419–422.

Komura, J. -I., Mitani, H., Nemoto, N., Ishikawa, T., and Shima, A., 1991, Preferential excision repair and non-preferential photoreactivation of pyrimidine dimers in the c-*ras* sequence of cultured goldfish cells, *Mutat. Res.* 254:191-198.

Leadon, S. A., and Hanawalt, P. C., 1983, Monoclonal antibody to DNA containing thymine glycol, *Mutat. Res.* 112:191-200.

Ley, R. D., 1984, Photorepair of pyrimidine dimers in the epidermis of the marsupial *Monodelphis domesticus, Photochem. Photobiol.* 40:141-143.

Ley, R. D., Sedita, B. A., and Grube, D. D., 1978, Absence of photoreactivation of pyrimidine dimers in the epidermal DNA of hairless mice exposed to ultraviolet light. *Photochem. Photobiol.* 27:483-485.

Lippke, J. A., Gordon, L. K., Brash, D. L., and Haseltine, W. A., 1981, Distribution of UV light-induced damage in a defined sequence of human DNA: Detection of alkaline-sensitive lesions at pyrimidine nucleoside-cytidine sequences, *Proc. Natl. Acad. Sci. U.S.A.* 78:3388-3392.

Menningmann, H. D., 1972, Pyrimidine dimers as premutational lesions in *Escherichia coli* WP2 Hcr⁻, *Mol. Gen. Genet.* 117:167-186.

Meyer, H. U., 1951, Photoreactivation of ultraviolet mutagenesis in the polar cap of *Drosophila, Genetics* 36:565.

Mitani, H., Yasuhira, S., Komura, J. -I., and Shima, A., 1991, Enhancement of repair of UV-irradiated plasmids in cultured fish cells by fluorescent light preillumination, *Mutat. Res.* 255:273-280.

Mitchell, D. L., 1988, The induction and repair of lesions produced by the photolysis of (6-4) photoproducts in normal and UV-hypersensitive human cells. *Mutat. Res.* 194:227-237.

Mitchell, D. L., and Cleaver, J. E., 1990, Photochemical alterations of cytosine account for most biological effects after ultraviolet radiation, in: *Trends in Photochemistry and Photobiology* 1:107-119.

Mitchell, D. L., and Hartman, P. S., 1990, The regulation of DNA repair during development, *BioEssays* 12:74-79.

Mitchell, D. L., and Nairn, R. S., 1988, The (6-4) photoproduct and human skin cancer, *Photodermatology* 5:61-64.

Mitchell, D. L., and Nairn, R. S., 1989, The biology of the (6-4) photoproduct. Annual review. *Photochem. Photobiol.* 49:805-819.

Mitchell, D. L., and Rosenstein, B. S., 1987, The use of specific radioimmunoassays to determine action spectra for the photolysis of (6-4) photoproducts, *Photochem. Photobiol.* 45:781-786.

Mitchell, D. L., Haipek, C. A., and Clarkson, J. M., 1985, (6-4) photoproducts are removed from the DNA of UV-irradiated mammalian cells more efficiently than cyclobutane pyrimidine dimers, *Mutat. Res.* 143:109-112.

Mitchell, D. L., Clarkson, J. M., Chao, C. C-K., and Rosenstein, B. S., 1986, Repair of cyclobutane dimers and (6-4) photoproducts in ICR 2A frog cells, *Photochem. Photobiol.* 43:595-597.

Mitchell, D. L., Brash, D. E., and Nairn, R. S., 1990a, Rapid repair of pyrimidine(6-4)pyrimidone photoproducts in human cells does not result from change in epitope conformation, *Nucl. Acids Res.* 18:963-971.

Mitchell, D. L., Applegate, L. A., Nairn, R. S., and Ley, R. D., 1990b, Photoreactivation of cyclobutane dimers and (6-4) photoproducts in the epidermis of the marsupial *Monodelphis domestica, Photochem. Photobiol.* 51:653-658.

Mitchell, D. L., Jen, J., and Cleaver, J. E., 1991a, Relative induction of cyclobutane dimers and cytosine hydrates in DNA irradiated *in vitro* and *in vivo* with UVC and UVB light, *Photochem. Photobiol.* 54:741-746.

Mitchell, D. L., Nguyen, T. D., and Cleaver, J. E., 1991b, Nonrandom induction of pyrimidine-pyrimidone(6-4)photoproducts in ultraviolet-irradiated human chromatin, *J. Biol. Chem.* 265:5353-5356.

Mitchell, D. L., Jen, J., and Cleaver, J. E., 1992, Sequence specificity of cyclobutane pyrimidine dimers in DNA treated with solar (ultraviolet B) radiation, *Nucl. Acids Res.* 20:225–229.

Mori, T., Nakane, M., Hattori, T., Matsunaga, T., Ihara, M., and Nikaido, O., 1991, Simultaneous establishment of monoclonal antibodies specific for either cyclobutane dimers of (6-4) photoproduct from the same mouse immunized with ultraviolet-irradiated DNA, *Photochem. Photobiol.* 54:225–232.

Nakamura, H., Kobayashi, J., and Hirata, Y., 1982, Separation of mycosporine-like amino acids in marine organisms using reversed-phase high-performance liquid chromatography, *J. Chromatogr.* 250:113–118.

Okaichi, K., Kajitani, N., Nakajima, K., Nozu, K., and Ohnishi, T., 1989, DNA damage and its repair in *Dictyostelium discoideum* irradiated by health lamp light (UV-B), *Photochem. Photobiol.* 50:69–73.

Patrick, M. H., 1977, Studies on thymine-derived UV photoproducts in DNA—I. Formation and biological role of pyrimidine adducts in DNA, *Photochem. Photobiol.* 25:357–372.

Patrick, M. H., and Rahn, R. O., 1976, Photochemistry of DNA and polynucleotides: Photoproducts, in: *Photochemistry and Photobiology of Nucleic Acids* (S. Y. Wang, ed.), Vol. II, Academic Press, New York, pp. 35–95.

Peak, M. J., and Peak, J. G., 1982, Single-strand breaks induced in *Bacillus subtilis* DNA by ultraviolet light: Action spectrum and properties, *Photochem. Photobiol.* 35:675–680.

Peak, M. J., and Peak, J. G., 1990, Hydroxyl radical quenching agents protect against DNA breakage caused by both 365-nm UVA and by gamma radiation, *Photochem. Photobiol.* 51:649–652.

Peleg, L., Raz, E., & Ben-Ishai, R., 1976, Changing capacity for DNA excision repair in mouse embryonic cells *in vitro, Exp. Cell Res.* 104:301–307.

Pfeiffer, G. P., Drouin, R., Riggs, A. D., and Holmquist, G. P., 1991, *In vivo* mapping of a DNA adduct at nucleotide resolution: Detection of pyrimidine(6-4)pyrimidone photoproducts by ligation-mediated polymerase chain reaction, *Proc. Natl. Acad. Sci. U.S.A.* 88:1374–1378.

Rasmussen, R. E., and Painter, R. B., 1964, Evidence for repair of ultraviolet damaged deoxyribonucleic acid in cultured mammalian cells, *Nature* 203:1360.

Regan, J. D., Snyder, R. D., Francis, A. A., and Olla, B. L., 1983, Excision repair of ultraviolet- and chemically-induced damage in the DNA of fibroblasts derived from two closely-related species of marine fishes, *Aquatic Toxicol.* 4:181–188.

Regan, J. D., Carrier, W. L., Samet, C., and Olla, B. L., 1982, Photoreactivation in two closely related marine fishes having different longevities. *Mech. of Ageing and Dev.* 18:59–66.

Renger, G., Völker, M., Eckert, H. J., Fromme, R., Hohm-Veit, S., and Gräber, P., 1989, On the mechanism of photosystem II deterioration by UV-B radiation, *Photochem. Photobiol.* 49:97–105.

Rosenstein, B. S., and Mitchell, D. L., 1987, Action spectra for the induction of pyrimidine(6-4)pyrimidone photoproducts and cyclobutane pyrimidine dimers in normal human skin fibroblasts, *Photochem. Photobiol.* 45:775–781.

Rupert, C. S., 1975, Enzymatic photoreactivation: Overview, in: *Molecular Mechanisms for Repair of DNA, Part A* (P. C. Hanawalt and R. B. Setlow, eds.), Plenum, New York, p. 73.

Sancar, A., and Rupp, W. D., 1983, A novel repair enzyme: UvrABC excision nuclease of *Escherichia coli* cuts a DNA strand on both sides of the damaged region, *Cell* 33:249–260.

Sancar, G. B., and Smith, F. W., 1989, Interactions between yeast photolyase and nucleotide excision repair proteins in *Saccharomyces cerevisiae* and *Escherichia coli, Mol. and Cell. Biol.* 9:4767–4776.

Sancar, A., Franklin, K. A., and Sancar, G. B., 1984, *Escherichia coli* DNA photolyase stimulates uvrABC excision nuclease *in vitro, Proc. Natl. Acad. Sci. U.S.A.* 81:7397–7401.

Setlow, R. B., and Carrier, W. L., 1966, Pyrimidine dimers in ultraviolet-irradiated DNAs, *J. Mol. Biol.* 17:237–254.

Setlow, R. B., Woodhead, A. D., and Grist, E., 1989, Animal model for ultraviolet radiation-induced melanoma: Platyfish–swordtail hybrid. *Proc. Natl. Acad. Sci. U.S.A.* 86:8922–8926.

Shibata, K., 1969, Pigments and a UV-absorbing substance in corals and a blue-green alga living in the Great Barrier Reef, *Plant Cell Physiol.* (Tokyo) 10:325–335.

Shick, J. M., Lesser, M. P., and Stochaj, W. R., 1991, Ultraviolet radiation and photooxidative stress in zooanthellate anthozoa: The sea anemone *Phyllodiscus semoni* and octoral *Clavularia* spp., *Symbiosis* 10:145–173.

Sisson, W. B., 1986, Effects of UV-B radiation on photosynthesis, in: *Stratospheric Ozone Reduction, Solar Ultraviolet Radiation and Plant Life,* (R. C. Worrest and M. M. Caldwell, eds.), NATO NSI Series G: Ecological Sciences Vol 8., Springer-Verlag, Berlin, pp. 161–170.

Smith, P. J., and Paterson, M. C., 1982, Abnormal responses to mid-ultraviolet light of cultured fibroblasts from patients with disorders featuring sunlight sensitivity, *Cancer Res.* 41:511–518.

Sterenborg, H. J. C. M., and Van der Leun, J. C., 1990, Tumorigenesis by a long wavelength UV-A source, *Photochem. Photobiol.* 51:325–330.

Takahashi, A., Takeda, K., and Ohnishi, T., 1991, Light-induced anthocyanin reduces the extent of damage to DNA in UV-radiated *Centaurea-cyanus* cells in culture, *Plant Cell Physiol.* 32:541–547.

Taylor, J. S., and Cohrs, M. P., 1987, DNA, light and Dewar pyrimidinones: The structure and biological significance of TpT3, *J. Am. Chem. Soc.* 109:2834–2835.

Taylor, J. S., Lu, H-F., and Kotyk, J. J., 1990, Quantitative conversion of the (6-4) photoproduct of TpdC to its Dewar valence isomer upon exposure to simulated sunlight, *Photochem. Photobiol.* 51:161–167.

Teramura, A. H., 1990, Implications of stratospheric ozone depletion upon plant production, *HortScience* 25:1557–1560.

Tevini, M., and Teramura, A. H., 1989, UV-B effects on terrestrial plants, *Photochem. Photobiol.* 50:479–487.

Thomas, D. C., Okumoto, D. S., Sancar, A., and Bohr, V. A., 1989, Preferential repair of (6-4) photoproducts in the dihydrofolate reductase gene of Chinese hamster ovary cells, *J. Biol. Chem.* 264:18005–18010.

Trosko, J. E., and Mansour, V. H., 1969, Photoreactivation of ultraviolet light-induced pyrimidine dimers in Ginkgo cells grown *in vitro, Mutat. Res.* 7:120–121.

Tyrrell, R. M., and Pidoux, M., 1986, Endogenous glutathione protects human skin fibroblasts against the cytotoxic action of UVB, UVA and near visible radiations, *Photochem. Photobiol.* 44:561–564.

Tyrrell, R. M., and Pidoux, M., 1987, Action spectra for human skin cells: Estimates of the relative cytotoxicity of the middle ultraviolet, near ultraviolet and violet regions of sunlight on epidermal keratinocytes, *Cancer Res.* 47:1825–1829.

Umlas, M. E., Franklin, W. A., Chan, G. L., and Haseltine, W. A., 1985, Ultraviolet light irradiation of defined-sequence DNA under conditions of chemical photosensitization, *Photochem. Photobiol.* 42:265–273.

van Zeeland, A. A., Natarajan, A. T., Verdegaal-Immerzeel, E. A. M., and Filon, A. R., 1980, Photoreactivation of UV induced cell killing, chromosome aberrations, sister chromatid exchanges, mutations, and pyrimidine dimers in *Xenopus laevis* fibroblasts, *Mol. Gen. Genet.* 180:495–500.

Wade, M. H., and Trosko, J. E., 1983, Enhanced survival and decreased mutation frequency after photoreactivation of UV damage in rat kangaroo cells, *Mutat. Res.* 112:231–243.

Woodhead, A. D., Setlow, R. B., and Grist, E., 1980, DNA repair and longevity in three species of cold-blooded vertebrates, *Exp. Geront.* 15:301–304.

Wu, J. H., Lewin, R. A., and Werbin, H., 1967, Photoreactivation of UV-irradiated blue-green algal virus LPP-1, *Virology* 31:657.

Yamamoto, K., Satake, M., and Shinagawa, H., 1984, A multicopy *phr*-plasmid increases the ultraviolet resistance of a *rec*A strain of *Escherichia coli, Mutat. Res.* 131:11–18.

Yasuhira, S., Mitani, H., and Shima, A., 1991, Enhancement of photorepair of ultraviolet damage by preillumination with fluorescent light in cultured fish cells, *Photochem. Photobiol.* 53:211–215.

13

Ultraviolet Radiation and Its Effects on Organisms in Aquatic Environments

Osmund Holm-Hansen, Dan Lubin, and E. Walter Helbling

1. Introduction

The problem of trying to determine the effect of solar ultraviolet radiation (UVR) on aquatic organisms is much more difficult than that of assessing the impact of UVR on terrestrial plants. The major reasons for this are that spectral irradiance changes dramatically with depth in the water column and that most aquatic organisms will be moving up and down in the upper water column, either through active motility processes or by physical mixing processes. It is thus not possible to determine the effect of UVR on planktonic organisms with any degree of certainty; the best one can do is to determine the effects under a wide variety of experimental techniques, and to estimate the potential damage to organisms when they are under completely natural conditions.

Osmund Holm-Hansen and E. Walter Helbling • Marine Research Division, Scripps Institution of Oceanography, University of California at San Diego, La Jolla, California 92093-0202. **Dan Lubin** • California Space Institute, Scripps Institution of Oceanography, University of California at San Diego, La Jolla, California 92093-0202.

Environmental UV Photobiology, edited by Antony R. Young *et al.* Plenum Press, New York, 1993.

There is a large literature on effects of UVR on plant and animal species living in aquatic environments. Most of these studies, however, have been done with organisms or cultures maintained in temperature-controlled incubators exposed either to solar radiation or to artificial illumination. Although most such studies attempt to simulate natural conditions, the spectral irradiance at various depths in the water column can not be duplicated very well, and hence there is always some degree of artificiality in such experimentation. The most reliable way to evaluate the ecological impact of UVR on aquatic organisms is to determine the *in situ* response as a function of depth in the upper water column. Such experiments, however, suffer from the fact that the samples are generally held at fixed depths in the water column, whereas in reality planktonic cells usually are being circulated within the upper mixed layer (UML) of the water column by physical mixing processes, and hence they are being exposed to continuously changing spectral irradiance regimes. Because the combined effects of varying both the dose rate (i.e., the irradiance incident upon the organism) and the duration of exposure at that irradiance are not known for natural populations of aquatic organisms, the interpretation of *in situ* data also involves some degree of uncertainty.

In this chapter, we are concerned with effects of solar radiation on aquatic organisms, and hence we will discuss both UV-B radiation (defined here as the spectral region from 280 to 320 nm) and UV-A radiation (320 to 400 nm), but the emphasis will be on UV-B effects, particularly the effect of enhanced UV-B radiation resulting from decreased ozone concentrations in the stratosphere. Laboratory studies utilizing lamps emitting UV-C radiation (190 to 280 nm) will not be included in the discussion. Although *in situ* experiments are desirable from the perspective of maintaining natural spectral irradiance incident upon the samples, it is difficult to vary experimental conditions sufficiently in such studies to determine dose-response relationships or to understand the mechanisms whereby UVR is affecting cellular processes. We will thus include both *in situ* and laboratory studies on effects of UVR in our discussion.

Emphasis will be placed on spectral solar irradiance incident upon organisms in the water column, the dose of UVR absorbed by organisms, the magnitude and action spectrum for any elicited physiological response, and adaptational mechanisms which minimize cellular damage caused by the incident UVR. It should be noted, however, that solar UVR can also affect biochemical dynamics in the environment surrounding the organisms, resulting in an altered chemical milieu which may have either deleterious or beneficial effects on aquatic organisms. As many aspects of such "extracellular" effects of UVR have recently been described (Blough and Zepp, 1990; Palenik *et al.*, 1991), they will not be included in this chapter. We are concerned with all aquatic environments, and hence our discussion will include the importance of latitudinal and seasonal aspects of incident solar radiation, as well as the importance of the depth distribution of organisms in the water column. Primary attention is given to studies involving phytoplankton, because they represent the base of the food web in natural waters and also because they have been the subject of the majority of UVR studies in aquatic environments. No attempt has been made to mention all past studies dealing with UVR effects on plankton, as many review articles summarize the older literature very well (e.g., Calkins, 1982; Worrest, 1982, 1986; Jagger, 1985; Urbach, 1989).

The problem of the effect of solar UVR on sea-ice assemblages and on populations in the water column beneath the ice will not be discussed in this chapter. It should be noted that during the time of the maximal development of the ozone hole in the Antarctic (September to October), the annual sea ice in the southern ocean is at approximately 75% of its maximal extent, and hence much of the solar UVR incident at that time will be absorbed/scattered by sea ice, and very much attenuated before it penetrates to the water beneath the ice. The reader is referred to the discussion and references found in recent studies by Trodahl and Buckley (1989, 1990), which give a good description of the attenuation of UVR by sea ice and the effects of UVR on the associated ice flora.

2. General Effects of UVR on Microbial Cells

Although UVR is often arbitrarily separated into the UV-B and UV-A regions, it should be recognized that such distinctions are generally specific to certain interests (e.g., the erythema action spectrum) and do not signify any general biological response specific to these spectral regions. Thus, UV-B radiation is often mentioned as being "harmful" and UV-A radiation as "beneficial" because of photorepair mechanisms. In regard to effects on microorganisms, it is more realistic to consider the electromagnetic spectrum as a continuum from 280 to 800 nm, with the elicited biological response being related to the wavelength of the radiation and to the absorption spectra of cellular chromophores. Any portion of UVR or visible radiation may have either deleterious or beneficial effects on certain organisms, depending upon the dose absorbed by the organism.

Most literature on photosynthesis arbitrarily cut off at 400 nm, but there are many studies with algae which show that UVR radiation (from 400 nm down to at least 300 nm) can be utilized in photosynthesis (McLeod and Kanwisher, 1962; Halldal, 1964, 1967). More recent studies (Neori et al., 1988) with a variety of unicellular and multicellular algae confirm that UVR with wavelengths down to 320 nm (the lowest wavelengths used in the investigation) is effective in providing energy required in photosynthesis. Under some conditions UVR can thus be utilized in photosynthesis, but data discussed in Section 5 show that both UV-A and UV-B generally decrease rates of primary production in natural waters. Studies with natural phytoplankton populations (Modert *et al.*, 1982) support the above results in that energy in both UV-B and UV-A radiation up to a threshold value apparently were used in photosynthesis, but at fluences above the threshold value, UVR decreased photosynthetic rates. In this chapter, we will not be concerned with any potentially beneficial aspects of UVR, but will concentrate on the deleterious effects that have been studied with organisms from aquatic environments.

Most studies of UVR effects on aquatic organisms have utilized artificial lamps, often in conjunction with natural solar radiation. Although results from such experiments cannot be extrapolated to the natural environment with complete confidence, they do provide the most comprehensive information on the sensitivity of cells to UVR and the diverse metabolic processes which may be affected. Such information is also of great value in regard to devising appropriate *in situ* or simulated *in situ* experiments. The following cellular processes have been described as being adversely affected by UVR. It should be noted that the chromophores, targets, action spectra, and mechanisms responsible for these diverse physiological effects are largely unknown. Some of the effects listed separately may have a common biochemical cause in regard to the target involved by the UV-induced damage. Studies discussed below have used ecologically relevant spectral ranges of UVR (wavelength > 280 nm); reference has not been made to earlier studies utilizing mercury lamps with peak emission close to 254 nm.

2.1. Photosynthetic Rates

Most studies on the effects of UVR on aquatic plants have determined the short-term rate of photosynthesis (measured either as the fixation of CO_2 or the production of O_2) in relation to the dose of UVR. The major reason for this emphasis on photosynthesis is related to the ecological importance of primary production in supporting the food web in natural waters. Reference to many of these studies which have documented the deleterious effects of UVR in laboratory experiments and in field studies with natural phytoplankton assemblages may be found in the general references listed toward the end of the Introduction. More recent data, much of which is concerned with effects of enhanced UV-B radiation, as well as discussion of the interpretation of such data, are found in section 5.

2.2. Effects on Photosynthetic Pigments

There is a large literature on the effects of solar UV radiation on degradation of photosynthetic pigments, with concomitant loss of photosynthetic capacity (e.g., Döhler, 1985; Nultsch and Agel, 1986; Häder and Häder, 1989a). There are some indications that the mechanism of UVR-induced damage to pigments is similar to that of visible radiation (Neale, 1987). Both laboratory and field tests have demonstrated the differential destruction of cellular pigments by ambient UVR (El-Sayed *et al.*, 1990). Long-term tests (1–2 days) with natural populations of Antarctic phytoplankton have also shown that concentrations of all major photosynthetic pigments were decreased by ambient levels of UVR; concomitant short-term (4 h) tests with ice algae, however, did not result in pigment loss (Bidigare, 1989). Hirosawa and Miyachi (1982/83) have shown that in the blue-green alga *Anacystis nidulans*, UV-A radiation resulted in lower levels of cellular chlorophyll *a*, presumably by inhibition of the synthesis of the pigment, rather than by a destructive effect on the pigment itself.

2.3. Photosystem II Sensitivity

UVR apparently can damage both the oxidizing side of photosystem II as well as the reaction centers (Bornman *et al.*, 1984), resulting in decreased rates of photosynthesis (Renger *et al.*, 1989). Photosystem I does not seem to be damaged by UVR (Bornman, 1989). UV-B radiation is most effective in photosystem II inhibition, presumably by degradation of essential proteins, with the chromophore possibly being a plastoquinone (Greenberg *et al.*, 1989). Hirosawa and Miyachi (1983), however, have also shown that inhibition of the Hill reaction is mediated through a chromophore absorbing in the UV-A region and that the effect is readily reversible by visible light.

2.4. Effect on Cell Growth and Division

Many workers (e.g., Calkins and Thordardottir, 1980; Ekelund and Björn, 1990) have reported the growth-limiting effect of UVR on algae. Vernet (1990) has shown that in Antarctic phytoplankton, UV-A and UV-B radiation inhibit cell growth about equally, which is similar to the inhibition often noted for photosynthetic rates (Jokiel and York, 1984; Holm-Hansen, 1990). Karentz *et al.* (1991a), working with cultures of Antarctic phytoplankton, have shown that there is a great range in sensitivity to UV-B radiation, with much of the variation in rates of cell survival being related to cell size and shape. The mechanisms responsible for such retardation in growth and cell division in marine phytoplankton have not been specifically identified. Direct effects on inhibition of cell division due to DNA damage certainly are of much importance in this regard, but other indirect effects (e.g., damage to essential enzymes or proteins involved in membrane transport processes) must also be considered.

2.5. Nucleic Acid Functioning

Direct or indirect effects of UVR on structure and function of DNA is thought to be one of the primary mechanisms responsible for cell injury and loss of viability (see Chapter 12 by Mitchell and Karentz, this volume). Damage to RNA by UVR might also affect many cellular processes through interfering with normal functioning of messenger RNA or transfer RNA (Döhler, 1990). Damage to nucleic acids can result either from direct absorption of UVR or indirectly through sensitizing mechanisms involving reactive oxygen transients such as hydroxyl radicals and superoxide.

2.6. Electron Transport

UVR is thought to result in damage to the electron transport system (Neale, 1987), which could have diverse effects on metabolic processes. The importance in electron transport systems of flavoproteins and cytochromes, which absorb in the UV region, suggests that respiration should be sensitive to damage by enhanced levels of UV-B radiation. The authors are not aware, however, of any studies demonstrating the effect of UVR directly on respiratory rates.

2.7. ATP Synthesis

Studies of Vosjan *et al.* (1990) have shown that ambient levels of solar UV-B in the Antarctic can have dramatic effects on the total concentration of adenosine triphosphate (ATP) in natural microbial assemblages from the upper 30 m of the water column. After 5 h of UV-B irradiation (1.35 W m^{-2}), ATP concentrations were reduced by an average of 75%. As cellular concentrations of adenosine phosphates play an important role in metabolic regulation (Atkinson, 1969), such an effect of ambient levels of solar UV-B on cellular ATP concentrations may have ramifications on many cellular processes and on cell viability.

2.8. Nitrogen Metabolism

Extensive work by Döhler (1985) and Döhler *et al.* (1991) has shown that ambient levels of UVR can have deleterious effects on many facets of nitrogen metabolism in microorganisms. This includes effects on rates of nitrogen fixation (Newton *et al.*, 1979), nitrate and ammonia utilization (Döhler and Biermann, 1987), amino acid metabolism (Döhler, 1989), and rates of protein synthesis (Döhler, 1985). In tests with blue-green algae and with the water fern *Azolla caroliniana,* Newton *et al.* (1979) have shown that nitrogenase activity, measured by acetylene reduction, is more sensitive to low doses of UV-B than is the CO_2 fixation system. Studies of Döhler *et al.* (1987) also show that nitrogen metabolism seems to be more sensitive to UVR than CO_2 fixation. Döhler (1989) has also shown that the patterns of incorporation of radiocarbon into amino acids associated with the Calvin–Benson carbon cycle were very different in diatoms exposed to UV-B radiation. This effect of UVR of altering the incorporation patterns of radiocarbon is in contrast, however, to the results of Zill and Tolbert (1958), who reported that gamma and UV radiation caused loss of viability in the green algae *Chlorella pyrenoidosa* without any detectable change in the distribution pattern of the radiocarbon.

2.9. Proteins and Cell Membranes

Many studies have also shown that protein structure and function can be adversely affected by UVR (Döhler and Bierman, 1987; Larson and Berenbaum, 1988). Because such cellular damage would include enzyme function, the indirect effects of this UV-induced damage might have far-reaching effects on general metabolic processes. Much attention has been paid to membrane proteins involved in active transport of solutes in or out of cells, because damage of these proteins would have many ramifications on general cellular processes. The proteins involved in the transport of inorganic carbon from outside the cell to the chloroplast have received particular attention, as such transport would relate to observed effects on photosynthetic rates (Miyachi, 1989) and on rates of photorespiration (Tolbert, 1989).

2.10. Tropisms: Cell and Organelle Movements

Extensive studies (Nultsch and Häder, 1988; Häder and Häder, 1989a,b; Häder and Worrest, 1991; Blakefield and Calkins, 1992) have shown that motility of organisms is very sensitive to UVR. This may have much significance to ability of organisms in nature to minimize damage incurred by solar UVR. There seems to be much variation in various phylogenetic groups of phytoplankton in degree of sensitivity to UVR. Ekelund and Björn (1990) have shown that some dinoflagellates are exceedingly sensitive to relatively low fluxes of UV-B radiation, in regard to both swimming speed and growth rate. They have speculated that the target for such UVR effects on motility may be the microtubules, which are involved in motility in dinoflagellates. Studies with a colored ciliate have shown that motility is very sensitive to both UV-B and UV-A radiation (Häder and Häder, 1991), with the suggested mechanism being a photodynamic action. There is not much known regarding the sensing mechanisms for UVR, but studies indicate that both visible and long-wavelength UV-A radiation are effective for the photoreceptor pigments involved in phototaxic movements (Foster and Smyth, 1980). If organisms have no mechanism with which to sense UV-B radiation, then enhanced UV-B resulting from ozone depletion might have especially deleterious effects on motility and orientation of organisms in the water column.

When phytoplankton are exposed to high irradiances, many species have the ability to minimize their cross-sectional area for absorption of radiation by coalescing and movement of their chloroplasts (Haupt and Schönbohm, 1970; Kiefer, 1973). Action spectra for these organelle movements were not published, so it is not known if this is a response just to photosynthetically available radiation (PAR) or to both PAR and UVR. In either event, it would serve to protect the cells from excessive UVR when solar radiation is high.

2.11. Effect on Metal Complexes

Many essential trace mineral elements (e.g., Fe, Mn, Cu, Co) exist in low concentrations in natural waters, and their activity (i.e., their availability for cells) may be affected by UV-catalyzed photochemical reactions. Such changes in the bioavailability of metals to plankton may have either deleterious effects (e.g., toxicity due to cuprous ion) or beneficial effects through increasing the availability of metals which are limiting to cellular growth. As the sensitivity of organisms to such chemical changes in the environment seems to vary considerably from the prokaryotic cyanobacteria to the eukaryotic picoplankton and to the larger eukaryotic phytoplankton (Moffett, 1990), enhanced UV-B may result in altering species composition of natural microbial assemblages in natural waters. An example of such an enhancement of primary production has been cited by Palenik *et al.* (1991), who suggest that enhanced UV-B irradiation may increase rates of primary production in the southern ocean by removing the limitation of Fe deficiency. This suggestion, however, assumes that Fe availability is limiting phytoplankton growth in the Antarctic as described by Martin *et al.* (1990), a scenario which is not adhered to by all researchers (e.g., Buma *et al.,* 1991; Mitchell *et al.,* 1991).

3. Downwelling Spectral UV Irradiance

Many of the studies referred to above report the illumination conditions to which the cells were exposed during the experiment. To be able to relate such data on dose-response relationships to effects on natural microbial assemblages in aquatic environments, it is necessary to know the ambient spectral irradiance regimes in aquatic environments in relation to latitude and season as outlined below. The magnitude of the impact of UVR on organisms will be a function of the total absorbed dose, which is dependent on both the incident spectral irradiance (dose rate) and the length of time of exposure to the radiation. It is thus necessary to examine both the spectral irradiance incident upon the water surface and the spectral attenuation of UVR with depth in the water column.

3.1. Latitudinal and Seasonal Variation in UVR

Figure 1 shows spectral irradiance at local noon as a function of latitude and season in the Southern Hemisphere when integrated over three regions of the electromagnetic spectrum: UV-B, 280–320 nm; UV-A, 320–400 nm; and PAR (photosynthetically available radiation), 400–700 nm. The enhanced UV-B radiation at vernal equinox under the severe 1987 ozone hole is also shown in Fig. 1A. Values for the integrated daily dose can be found in Holm-Hansen and Lubin (1992). At no time does either the irradiance of UV-B at local noon or the integrated daily dose of UV-B in the Antarctic (defined here as south of 55°) reach those levels normally prevailing between 0 to 30°S. Hence, even under maximal development of the ozone hole, the fluence of UV-B is still less than that normally encountered in tropical and temperate waters. With the exception that there is no pronounced ozone hole, the same general trends in UVR also apply to the Northern Hemisphere, but the chapter by Madronich (Chapter 1, this volume) should be consulted for a more detailed discussion of the interaction between the atmosphere and incident UVR.

Although the fluence of UV-B under the ozone hole in the oceanic areas of the Antarctic increases by less than a factor of 2, the most dramatic changes are at the shorter UV-B wavelengths, as seen by the data in Fig. 2A. The magnitude of this effect is most evident by plotting the ratio of the spectral irradiance on a low-ozone day to that on a normal-ozone day (Fig. 2B). This ratio is >10 at 304 nm, and increases exponentially at progressively shorter wavelengths. Although this effect is dramatic, the biological impact of this enhancement of shorter wavelength UV-B radiation is mitigated by (1) very low absolute energy levels at these wavelengths, as can be seen in Fig. 2A, and (2) rapid attenuation of these shorter UV-B wavelengths with depth in the water column.

More detailed descriptions of incident spectral irradiance in Antarctica are available in Lubin and Frederick (1991), Lubin *et al.* (1992), and Smith and Wan (1992).

3.2. Attenuation of UVR in the Water Column

For many decades it was thought that UVR was so quickly attenuated by water that it would not have any significant effect on aquatic organisms.

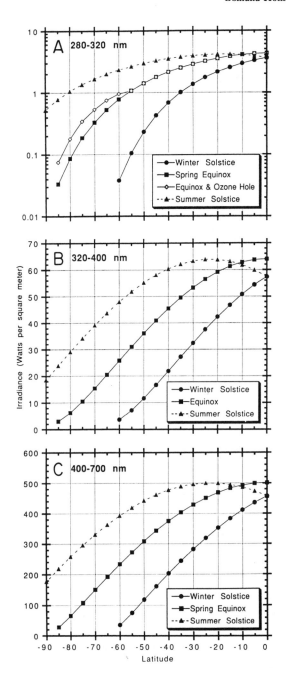

Early work by Jerlov (1950) and more recent studies (e.g., Smith and Baker, 1981) have shown, however, that UVR can penetrate to tens of meters in natural waters, and hence is of considerable ecological significance. In evaluating the impact of UVR on overall rates of primary production, it is necessary to consider the attenuation of both UVR and PAR with depth in the water column. Figure 3 shows the spectral irradiance at the spring equinox from just below the surface to 50 m depth for clear ocean waters, both in the tropics (Fig. 3A) and in the Antarctic at 65°S (Fig. 3b). Both UV and red–infrared radiation are relatively rapidly attenuated, while maximal transmission is close to 450 nm in the blue region of the visible spectrum. Data in Fig. 4 show a more detailed picture of the attenuation of UVR during the spring equinox at the equator (Fig. 4A), at 65°S under normal ozone (Fig. 4B), and at 65°S under an ozone hole (Fig. 4C). It is seen that the flux of short UV-B wavelengths is considerably greater in tropical waters than in polar waters, even under a well-developed ozone hole.

The fluence and rates of attenuation of UV-B radiation (with and without an ozone hole) in the upper water column (0 to 50 m) at 65°S during the spring equinox are shown in Fig. 5, together with values at the equator. The curved portions (0 to 5 m depth) of the PAR lines reflect chromatic changes due to the rapid attenuation of wavelengths between 600 to 700 nm as evident in Fig. 3. It is seen that the UV-B irradiance in Antarctic waters, even when the ozone hole is present, is much lower than at low latitudes, and that it is attenuated much faster than UV-A radiation or PAR. Calculated values for rate of attenuation of UV-B radiation at 305 and 320 nm (with and without an ozone hole) for clear ocean waters are shown in Fig. 6. The actual rate of attenuation of UVR in Antarctic waters is considerably faster, as indicated by actual field data shown in Fig. 7. It is seen that UV-B radiation at 320 nm is measurable to at least 30 m, while UV-A radiation at 380 nm is measurable to over 60 m (Fig. 7A); the profile of water density (sigma-t) and the chlorophyll-a concentration with depth at that station are shown in Fig. 7B.

In addition to absorption by water molecules, the rate of spectral attenuation of UVR with depth will be related to the concentrations of dissolved organic material, size distribution and total biomass of microbial cells, and also the amount of pigmented detrital material. Some of the

---

Figure 1. Results of atmospheric radiative transfer calculations showing local noon solar irradiance incident upon the earth as a function of latitude and season in the Southern Hemisphere, in addition to increased UV-B radiation at vernal equinox under the severe Antarctic ozone depletion of 1987. Note the change in ordinate scales for the integrated values for the three spectral regions. (A) UV-B radiation; (B) UV-A radiation; (C) photosynthetically available radiation.

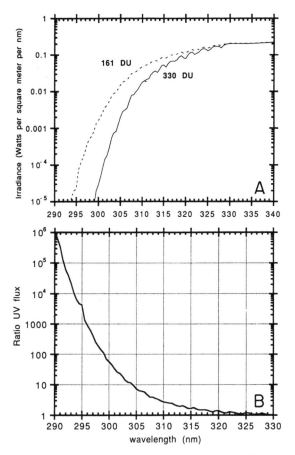

Figure 2. Spectral UVR irradiance at local noon at McMurdo Station (77° 51′ S) in the Antarctic as influenced by ozone concentrations in the atmosphere. (A) Solid line: spectral UV-B radiation measured on October 20, 1988, with a "normal" ozone concentration of 330 Dobson Units (DU); Dashed line: spectral UV-B radiation measured under the ozone hole (161 DU) on October 20, 1989, showing enhanced flux of shorter wavelengths. Data from the 2 days have been normalized at 340 nm. Note that ordinate values are on a log scale. (B) Ratio of spectral UV-B radiation for the 2 days shown above (values under the ozone hole divided by values under normal ozone concentrations). Data from C. R. Booth.

difficulties of measuring the diffuse attenuation coefficients for spectral UVR have been discussed by Mitchell (1990), while Nelson (1990) has discussed the importance of pigmented detrital material in oceanic waters. Many coastal ecosystems have high concentrations of "Gelbstoff," which will attenuate UVR very rapidly. As yet we do not have sufficient data

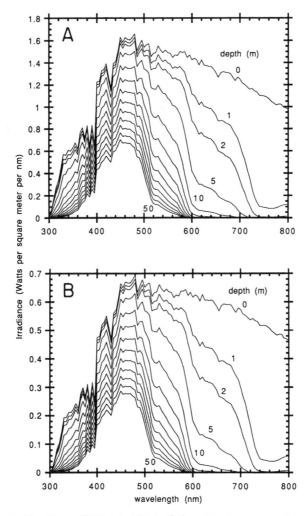

Figure 3. Spectral irradiance of UV and visible radiation at local noon on a clear day from just below the surface to 50-m depth at the time of the spring equinox. (A) At the equator with ozone at 261 DU. (B) At 65°S with ozone at 275 DU. The rate of attenuation with depth is based on diffuse attenuation coefficients for clearest ocean water given by Smith and Baker (1979, 1981).

from diverse aquatic environments to adequately describe the underwater UV regime in terms of concentrations of particulate and dissolved organic materials. With the recent advent of commercial UV-spectroradiometers suitable for field studies, such data should be available in the near future.

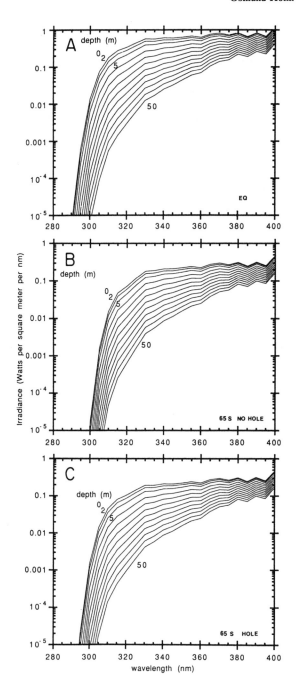

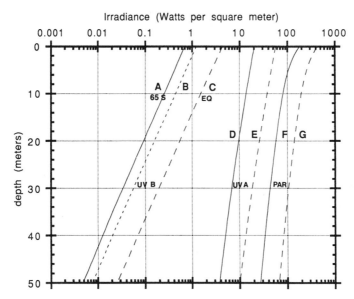

Figure 5. Local noon irradiance values during the spring equinox of UV-B, UV-A, and PAR in the upper 50 m of the water column at the equator and at 65°S (with and without an ozone hole). Values are based on clearest ocean waters with diffuse attenuation coefficients as described by Smith and Baker (1979, 1981). (A) UV-B at 65°S with ozone at 377 DU. (B) UV-B at 65°S under an ozone hole of 275 DU. (C) UV-B at the equator. (D) UV-A at 65°S. (E) UV-A at the equator. (F) PAR at 65°S. (G) PAR at the equator.

4. Importance of Nutritional Mode

As is evident from Figs. 3 and 4, water acts as a selective filter which attenuates UVR rather rapidly with depth, so that organisms can lessen any UV exposure by moving down the water column. All photoautotrophic organisms (e.g., photosynthetic bacteria, cyanobacteria, and eukaryotic algae) are dependent upon solar radiation as an energy source for photosynthetically fixing CO_2, and hence must be close enough to the surface so that irradiance levels are sufficient to result in net CO_2 fixation. This re-

---

Figure 4. Spectral irradiance of UVR (local noon) at the time of the spring equinox in the upper 50 m of the water column as a function of latitude and atmospheric ozone concentrations. (A) At the equator with ozone of 260 DU. (B) At 65°S with normal ozone of 377 DU. (C) At 65°S with ozone hole of 275 DU. Spectral attenuation of radiation with depth is based on diffuse attenuation coefficients for clearest ocean water given by Smith and Baker (1979, 1981).

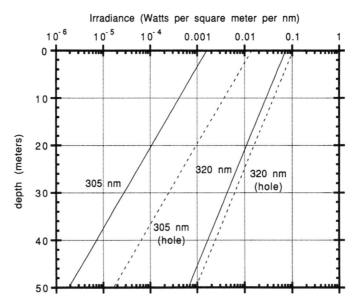

Figure 6. Spectral attenuation of UV-B radiation (305, 320 nm) at local noon in the upper water column at 65°S during the spring equinox under normal ozone concentrations (377 DU) and under an ozone hole with ozone of 170 DU. Rate of attenuation is based on diffuse attenuation coefficients described by Smith and Baker (1979, 1981).

quirement for phytoplankton to remain in well-illuminated surface waters has led to the general belief that all photoautotrophic organisms are surviving under continual UVR stress (Häder and Worrest, 1991). Heterotrophic organisms need not inhabit surface waters where the flux of UVR is high, because both particulate and dissolved organic carbon concentrations are high throughout the euphotic zone (defined here as extending to that depth to which 0.1% of noon solar irradiance penetrates) and often for many meters below the 0.1% light level. Many microorganisms, such as bacteria and unicellular algae, have limited movement by cilia, flagella, or changing cell buoyancy, but such motility is generally limited to less than 1–2 m per hour. Such rates of movement are generally not sufficient to overcome physical mixing forces induced by wind stress, and hence such organisms will on occasion be exposed to high fluxes of UVR. Other organisms, such as young fish and macrozooplankton, have sufficient motility to minimize UVR stress, but it is not known if they possess the capability to sense UVR, which would be needed to direct their motion toward deeper water.

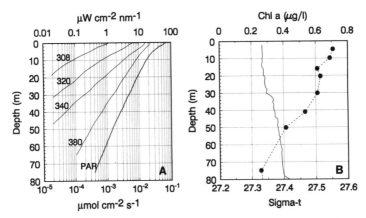

Figure 7. Attenuation of UVR and PAR (close to local noon) in the upper water column at 61°S in January 1992 at a station close to Elephant Island, Antarctica, (A) Recorded values for four UV wavelengths (use upper scale) and for PAR (lower scale), using a profiling UV-radiometer (model PUV-500, Biospherical Instruments Inc.). (B) Profile of water density (continuous line) and of chlorophyll-a concentrations (dashed line) at the same station.

5. Photosynthetic Studies with Phytoplankton

Primary attention will be given to UVR effects on phytoplankton because they constitute the primary producers in most aquatic environments and hence are responsible for production of most of the organic carbon reservoirs which support the entire food web. There has also been more studies with them than with other plant or animal groups.

Although there were scattered reports in the older literature regarding deleterious effects of UVR on phytoplankton (e.g., Steeman Nielsen, 1964), there were relatively few studies on this subject until the late 1970s, when there was concern over the effect of operating supersonic aircraft, and again in the late 1980s, when the magnitude of the ozone hole in the Antarctic became obvious. These studies suggested that ambient levels of UV-B radiation decreased photosynthetic rates by marine phytoplankton by up to 50% close to the surface, with effects detectable down to tens of meters (Lorenzen, 1979; Smith *et al.,* 1980). The overall decrease of primary production due to UV-B radiation, however, was estimated to be less than 2% by Lorenzen (1979). *In situ* studies in Canadian coastal waters showed UVR decreased primary production by 8–30% during spring (April–May) at depths down to 1 m; no inhibitory effect was evident, however, during the summer experiments when irradiance was much higher (Hobson and Hartley, 1983). Similar studies in fresh water (Lake Michigan) showed a 13% suppression of rates of primary production, but it was restricted to the upper 6 m of the water (Gala and Giesy, 1991). These authors reported that the dose of UV-B required to produce a 50% decrease in photosynthetic rates increased from 17.6 kJ m^{-2} in spring to 132 kJ m^{-2} in summer, and concluded that enhanced UV-B radiation from decreased atmospheric ozone concentrations would not increase the loss of primary production to any significant extent in Lake Michigan.

Studies of Bühlmann *et al.* (1987) concluded that UV-B was of relatively minor importance in photoinhibition in natural waters as compared to the effect of UV-A radiation. These results are quite similar to those of Maske (1984), who studied UVR inhibition of natural populations of phytoplankton and of cultures incubated *in situ* in the Kiel-fjord (Germany). Cutting off all UVR below 360 nm in wavelength increased photosynthetic rates 40% to 150%, with most of the inhibition being caused by UVR with wavelengths between 320 to 360 nm. Recent studies in the Antarctic (R. C. Smith *et al.*, 1992) utilizing *in situ* techniques also showed that the magnitude of inhibition caused by UV-A radiation was at least twice that caused by UV-B radiation.

5.1. Studies in Antarctica

Because the most rapid changes in incident UV-B radiation occur under the ozone hole in the Antarctic (Lubin and Frederick, 1991), there have been extensive studies to ascertain whether or not enhanced UV-B radiation resulting from ozone depletion in the stratosphere will have any calamitous effects on the southern ocean ecosystem. The major findings from these studies are outlined below.

5.1.1. Incubator Experiments

The first studies in Antarctic waters which dealt directly with possible effects of enhanced UV-B radiation due to formation of the ozone hole showed that UVR significantly decreased primary production rates and also produced drastic changes in phytoplankton pigmentation and viability (El-Sayed, 1988; El-Sayed *et al.*, 1990). In many of the experiments UV-A radiation was responsible for more than 50% of the inhibitory effects, with UV-B radiation accounting for less than 50%. These authors also speculated that differential sensitivity of phytoplankton species might be important in altering the species composition of the phytoplankton crop, but no quantitative data were presented to support that statement.

In extensive studies during 1988–89 (Holm-Hansen, 1990) and 1990–92 (Helbling *et al.*, 1992), the magnitude of the increase of photosynthetic rates was determined when natural phytoplankton assemblages were exposed to ambient solar radiation, but with various portions of the UVR blocked by a variety of sharp-cutoff plastic and glass filters. Samples were contained in quartz tubes (50–100 ml) or round vessels (225 ml) with ground-glass joints to eliminate any possibility of toxic materials in the stopper. The results (Fig. 8) indicate that energy in the UV-B portion of the spectrum accounted for approximately 50% of the total inhibition caused by all wavelengths less than 378 nm. The shorter UV-B wavelengths (around 300 nm) had relatively little impact on photosynthetic rates.

An action spectrum showing the relative effectiveness of spectral UVR for inhibition of photosynthesis of natural Antarctic phytoplankton assemblages is shown in Fig. 9. It is seen that the shorter the wavelength of UVR, the greater is its effectiveness in inhibiting CO_2 fixation rates when expressed as response per unit energy. The relative effectiveness of energy at a wavelength of 296 nm is almost two orders of magnitude greater than the effectiveness of energy at the longer UV-A wavelengths. Although the shorter UV-B wavelengths are potentially the most effective in reducing photosynthetic rates, wavelengths < 300 nm apparently are not responsible for much of the total inhibition of photosynthesis by solar UVR (see Fig. 8). The reason for this is that the flux of incident solar UVR decreases very rapidly below 320 nm; the

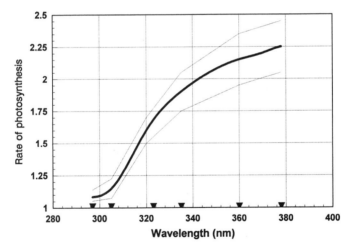

Figure 8. Magnitude of enhancement of photosynthesis by Antarctic phytoplankton when solar UVR was selectively "cut off" at various wavelengths by use of sharp-cutoff plastic or glass filters. Wavelengths at which the filters showed 50% transmission are indicated by the solid triangles on the abscissa. The rate of photosynthesis in quartz control vessels has been set at 1.0. The solid dark line, which has been generalized from all our data, represents the increase in photosynthetic rate relative to that in the quartz vessels; the dotted lines represent one standard deviation. From Holm-Hansen (1990). Data have been amended slightly by inclusion of additional data obtained during 1990 to 1992.

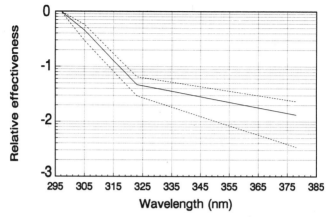

Figure 9. Estimation of the action spectrum for UVR inhibition of photosynthesis in Antarctic marine phytoplankton by comparison of photosynthetic enhancement results with data on incident spectral irradiance obtained with the UV-spectroradiometer maintained at Palmer Station by the U.S. National Science Foundation. Solid line shows the relative effectiveness of UVR in photoinhibition (expressed as response per unit energy), where the effectiveness is set at 1.0 at 296 nm. Note that the ordinate is a log scale. Dashed lines represent minimum and maximum values. From Helbling *et al.* (1992).

fluence of radiation at 320 nm is almost four orders of magnitude greater than that at 300 nm as seen by the data in Fig. 2A.

Generalized results from many incubator experiments using Antarctic phytoplankton exposed to either direct or attenuated (by neutral density screening) solar radiation to simulate conditions in the upper water column and cutting off UVR below 305, 323, and 360 nm (Pyrex, Mylar, and Plexiglas filters, respectively) are shown in Fig. 10. It is seen that the use of Mylar increased the mean photosynthetic assimilation number from 0.38 (samples in quartz glass vessels) to 0.56, a mean increase of 47%; use of a Plexiglas

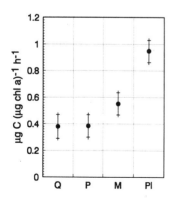

Figure 10. Photosynthetic assimilation numbers of Antarctic phytoplankton when incident solar radiation is filtered through quartz (Q), Pyrex (P), Mylar (M), or Plexiglas (Pl). Solid circles represent the mean value of all samples and the vertical bars the 95% confidence intervals. The thickness of the Mylar and Plexiglas filters was 0.1 and 3.2 mm, respectively. All samples (132) were collected within the upper mixed layer. From Helbling *et al.* (1992).

filter increased the assimilation number to 0.96, a mean increase of 146% as compared to the controls in quartz. The results from incubations with Mylar and Plexiglas filters were significantly different from those with Pyrex and quartz vessels and also different between themselves (Tukey test). There was no significant difference between the samples incubated in Pyrex and quartz vessels.

The question of whether or not there is a threshold value for inhibition of photosynthesis by UVR is important, as it would determine the depth to which UVR has any significant deleterious effects on primary production. The effect on photosynthetic rates of Antarctic phytoplankton when the incident solar UVR varies from low to high is shown in Fig. 11. Photosynthetic rates in the three treatments (Pyrex, Mylar, and plastic film) did not show any significant differences as compared to the control samples in quartz when the ambient UVR (295–385 nm) was less than 5 W m^{-2}. Above this threshold value, the magnitude of enhancement of photosynthetic rates was directly related to the elevated levels of UVR irradiance. The maximum enhancement

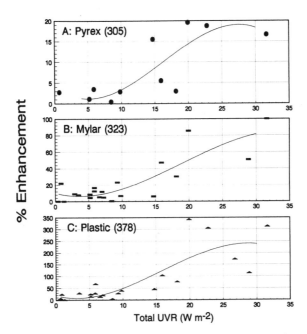

Figure 11. Percentage of enhancement of photosynthetic rates as a function of mean incubation values of UVR incident upon the samples in quartz vessels and when various spectral regions of UVR are removed by Pyrex, Mylar, or a plastic film (thickness 0.27 mm) with 50% transmission at 378 nm. Note the different ordinate values in A, B, and C. The lines through the data points were computer drawn, using a third-degree polynomial function. From Helbling *et al.* (1992).

values obtained by screening off wavelengths less than 305, 323, and 378 nm were approximately 17%, 80%, and 250%, respectively.

5.1.2. In Situ *Incubations*

The authors are aware of few studies (which are discussed below) using *in situ* techniques to determine the effect of UVR on Antarctic phytoplankton. The experiments of Holm-Hansen and colleagues were all done close to Palmer Station (64°S); natural water samples were obtained from the upper 20 m of the water column, inoculated with radiocarbon, and incubated under three different spectral regimes for 8–10 h centered at local noon at the same depth from which they were obtained. Rates determined in the quartz vessels (with no supplemental filter) are referred to as 1.0 for all samples, and the rates determined in other treatments, where wavelengths shorter than 305 nm (Pyrex filter) or 360 nm (Plexiglas filter) were removed, are shown as enhancement factors as compared to the rates in the quartz vessels (Fig. 12). Elimination of the shorter UV-B wavelengths (samples with Pyrex filter) resulted in approximately 20% higher rates (enhancement factor of 1.2) of photosynthesis in samples collected and incubated close to the surface, with the effect diminishing rapidly with depth; by 10 m depth there was no difference in the samples contained within quartz or Pyrex vessels. Effects of removing all energy below 360 nm (Plexiglas filter) resulted in much higher rates of incorporation of radiocarbon. As compared to data from quartz vessels, the rates were approximately doubled in samples close to the surface, approximately 10% higher at 10 m depth, and showed no detectable differences at a

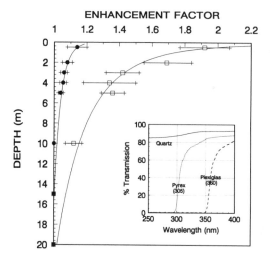

Figure 12. Relative rates of *in situ* photosynthesis in the upper 20 m of the water column when natural phytoplankton samples are exposed to varying proportions of UV radiation. Samples were incubated in (a) quartz glass, which transmits nearly all UV radiation (photosynthetic rates for these samples are assumed to be "1.0" for all depths), (b) Pyrex glass (●), with 50% transmission at about 305 nm, or (c) quartz glass screened with a Plexiglas filter (□), with 50% transmission at 360 nm. Samples were collected close to Anvers Island during November–December, 1988. Insert shows transmission characteristics of the filters used.

depth of 20 m. These data suggest that the shorter wavelengths (280 to 306 nm) of UV-B radiation depress photosynthetic rates much less than the suppression caused by longer UVR wavelengths, results consistent with all our incubator experiments discussed above. Holm-Hansen *et al.* (1993) have estimated, on the basis of the above data, that UV-B radiation under normal ozone concentrations is responsible for a loss of approximately 9% of primary production in the upper 20 m of the water column and 5.4% for the entire euphotic zone.

Smith *et al.* (1992) have also used *in situ* incubations to estimate the effect of UVR on primary production within the marginal ice zone in the Bellingshausen Sea. Data in Fig. 7 of their paper show that the total carbon loss due to UV-B radiation in the upper 25 m of the water column ranges from about 2.5% to 12%, depending upon ozone concentrations. Although these estimates seem reasonable, there is some concern regarding the validity of their radiocarbon data, as they employed polyethylene whirlpak bags for incubating their samples, and these bags have been shown to absorb significant amounts of UV-B radiation (El-Sayed *et al.,* 1990) and also to be toxic to phytoplankton after exposure to solar radiation (Holm-Hansen and Helbling, 1993).

One criticism of *in situ* experiments is that the sample bottles are held at one depth throughout the incubation period, whereas in reality phytoplankton cells are generally moving up and down (at unknown rates) vertically throughout the depth of the UML. The rates at which phytoplankton circulate within the UML are generally not known, but Denman and Gargett (1983) have estimated time scales ranging from 0.5 h to hundreds of hours for a UML of 10 m. The light regimes experienced by cells in the bottles will be fairly constant (as a percentage of incident radiation), in contrast to the phytoplankton circulating in the water column, which are experiencing fluctuations ranging from low to high irradiance depending upon the depth to which they are mixed. As the ratio of UV-B to UV-A is artificially high in some of the incubator experiments, it would thus tend to maximize the net inhibition by UVR. Neale (1987) has discussed the significance of such mixing in studies of photoinhibition caused by high irradiance of visible light, while Cullen and Lesser (1991) have commented that the kinetics and mechanism of photoinhibition by UV-B are similar to that resulting from excess visible light. Not only is the total irradiance varying with depth, but the ratio of UV-B to UV-A (which is of importance in regard to photorepair mechanisms) decreases rapidly with depth as shown in Fig. 13. Cullen and Lesser (1991) have discussed these effects and emphasized the variable impacts of solar UVR which might be expected, depending upon the dose to dose-rate relationship involved in the physiological response. Unfortunately, we do not know much about the mechanisms involved in damage by UVR, so at the present time we do not

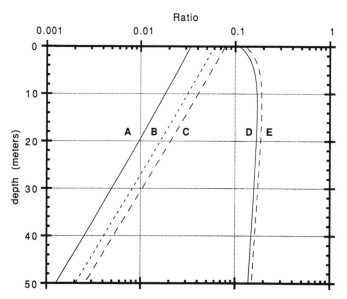

Figure 13. Ratios of integrated UV-B, UV-A, and PAR with depth at 65°S (with and without an ozone hole) and at the equator at local noon during spring equinox. (A) UV-B/UV-A at 65°S. (B) UV-B/UV-A at 65°S under the ozone hole. (C) UV-B/UV-A at the equator. (D) UV-A/PAR at 65°S. (E) UV-A/PAR at the equator. Spectral attenuation of radiation with depth is based on diffuse attenuation coefficients for clearest ocean water given by Smith and Baker (1979, 1981).

know if the *in situ* incubation techniques represent the worst or the best scenario in regard to the magnitude of loss of primary production due to solar UVR. Marra (1978) attempted to simulate natural vertical movements using *in situ* techniques and found that vertically cycled samples gave considerably higher rates of primary production (by 19% to 87%) than the samples at fixed depth. Another approach to this problem is to use a rotating incubator, which exposes samples to variable fluences of solar radiation to mimic the illumination regimes they might encounter by circulating in the UML, and to compare the integrated production with samples which are in an identical incubator which does not rotate (i.e., samples receive a constant percentage of incident solar radiation and hence correspond to the fixed samples in a standard *in situ* experiment). Results from preliminary tests with a set of such incubators are shown in Fig. 14. It is seen that on overcast days the integrated production in the rotating incubator exceeds that in the fixed incubator by amounts up to 24%; on very sunny days, however, the production in the rotating incubator is less than that in the fixed incubator by values up to 15%. The time scales of physical and biological processes which interact on cells mixing in the

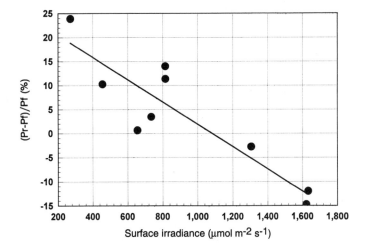

Figure 14. Effect on integrated primary production when samples are rotated (Pr) to simulate vertical movement in the UML as a percentage of that when the samples are fixed (Pf) and receive a constant percentage of incident solar radiation. Values for surface irradiance represent the mean irradiance during the 6 h of incubation centered at local noon.

upper water column are further discussed by Lewis *et al.* (1984), Neale (1987), and Cullen and Lewis (1988).

5.1.3. Effect of Enhanced UV-B Radiation

If one convolves the action spectrum for UVR inhibition of photosynthesis (Fig. 9) with the varying UVR spectral irradiance as a function of atmospheric ozone concentrations (e.g., Fig. 2), one obtains the curve showing percentage enhancement of UV-B-induced inhibition of photosynthesis as a function of column ozone values (Fig. 15). It is seen that a decrease of column ozone to 150 DU would increase inhibition of photosynthesis by approximately 80% as compared to the inhibition when column ozone was 350 DU. A fairly similar value (about 60%) has been estimated by Lubin *et al.* (1992) on the basis of a different data set. The recent work by R. C. Smith *et al.* (fig. 7, 1992) suggests that the carbon loss in the upper 25 m of the water column due to UV-B inhibition would be 12% (maximal value) "inside" the ozone hole, and about 5% "outside" the hole; the additional loss due to a well-developed ozone hole would thus be 140% of that prevailing under normal ozone values. This estimate is higher than the two estimates mentioned above, but the reader should be aware of the possible artifact associated with the Smith *et al.* (1992) work as mentioned in Section 5.1.2.

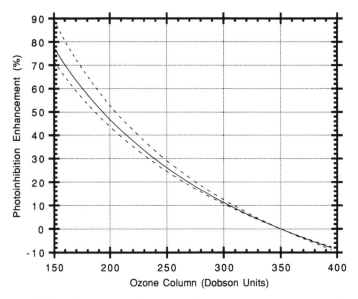

Figure 15. Inhibition of photosynthesis by UVR in antarctic phytoplankton as a function of column ozone values. Dashed lines indicate minimal and maximal estimates. A column ozone value of 350 DU is considered 'normal.'

5.2. Studies in Tropical Waters

It is commonly assumed that phytoplankton are continually living under UVR stress, even under normal ozone conditions (Worrest, 1983). The data discussed above showed that photosynthetic rates of Antarctic phytoplankton are inhibited significantly by both UV-A and UV-B radiation, even though solar radiation levels in Antarctic waters are much lower than in tropical waters (see Figs. 1 and 4). The results of incubator experiments done on board ship between 20°S and 10°N in the eastern Pacific Ocean (80–110°W) are shown in Fig. 16. During studies done in 1991, there were no significant differences in photosynthetic assimilation numbers between the controls (in quartz) and the treatments which had UVR below 305, 323, or 360 nm blocked by appropriate filters (ANOVA test). Results of ANOVA and Tukey tests indicated that the increase in assimilation numbers between the Pl and Pl400 treatments (1992 tests) was significant, suggesting that UV-A radiation between 360 and 400 nm was depressing photosynthetic rates of these tropical phytoplankton. This is consistent with data in the literature indicating that long-wavelength UV-A radiation inhibits cellular reactions mediated by blue light. It should be noted that the species composition of the phytoplankton assem-

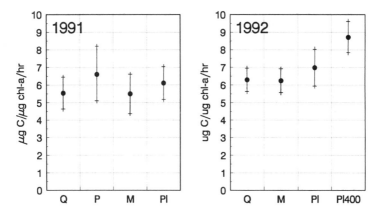

Figure 16. Photosynthetic assimilation numbers of tropical phytoplankton when incident solar radiation is filtered through quartz (Q), Pyrex (P), Mylar (M), Plexiglas (Pl), or Plexiglas-2 (Pl400). The P, M, Pl, and Pl400 filters cut off at 305, 323, 360, and 400 nm, respectively. Experiments were done in March–April of 1991 and 1992. All samples used in these tests were from the UML of the water column. Solid circles represent the mean value of all samples and the vertical bars the 95% confidence intervals. The 1991 data are from Helbling *et al.* (1992).

blages in tropical waters is generally different from that of polar waters. In tropical waters the phytoplankton are dominated by coccolithophorids, flagellates, and dinoflagellates, whereas in polar waters diatoms generally compose a significant portion of the total phytoplankton crop. The lack of sensitivity of phytoplankton in tropical waters (which are characterized by low nutrient concentrations in the UML) to UV-B radiation is surprising in view of laboratory findings that nutrient stress exacerbates cellular damage by UVR (Kiefer, 1973; Prézelin *et al.*, 1986).

5.3. Action Spectra for UV Inhibition of Photosynthesis

To be able to predict the effect of increasing UVR on rates of primary production, it is necessary to combine spectral irradiance data with the action spectrum for UV inhibition of photosynthesis. The difficulty of deriving an action spectrum for phytoplankton populations under natural conditions has been discussed by Coohill (1989) and Rundell (1983), who suggested that polychromatic action spectra are most realistic for natural field conditions. The action spectrum shown in Fig. 9 was derived by such a procedure, and it is quite similar to one described by Lubin *et al.* (1992) for Antarctic phytoplankton. These action spectra cannot, however, be extrapolated to other geographic regions, as can be seen by our above discussion regarding the lack of inhibition by UV-B radiation on tropical phytoplankton. The action spectra for Antarctic phytoplankton do not correspond very well to other well-publicized plant action spectra, such as that for chloroplast Hill reaction inhibition (Jones and Kok, 1966), generalized plant damage (Caldwell, 1971), inhibition of photosynthesis in higher plants (Caldwell *et al.*, 1986), or DNA damage (Setlow, 1974).

6. DNA and Cell Viability

The effect of enhanced UVR on nucleic acids is perhaps more serious than its effects on photosynthetic rates. When the spectral changes in UV-B in the Antarctic are weighted by Setlow's (1974) DNA weighting function, the resulting curve showing the amplification of DNA-effective irradiance shows an increase of over 400% at 150 DU as compared to that occurring at 350 DU (Lubin *et al.*, 1992). The net effect of such an enhancement of radiation absorbed by DNA can not be estimated, as most organisms have many enzymatic mechanisms to repair DNA lesions (see Chapter 12 by Mitchell and Karentz, this volume). Whether or not the cell loses viability will depend upon many factors, including the dose and dosage rate of the UVR and the rate of synthesis of repair enzymes. It is of interest to note that lethal effects of UVR on cell division in a repair-deficient strain of *Escherichia coli* could be detected down to 30 m in Antarctic waters during a well-developed ozone hole on October 6, 1988, at Palmer Station (Karentz and Lutze, 1990). It should be noted, however, that effects of UV-B radiation could be detected only to 10 m depth. UVR from 320 to 365 nm was responsible for considerably more loss of cell viability in *E. coli* than that resulting from UV-B radiation.

These depths to which UVR penetrates and causes loss of viability are considerably greater than the depth to which dimer formation can be detected in solubilized DNA. Regan *et al.* (1992) have described the use of purified DNA in solution as a solar dosimeter for measuring actual DNA damage in aquatic environments. Using *in situ* techniques and quartz glass vessels, they could, however, measure DNA damage only to 3 m depth in clear ocean water. This is surprising in view of the data of Karentz and Lutze (1990) showing that UVR down to 30 m in Antarctic waters decreased the survival rate of *E. coli*. Such results emphasize the problems inherent in applying a simple weighting function based on DNA absorption to estimate cellular damage by UVR.

The degree to which viability of phytoplankton cells is affected by UVR *in situ* has not been determined. Using a UV lamp, however, Karentz *et al.* (1991a) have determined the average fluence of UV-B required to kill various species of Antarctic diatoms. There was much variability among the 12 species used in the degree of sensitivity to UV-B, as the average fluence needed to kill cells ranged from 681 J m^{-2} to over 25 kJ m^{-2}. Their data also suggested that smaller cells were more sensitive to UVR than larger cells, which is contrary to the assumption often made that large diatoms in Antarctic waters might be more sensitive to UVR than the nanoplankton (El Sayed, 1988).

7. Species-Specific Effects of UVR

It is apparent from the literature on effects of UVR on organisms that there is much variation between species as well as between phytoplankton groups in regard to their sensitivity to UVR and to the cellular manifestations of UVR-induced damage. This has led to the general concern that the floristic composition of the phytoplankton community may be significantly altered by UVR, resulting in disruption of the normal feeding interactions between autotrophic and heterotrophic organisms (Worrest *et al.*, 1981; El-Sayed, 1988). Karentz (1991), on the basis of extensive studies in Antarctica, has suggested that enhanced UV-B levels will most likely cause changes in the species composition of the phytoplankton assemblages. El-Sayed (1988) noted that larger diatoms seemed to be killed by ambient UVR. Replacement of chain-forming diatoms by a nanoplankton-size flagellate population with diameters <20 μm might have serious consequences on the overall cycling of organic carbon in the upper water column. This would be of most concern in upwelling areas (e.g., off the west coasts of the continents and in Antarctic waters) where the food chain is characterized by being short, meaning that fairly large heterotrophic organisms (e.g., the anchoveta off Peru, and krill in the Antarctic) feed directly on phytoplankton. Any UV-induced change whereby the "steps" in the food chain become more numerous not only would decrease the amount of food resources that can be harvested (Ryther, 1969) but also might decrease the amount of organic material which settles to deeper water, where it supports demersal and benthic populations of animals.

8. Aspects of Photoadaptation

The major ways which organisms can minimize UVR-induced damage include: (1) movement or organization of organelles to minimize absorptional cross-sectional area, as well as movement of the organism away from the source of the radiation; (2) synthesis of screening pigments which absorb and dissipate UVR; (3) synthesis of accessory photosynthetic pigments which may utilize UV radiation for photosynthesis or for other energy-requiring metabolic pathways; (4) absorption by pigments which dissipate the energy as fluorescence; (5) synthesis of antioxidant enzymes (Lesser and Stochaj, 1990) or accessory pigments (Bidigare, 1989) to protect the cell from active forms of oxygen; and (6) both dark- and light-activated mechanisms to repair UVR-induced damage to nucleic acids (see Chapter 12 by Mitchell and Karentz, this volume).

8.1. Screening Pigments

The historical significance of screening pigments in marine phytoplankton has been discussed by Yentsch and Yentsch (1982), with emphasis on the photosynthetic pigments. Garcia-Pichel *et al.* (1992) have recently suggested that scytonemin, an extracellular pigment occurring in the outer sheath of blue-green algae, also serves a protective function against deleterious effects of UV-A radiation. The synthesis of pigments which have high-absorbance peaks in the UV-B and short-wavelength UV-A portions of the spectrum is prevalent in most marine plants and animals investigated. Chalker and Dunlap (1990) cite occurrence of such water and alcohol-soluble pigments in a variety of macroalgae, phytoplankton, corals, coelenterates, and ascidians. Pioneering studies on UV-B-absorbing compounds in corals and blue-green algae were done by Shibata (1969), who termed the group of structurally related compounds "S-320" and postulated that they had a functional role in protecting the organisms from UVR.

The strongest evidence that such pigments play a protective role against UV damage is afforded by the extensive studies of mycosporine-like amino acid compounds (MAAs) by Dunlap and Chalker (1986) and Dunlap *et al.* (1986) in corals and associated zooxanthellae on barrier reefs in Australia. The concentrations of MAAs in corals is directly related to the UVR fluence they experience; corals in shallow water have high concentrations of MAAs and those in deep water have low concentrations of MAAs. Neither the intracellular localization of these MAAs nor the sensing mechanism which initiates their synthesis, however, is known.

Cellular concentrations of MAAs in Antarctic phytoplankton also seem to be related to the immediate light-history of the cells (Vernet, 1990). Cells in deeper water have low or no discernable concentrations of MAAs, whereas cells maintained in high-light environments develop high concentrations of MAAs, as seen from Fig. 17. The occurrence of MAAs is widespread in Antarctic marine plants and animals as described by Karentz *et al.* (1991b). In studies by Carreto *et al.* (1989), it was shown that some species require only hours to synthesize UV-absorbing pigments when transferred from low light to high light. These studies were with dinoflagellates, which typically make a

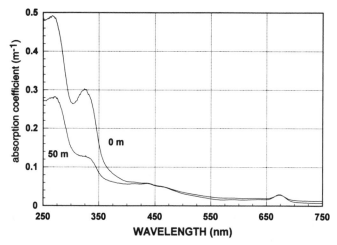

Figure 17. Spectral absorption coefficients of natural phytoplankton assemblages as a function of depth (0 and 50 m) in the water column at a station close to Elephant Island, Antarctica, in January of 1991. Samples were concentrated onto glass-fiber filters, and the spectral absorption coefficients determined as described by Kishino *et al.* (1985).

daily migration to the near surface, where they will be exposed to high UVR fluences. These photoprotective pigments (MAAs) consist of a variety of compounds which have absorption maxima ranging from 310 nm to 359 nm (Carreto *et al.*, 1990).

Studies of Lesser and Stochaj (1990) with the symbiotic prokaryote *Procholoron* sp. and its ascidian host have shown that cellular concentrations of superoxide dismutase, ascorbate peroxidase, and catalase activities were directly proportional to irradiance upon the organisms, whereas concentrations of MAAs were inversely proportional to irradiance. They speculated that both these major mechanisms to protect cells from UVR are required by these organisms, and serve as adaptive responses in different photic environments.

Possibly the most extreme marine UV environment for an organism is the surface film in tropical waters. An organism living on the surface during its entire existence, such as the wingless marine ocean-skaters (genus *Halobates*), must be able to cope with direct solar radiation. Such organisms (and other microorganisms living within the surface film) will be exposed to much higher fluences of UV-B radiation than any organism living within the water column. Apparently these marine insects are adequately protected from solar UVR by synthesis of black pigments which have highest absorbance in the UV regions and which are deposited in the dorsal cuticle (Cheng *et al.*, 1978). They also report that other species of water-skaters which inhabit mangrove swamps or streams have much lower concentrations of these UV-absorbing pigments than the open-ocean forms.

8.2. Dark and Light Repair Mechanisms

Mitchell and Karentz (Chapter 12, this volume) have described both dark and light mechanisms whereby cells can repair UV-induced damage to nucleic acids. In regard to photorepair processes, the relative ratios of UV-B to UV-A or to PAR is of importance, because the relative proportion of "damaging" radiation to "repair" radiation will affect the degree to which cells will be damaged. Data in Fig. 13 show the ratios of UV-B to UV-A in the upper 50 m of the water column at 65°S with and without the ozone hole and at the equator, and also the ratios of UV-A to PAR at 65°S and at the equator. The photoregimes in the upper water column at the equator would appear to be more damaging to organisms than at 65°S (with or without the ozone hole), not only because of the higher absolute fluences of UVR, but also because of the higher proportionate amounts of UV-B to UV-A and also of UV-A to PAR.

8.3. Importance of Vertical Mixing Processes

It was seen above that tropical phytoplankton apparently show no loss of photosynthetic ability due to solar UV-B radiation, whereas Antarctic phytoplankton are relatively sensitive to UV-B radiation, and to a lesser extent to UV-A radiation. This would be in keeping with the suggestion sometimes made that Antarctic phytoplankton, which are normally exposed to much lower seasonal fluences of UVR than are phytoplankton at lower latitudes (see Fig. 1), have not developed the genetic capability to deal with high fluences of UVR and hence might be particularly sensitive to enhanced UV-B resulting from ozone depletion. It is known, however, that Antarctic phytoplankton possess both dark and light DNA-repair mechanisms (Karentz *et al.,* 1991a) and that they can synthesize screening pigments such as MAAs to minimize damage from UVR. Both these types of photoadaptation, however, require significant time periods (hours to days) under high-light conditions to maximize the cellular protective mechanisms against UVR damage. This requires that the organisms remain under high illumination sufficiently long to permit the synthesis of needed enzymes and pigments.

As mentioned previously, the depth distribution of plankton in the upper water column will often be controlled by physical mixing processes which circulate the cells vertically throughout the depth of the UML. The oceanographic conditions in tropical waters are very different from those in Antarctic waters in regard to the depth and stability of the UML, factors which will control the mean irradiance experienced by phytoplankton cells. Data in Fig. 18 show typical profiles of water density (sigma-t) in the Antarctic and in tropical waters where UV studies have been done. The profile for tropical waters (station UV07, Fig. 18) shows a relatively shallow (about 12 m) and very stable UML. The water within that layer will mix vertically, depending upon wind stress, but it is difficult for that layer of low-density water to mix against the strong gradient of denser water immediately below it, which extends over 3.5 sigma-t units. As tropical waters are very low in inorganic nutrients, phytoplankton biomass is usually low (<0.5 μg chl-a liter^{-1}), with the result that the euphotic zone generally extends to more than 60 m depth. The phytoplankton in the UML will thus be exposed to high-light levels throughout the day.

Pelagic Antarctic waters (stn. D05, Fig. 18), in contrast, are characterized by very deep UMLs (>50 m) which are relatively unstable because of the relatively little change in water density with depth below the UML (for more discussion on this point, see Mitchell *et al.*, 1991). The result of this is that phytoplankton in pelagic Antarctic waters are generally deeply mixed in the upper 50 m of the water column, resulting in a low mean irradiance experienced by the cells. If such vertical mixing processes are fast enough, then the cells might not have sufficient time to fully photoadapt to the high-illumination conditions existing in the upper 5–10 m. Coastal Antarctic waters in sheltered

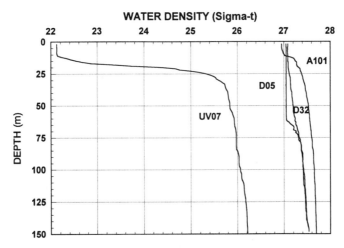

Figure 18. Representative profiles of water density (sigma-t) in tropical waters (stn. UV07) and in Antarctic waters showing three representative profiles of water density (stn. D05, a deep UML; stn. A101, a shallow UML; stn. D32, no UML, but a continuous increase in water density with depth).

areas such as the Gerlache Strait often show formation of relatively shallow UMLs (e.g., stn. A101, Fig. 18), which often develop phytoplankton blooms with chlorophyll-a concentrations of over 20 μg l^{-1} (Holm-Hansen *et al.*, 1989; Holm-Hansen and Mitchell, 1991). The attenuation of UVR and PAR is so rapid under such bloom conditions that phytoplankton within the UML are often light-limited in spite of the shallowness of the UML. Under some conditions, especially in the area around Elephant Island, physical mixing processes are such that sharply defined UMLs are not formed, but instead there is a continual increase in water density with depth (e.g., stn. D32, Fig. 18). As the gradient in density is so small in the upper 50 m of the water column, it can be expected that periodic storms will mix such a water column to 40–50 m depth and hence prevent the phytoplankton from adapting to high-light conditions existing in surface waters.

It was seen above (Section 5.2) that Antarctic phytoplankton are relatively sensitive to solar UVR as compared to phytoplankton from tropical waters. As it takes considerable time for cells to synthesize the various enzymes and pigments involved in photoprotection, the question arises as to the nature of the sensitivity of Antarctic phytoplankton to UVR; does the sensitivity reflect genetically controlled limitation of the ability of Antarctic phytoplankton to respond to UVR, or does it merely reflect the fact that mixing processes in Antarctic waters do not permit sufficient time under high-light conditions for the cells to become fully photoadapted? This question has been discussed previously by Cullen and Lewis (1988), but further studies are needed to determine the answer to this question for Antarctic phytoplankton.

8.4. Avoidance Reactions

Most plants and animals have means of both sensing potentially deleterious levels of high light and the ability to either move away from the source of light or minimize their cross-sectional absorbance area. As mentioned in Section 2.10, many studies have shown that motility and phototaxic orientation in various phytoplankton species are diminished by UVR. It has not been established, however, whether these organisms have special sensory mechanisms for perception of UVR, or if they are just responding to high levels of visible light (Häder and Worrest, 1991). The latter situation is suggested by studies by Damkaer and Dey (1983), who determined that adult euphausiids respond to levels of visible light and apparently can not respond to enhanced levels of UV-B radiation. If organisms do not have any means of sensing UVR, they might be at increased danger from enhanced UV-B radiation resulting from ozone depletion.

Some organisms, however, apparently do have the ability to sense and to respond to UVR. Rüdiger and López-Figueroa (1992) have reviewed photoreceptors in algae, including those effective in the UV-B and blue-light regions. Galland and Senger (1991) cite the possible importance of flavins and pterins as photoreceptors for various cellular processes in microorganisms for both the UV-B and UV-A spectral regions. The retinas of some fish have specialized cone cells that are sensitive to UVR and show maximal absorption in the UV-A region, with a peak at 380 nm (Hawryshyn, 1991, 1992). Additional discussion of photoreceptors in diverse plants and animals is available in Holmes (1991).

9. Studies with Other Marine Organisms

9.1. Macroalgae

There is a paucity of experimental data on the effects of UVR on macroalgae (Rhodophyta, Phaeophyta, and some Chlorophyta). Some of these forms live at depths where UVR would not be expected to be significant (> 30 m depth), but the plants living in the intertidal zone are in a most vulnerable location for incurring damage by UVR. It has been known for a long time that many such algae produce UV-absorbing compounds ("Gelbstoff") that supposedly serve as protective screening pigments against UVR (Fogg and Boalch, 1958; Craig and McLachan, 1964; Kroes, 1970; Sieburth and Jensen, 1969) for the mature plants. Wood (1987) has shown that mature kelp fronds apparently are not adversely affected by high fluences of solar radiation, but that the growth and survival of the early sporophyte stages are damaged by UVR. Enhanced UV-B radiation thus might be a problem with settlement and growth of the young plants. Polne and Gibor (1982) have shown that the fluorescence spectra of a variety of marine benthic algae change dramatically after exposure to fairly high levels of UV-A radiation, indicating that the photosynthetic process was being damaged. They also found that plants living at depth were much more sensitive to UVR than those from the intertidal zone, suggesting that these algae possess photoadaptive mechanisms to minimize damage by solar UVR. We are not aware, however, of any publications demonstrating deleterious effects of solar UVR on photosynthetic rates of macroalgae when measured by either CO_2 fixation or oxygen liberation. According to Kirk (1983), available data indicate that maximal photosynthetic rates in macroalgae are generally not decreased by direct sunlight. Hanelt (1992), however, has claimed to have demonstrated photoinhibition in a variety of macroalgae growing in the littoral zone. Although he did not study UVR *per se,* but instead determined rates of photosynthesis (by measurement of oxygen and *in vivo* fluorescence) in relation to PAR, his results suggest that effects of UVR in these algae should be further explored.

9.2. Bacterioplankton

Although there have been many laboratory studies with bacteria, particularly with *Escherichia coli* (Witkin, 1976; Peak and Peak, 1982), there have not been many investigations dealing with effects of UVR on natural populations of bacteria. In recent years the importance of bacterioplankton in marine waters has been recognized, not only for their role in regeneration of nutrients, but also for "repackaging" of organic material, which has many consequences for feeding preferences by zooplankton as well for altering both the amount and the composition of organic matter transported to deep water and to the benthos (Azam *et al.,* 1983). It has been estimated that approximately 50% of total primary production passes through the "microbial loop" in oceanic waters (Fuhrman, 1992), and hence the effect of solar UVR on this aspect of carbon cycling in natural waters has to be taken into consideration. At present we do not have sufficient experimental data with natural bacterial assemblages to assess the possible impact of enhanced UV-B on the functioning of the microbial loop.

UVR in bacteria apparently can damage general membrane permeability as well as specific permease mechanisms (Peak and Peak, 1982). This would have serious consequences for small heterotrophic organisms such as bacteria which subsist on the very low concentrations of dissolved organic matter which are found in ocean water. It should be noted, however, that heterotrophic organisms need not inhabit the upper portion of the water column (as photoautotrophic organisms must), and hence enhanced UVR most likely would not have any dramatic effect on the overall functioning of the ecosystem. Relatively high bacterial populations ($> 10^4$ cells per ml) are found throughout the entire water column in the oceans, with much higher concentrations ($\geq 10^6$ cells per ml) within the euphotic zone. Injurious effects to bacteria by UVR in the upper 5–10 m of the water column would thus not appear to be of serious consequence when considering the entire ecosystem.

An indication of the variability in sensitivity to UVR encountered in marine bacterioplankton is afforded by the studies of Yayanos (1989), who compared the loss of viability resulting from UVR in a bacterial isolate from the deep sea as compared to a mesophile isolate from the sea surface. A dose of 100 J m^{-2} reduced the surviving fraction of the mesophile isolate to approximately 4%, whereas in the barophilic isolate the surviving fraction was approximately 0.2% after a UV dose of only 2 J m^{-2}.

9.3. Invertebrates

There have not been many intensive studies dealing with the effects of UVR on invertebrates, but there are scattered reports which show that invertebrates can be damaged by ambient UVR (Worrest, 1982). Jokiel (1980) has found that ambient levels of UVR damage coral reef epifauna, with the effects including discoloration, necrosis, and death. Ambient levels of PAR did not cause such damage. Working with various corals from the Great Barrier Reef, Siebeck (1988), found that exposure of corals to UVR resulted in visible cellular damage and death in all species studied, but that there was much variation in sensitivity between species, with the LD_{50} values ranging from <40 to 240 kJ m^{-2} when kept in darkness following UVR exposure, and from <500 to 1,150 kJ m^{-2} when given visible light following the exposure to UVR.

As mentioned in Section 8.1, studies by Cheng *et al.* (1978) on marine insects (*Halobates,* the wingless ocean-skater) have shown that these insects develop high concentrations of UV-absorbing pigments in their dorsal cuticle, which apparently protects them from the harsh photoenvironment of the sea surface.

9.4. Zooplankton

Many researchers have reported that UVR causes increased mortality in a variety of larval crustaceans (Damkaer *et al.,* 1981) as well as in adult planktonic crustaceans such as copepods and euphausiids (Karanas *et al.,* 1981; Damkaer and Dey, 1983; Dey *et al.,* 1988). Fecundity of the adult stages is also much reduced by UVR. The question as to whether or not these organisms are being damaged by UVR in their natural habitats is not clear. Dey *et al.* (1988) found that the UV-B threshold values for mortality for five abundant copepod species were slightly less than present-day radiation levels. The results of Damkaer *et al.* (1981), however, showed that the threshold values for UV-B-induced mortality in larval and adult crustaceans were slightly higher than the present-day UV-B fluxes incident upon the sea surface. They concluded that the species investigated apparently are not being damaged by present-day levels of UV-B radiation to any significant extent. An important consideration in this respect is the behavioral strategies of these organisms in their natural habitats, which is a problem that is difficult to approach experimentally.

9.5. Fish Larvae

Through the extensive investigations by Hunter and colleagues, it is clear that ambient solar radiation has the potential to have severe effects on viability of fish larvae (Hunter *et al.,* 1979). These studies have been done using both natural solar radiation as well as artificial illumination whose spectral characteristics have been carefully determined. No *in situ* studies have been done, however, so the problem of the effects of vertical mixing in the upper water column (as discussed in Section 8.3) must ultimately be taken into account when considering the effect of enhanced UV-B radiation on fish larvae. The LD$_{50}$ dose (weighted for biological effectiveness) for northern anchovy larvae was approximately 50 J m^{-2} day^{-1}, a value which is generally exceeded at 1 m depth off southern California (approx. 33°N) from March through September (Hunter *et al.*, 1981). Taking into consideration the seasonality of spawning and the distribution within the water column of the larvae for various fish species, the authors concluded that northern anchovy stocks may not be greatly affected by enhanced UV-B resulting from ozone depletion. This suggestion was strengthened by the studies of Kaupp and Hunter (1981), who showed the fluence of radiation effective in photorepair (320 to 500 nm) was more than sufficient throughout the upper water column to ensure maximal photorepair of UV-induced damage. The question of assessing the future impacts of UVR on other fish stocks, taking into consideration additional factors such as time of larval development, depth distribution of the larvae, pigmentation, and DNA-repair mechanisms, has been discussed by Hunter *et al.* (1982) and Dunlap *et al.* (1989).

10. Concluding Comments

Our knowledge concerning the effects of solar UVR on aquatic organisms has advanced considerably during the past decade, but it is obvious that there are still enormous gaps in our understanding which make it difficult to predict, with any degree of assurance, what the future impact of enhanced UV-B will be on aquatic ecosystems. Some areas that need more focused research include the following.

10.1. Long-Term Effects

Most studies dealing with UVR and phytoplankton have concentrated on short-term effects of UVR on rates of photosynthesis, on pigmentation, and on rate of formation of DNA lesions. In the absence of knowing the degree to which effects of these processes are reversible under natural conditions, it is not realistic to extrapolate such data to long-term functioning of the ecosystem. It would appear that the most serious consequence of enhanced UV-B radiation might be damage to DNA and hence to viability of plankton. In order to determine the magnitude of impairment to mitosis or meiosis, long-term (days to weeks) data sets are required. Some researchers have started to address this problem by use of isolated cultures, but one limitation of this approach is that it is often difficult to maintain in culture those species which are most dominant in natural water samples, due to the many problems (e.g., altering normal grazing relationships) inherent in putting water samples containing microbial assemblages in containers. Hopefully, future studies will combine the usual "bottle" experiments with new technology designed to document changes in species composition and metabolic activity of plankton in the upper water column by *in vivo* profiling techniques.

10.2. Action Spectra

Compared to the data which serve as the basis for the erythema weighting function, it is obvious that our knowledge of appropriate action spectra for various cellular processes in aquatic plankton is extremely limited. To be able to predict the impact of enhanced UV-B radiation on aquatic ecosystems, it is essential that more detailed action spectra be obtained for effects on photosynthesis, cell viability, fecundity, and the like. This will most likely require continued studies using polychromatic radiation to derive action spectra for field samples, in addition to laboratory studies utilizing cultures and traditional monochromatic irradiation techniques. A combination of such approaches should yield information on the chromophores and targets involved in UVR-induced damage.

10.3. Variability and UV Stress

The assumption is generally made that most plants and animals are living under UVR stress, and that any additional UV (e.g., enhanced UV-B from ozone depletion) will cause additional cellular damage or death of organisms (Voytek, 1989). Data discussed in Section 5, however, show that while some phytoplankton assemblages are damaged by ambient UVR, assemblages in tropical waters seem not be harmed by solar UV-B radiation. An intriguing example of the great variability of organisms to various forms of radiation is afforded by the studies of bacteria belonging to the family Deinococcaceae (Mosely, 1983; M. D. Smith *et al.*, 1992). These bacteria are extremely resistant to far UVR (200 to 290 nm) and to ionizing radiation, but they are extremely sensitive to near UV (290 to 400 nm). Their resistance to "sterilizing" far UVR is believed to be due to a combination of DNA-repair mechanisms and to genome multiplicity. The fact that such extreme resistance to far UVR can evolve in bacteria suggests that planktonic organisms may also have the capability to adapt so as to minimize any deleterious effects of ambient UVR. It should be recognized, however, that the conditions imposed upon a planktonic organism, which is at the mercy of physical mixing processes in aquatic environments, may limit the degree to which such organisms can adapt to high-irradiance conditions. Another characteristic of polar environments (which will experience the greatest changes in ambient UV-B radiation due to ozone depletion) is the temporal aspects of ontogenetic development of plant and animal species, which are keyed to the very pronounced seasonal cycles in environmental conditions. The autecological approach to UV-induced damage to natural planktonic populations should be addressed in future studies. Until we know more about variability and adaptability of natural microbial populations to UVR, we can not assess the degree to which they are currently under UVR-stress or what the effect will be under conditions of enhanced UV-B radiation.

ACKNOWLEDGMENTS. Material in this chapter is based on studies supported by the U.S. National Science Foundation, the U.S. Antarctic Marine Living Resources Program (NOAA), and the Alternative Fluorocarbon Environmental Acceptability Study (AFEAS).

References

Atkinson, D. E., 1969, Regulation of enzyme function, *Ann. Rev. Microbiol.* 23:47–68.

Azam, F., Fenchel, T., Field, J. G., Gray, J. S., Meyer-Reil, L. A., and Thingstad, F., 1983, The ecological role of water column microbes in the sea, *Mar. Ecol. Prog. Ser.* 10:257–263.

Bidigare, R. R., 1989, Potential effects of UV-B radiation on marine organisms of the Southern Ocean: Distributions of phytoplankton and krill during austral spring, *Photochem. Photobiol.* 50:469–477.

Blakefield, M. K., and Calkins, J., 1992, Inhibition of phototaxis in *Volvox aureus* by natural and simulated solar ultraviolet light, *Photochem. Photobiol.* 55:867–872.

Blough, N. V., and Zepp, R. G., 1990, *Effects of Solar Ultraviolet Radiation on Biogeochemical Dynamics in Aquatic Environments,* WHOI-90-09, Woods Hole Oceanographic Institution, Woods Hole, MA.

Bornman, J. F., 1989, Target sites of UV-B radiation in photosynthesis of higher plants, *Photochem. Photobiol.* B4:145–158.

Bornman, J. F., Björn, L. O., and Åkerlund, H., 1984, Action spectrum for inhibition by UV radiation of photosystem II activity in spinach thylakoids, *Photobiochem. Photobiophys.* 8:305–313.

Bühlmann, B., Bossard, P., and Uehlinger, U., 1987, The influence of longwave ultraviolet radiation (UV-A) on the photosynthetic activity (^{14}C-assimilation) of phytoplankton, *J. Plankton Res.* 9:935–943.

Buma, A. G. J., de Baar, H. J. W., Nolting, R. F., and van Bennekom, A. J., 1991, Metal enrichment experiments in the Weddell-Scotia Seas: Effects of iron and manganese on various plankton communities, *Limnol. Oceanogr.* 36:1865–1878.

Caldwell, M. M., 1971, Solar UV irradiation and the growth and development of higher plants, in: *Photophysiology* (A. C. Giese, ed.), Vol. 6, Academic Press, New York, pp. 131–177.

Caldwell, M. M., Camp, L. B., Warner, C. W., and Flint, S. D., 1986, Action spectra and their key role in assessing biological consequences of solar UV-B radiation, in: *Stratospheric Ozone Reduction, Solar Ultraviolet Radiation and Plant Life* (R. C. Worrest and M. M. Caldwell, eds.), Springer, Heidelberg, pp. 87–111.

Calkins, J. C., 1982, *The Role of Solar Ultraviolet Radiation in Marine Ecosystems,* Plenum Press, New York.

Calkins, J., and Thordardottir, T., 1980, The ecological significance of solar UV radiation on aquatic organisms, *Nature* 283:563–566.

Carreto, J. I., De Marco, S. G., and Lutz, V. A., 1989, UV-absorbing pigments in the dinoflagellates *Alexandrium excavatum* and *Prorocentrum micans,* Effects of light intensity, in: *Red Tides* (T. Okaichi, D. M. Anderson, and T. Nemoto, eds.), Elsevier, New York, pp. 333–339.

Carreto, J. I., Carignan, M. O., Daleo, G., and De Marco, S. G., 1990, Occurrence of mycosporine-like amino acids in the red-tide dinoflagellate *Alexandrium excavatum:* UV-photoprotective compounds? *J. Plank. Res.* 12:909–921.

Chalker, B. E., and Dunlap, W. C., 1990, UV-B and UV-A light absorbing compounds in marine organisms, in: *Response of Marine Phytoplankton to Natural Variations in UV-B Flux* (B. G. Mitchell, O. Holm-Hansen, and I. Sobolev, eds.), FC138-088, Scripps Institution of Oceanography, La Jolla, pp. (J)1–21.

Cheng, L., Douek, M., and Goring, D. A. I., 1978, UV absorption by gerrid cuticles, *Limnol. Oceanogr.* 23:554–556.

Coohill, T. P., 1989, Ultraviolet action spectra (280 to 380 nm) and solar effectiveness spectra for higher plants, *Photochem. Photobiol.* 50:451–457.

Craig, J. S., and McLachlan, J., 1964, Excretion of colored ultraviolet absorbing substances by marine algae, *Can. J. Bot.* 42:23–33.

Cullen, J. J., and Lesser, M. P., 1991, Inhibition of photosynthesis by ultraviolet radiation as a function of dose and dosage rate: Results for a marine diatom, *Mar. Biol.* 111:183–190.

Cullen, J. J., and Lewis, M. R., 1988, The kinetics of algal photoadaptation in the context of vertical mixing, *J. Plankton Res.* 10:1039–1063.

Damkaer, D. M., and Dey, D. B., 1983, UV damage and photoreactivation potentials of larval shrimp, *Panalus platyceros,* and adult euphausiids, *Thysanoessa raschii, Oecologia* 60:169–175.

Damkaer, D. M., Dey, D. B., and Heron, G. A., 1981, Dose/dose-rate responses of shrimp larvae to UV-B radiation, *Ecologia* (Berl.) 48:178–182.

Denman, K. L., and Gargett, A. E., 1983, Time and space scales of vertical mixing and advection of phytoplankton in the sea, *Limnol. Oceanogr.* 28:801–815.

Dey, D. B., Damkaer, D. M., and Heron, G. A., 1988, UV-B dose/dose-rate responses of seasonally abundant copepods of Puget Sound, *Oecologia* 76:321–329.

Döhler, G., 1985, Effect of UV-B radiation (190–320 nm) on the nitrogen metabolism of several marine diatoms, *J. Plant Physiol.* 118:391–400.

Döhler, G., 1989, Influence of UV-B (290–320 nm) radiation on photosynthetic $^{14}CO_2$ Fixation of *Thalassiosira rotula* Meunier, *Biochem. Physiol. Pflanz.* 185:221–226.

Döhler, G., 1990, Impact of UV-B (290–320 nm) radiation on metabolic processes of marine phytoplankton, in: *Effects of Solar Ultraviolet Radiation on Biogeochemical Dynamics in Aquatic Environments* (N. V. Blough and R. G. Zepp, eds.), WHOI-90-09, Woods Hole Oceanographic Institution, Woods Hole, Massachusetts.

Döhler, G., and Biermann, 1987, Effect of UV-B irradiance on the response of ^{15}N-nitrate uptake of *Lauderia annulata* and *Synedra planctonica, J. Plankton Res.* 9:881–890.

Döhler, G., Worrest, R. C., Biermann, I., and Zink, J., 1987, Photosynthetic $^{14}CO_2$ fixation and ^{15}N-ammonia assimilation during UV-B radiation of *Lithodesmium variabile, Physiol. Plant.* 70:511–515.

Döhler, G., Hagmeier, E., Grigoleit, E., and Krause, K.-D., 1991, Impact of solar UV radiation on uptake of ^{15}N-ammonia and ^{15}N-nitrate by marine diatoms and natural phytoplankton, *Biochem. Physiol. Pflanz.* 187:293–303.

Dunlap, W. C., and Chalker, B. E., 1986, Identification and quantitation of near-UV absorbing compounds (S-320) in a hermatypic scleractinian, *Coral Reefs* 5:155–159.

Dunlap, W. C., Chalker, B. E., and Oliver, J. K., 1986, Bathymetric adaptations of reef-building corals at Davies Reef, Great Barrier Reef, Australia. III. UV-B absorbing compounds, *J. Exp. Mar. Biol. Ecol.* 104:239–248.

Dunlap, W. C., Williams, D. McB., Chalker, B. E., and Banaszak, A. T., 1989, Biochemical photoadaptation in vision: UV-absorbing pigments in fish eye tissues, *Comp. Biochem. Physiol.* 93B:601–607.

Ekelund, N. G. A., and Björn, L. O., 1990, Ultraviolet radiation stress in dinoflagellates in relation to targets, sensitivity and radiation climate, in: *Response of Marine Phytoplankton to Natural Variations in UV-B Flux* (B. G. Mitchell, O. Holm-Hansen, and I. Sobolev, eds.), FC138-088, Scripps Institution of Oceanography, La Jolla, pp. (F)1–10.

El-Sayed, S. Z., 1988, Fragile life under the ozone hole, *Natural History* 97(10):72–80.

El-Sayed, S. Z., Stephens, F. C., Bidigare, R. R., and Ondrusek, M. E., 1990, Effect of ultraviolet radiation on antarctic marine phytoplankton, in: *Antarctic Ecosystems. Ecological Change and Conservation* (K. R. Kerry and G. Hempel, eds.), Springer, Heidelberg, pp. 379–385.

Fogg, G. E., and Boalch, G. T., 1958, Extra-cellular products in pure cultures of brown algae, *Nature* 181:789–790.

Foster, K. W., and Smyth, R. D., 1980, Light antennas in phototactic algae, *Microbiol. Rev.* 44: 572–630.

Fuhrman, J., 1992, Bacterioplankton roles in cycling of organic matter: The microbial food web, in: *Primary Production and Biogeochemical Cycles in the Sea* (P. G. Falkowski and A. D. Woodhead, eds.), Plenum Press, New York, pp. 361–383.

Gala, W. R., and Giesy, J. P., 1991, Effects of ultraviolet radiation on the primary production of natural phytoplankton assemblages in Lake Michigan, *Ecotoxicology and Environ. Safety* 22:345–361.

Galland, P., and Senger, H., 1991, Flavins as possible blue light photoreceptors, in: *Photoreceptor Evolution and Function* (M. G. Holmes, ed.), Academic Press, New York, pp. 65–124.

Garcia-Pichel, F., Sherry, N. D., and Castenholz, R. W., 1992, Evidence for an ultraviolet sunscreen role of the extracellular pigment scytonemin in the terrestrial cyanobacterium *Chlorogloeopsis* sp. *Photochem. Photobiol.* 56:17–23.

Greenberg, B. M., Gaba, V., Canaani, O., Malkin, S., Mattoo, A. K., and Edelman, M., 1989, Separate photosensitizers mediate degradation of the 32-kDa photosystem II reaction center protein in the visible and UV spectral regions, *Proc. Natl. Acad. Sci. U.S.A.* 86:6617–6620.

Häder, D.-P., and Häder, M. A., 1989a, Effects of solar and artificial radiation on motility and pigmentation in *Cyanophora paradoxa*, *Arch. Microbiol.* 152:453–457.

Häder, D.-P., and Häder, M. A., 1989b, Effects of solar UV-B irradiation on photomovement and motility in photosynthetic and colorless flagellates, *Environ. Exper. Bot.* 29:273–282.

Häder, D.-P., and Häder, M. A., 1991, Effects of solar radiation on motility in *Stentor coeruleus*, *Photochem. Photobiol.* 54:423–428.

Häder, D.-P., and Worrest, R. C., 1991, Effects of enhanced solar ultraviolet radiation on aquatic ecosystems, *Photochem. Photobiol.* 53:717–725.

Halldal, P., 1964, Ultraviolet action spectra of photosynthesis and photosynthetic inhibition in a Green and a Red alga, *Physiol. Plant.* 17:414–421.

Halldal, P., 1967, Ultraviolet action spectra in algology, *Photochem. Photobiol.* 6:445–460.

Hanelt, D., 1992, Photoinhibition of photosynthesis in marine macrophytes of the South China Sea, *Mar. Ecol. Prog. Ser.* 82:199–206.

Haupt, W., and Schönbohm, E., 1970, Light-oriented chloroplast movements, in: *Photobiology of Microorganisms* (P. Halldal, ed.), Wiley-Interscience, New York, pp. 283–307.

Hawryshyn, C. W., 1991, Light-adaptation properties of the ultraviolet-sensitive cone mechanisms in comparison to the other receptor mechanisms of goldfish, *Visual Neurosci.* 6:293–301.

Hawryshyn, C. W., 1992, Polarization vision in fish, *Am. Sci.* 80:164–175.

Helbling, E. W., Villafañe, V., Ferrario, M., and Holm-Hansen, O., 1992, Impact of natural ultraviolet radiation on rates of photosynthesis and on specific marine phytoplankton species, *Mar. Ecol. Prog. Ser.* 80:89–100.

Hirosawa, T., and Miyachi, S., 1982/83, Effects of long-wavelength ultraviolet (UV-A) radiation on the growth of *Anacystis nidulans, Plant Science Lett.* 28:291–298.

Hirosawa, T., and Miyachi, S., 1983, Inactivation of Hill reaction by long-wavelength ultraviolet radiation (UV-A) and its photoreactivation by visible light in the cyanobacterium, *Anacystis nidulans, Arch. Microbiol.* 135:98–102.

Hobson, L. A., and Hartley, F. A., 1983, Ultraviolet irradiance and primary production in a Vancouver Island fjord, British Columbia, Canada, *J. Plank. Res.* 5:325–331.

Holmes, M. G., 1991, *Photoreceptor Evolution and Function,* Academic Press, New York.

Holm-Hansen, O., 1990, Effects of ultraviolet-B and ultraviolet-A on photosynthetic rates of antarctic phytoplankton, *Antarctic Journal U.S.* 25:176–177.

Holm-Hansen, O., and Helbling, E. W., 1993, Polyethylene bags and solar ultraviolet radiation, *Science* 259:534.

Holm-Hansen, O., and Lubin, D., 1993, Solar ultraviolet radiation: Effect on rates of CO$_2$ fixation in marine phytoplankton, in: *Photosynthetic Carbon Metabolism and Regulation of Atmospheric CO$_2$ and O$_2$* (N. E. Tolbert and J. Preiss, eds.), Oxford University Press, New York, in press.

Holm-Hansen, O., and Mitchell, B. G., 1991, Spatial and temporal distribution of phytoplankton and primary production in the western Bransfield Strait region, *Deep-Sea Res.* 38:961–980.

Holm-Hansen, O., Mitchell, B. G., Hewes, C. D., and Karl, D. M., 1989, Phytoplankton blooms in the vicinity of Palmer Station, Antarctica, *Polar Biol.* 10:49–57.

Holm-Hansen, O., Helbling, E. W., and Lubin, D., 1993, Ultraviolet radiation in Antarctica: Inhibition of primary production, *Photochem. Photobiol.,* in press.

Hunter, J. R., Taylor, J. H., and Moser, H. G., 1979, Effect of ultraviolet irradiation on eggs and larvae of the northern anchovy, *Engraulis mordax,* and the Pacific mackerel, *Scomber japonicus,* during the embryonic stage, *Photochem. Photobiol.* 29:325–338.

Hunter, J. R., Kaupp, S. E., and Taylor, J. H., 1981, Effects of solar and artificial ultraviolet-B radiation on larval northern anchovy, *Engraulis mordax, Photochem. Photobiol.* 34:477–486.

Hunter, J. R., Kaupp, S. E., and Taylor, J. H., 1982, Assessment of effects of UV radiation on marine fish larvae, in: *The Role of Solar Ultraviolet Radiation in Marine Ecosystems* (J. Calkins, ed.), Plenum Press, New York, pp. 459–497.

Jagger, J., 1985, *Solar-UV Actions on Living Cells,* Praeger, New York.

Jerlov, N. G., 1950, Ultraviolet radiation in the sea, *Nature* 166:111–112.

Jokiel, P. L., 1980, Solar ultraviolet radiation and coral reef epifauna, *Science* 207:1069–1071.

Jokiel, P. L., and York, R. H., 1984, Importance of ultraviolet radiation in photoinhibition of microalgal growth, *Limnol. Oceanogr.* 29:192–199.

Jones, L. W., and Kok, B., 1966, Photoinhibition of chloroplast reactions: 1. Kinetics and action spectra, *Plant Physiol.* 41:1037–1043.

Karanas, J. J., Worrest, R. C., and Van Dyke, H., 1981, Impact of UV-B radiation on the fecundity of the copepod *Acartia clausii, Mar. Biol.* 65:125–133.

Karentz, D., 1991, Ecological considerations of antarctic ozone depletion, *Antarctic Science* 3:3–11.

Karentz, D., and Lutze, L. H., 1990, Evaluation of biologically harmful ultraviolet radiation in Antarctica using a biological dosimeter designed for aquatic environments, *Limnol. Oceanogr.* 35:549–561.

Karentz, D., Cleaver, J. E., and Mitchell, D. L., 1991a, Cell survival characteristics and molecular responses of antarctic phytoplankton to ultraviolet-B radiation, *J. Phycol.* 27:326–341.

Karentz, D., McEuen, F. S., Land, M. C., and Dunlap, W. C., 1991b, Survey of mycosporine-like amino acid compounds in antarctic marine organisms: Potential protection from ultraviolet exposure, *Mar. Biol.* 108:157–166.

Kaupp, S. E., and Hunter, J. R., 1981, Photorepair in larval Anchovy, *Engraulis mordax, Photochem. Photobiol.* 33:253–256.

Kiefer, D. A., 1973, Chlorophyll-a fluorescence in marine centric diatoms: Responses of chloroplasts to light and nutrient stress, *Mar. Biol.* 23:39–46.

Kirk, J. T. O., 1983, *Light and Photosynthesis in Aquatic Ecosystems,* Cambridge University Press, Cambridge.

Kishino, M., Takahashi, M., Okami, N., and Ichimura, S., 1985, Estimation of the spectral absorption coefficients of phytoplankton in the sea, *Bull. Mar. Sci.* 37:634–642.

Kroes, H. W., 1970, Excretion of mucilage and yellow-brown substances by some brown algae from the intertidal zone, *Bot. Mar.* 13:107–110.

Larson, R. A., and Berenbaum, M. R., 1988, Environmental phototoxicity, *Environ. Sci. Technol.* 22:354–360.

Lesser, M. P., and Stochaj, W. R., 1990, Photoadaptation and protection against active forms of oxygen in the symbiotic procaryote *Prochloron* sp. and its ascidian host, *Appl. Environ. Microbiol.* 56:1530–1535.

Lewis, M. R., Cullen, J. J., and Platt, T., 1984, Relationships between vertical mixing and photoadaptation of phytoplankton: Similarity criteria, *Mar. Ecol. Prog. Ser.* 15:141–149.

Lorenzen, C. J., 1979, Ultraviolet radiation and phytoplankton photosynthesis, *Limnol. Oceanogr.* 24:1117–1120.

Lubin, D., and Frederick, J. E., 1991, The ultraviolet radiation environment of the antarctic peninsula: The roles of ozone and cloud cover, *J. Applied Meteorology* 30:478–493.

Lubin, D., Mitchell, B. G., Frederick, J. E., Alberts, A. D., Booth, C. R., Lucas, T., and Neuschuler, D., 1992, A contribution toward understanding the biospherical significance of antarctic ozone depletion, *J. Geophys. Res.* 97(D8):7817–7828.

Marra, J., 1978, Phytoplankton photosynthetic response to vertical movement in a mixed layer, *Mar. Biol.* 46:203–208.

Martin, J. H., Gordon, R. M., and Fitzwater, S. E., 1990, Iron in Antarctic waters, *Nature* 345: 156–158.

Maske, H., 1984, Daylight ultraviolet radiation and the photoinhibition of phytoplankton carbon uptake, *J. Plankton Res.* 6:351–357.

McLeod, G. C., and Kanwisher, J., 1962, The quantum efficiency of photosynthesis in ultraviolet light, *Physiol. Plant.* 15:581–586.

Mitchell, B. G., 1990, Action spectra of ultraviolet photoinhibition of antarctic phytoplankton and a model of spectral diffuse attenuation coefficients, in: *Response of Marine Phytoplankton to Natural Variations in UV-B Flux* (B. G. Mitchell, O. Holm-Hansen, and I. Sobolev, eds.), FC138-088, Scripps Institution of Oceanography, La Jolla, pp. (H)1–10.

Mitchell, B. G., Brody, E. A., Holm-Hansen, O., McClain, C., and Bishop, J., 1991, Light limitation of phytoplankton biomass and macronutrient utilization in the Southern Ocean, *Limnol. Oceanogr.* 36:1662–1677.

Miyachi, S., 1989, CO_2 concentrating mechanism in microalgae and cyanobacteria, in: *Marine Biotechnology* (S. Miyachi, I. Karube, and Y. Ishida, eds.), Fuji Press, Tokyo, pp. 21–24.

Modert, C. W., Norris, D. R., Blatt, J. H., and Petrilla, R. D., 1982, Effects of UV radiation on photosynthesis of natural populations of phytoplankton, in: *The Role of Solar Ultraviolet Radiation in Marine Ecosystems* (J. Calkins, ed.), Plenum Press, New York, pp. 563–571.

Moffett, J. W., 1990, Chemical reactions affected by UV irradiation in the oceans and their influence on primary productivity: Some general considerations, in: *Effects of Solar Ultraviolet Radiation on Biogeochemical Dynamics in Aquatic Environments,* WHOI-90-09, Woods Hole Oceanographic Institution, Woods Hole, pp. 68–69.

Mosely, B. E. B., 1983, Photobiology and radiobiology of *Micrococcus (Deinococcus) radiodurans,* in: *Photochemical and Photobiological Reviews* (K. C. Smith, eds.), Vol. 7, Plenum Press, New York, pp. 223–274.

Neale, P. J., 1987, Algal photoinhibition and photosynthesis in the aquatic environment, in: *Photoinhibition* (D. J. Kyle, C. B. Osmond and C. J. Arntzen, eds.), Elsevier, New York, pp. 39–65.

Nelson, J. R., 1990, Sunlight-dependent changes in the pigment content and spectral characteristics of particulate organic material derived from phytoplankton, in: *Effects of Solar Ultraviolet Radiation on Biogeochemical Dynamics in Aquatic Environments,* WHOI-90-09, Woods Hole Oceanographic Institution, Woods Hole, pp. 108–109.

Neori, A., Vernet, M., Holm-Hansen, O., and Haxo, F. T., 1988, Comparison of chlorophyll far-red and red fluorescence excitation spectra with photosynthetic oxygen action spectra for photosystem II in algae, *Mar. Ecol. Prog. Ser.* 44:297–302.

Newton, J. W., Tyler, D. D., and Slodki, M. E., 1979, Effect of ultraviolet-B (280 to 320 nm) radiation on blue-green algae (Cyanobacteria), possible biological indicators of stratospheric ozone depletion, *Appl. Environ. Microbiol.* 37:1137–1141.

Nultsch, W., and Agel, G., 1986, Fluence rate and wavelength dependence of photobleaching in the cyanobacterium, *Anabaena variabilis, Arch. Microbiol.* 144:268–271.

Nultsch, W., and Häder, D.-P., 1988, Photomovement in motile microorganisms—II, *Photochem. Photobiol.* 47:837–869.

Palenik, B., Price, N. M., and Morel, F. M. M., 1991, Potential effects of UV-B on the chemical environment of marine organisms: A review, *Environ. Pollution* 70:117–130.

Peak, M. J., and Peak, J. G., 1982, Lethal effects on biological systems caused by solar ultraviolet light: Molecular considerations, in: *The Role of Solar Ultraviolet Radiation in Marine Ecosystems* (J. Calkins, ed.), Plenum Press, New York, pp. 325–336.

Polne, M., and Gibor, A., 1982, The effect of high intensity UV radiation on benthic marine algae, in: *The Role of Solar Ultraviolet Radiation in Marine Ecosystems* (J. Calkins, ed.), Plenum Press, New York, pp. 573–579.

Prézelin, B. B., Samuelsson, G., and Matlik, H. A., 1986, Photosystem II inhibition and altered kinetics of photosynthesis during nutrient-dependent high-light photoadaptation in *Gonyaulax polyedra, Mar. Biol.* 93:1–12.

Regan, J. D., Carrier, W. L., Gucinski, H., Olla, B. L., Yoshida, H., Fujimura, R. K., and Wicklund, R. I., 1992, DNA as a solar dosimeter in the ocean, *Photochem. Photobiol.* 56:35–42.

Renger, G., Völker, M., Eckert, H. J., Fromme, R., Holm-Veit, S., and Gräber, P., 1989, On the mechanism of photosystem II deterioration by UV-B irradiation, *Photochem. Photobiol.* 49:97–105.

Rüdiger, W., and López-Figueroa, F., 1992, Photoreceptors in algae, *Photochem. Photobiol.* 55:949–954.

Rundel, R. D., 1983, Action spectra and estimation of biologically effective UV radiation, *Physiol. Plant.* 58:360–366.

Ryther, J. R., 1969, Photosynthesis and fish production in the sea, *Science* 166:72–76.

Setlow, R. B., 1974, The wavelengths in sunlight effective in producing skin cancer. A theoretical analysis, *Proc. Nat. Acad. Sci. U.S.A.* 71:3363–3366.

Shibata, K., 1969, Pigments and a UV-absorbing substance in corals and a blue-green alga living in the Great Barrier Reef, *Plant & Cell Physiol.* 10:325–335.

Siebeck, O., 1988, Experimental investigation of UV tolerance in hermatypic corals (Scleractinia), *Mar. Ecol. Prog. Ser.* 43:95–103.

Sieburth, J. McN., and Jensen, A., 1969, Studies on algal substances in the sea. II. The formation of Gelbstoff (humic material) by exudates of Phaeophyta, *J. Exp. Mar. Biol. Ecol.* 3:275–289.

Smith, M. D., Masters, C. I., and Mosely, B. E. B., 1992, Molecular biology of radiation-resistant bacteria, in: *Molecular Biology and Biotechnology of Extremophiles* (R. A. Herbert and R. J. Sharp, eds.), Blackie, Glasgow, pp. 259–280.

Smith, R. C., and Baker, K. S., 1979, Penetration of UV-B and biologically effective dose-rates in natural waters, *Photochem. Photobiol.* 29:311–323.

Smith, R. C., and Baker, K. S., 1981, Optical properties of the clearest natural waters (200–800 nm), *Appl. Opt.* 20:177–184.

Smith, R. C., and Wan, Z., 1992, Ozone depletion in Antarctica: Modeling its effect on solar UV irradiance under clear-sky conditions, *J. Geophys. Res.* 97(C5):7383–7397.

Smith, R. C., Baker, K. S., Holm-Hansen, O., and Olson, R., 1980, Photoinhibition of photosynthesis in natural waters, *Photochem. Photobiol.* 31:585–592.

Smith, R. C., Prézelin, B. B., Baker, K. S., Bidigare, R. R., Boucher, N. P., Coley, T., Karentz, D., MacIntyre, S., Matlick, H. A., Menzies, D., Ondrusek, M., Wan, Z., and Waters, K. J., 1992, Ozone depletion: Ultraviolet radiation and phytoplankton biology in antarctic waters, *Science* 255:952–959.

Steeman Nielsen, E., 1964, On a complication in marine productivity work due to the influence of ultraviolet light, *J. Cons. Cons. Perm. Int. Explor. Mer.* 29:130–135.

Tolbert, N. E., 1989, Role of photosynthesis and photorespiration in regulating atmospheric CO_2, in: *Marine Biotechnology* (S. Miyachi, I. Karube, and Y. Ishida, eds.), Fuji Press, Tokyo, pp. 31–34.

Trodahl, H. J., and Buckley, R. G., 1989, Ultraviolet levels under sea ice during the antarctic spring, *Science* 245:194–195.

Trodahl, H. J., and Buckley, R. G., 1990, Enhanced ultraviolet transmission of antarctic sea ice during the austral spring, *Geophys. Res. Lett.* 17:2177–2179.

Urbach, F., 1989, The biological effects of increased ultraviolet radiation: An update, *Photochem. Photobiol.* 50:439–441.

Vernet, M., 1990, UV radiation in antarctic waters: Response of phytoplankton pigments, in: *Response of Marine Phytoplankton to Natural Variations in UV-B Flux* (B. G. Mitchell, O. Holm-Hansen, and I. Sobolev, eds.), FC138-088, Scripps Institution of Oceanography, La Jolla, pp. (I)1–12.

Vosjan, J. H., Döhler, G., and Nieuwland, G., 1990, Effect of UV-B irradiance on the ATP content of microorganisms of the Weddell Sea (Antarctica), *Netherlands J. Sea Res.* 25:391–393.

Voytek, M. A., 1989, *Ominous Future under the Ozone Hole,* Environmental Defense Fund, Washington.

Witkin, E. M., 1976, Ultraviolet mutagenesis and inducible DNA repair in *Escherichia coli,* *Bacteriol. Rev.* 40:869–907.

Wood, W. F., 1987, Effect of solar ultraviolet radiation on the kelp *Ecklonia radiata, Mar. Biol.* 96:143–150.

Worrest, R. C., 1982, Review of literature concerning the impact of UV-B radiation upon marine organisms, in: *The Role of Solar Ultraviolet Radiation in Marine Ecosystems* (J. Calkins, ed.), Plenum Press, New York, pp. 429–457.

Worrest, R. C., 1983, Impact of solar ultraviolet-B radiation (290–320 nm) upon marine microalgae, *Physiol. Plant.* 58:428–434.

Worrest, R. C., 1986, The effect of solar UV-B radiation on aquatic systems: An overview, in: *Effects of Changes in Stratospheric Ozone and Global Climate. Overview* (J. G. Titus, ed.), U.S. Environmental Protection Agency and United Nations Environmental Program 1, 175–191.

Worrest, R. C., Wolniakowski, K. U., Scott, J. D., Brooker, D. L., Thomson, B. E., and Van Dyke, H., 1981, Sensitivity of marine phytoplankton to UV-B radiation: Impact upon a model ecosystem, *Photochem. Photobiol.* 33:223–227.

Yayanos, A. A., 1989, Physiological and biochemical adaptation to low temperatures, high pressures, and darkness in the deep sea, in: *Recent Advances in Microbial Ecology* (T. Hattori, Y. Ishida, Y. Maruyama, R. Y. Morita, and A. Uchida, eds.), Japan Scientific Society Press, Tokyo, pp. 38–42.

Yentsch, C. S., and Yentsch, C. M., 1982, The attenuation of light by marine phytoplankton with specific reference to the absorption of near-UV radiation, in: *The Role of Solar Ultraviolet Radiation in Marine Ecosystems* (J. Calkins, ed.), Plenum Press, New York, pp. 691–700.

Zill, L. P., and Tolbert, N. E., 1958, Effect of ionizing radiation and ultraviolet radiations on photosynthesis, *Arch. Biochem.* 76:196–203.

14

Effects of Ultraviolet-B Radiation on Terrestrial Plants

Janet F. Bornman and Alan H. Teramura

1. Penetration and Measurement of UV-B Radiation in Plants

1.1. Significance of Changes within Plant Tissue

Determination of spectral radiation within plant tissues is normally complicated by the intricate path taken by photons of light through the tissue as a result of scattering, internal reflection, and absorption, none of which occurs in a uniform way. Methods which have been employed to estimate the penetration of radiation into plant tissues have included theoretical calculations, measurements of whole leaf or epidermal reflectance and absorptance, and the direct measurement of the internal spectral regime using fiber optics.

Exposure to UV radiation may cause alterations in morphology, ultrastructure, and pigments, which in turn may result in many indirect effects on cell processes and subsequent growth. One example of this is the altered distribution of photosynthetically active radiation (PAR, 400–700 nm) with depth in a leaf after exposure of plants to UV-B radiation (Bornman and Vogelmann, 1991), with concomitant change of the microenvironment for processes such as photosynthesis. It was noted that in *Brassica campestris* the scattered 400-

Janet F. Bornman • Plant Physiology, Lund University, S-220 07 Lund, Sweden.   Alan H. Teramura • Department of Botany, University of Maryland, College Park, Maryland 20742.

Environmental UV Photobiology, edited by Antony R. Young *et al.* Plenum Press, New York, 1993.

to 700-nm light increased both in the palisade and spongy mesophyll tissues after exposure to UV-B radiation (Bornman and Vogelmann, 1991). This was likely a result of ultrastructural and pigment changes. In addition, in the spongy mesophyll, collimated light was decreased after UV radiation, which seems to have been compensated for by the increased contribution of the scattered component.

1.2. Theoretical and Experimental Determination of UV Radiation within Plant Tissue

For the theoretical estimation of internal radiation in plant organs, light gradients are determined from reflectance and transmittance values using the light-scattering model based on Kubelka–Munk calculations (Seyfried and Fukshansky, 1983). This approach is strictly valid only for conditions in which the sample is irradiated with diffuse light, and not for material exposed to collimated light or direct sunlight.

The epidermis of plant leaves, which forms the first filter for radiation, is particularly effective in reducing the penetration of ultraviolet-B (UV-B, 280–315 nm) radiation, while transmitting a large portion of PAR. Measurements through detached epidermal peels of a variety of plant species using an integrating sphere/monochromator system have shown that transmittance of UV radiation was generally less than 10%. In many of the species tested, it was also estimated that UV-absorbing pigments accounted for 20% to 57% of the attenuation of the UV radiation (Robberecht and Caldwell, 1978), providing an effective screen for protecting the photosynthetic apparatus in underlying cells (Flint et al., 1985). Using the same integrating sphere/monochromator system, Flint and Caldwell (1983) found that transmittance of UV radiation through corollas and anther walls was very small, so that pollen would be well shielded. However, the period between dehiscence and penetration of the pollen tube into stigmatic tissues might be a particularly vulnerable stage (Flint and Caldwell, 1983).

The penetration of UV radiation can be measured directly by inserting fiber optic microprobes of μm tip size into plant tissue (Bornman and Vogelmann, 1988; Day et al., 1992; Cen and Bornman, 1993). In addition, anatomical and biochemical changes resulting from exposure to UV-B radiation can be measured using other wavelengths of light, e.g., in the range 400–700 nm (PAR) (Bornman and Vogelmann, 1991). The advantage of using the fiber optic technique is that direct, in situ spectral information can be obtained in anatomically complex material. Since the fiber used is a directional sensor, both collimated and diffuse radiation can be measured by varying the angle of insertion with respect to the plant organ. As is to be expected, UV-A penetrates further into leaves than UV-B radiation (Bornman and Vogelmann, 1988). Although not as biologically effective as UV-B radiation in inducing a direct response, UV-A radiation may still contribute to adverse effects in other ways, some of which are briefly discussed in the following sections. Penetration and internal distribution of UV-B radiation varies among plant species, and is strongly affected by leaf anatomy, pigments, and other physiological changes resulting from UV radiation. In a study of 22 different plant species, it was found that UV-B radiation penetrated deepest into leaves of herbaceous dicotyledons and least into conifer needles, while penetration into leaves of woody dicotyledons and grasses was intermediate (Day *et al.,* 1992). Even in conifers, where attenuation is high, UV-B radiation may be particularly effective during the period shortly following needle elongation and emergence past the highly UV-protective bud scales. This growth period coincides with the annual solar UV-B radiation maximum. Although 300-nm UV-B radiation is strongly attenuated in the epidermal layer, a small proportion of this radiation reaches the mesophyll. In mature needles, virtually all UV-B radiation is absorbed in the cell walls of the epidermis.

UV-B screening pigments are primarily located toward the adaxial leaf surfaces in predominantly horizontally inclined leaves. The effectiveness of these pigments is reflected in light gradients when the fiber optic microprobe is driven through a leaf from the shaded side with the adaxial leaf surface facing the light source (Fig. 1). Despite the different protective mechanisms, it is also possible that microsites of relatively high amounts of UV-B radiation occur where screening is imperfect.

2. Heliotropism and UV-B Radiation

The fact that leaves of some plants orient themselves to incident light has been known for a long time. Rapid and reversible leaf movements in response to direct solar radiation have been called heliotropism (Darwin, 1880). Two types of heliotropic leaf movement can be distinguished. *Diaheliotropism* is the ability of leaves to track the sun, keeping leaf lamina perpendicular to the sun's direct rays. *Paraheliotropism* is the ability of leaves to avoid the sun, maintaining leaf lamina parallel to the sun's direct rays. Heliotropic leaf movements have been observed in plant species from 16 different families. The tracking ability is independent of photosynthetic pathway and taxonomic affinity, since species representing both C_3 and C_4 photosynthetic pathways and families from diverse plant orders demonstrate solar tracking (Ehleringer and Forseth, 1980).

The most apparent consequence of heliotropism is a modification of the direct solar irradiance on leaves. However, this regulation of incident radiation

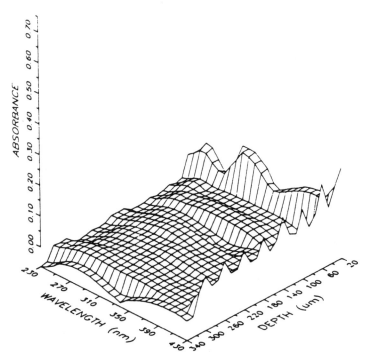

Figure 1. Difference plot showing the content of UV-screening pigments from paradermal sections of *Brassica napus,* expressed as absorbance. Values from control plants were subtracted from those of UV-treated plants. Control plants received 740 μmol m^{-2} s^{-1} photosynthetically active radiation, 400–700 nm. In addition, treated plants were exposed to 8.9 kJ m^{-2} day^{-1} biologically active UV-B radiation, calculated for the 15th of June on a cloudless day and at aerosol level zero. Plants were grown under the experimental conditions for 14 days. (Unpublished data from C. Ålenius, J. F. Bornman, and T. C. Vogelmann.)

has far-reaching ramifications on many aspects of plant–environment interactions, including leaf radiant energy balance, leaf temperature, transpirational water loss, photosynthetic carbon gain, water use efficiency, high-light-associated damage to the photosynthetic apparatus, and carbon return on investments of limiting resources into photosynthesis (Forseth, 1990).

It has been hypothesized that the differential cultivar sensitivity in soybeans to UV-B radiation may be linked to differences in the ability of some cultivars to avoid high leaf temperatures (outside the optimal range for enzyme activity) by orienting leaves away from the sun at the times when air temperature and other factors are at stressful levels (M. Rosa and I. N. Forseth, personal communication, July 1992). Preliminary experiments with soybean cultivars Forrest, Cumberland, and Essex (UV-B tolerance in Forrest > Cumberland > Essex) tested for UV-B radiation effectiveness under conditions of decreasing soil water potential have shown variability among cultivars in their leaf-angle response to UV-B radiation under water stress conditions. Under well-watered conditions no difference could be found between UV-B irradiated plants and controls for the cultivars used. However, as soil water availability decreases, leaves of UV-B irradiated cultivars Forrest and Cumberland tended to move to lower angles than those of control plants. Cultivar Essex did not show marked differences between UV-B treated plants and controls. This difference in leaf behavior may be correlated to the differential sensitivity of these soybean cultivars to UV-B radiation.

3. Some Important Targets of UV-B Radiation

Some of the studies cited in this section used very low levels of PAR with unnaturally high amounts of supplemental UV-B radiation. Nonetheless, the work is still clearly important in identifying key targets of UV radiation, but further work is necessary before much of the information can be extrapolated to ecologically relevant conditions.

3.1. Nucleic Acids

UV-B radiation seems to exert a direct effect on DNA (acting as both main chromophore and target) in contrast to the action of UV-A (315–400 nm), which appears to be indirect via stimulation of hydroxyl radical formation (Peak and Peak, 1990). However, this has apparently not yet been experimentally demonstrated in higher plants. Apart from UV-A radiation, there is also evidence that H_2O_2 plays a role in DNA damage through mediation of Fenton reactions (Imlay and Linn, 1988), whereby iron catalyzes the reduction of H_2O_2, giving rise to $\cdot OH$ and OH^-. Mechanisms other than Fenton reactions are of course also involved in the generation of free-radical species, which can alter DNA. Although most work has concentrated on nuclear DNA, there is evidence that mitochondrial DNA may be more sensitive to oxidative stress (Hruszkewycz, 1988), which makes this genome of particular interest with respect to increasing levels of UV-B radiation.

So far, relatively little research on the effect of UV radiation on nucleic acids has been carried out on higher plant systems. In general, it is assumed that the mechanisms found for prokaryotes and single-cell eukaryotes are similar to those of higher plants. Evidence for pyrimidine dimer formation in plants comes from a study on *Medicago sativa,* which was irradiated with UV-B radiation. Dimers were also found with the UV-B component removed, although levels were low (Quaite *et al.*, 1992a). An action spectrum for the induction of cyclobutyl pyrimidine dimers in the same material showed that damage occurred by wavelengths as long as 365 nm (Quaite *et al.*, 1992b). An important observation was that the action spectrum of Quaite *et al.* (1992b) for intact leaves has a maximum near 280 nm, instead of near 260 nm. This discrepancy may be due to self-screening within the plant.

3.2. Proteins

The main CO_2-fixing enzyme in C_3 plants, ribulose 1,5-bisphosphate carboxylase (Rubisco), comprises about 50% of the total soluble plant protein and consists of a large and small subunit (55 and 14 kDa, respectively), the former of which is encoded by the chloroplast and the latter by the nucleus. Rubisco activity may decline with enhanced levels of UV-B radiation (Vu *et al.*, 1984, Strid *et al.*, 1990). The levels of both the subunits of Rubisco are decreased by UV-B radiation. This was correlated to a reduction in the RNA transcript levels for these subunits in UV-B-treated material (Jordan *et al.*, 1992). Other proteins, such as the chloroplast-encoded D1 protein and the nuclear-encoded chlorophyll *a/b* binding protein have been investigated also at the transcription level (*psb* A and *cab* transcripts; Jordan *et al.*, 1991). The steady-state mRNA transcript levels for the *cab* sequence and the *psb* A RNA transcripts were reduced. However, the latter reduction did not occur as rapidly nor to the same extent as that for *cab*. The supplementary UV-B radiation did not cause an irreversible change of the regulatory control involved in maintaining the normal RNA transcript levels. Increased visible irradiance during UV-B exposure lessened the effect on the *cab* transcript levels to some extent. The RNA transcripts for Rubisco (*rbc* S) show the same response as *cab*, suggesting a common UV-B effect on nuclear-encoded biosynthesis of chloroplast proteins. It is likely that the perception of UV-B radiation leads to the transduction of a signal that rapidly inhibits the transcription of the *cab* gene and subsequently to the reduction in the RNA transcript level. Inactivation of DNA by UV-B radiation does not seem to be only by direct photochemical action, since transcription can be switched on or off for the different nuclear-encoded genes (Jordan *et al.*, 1991).

3.3. Membrane Lipids

Lipids and proteins are the main constituents of cell membranes, with the lipid/protein ratio usually being 1:1 in plant cells. Apart from, for example, ozone and chilling, UV radiation is one of the causes of membrane alteration and damage (Bornman *et al.*, 1983; Murphy, 1983; Kramer *et al.*, 1991). One of the main processes involved appears to be lipid peroxidation, which results in the oxidation of polyunsaturated fatty acids, with reactions that involve molecular oxygen. Lipid peroxidation is stimulated not only by UV-B (Panagopoulos et al., 1990; Kramer et al., 1991), but also by UV-A radiation. However, with supplemental UV radiation, the peroxidation appears to be affected by the level of PAR, with higher irradiances having little effect on total peroxidase activity (Murali et al., 1988). Singlet oxygen also appears to be involved, as was found by Chamberlain and Moss (1987) in E. coli. They suggested that singlet oxygen ($*O_2$) attacks unsaturated double carbon bonds. In chloroplast membrane lipids, the total content of the polar lipids, the mono- and digalactosyldiglycerides (MGDG, DGDG), may show a differential decrease resulting in a change in the ratio, whereby the content of MGDG is more reduced. Levels of phosphatidylglycerol have also been reported to decrease (Tevini et al., 1981). Since a high degree of unsaturation of MGDG is necessary for stability of the chloroplast membrane structure (Hugly et al., 1989), the decreases in the MGDG/DGDG ratio with UV-B radiation may jeopardize this stability (Kramer et al., 1991). Alterations to chloroplast membranes are likely to be reflected in changes in photosynthesis upon exposure to enhanced levels of UV-B radiation, and are therefore of particular significance.

3.4. Cytoskeleton

The plant cytoskeleton is composed of a filamentous network consisting of microtubules, actin filaments, and intermediate filaments. Microtubules are made up of partially homologous α- and β-polypeptides. The cytoskeleton can also be described as that part of the cell which remains after extraction with detergent.

The involvement of the cytoskeleton is of considerable importance in cell growth and morphology, since microtubules are essential, dynamic structures of the preprophase band, mitotic and meiotic spindles, phragmoplast and interphase arrays. However, most of our information on the potential damaging effect of UV-B radiation on the cytoskeleton comes from experiments conducted on animal cells, where, in addition to other effects, the disassembly of the cortical microtubules (Zamansky and Chou, 1987) as well as inhibition of the reassembly (polymerization) from the pool of cytoplasmic tubulin dimers during interphase has been shown (Zamansky et al., 1991). In plant systems, subjecting Petunia hybrida protoplasts to UV radiation (UV-B and UV-A) altered cortical microtubule organization, resulting in shortened cortical microtubules (Fig. 2). Tubulin may be a particularly sensitive target, since it has a high content of amino acids with aromatic side chains. UV radiation delayed progression through the cell cycle, with the phases, S_1, G_1 and G_2, being affected. There may be a correlation between cortical micro-

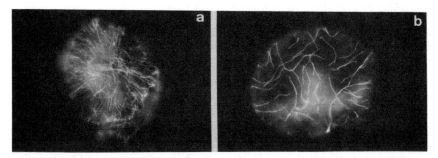

Figure 2. Mesophyll protoplasts from *Petunia hybrida* were irradiated 30 min after isolation and cultured for 48 h. In the nonirradiated protoplast (a), the cortical microtubules are abundant and run parallel to each other. In the protoplasts which had received a dose of 8 mmol photons m^{-2} UV radiation (b), there are fewer microtubules, and the parallel orientation of the cortical microtubules is much less apparent. (Data from I. Staxén.)

tubule disruption and the delay in the onset of DNA synthesis (Staxén *et al.,* 1993). Further evidence from animal systems also suggests that delay of the cell cycle is not solely due to a delay in the S phase, during which time DNA repair is thought to occur. Prolongation of the G_2 phase has also been reported, which suggests that there are other mechanisms involved apart from DNA repair (Carlson, 1976a,b). These may include loss of sulfhydryl groups from the tubulin dimer, which affects its stability and hence polymerization, as well as conformational change and apparent cross-linking of dimers after exposure to UV-B radiation (Zaremba *et al.,* 1984). Since tubulin absorbs maximally at 280 nm, the structural network of cells, which is involved in both the architecture and regulation of many cell functions, is an obvious target of UV radiation.

3.5. Photosystem II and Net Photosynthesis

Productivity and ecosystem structure are in large part affected by changes in photosynthesis, making the latter one of the key physiological processes of prime importance with regard to anthropogenic disturbances to the environment. The composition of photosystems (PS) I and II suggests that UV radiation would be effective in inducing changes in both systems; however, this does not seem to be the case. Research to date has shown that PS II is generally more affected (Noorudeen and Kulandaivelu, 1982; Kulandaivelu and Noorudeen, 1983; Iwanzik *et al.,* 1983; Bornman 1989), with few reports on the weak inhibition of PS I (Brandle *et al.,* 1977; Van *et al.,* 1977). The different components of PS II are variably influenced by UV radiation. The PS II reaction center (RC) itself may be affected by enhanced UV-B radiation. This center consists of a chlorophyll-binding complex of two polypeptides (D1 and D2), which is assumed to contain four molecules of chlorophyll a, two molecules of pheophytin, a β-carotene, and one or two nonheme iron atoms (Barber, 1987). The oxidizing as well as the reducing side of PS II can also be affected by UV-B radiation. On the oxidizing side, the functional connection between the oxygen-evolving complex and P680 has been implicated, as well as impairment of the reaction center (Renger $et\ al.$, 1989). The D1 and D2 polypeptides on the acceptor or reducing side were also found to be modified by UV-B radiation, such that the number and activity of quinone binding sites were altered (Renger $et\ al.$, 1989). The pattern of excitation energy transfer between different photosynthetic units may also be influenced by UV radiation (Renger $et\ al.$, 1986). Changes in the decay of fluorescence, which appears to have more than one component, indicate that exciton lifetimes are altered such that the fast components are accelerated and the slow ones retarded (Renger $et\ al.$, 1991). This observation taken together with the fact that UV-B radiation also decreases the total fluorescence decay, points to the formation of additional quenchers of exciton energy (Renger $et\ al.$, 1991).

The turnover of the D1 polypeptide is fastest with UV-B irradiation (Greenberg $et\ al.$, 1989a). The degradation of D1 in the UV, visible, and far-red spectral regions involves a 23.5 kDa polypeptide (Greenberg $et\ al.$, 1989b). However, it appears that the photoreceptor for the D1 degradation is different from that in the visible and far-red regions. The UV photoreceptor is probably a semiquinone anion radical (Q_B^-), while for the visible and far-red the photoreceptor is thought to be bulk photosynthetic pigments (mainly chlorophyll and carotenoids; Greenberg $et\ al.$, 1989a). In another study, high levels of UV-B radiation inhibited formation of the semiquinone anion at Q_A (a primary quinone electron acceptor of PS II), and decreased the overall photoreduction of plastoquinone (PQ) (Melis $et\ al.$, 1992). However, as these latter authors point out, the decrease in photoreduction of PQ cannot solely be accounted for by the inhibition at Q_A, since different PS II reaction centers can donate electrons independently into a common pool of PQ molecules (Siggel $et\ al.$, 1972). Consequently, it was suggested that UV-B radiation also causes damage to the PQ molecules of the pool itself. That quinones are involved seems probable, since the quinone, semiquinone anion radical, and the quinol, comprising the three redox states of plastoquinone, all absorb in the UV.

Net photosynthesis, measured as the rate of CO_2 assimilation, does not always appear to follow the sometimes larger changes due to UV radiation found from studies on the partial reactions. Disturbances of these partial reactions may also reflect more subtle changes of the carbon cycle, which, if of sufficient magnitude, are likely to affect productivity. Even primary carbon metabolism appears to be influenced by UV-B radiation, as is seen in the reduction of organic acids and soluble sugars (Takeuchi $et\ al.$, 1989). Fluorescence induction is influenced by photochemical events related to the redox state of Q_A, whereby a decrease in fluorescence reflects oxidized Q_A, or photochemical quenching (q_p). Q_A will be kept in the oxidized state by the transfer of electrons to NADPH and finally to CO_2. Nonphotochemical quenching processes (q_{NP}), thought to mainly reflect energization of thylakoid membranes due to movement of H^+ ions and subsequent production of ATP, also influence chlorophyll fluorescence. Thus closer investigation of this link between carbon assimilation and the redox state of Q_A may provide more integrated information on the involvement of UV-B radiation in photosynthesis. It seems that there is a tendency for q_P to decrease and q_{NP} to increase after exposure of plants to enhanced levels of UV-B radiation, which may partially explain some of the reductions in fluorescence after UV exposure.

Reductions in net photosynthesis, measured as the rate of CO_2 assimilation, have been noted in the greenhouse and in the field, although under certain conditions no effect or apparent increases in photosynthesis have also been documented. These differences are likely the outcome of many factors, such as inherent plant adaptation, differences in sensitivity with ontogenetic stage, modification of UV effectiveness by interaction with natural climatic changes with season, and different experimental conditions and protocols. Where reduced photosynthetic rates have been observed, stomatal closure, decreases in photosynthetic pigments, and altered Rubisco activity, as well as impairment to PS II, are some of the likely causes (Teramura *et al.*, 1983; Negash, 1987; Teramura, 1983; Vu *et al.*, 1984; Strid *et al.*, 1990; Bornman, 1989).

4. Effects of UV-B Radiation on Growth and Morphology

Often, plant response to UV radiation differs qualitatively and quantitatively among both species and cultivars, so that the idea of a general plant response to UV radiation may be too simplistic. UV radiation may either inhibit or stimulate plant growth through various physiological and morphological processes. Photosynthesis may decrease in early stages of leaf development or ontogeny, which suggests that plants may be particularly sensitive to UV radiation before full leaf expansion (Teramura and Caldwell, 1981). Independent studies have shown that leaf expansion was also reduced after exposure of plants to enhanced UV-B radiation (Ambler *et al.*, 1975).

Among the most commonly observed changes after UV exposure are changes in biomass and biomass allocation, flowering pattern, plant height, and leaf volume. Other structural changes may include reduced internodal length and sometimes increased epicuticular wax (Basiouny *et al.*, 1978; Tevini *et al.,* 1986). Decreases in height and changes in leaf volume reflect alterations in both cell division and cell expansion. Reduced leaf area may also accompany increases in leaf thickness (Murali *et al.,* 1988). It is therefore not surprising that the number of stomata also decreases (Tevini *et al.,* 1986). Seed lipid and protein content may also decrease in sensitive cultivars (Teramura *et al.,* 1990a).

At the level of the cell, changes in volume, cytoplasmic streaming, and organelle morphology, as well as increases in water permeability and reduced viscosity, may be found after irradiation with short-wavelength UV-B radiation (Lichtscheidl-Schultz, 1985). At the ultrastructural level, changes occurring in organelle morphology include, for example, rounding of mitochondria and swelling of endoplasmic reticulum. In other respects, several variations among different species are seen. For example, in *Phaseolus vulgaris* pronounced elongation of the columnar cells of the palisade was seen in comparison to those leaves which did not receive UV radiation (Cen and Bornman, 1990). In leaves of *Brassica carinata* and *Medicago sativa,* increased thickness was a consequence of additional spongy mesophyll cells, while in *B. campestris,* palisade cells increased in number (Bornman and Vogelmann, 1991). At the level of the chloroplast, UV radiation can cause dilation of the thylakoid membranes (Brandle *et al.,* 1977) and apparent disruption to the double membrane of the chloroplast (Brandle *et al.,* 1977; Bornman *et al.,* 1983).

Morphological changes may also result in shifts in the balance of competition among different species. Changes can occur in the competitive ability within plant communities as a result of differential UV-B resistance of the competing species (Caldwell, 1977). This subject is expanded further in Section 7.

5. Reproduction

UV-B radiation can influence plants in many different ways during seedling growth. This is partially due to inherent ontogenetic differences in UV-B sensitivity and also to modifications in protective mechanisms, since these may change during adaptation to enhanced levels of UV-B radiation. For example, germination and early seedling stages are likely to be particularly sensitive periods in development. Although pollen is fairly well protected from UV radiation, after transfer to the stigma the pollen tube itself may be susceptible (Flint and Caldwell, 1986). UV radiation may also inhibit pollen germination (Campbell *et al.,* 1975; Caldwell *et al.,* 1979; Flint and Caldwell, 1984). Flowering may be repressed in plants receiving UV-B radiation as compared to those where the radiation is excluded by Mylar film or glass

(Kasperbauer and Loomis, 1965; Caldwell, 1968), although this effect is not always found. More important with respect to pollination and pollinators, is that the timing of flowering may be altered by UV-B radiation. It has also been found that early stages of development and the transition period between vegetative and reproductive growth are sensitive to enhanced levels of UV-B radiation in soybeans (Murali and Teramura, 1986a; Teramura and Sullivan, 1987).

6. Protective Mechanisms

Plants have evolved several mechanisms to deal with stress. Apart from light, other factors such as temperature, water, nutrient supply, wounding, herbivore attack, and also atmospheric pollutants induce the plant to respond. This response ranges from physiological, molecular, and biochemical changes at the cell level to altered morphology at the whole plant level. Apparent protective measures specifically induced by UV radiation at the whole plant level are sometimes difficult to characterize. However, decreased leaf area and increases in leaf thickness and epicuticular wax are generally believed to be protective modifications. At the cell level, induction of specific proteins and enzymes (especially of the isoflavonoid pathway and repair enzymes such as DNA photolyase, see Chapter 12, this volume) increase upon irradiation of plant material with UV radiation.

6.1. Regulation of Flavonoid and Isoflavonoid Production

Flavonoids, a class of water-soluble phenolic derivatives, include the colorless pigments which effectively absorb in the UV region of the spectrum. UV-B radiation specifically induces the accumulation of these UV-absorbing pigments, with a peak showing maximum effectiveness at around 295 nm (Wellmann, 1975). However, UV-A and visible radiation are also involved in induction and in the quantitative and qualitative pattern of flavonoids. Thus apart from a UV-B receptor, the response may also be modified by a blue-light receptor as well as phytochrome, for example, at the level of regulation of chalcone synthase mRNA (Bruns et al., 1986). The effect of UV-B radiation at the transcription level has also been observed in the increase in certain key enzymes of the flavonoid pathway (Schulze-Lefert et al., 1989). Chalcone synthase is one of the main regulatory enzymes which must be induced for synthesis of flavonoids (Hahlbrock and Grisebach, 1979; Hahlbrock and Scheel, 1989). Other key enzymes of flavonoid biosynthesis that have been shown to be induced by UV radiation include PAL (phenylalanine ammonia-lyase), chalcone-flavonone isomerase (Chappell and Hahlbrock, 1984), and 4-coumarate-CoA-ligase (Douglas *et al.,* 1987).

Evidence for a possible UV photoreceptor points to a flavin (Ensminger and Schäfer, 1992), although other compounds may also be involved. Using cell cultures, these authors found that cells supplied with UV-B radiation, visible light, and riboflavin produced higher amounts of chalcone synthase and flavonoids than control cells. Other factors may also be involved in the regulation or modification of flavonoid accumulation. For example, a role for reduced glutathione (GSH) as a signal compound in the UV-induced induction of the enzymes required for flavonoid biosynthesis has been suggested (Wingate *et al.,* 1988).

Anthocyanins, which also belong to the flavonoid group, may afford protection. Takahashi *et al.* (1991) reported that with increasing anthocyanin levels, there was a reduction in formation of pyrimidine dimers upon exposure of cultured cells to UV-B radiation. The action spectrum for anthocyanin formation in sorghum shows a prominent peak around 290 nm, with lesser peaks in the UV-A, blue, and red regions of the spectrum (Yatsuhashi *et al.,* 1982). However, the additional presence of phenylpropanoids also plays a protective role, and thus the individual effects are difficult to separate. Yatsuhashi *et al.* (1982) found that the action of the 385-, 480-, and 650-nm anthocyanin peaks in sorghum was mediated by phytochrome, but not that of the 290-nm peak. Synthesis of anthocyanin and flavone glycoside appears to be mediated by UV-B and phytochrome photoreceptors (Wellmann, 1974; Drumm-Herrel and Mohr, 1981; Duell-Pfaff and Wellmann, 1982). Although both types of photoreceptor sometimes act synergistically, they may also act independent of one another (Yatsuhashi and Hashimoto, 1985). Takahama *et al.* (1989) found a stimulation of flavonoid synthesis in *Vicia faba* by applying H_2O_2. This suggests that H_2O_2 is involved in the oxidation of flavonols. As Takahama *et al.* (1989) point out, radical-triggered induction of flavonoid synthesis may be very rapid, especially if UV-A radiation induces radical formation, as reported by Peak and Peak (1990).

Flavonoids are found in leaves, pollen, petals, stems, and bark, mainly in vacuoles and cell walls. In leaves, the epidermis usually forms the first, most effective UV-B filter, with high concentrations of flavonoids, varying from over 1 mM up to 10 mM in this epidermal layer (Vierstra *et al.,* 1982). This filtering effect would seem to be largely due to the optical properties of the pigments, although anatomical features will also play a role. Both these features will affect physiological processes in deeper-lying tissue. The attenuation of UV-B radiation in deeper-lying tissues is seen, for example, in the apparent reduction of changes in the photosynthetic transport system upon accumulation of these compounds (Tevini *et al.,* 1991).

6.2. Effect of UV-Absorbing Pigments, Pubescence, and Wax Layer

In higher plants, flavonoids are mainly located in the epidermis (Robberecht and Caldwell, 1978) and upper mesophyll tissues in vacuoles (Jahnen
and Hahlbrock, 1988) and to a lesser extent also in chloroplasts (McClure,
1976; Weissenböck et al., 1976).

It seems logical to assume that there should be a relationship between
tolerance to UV radiation and the absorption capacity of the upper leaf tissue.
The attenuation of UV radiation with depth in a leaf follows the pattern of
the UV-absorbing pigments (compare Figs. 1 and 3). Apart from the optical,
protective role of flavonoids, herbivore resistance may increase as a result of
increases in certain, specific compounds, such as the furanocoumarins and
terpenoids, which are toxic to many organisms (Tevini and Teramura 1989,
and refs. therein), although in some cases these may serve as chemical at-
tractants.

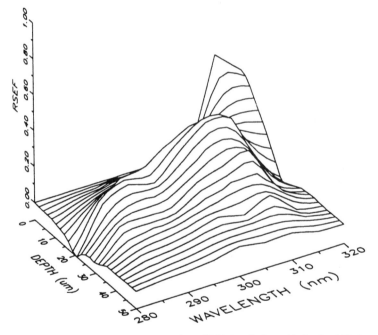

Figure 3. Difference plot showing the penetration of UV-B radiation with depth within leaves of
Brassica napus. Values from UV-treated plants were subtracted from those of control plants.
Control plants received 740 μmol m^{-2} s^{-1} photosynthetically active radiation, 400–700 nm. In
addition, treated plants were exposed to 8.9 kJ m^{-2} day^{-1} biologically active UV-B radiation,
calculated for the 15th of June on a cloudless day and at aerosol level zero. Plants were grown
under the experimental conditions for 14 days. RSEF, relative steric energy flux. (Unpublished
data from C. Ålenius, J. F. Bornman, and T. C. Vogelmann.)

Epicuticular wax, a mixture of highly nonpolar lipids, does not absorb strongly in the UV-B region, although it may serve to reflect and protect from high-UV irradiances and visible light. This would be of ecological importance for species growing in high elevations (Clark and Lister, 1975). Although reflectance is generally lower in the UV wavebands compared to that for the visible region, 10–20% of UV radiation may be reflected (Gausman *et al.*, 1975; Robberecht *et al.*, 1980). Despite the low absorption by epicuticular wax, UV radiation may still have a regulatory effect on its biosynthesis, which may result in increased wax cover or changes in its structure or composition (Basiouny *et al.*, 1978; Steinmüller and Tevini, 1985). At the same time, an increase in epicuticular wax will reduce transpiration. Just as stimulation of flavonoid biosynthesis may differ widely among species, so do the effects of UV-B radiation on epicuticular wax formation. In a comparative study using bean, cucumber, and barley leaves of plants subjected to UV-B radiation, the wax layer increased most on barley leaves, followed by bean and cucumber (Steinmüller and Tevini, 1985). In cucumber, where the main wax components are the alkane-1-ols and alkanes, there was a shift toward shorter-chain homologues with concomitant reduction in long-chain alkanes with enhanced levels of UV-B radiation (Tevini and Steinmüller, 1987). It was also shown by these authors that the influence of UV-B radiation was different from that of visible light, making UV-B radiation a potentially important factor in altered wax composition. Since the epicuticular wax is influenced by other environmental factors (Hull *et al.*, 1975), additional UV-B radiation may alter the natural balance in response to factors such as temperature, visible radiation, and humidity.

6.3. Phytoalexins and Related Compounds

The induction of stress metabolites is caused by infection or other cellular injury as well as by abiotic elicitors. Together with other coordinated responses such as increased lignification and various enzyme inhibitors, these stress metabolites generally increase plant resistance (Kuć, 1984). Apart from the flavonoid UV-screening pigments discussed in Sections 6.1 and 6.2, an interesting group of metabolites are the phytoalexins, which are low-molecular-weight, phenolic compounds that are induced by biotic and abiotic elicitors and are found in several classes of compounds, e.g., the isoflavonoids, terpenoids, stilbenes, polyacetylenes, and dihydrophenanthrenes. The most studied class has been the isoflavonoids. Of the abiotic elicitors, fungicides, heavy metals (e.g., $CuCl_2$, $HgCl_2$), surfactants, and UV radiation, among others, induce phytoalexin synthesis (Dixon *et al.*, 1983, and refs. therein).

To date, most of the information on UV radiation and induction of secondary metabolites comes from studies using UV-C radiation (<280 nm), where correlations have been found between high levels of phytoalexins and decreased susceptibility to disease attack after exposure to UV-C radiation (see, e.g., Bridge and Klarman, 1973). A study using UV-B radiation showed increases in the secondary metabolite tetrahydrocannabinol in *Cannabis sativa*. However, in the plants studied, the increased levels of cannabinoid did not appear to contribute to further UV tolerance. This might have been a result of already high levels in combination with flavonoids, making the plant insensitive to UV-B radiation (Lydon *et al.*, 1987). However, since many phytoalexins are flavonoid-related compounds, this research area may be worth exploring further using solar ultraviolet radiation.

6.4. Protection against Active Oxygen Metabolites

Many key components of the plant cell are susceptible to damage by oxygen species and free radicals, i.e., membrane lipids and unsaturated fatty acids, proteins, carbohydrates, and nucleic acids. The level of active oxygen metabolites may increase during stress conditions such as exposure to UV radiation, drought, high PAR, certain herbicides (paraquat and diquat; Halliwell, 1984a), and air pollutants. Some of the physiological events during senescence (Dhindsa *et al.*, 1981) and those seen upon irradiation with UV appear to be similar in some respects, and may involve oxidative damage. These include increased levels of membrane permeability, lipid peroxidation, and generation of free-radical species, all of which may be closely linked.

Singlet oxygen ($*O_2$) and the hydroxyl free radical ($OH^{\bullet}$) may be produced by $O_2^{\bullet-}$ and H_2O_2. Other sources of oxygen-derived free radicals include many enzymes, e.g., lipoxygenases, peroxidases, NADPH oxidase, and xanthine oxidase. Using ultraweak luminescence (UL), an indirect measure of the production of electronically excited states can be obtained, which is useful for estimating the oxidation of certain chemical reactions, from which UL derives (Abeles, 1986). An increase in UL occurs after exposure to many types of stress, for example, wounding, disease, temperature, and radiation, including UV-B radiation (Panagopoulos *et al.*, 1989, 1992; Levall and Bornman, 1993).

Protection against active oxygen metabolites can be achieved by several different systems, and thus their potential protective roles after exposure to enhanced UV radiation is of considerable interest. Besides superoxide dismutase, which is particularly prevalent in mitochondria (Rich and Bonner, 1978) and chloroplasts (Halliwell, 1984a) and which plays a key role as an antioxidant, other systems include: (1) alpha-tocopherol, or vitamin E, a lipid-soluble phenol derivative which is one of the main scavangers of membrane free radicals, and an efficient quencher of $^1O_2$ (Larson, 1988); (2) hydrogen peroxide, which is removed by both catalase and peroxidases; and (3) glutathione (GSH), a tripeptide containing a thiol group which may become oxidized (GSSG) as a result of environmental stress. This oxidation to the disulfide occurs as a result of its reaction with H_2O_2. GSH also removes oxidants such as $OH^{\cdot}$, $O_2^{-\cdot}$ and $^1O_2$ (Larson, 1988). At the molecular level GSH induces transcription of defense genes encoding for PAL and chalcone synthase (Wingate *et al.*, 1988), which makes GSH a possible stress marker involved in the flavonoid pathway.

Changes in the pattern of peroxidase activity, as visualized by electrophoresis, suggests that different isozymes may be induced by enhanced levels of UV-B radiation (Murali *et al.*, 1988). It was tentatively suggested that the presence of these isozymes as a response to UV-B radiation may have a role in free-radical scavenging. Ascorbate, β-carotene, uric acid, and several flavonoids also serve as antioxidants, with β-carotene being particularly effective in quenching $^1O_2$ and thus preventing reactions leading to peroxidation of membrane lipids (Larson, 1988). Flavonoids with several hydroxyl groups appear to have the highest antioxidant activity (see refs. in Larson, 1988). It also seems that flavonoids may inhibit $O_2^{-\cdot}$-promoted redox reactions in the chloroplast (Takahama, 1983), for example, in lipid photoperoxidation reactions, as well as quenching $^1O_2$ (Larson, 1988, and refs. therein). Photobleaching of carotenoids can also be prevented by kaempferol and quercetin (Takahama, 1982), which are both flavonoids.

Thus together with the UV-B-screening function of flavonoids, these compounds in their additional role as antioxidants are likely to contribute considerably in protecting plants from deleterious effects of UV radiation.

6.5. Structural Response

Many plants respond to UV-B radiation by an increase in leaf thickness (Murali and Teramura, 1986a; Murali *et al.*, 1988; Cen and Bornman, 1990; Bornman and Vogelmann, 1991). This response may serve as a direct protective measure with regard to penetration of UV radiation to lower-lying tissues. However, the outcome is a little more complex, since scattered light (400–700 nm, PAR) increases markedly in thicker leaves exposed to UV radiation. At the same time, parallel, or collimated, light is more reduced, especially in the spongy mesophyll of these leaves (Bornman and Vogelmann, 1991). For further discussion of light penetration, see Section 1. Reduced leaf area is a further change which contributes to protection against UV-B radiation. Plant adaptation can therefore arise from changes in molecular, biochemical, anatomical, and morphological characteristics.

6.6. Polyamines

With respect to stress, polyamines, simple aliphatic bases of intermediary pathways of nitrogen metabolism, generally serve a protective function in situations of nutrient deficiency, osmotic changes, and drought. Polyamines also increase during senescence (Flores, 1990, and refs. therein). In addition, polyamines, such as spermidine, spermine, and their diamine precursor, putrescine, play a role in reducing lipid peroxidation and are generally stimulated upon exposure of plants to enhanced UV-B radiation (Kramer *et al.,* 1991). This stimulation peaks and then declines. Thus polyamines also seem to contribute as protectants against some of the harmful effects of UV-B radiation.

In addition to levels of polyamines being dose dependent with respect to UV-B radiation, higher levels of PAR also increase polyamine concentrations (Kramer *et al.,* 1992), thereby affording added protection (UV-induced synthesis is also favored by high PAR). Lower polyamine levels may be a contributing factor to increased sensitivity, usually seen in experiments under low PAR and relatively high UV-B radiation.

Using *in vitro* and animal systems, it has been suggested that the mechanism underlying the decreases in lipid peroxidation by polyamines involves binding of these compounds to both phosphorus-containing compounds and the Fe^{3+} catalyst of Fe^{2+} oxidation (Kitada *et al.,* 1979; Tadolini, 1988). Since Fe^{2+} takes part in the Fenton reaction, generation of free-radical species is thus decreased.

7. Species and Cultivar Differences: Consequences for Plant Competition

Different species and cultivars often exhibit a wide range of sensitivity to enhanced UV-B radiation, which may result in both inter- and intraspecific competition being altered. In natural ecosystems, an increase in a population may be at the expense of another population, and this might occur either inter- or intraspecifically. Apparent "stimulation" of growth by UV-B radiation may actually reflect decreased growth of the competitor, although exceptions have been found (Fox and Caldwell, 1978). In agricultural systems interspecific competition could also be important between crops and weedy species. The competitive ability of a species depends on the sensitivity of the competitor and the strength of the stress. However, decreases in total biomass production are not commonly observed. Instead, morphological changes and shifts in biomass allocation may occur (Gold and Caldwell, 1983; Barnes *et al.,* 1988).

In general, changes in morphological characters may be enough to trigger shifts in competitive balance. This is particularly true where shoot morphology changes, since this will alter light interception. For example, in field experiments with enhanced UV-B radiation, where more than one species shares the same habitat, leaf insertion height and leaf length of one of the species may decrease or increase. In wheat this was found to be due to stimulation in height, resulting in shading of the other species (wild oat) (Barnes *et al.*, 1988), probably resulting in altered growth allocation patterns. Grasses are generally more responsive than broad-leaved plants with regard to changes in growth form under conditions of enhanced UV-B radiation. These morphological changes include increased branching, reduced leaf area, and increased leaf number (Barnes *et al.*, 1990). When the same two species were grown in a greenhouse, similar results were found, and, in addition, wheat responded with an increase in number of leaves (increased tillering) under elevated UV-B radiation.

Current levels of UV radiation may already affect interspecific competition among various native plant species, since solar UV exclusion studies in western Germany by Bruzek (cited by Gold and Caldwell, 1983) using two pairs of naturally competing species showed large differences in response to present levels of ambient solar UV radiation. This was characterized by a reduction in shoot biomass.

Other effects of enhanced UV-B radiation resulting in interspecific competition have been noted for field grown, competing pairs from (1) agricultural crops and associated weeds, (2) montane forage crops, and (3) disturbed weedy associates (Fox and Caldwell, 1978; Gold and Caldwell, 1983). The results are summarized in Table 1. To measure the competitive ability of one species when grown in a mixture with a second species, relative crowding coefficients (RCC) were determined. An RCC of 1.0 indicates that both species have a similar competitive ability. Species 1 has a competitive advantage when the RCC is greater than 1.0. As shown in Table 1, the RCCs based upon total above-ground biomass indicate that there was a significant shift in the competitive balance in four of the species pairs: *Medicago sativa/Amaranthus retroflexus; Triticum aestivum/Avena fatua; Triticum aestivum/Aegilops cylindrica; Geum macrophyllum/Poa pratensis.* However, the fact that in the previous year the competitive balance for *Avena* and *Triticum* shifted in favor of *Avena* suggests that the modification of competitive balance induced by UV-B radiation can be altered by other prevailing microclimate conditions. Also, the extent of dominance can vary with UV-B irradiance. In the *Setaria glauca/Trifolium pratense* pairing, *Setaria* was dominant both under ambient and enhanced UV-B radiation levels, but under enhanced levels the degree of dominance was expanded to a much greater extent. By contrast, in the *Bromus tectorum/Alyssum alyssoides* pairing, although *Bromus* was dominant under both ambient and enhanced UV-B radiation levels, this dominance was greatly reduced with enhanced UV-B radiation.

Thus, enhanced levels of UV-B radiation can alter the competitive interactions of some species pairs, with the competitive advantage of one species over the other depending upon the species pairing and the level of UV-B

Table 1. Relative Crowding Coefficients Based upon Shoot Biomass under Ambient and Enhanced Levels of UV-B Radiation[a]

Plant association	Competing species pair		Simulated ozone depletion[c] (%)	Relative crowding coefficient[b]	
	Species 1	Species 2		Ambient UV	Enhanced UV[d]
Agricultural crops and associated weed species	Alyssum alyssoides	Pisum sativum	40	0.34	0.25
	Amaranthus retroflexus	Medicato sativa	40	3.56	0.73
	Amaranthus retroflexus	Allium capa	40	1.89	2.01
	Setaria glauca	Trifolium pratense	40	2.06	18.74
	Triticum aestivum	Avena fatua	16	1.08	1.28
	Triticum aestivum	Avena fatua	40	1.08	1.69*
	Triticum aestivum	Aegilops cylindrica	16	0.48	1.57*
Montane forage species	Poa pratensis	Geum macrophyllum	40	0.85	2.28*
Disturbed area weedy associates	Bromus tectorum	Alyssum alyssoides	40	6.35	1.63
	Plantago patagonica	Lepidium perfoliatum	40	0.75	0.68

[a] Data from Gold and Caldwell (1983) and Fox and Caldwell (1978).
[b] Relative crowding coefficient of 0 means neither species has competitive advantage, more than 1 means species 1 has competitive advantage, less than 1 means species 2 has advantage.
[c] Simulated ozone depletion based on the generalized plant action spectrum (Caldwell, 1971), calculated at Logan, Utah (40°N).
[d] Asterisk denotes a significant difference (p less than .05) between control and enhanced-UV treatment.

irradiation. Since there are a large number of weeds typically associated with various crop plants, the impact of enhanced levels of UV-B radiation upon agro-ecosystems is quite complex, but could potentially have serious consequences if weeds generally have a competitive advantage over crop plants. Total harvestable yield, as well as its quality, can be altered by the presence of weeds (Bell and Nalewaja, 1968; McWhorter and Patterson, 1980), even in the absence of UV-B radiation.

With interspecific competition affected by enhanced levels of UV-B radiation, the relative species composition of many natural plant communities could possibly change as a consequence of increased levels of UV-B radiation. Consequently, because of the subtle nature of UV-B radiation stress, an enhancement of solar UV-B radiation may more likely alter the competitive balance of plants rather than directly affect ecosystem primary productivity (Caldwell, 1977; Gold and Caldwell, 1983; Fox and Caldwell, 1978).

Thus, it is to be expected that at the whole-plant level, changes in leaf morphology, surface reflectance, physiology, and leaf biochemistry will alter competitive balance and change canopy dynamics. At the same time, the response to UV-B radiation will be modified by developmental stage of the plant and by prior mechanisms of adaptation and changes in the microclimate with respect to moisture and temperature.

8. Crop Yield

One of the major concerns of increasing UV-B radiation is its potential impact on global food production. However, interpretation and intercomparison of results are sometimes difficult for experiments performed outdoors in field trials, which are subject to large, natural variations in temperature, precipitation, and the like, and therefore require well-designed and replicated experiments.

A reduction in yield under enhanced UV-B radiation has been reported for more than half of the crop species studied outdoors. At the same time, contradictory observations have been made, which are not always easy to explain. For example, in solar exclusion studies, yields in bean and corn were reduced by 11% and 28%, respectively. Tomato fruits matured earlier when ambient UV-B radiation was excluded (Bartholic et al., 1975). In another study, Becwar et al. (1982) found that exclusion of ambient levels of UV-B radiation had no affect on the growth of pea, potato, radish, or wheat, while ambient levels of UV-B radiation apparently were responsible for an 8–19% stunting in height of the wheat plants.

Ozone depletions ranging from 16% to 32% may cause large reductions in yield for some plants (e.g., squash, tomato, mustard and black-eyed pea;

Biggs and Kossuth, 1978; Biggs *et al.*, 1984). However, final seed yield and vegetative characteristics in six soybean cultivars under similar conditions showed no significant UV-B radiation effect (Sinclair *et al.*, 1990). Grain yield of corn does not usually show decreases due to enhanced UV-B radiation, although Eisenstark *et al.* (personal communication, January 1992) did report one instance of a significant reduction in grain yield over a 3-year period.

In a 6-year study, wheat grown in the field in competition with a weedy competitor, wild oat, with supplemental levels of UV-B radiation was more competitive than the wild oat (Barnes *et al.*, 1988).

The effects of a 16% and 25% ozone depletion (College Park, Maryland, 39°N latitude) on two soybean cultivars, Essex (sensitive to enhanced UV-B radiation) and Williams (tolerant), over six growing seasons, May through October, have been examined (Teramura *et al.*, 1990a). Seeds were sown directly into the soil at a density commonly used by growers in the region. Filtered Westinghouse FS-40 sunlamps were arranged perpendicularly to the planted rows to minimize the effects of the physical gradient of UV-B irradiance along the long axis of the tubes. In controls, lamps were filtered with Mylar Type S plastic films, which absorb all radiation below 320 nm. Several important conclusions were drawn from this study. First, under a full solar spectrum, there were clear UV-B radiation-induced reductions in seed yield in sensitive soybeans. Evaluated over the entire study period, yield in the Essex soybean was reduced by 19–25% in 4 of the 6 years when grown under a simulated 25% ozone depletion. Second, there were clear intraspecific (cultivar) differences in UV-B radiation sensitivity for soybean. Unlike Essex, the more UV-B radiation tolerant cv. Williams showed yield increases ranging from 4% to 22% when grown under a simulated 25% ozone depletion. Such intraspecific variability may be important to plant breeders, since it suggests that a degree of resistance to UV-B radiation is already present in the soybean germplasm. Finally, the study demonstrated the importance and need for multiyear field studies, since UV-B radiation effectiveness can be modified by other prevailing stresses. In the sensitive cv. Essex, seed yield was reduced only in years where water availability was high. In dry years, when plants experienced drought stress, such as in 1983 and 1984, the effects of UV-B radiation on yield were masked by reductions in growth due to drought (Murali and Teramura, 1986b).

In addition to quantitative modifications on crop yield, there is some evidence that yield quality might also be modified by UV-B radiation. For example, the number of abnormally shaped tomato fruit and the number of culled potato tubers due to rot, cracks, and other undesirable characteristics apparently decrease with enhanced levels of UV-B radiation (Biggs and Kossuth, 1978). Seed protein can also be affected by enhanced levels of UV-B radiation, as shown in the UV-B sensitive cv. Essex, where seed protein decreased by as much as 5%, although total seed lipid concentrations were reduced by only 1–2% (Teramura *et al.,* 1990a).

By using somaclonal variation and selection of plants *in vitro,* tolerance to UV radiation can potentially be improved (Levall and Bornman, 1993). Selection in this study, using *Beta vulgaris,* started at the callus stage, when high levels of UV radiation were applied, and surviving material was further subcultured. After shoot and root development, plants were subjected to supplementary UV-B radiation and several physiological parameters compared for clones selected under high-UV radiation with those cultivated without the selection pressure. Results suggested that increased levels of UV-B screening compounds and carotenoids may have contributed to increased tolerance in selected material. This kind of approach may be useful in other studies for creating tolerant clones or cultivars.

Despite the wide range of differences in experimental protocols and dosimetry used in these studies, there were still more instances of modifications to yield than reports of no effect at all. The uncertainties associated with this generalization are quite large, and presently it is not possible to make any quantitative predictions on how agricultural productivity will be affected by potential future increases in UV-B radiation reaching the surface of the earth.

9. Forests

Forests contain nearly 80% of total terrestrial plant biomass (Whittaker 1975). However, not much information is available on UV effects on forests. Tropical forests account for over one half of total plant productivity and by far the bulk of species diversity, but no study has yet included any tropical forest species. Long-lived perennial tree species provide a unique opportunity to observe the possible cumulative effects of protracted UV exposures, which are not possible to investigate in annual crop species. It is presently unknown whether natural protective or repair mechanisms may fully or partially restore physiological or biochemical damage caused by UV-B radiation during periods of low UV-B irradiation.

In UV-B exclusion studies where ambient levels of UV-B radiation were filtered out with Plexiglas, Bogenrieder and Klein (1982) found that growth in *Fraxinus excelsior, Carpinus betulus, Fagus sylvatica,* and *Acer platanoides* was significantly increased in terms of total dry matter accumulation compared with trees grown under a full solar spectrum including UV-B radiation. These observations suggested that present-day levels of UV-B radiation may already be deleterious to tree seedling growth and survival. In contrast, two high-elevation species of conifers (*Picea engelmannii* and *Pinus contorta*) grown in the field with exclusion of UV-B radiation or supplemental UV-B radiation from sunlamps did not show any visual symptoms or morphometric changes after one growing season (Kaufmann, 1978). In greenhouse studies Basiouny and Biggs (1975) reported visual stunting in 1-month-old peach seedlings, although Semeniuk (1978) observed no visual symptoms in seedlings of six ornamental tree species. In studies where 15 conifer species were subjected to a range of elevated levels of UV-B radiation in growth chambers and greenhouses (Kossuth and Biggs, 1981; Sullivan and Teramura, 1988), nearly half (47%) of the species tested showed reductions in total dry matter production. However, biomass increased in *Picea engelmannii* and *Abies concolor*, two high-elevation species, and was unaffected in *Abies fraseri, Pinus edulis, Pinus nigra*, and *Pseudotsuga menziesii*. In general, as in natural ecosystems, forest species native to high elevations, where natural levels of UV-B radiation are higher, appear to be more resistant to UV-B radiation than those from lower elevations. *Pinus taeda* displayed the largest UV-B responsiveness, with significant reductions in seedling height (16%) and total dry biomass (>25%) when grown in a UV-B environment simulating a 40% ozone depletion. *Pinus taeda* is found at elevations below 300 m and is the dominant forest species in the southeastern United States, occupying nearly 16% of the total area over its range (Gjerstad and Barber, 1987). It is the leading commercial species in this region, accounting for nearly 65% of U.S. pulp production.

In contrast to the use of newly germinated seedlings and relatively short exposure times of 2 to 3 months or less, a greenhouse study, where 1-year-old *P. taeda* seedlings were grown for 7 months under elevated levels of UV-B radiation (simulating 16%, 25%, and 40% ozone depletions at midlatitudes), showed that UV-B radiation significantly affected growth, photosynthesis, and foliar concentrations of UV-B-absorbing compounds (Sullivan and Teramura, 1989). It also appears that the effects of UV-B radiation may be cumulative throughout the growing season.

In a field validation study, *P. taeda* was grown for 3 years under supplemental UV-B irradiances, which simulated 16% and 25% stratospheric ozone reductions (Sullivan and Teramura, 1992). Seeds were obtained from seven locations in the United States ranging from 31° to 39°N latitude. The substantial differences in UV-B sensitivity among the seed sources tested suggested the presence of large genetic variability similar to that found in crops. Nonetheless, at the end of the 3-year irradiation period, plant biomass was reduced by 12–20% in all the seed sources tested, indicating that for field-grown trees the effects of UV-B radiation over an extended period are deleterious to growth and suggesting again that these effects may be cumulative. Furthermore, reductions in growth appear to result from an overall reduction in carbon accumulation, rather than merely an alteration in biomass partitioning.

10. Natural Ecosystems

The effects of UV-B radiation on wild plants and natural ecosystems indicate that species from high elevations, where natural levels of UV-B radiation are higher, generally tend to be more resistant than their counterparts from low elevations. Several studies illustrate this, although not all growth responses are clearly separated. For example, two species of columbines, one from an alpine environment (*Aquilegia caerulea*) and the other from lower elevations (*A. canadensis*), grown in a greenhouse under supplemental UV-B radiation [0 and 19 kJ m^{-2} day^{-1} biologically effective UV-B (UV-B$_{BE}$)], showed reduced growth and an increased number of leaves in response to UV-B radiation. However, the degree of growth reduction was greater in the low-elevation species, while leaf number increased faster in the high-elevation species. Flavonoid concentrations increased significantly in both species when exposed to UV-B radiation, but alkaloid concentrations declined in the high-elevation species (Larson *et al.*, 1990).

Further evidence of increasing tolerance of plants from high-elevation areas comes from a study of 33 plant species collected from their natural habitats in Hawaii and grown in greenhouse conditions under supplemental levels of UV-B radiation (0, 15.5, and 23.1 kJ m^{-2} day^{-1} UV-B$_{BE}$). These plants differed widely in their responsiveness to UV-B radiation. Height was significantly reduced in 42% of the species and biomass in 24%, although the overall trend of increasing tolerance with increasing elevation was evident (Sullivan *et al.*, 1992; Ziska *et al.*, 1992). Furthermore, these Hawaiian plants generally were more tolerant to UV-B radiation than crop plants, with biomass increasing in 12% of the plants when grown under relatively large UV-B enhancements. In terms of biomass alteration, endemic or native species were less responsive than introduced species. However, in terms of plant height, about half the species examined in each group were affected. Therefore, even though UV-B radiation had less of an effect on total biomass in endemics, morphological alterations were evident. These subtle morphological changes, even in the absence of biomass changes, could lead to shifts in competitive balance in mixed communities (Barnes *et al.*, 1990, 1988). The more direct UV-B-induced reductions in productivity could lead to changes in ecosystem biodiversity by virtue of the fact that more UV-resistant species would replace the UV-sensitive ones.

Differences in high- and low-elevation species are also seen from results of epidermal transmittance of UV-B radiation of plants collected along a latitudinal gradient from 9° to 71°N latitude (Robberecht and Caldwell, 1978; Robberecht *et al.*, 1980). In plants collected from equatorial and tropical regions, epidermal transmission averaged less than 2%, while at higher latitudes, transmittance was 5% or greater. Flavonoids and related pigments found in the epidermis attenuated 20–57% of the UV-B radiation incident on leaves, making them effective solar screens against the damaging effects of UV-B radiation.

Contrasting results have been found for the sensitivity of indigenous monocotyledonous and dicotyledonous plants. Two species of salt-marsh plants, a dicotyledon (*Aster tripolium*) and a monocotyledon (*Spartina anglica*), were grown in a greenhouse with supplemental UV-B radiation (Van de Staaij *et al.,* 1990). After 4 weeks of growth, total biomass and net photosynthesis were significantly reduced by UV-B treatment in the dicotyledon and unaffected in the monocotyledon. However, in another study where three monocotyledonous and three dicotyledonous weeds were grown in a greenhouse under supplemental levels of UV-B radiation, monocotyledons were morphologically more responsive to UV-B radiation than dicotyledons (Barnes *et al.,* 1990). It was suggested that changes in competitive balance resulting from increased UV-B radiation may be more frequent when monocotyledons are involved in mixtures, compared with dicotyledons.

The alteration and stimulation of specific biochemical pathways by UV-B radiation, such as the phenylpropanoid pathway, which is responsible for flavonoid and lignin production, can modify the chemical defense against herbivores and pathogens, rate and extent of tissue decomposition in the soil, and plant sensitivity to UV-B radiation (Caldwell *et al.,* 1989). These changes could influence the competitive balance among species, and species diversity, ultimately leading to alterations in community or ecosystem productivity.

11. Interaction of UV-B Radiation with Other Environmental Factors

Visible light, temperature, water supply, soil nutrient composition, toxic pollutants, infection, wounding, and the like, can all cause plant stress and may already limit plant productivity. It is therefore important to understand how an additional stress, namely, enhanced levels of UV-B radiation, may interact with these other environmental constraints. Depending on the specific combination, UV-B radiation may ameliorate or aggravate a stress, or alternatively, not interact at all.

11.1. PAR and UV-A Radiation

A number of studies (e.g., Teramura, 1980; Warner and Caldwell, 1983; Mirecki and Teramura, 1984; Cen and Bornman, 1990) have demonstrated that photosynthetically active radiation (PAR, 400–700 nm) has a direct influence on the manner in which a plant responds to UV-B radiation. For example, the response of plants to UV-B radiation during growth is a function of incident PAR (Teramura, 1980). Therefore, one generally observes reductions in total biomass, shifts in biomass allocation patterns, and reduction in plant height with decreasing PAR and increasing UV radiation. However, different plant species do not always respond in the same way. For example, dry matter decreased significantly in wheat when these plants were exposed to high UV-B irradiances and relatively high PAR, with little effect under low PAR and UV radiation (Teramura, 1980). Equally important is the composition or quality of PAR, in particular with respect to blue light, which together with UV-A radiation is a necessary component for photoreactivation. It has also been suggested that the accumulation of polyamines (Section 6.6), which are protective compounds induced during stress conditions, is sensitive to light quality, especially blue and red/far-red, implicating the involvement of blue/UV photoreceptors and phytochrome (Kramer *et al.*, 1992; Dai and Galston, 1981).

Interactions between UV-B radiation and PAR can also be seen with respect to physiological processes such as net photosynthesis, dark respiration, transpiration, and stomatal diffusive resistance, with UV-B radiation having a distinctly negative effect in combination with low levels of PAR (Teramura *et al.*, 1980; Mirecki and Teramura, 1984).

In natural environments exposure to low-light conditions, e.g., shade, can be important if plants have acclimatized under these conditions and are then suddenly exposed to both high PAR and UV-B radiation in sunflecks or canopy gaps. The lack of fully developed protective pigments and wax cover, and different photosynthetic pigment ratios together with the shade-type, structural arrangement of the photosynthetic apparatus, are likely to result in photoinhibition. Warner and Caldwell (1983) found that although high PAR tends to make plants less sensitive to UV-B radiation, this may be a somewhat indirect effect involving changes in morphology and physiology, such as thicker leaves and higher levels of flavonoids. At the same time their results indicated that high PAR together with UV-B radiation depresses photosynthesis. It seems that protective mechanisms of photoinhibition may also be influenced by concomitant UV-B radiation. One such mechanism involves carotenoids, and specifically the xanthophyll cycle, which contributes to photoprotection under visible light by dissipating excess excitation energy (Demmig-Adams, 1990). The cyclic epoxidation and de-epoxidation of the xanthophylls occur across the thylakoid membrane and include enzymes operating in the stroma and inner lumenal matrix. The violaxanthin de-epoxidase of the lumen is activated by light-dependent proton pumping. There is now evidence that this xanthophyll cycle may be a target of UV-B radiation, with the de-epoxidation of violaxanthin to zeaxanthin being inhibited (Pfündel *et al.*, 1992). This was shown for chloroplasts irradiated *in vitro* as well as for those isolated from UV-B-treated plants. Changes in the cycle can be followed chromatographically as well as spectrophotometrically, the latter by measuring changes in absorbance at 505 nm, which results from the enzymatic formation of zeaxanthin from violaxanthin. The redox state of PQ or an electron carrier near PQ seems to determine the availability of violaxanthin (Siefermann and Yamamoto, 1975), such that a more oxidized PQ pool results in a decrease in violaxanthin availability. Pfündel *et al.* (1992) found a drop in the apparent availability of violaxanthin after UV-B irradiation. Consequently, any decrease in the reduction potential of PQ by UV radiation would affect the xanthophyll cycle and its protective properties.

However, photoinhibition by visible light and UV inhibition may not always reflect damage via similar mechanisms. For example, both Q_A and the photoreduction of pheophytin are adversely affected by high levels of visible light (Demeter *et al.*, 1987), although photoreduction of pheophytin is not affected by high levels of UV-B radiation (Melis *et al.*, 1992).

11.2. Water Stress and UV-B Radiation

In the field and in nature, drought is very common, resulting in plant water stress. The effects of water deficits on plant growth and development have been well documented (for reviews see Schulze, 1986). Since diurnal and seasonal patterns of water stress in plants broadly overlap with periods of solar UV-B radiation maxima, research into the combined effects of these two environmental stresses is important. It has been shown that certain species are more sensitive to UV-B radiation regardless of the level of water availability. For example, cucumber (*Cucumis sativus*) was much more sensitive than radish (*Raphanus sativus*) when grown in a growth chamber under a UV-B level similating a 12% ozone depletion over midlatitudes (Teramura *et al.*, 1983; Tevini *et al.*, 1983). Water stress appeared to further diminish radish sensitivity to UV-B radiation by inducing the formation of UV-B-absorbing compounds in the leaves (Tevini *et al.*, 1983). Under well-watered conditions, UV-B radiation had little effect on stomatal conductance in either species, but under water stress, the pattern of stomatal conductance was substantially altered in cucumber, which also contributed to the susceptibility of this species to drought.

In other studies the effect of UV-B radiation on the photosynthetic recovery from water stress (Teramura *et al.*, 1984a) and on the various components of internal water relations (Teramura *et al.*, 1984b) in soybean (*Glycine max* cv. Essex) was investigated. Both studies were conducted in greenhouses using two levels of UV-B radiation (no UV-B radiation and an amount of

UV-B radiation simulating a 25% ozone depletion over midtemperate latitudes). The results of the first study clearly demonstrated an adverse, additive effect of the combination of UV-B radiation and water stress on photosynthetic recovery from water stress. This additive effect was not simply a short-term stomatal response. On a relative basis, photosynthesis in UV-B-irradiated plants recovered more quickly from water stress and to a greater extent. The mechanism for such UV-B radiation effects on photosynthetic recovery from water stress are presently unknown. However, there is evidence that UV-B radiation and water stress inhibit photosynthesis by affecting different processes. In a follow-up study, Teramura *et al.* (1984b) showed that UV-B radiation had no additional direct effects on leaf water potential or its components. It should be noted that despite an absence of any UV-B radiation effect on the water relations of soybean, this species generally lacks the ability to osmotically adjust. In active osmotic adjusters such as sorghum and cotton, UV-B radiation may still interfere with this physiological adjustment.

Under field conditions, the combination of stresses does not result in additive effects on total plant growth or seed yield as was observed in greenhouse studies (Teramura *et al.,* 1984a). In two field studies, Essex soybean was grown under two levels of UV-B radiation (ambient and an amount simulating a 25% ozone depletion over midlatitudes) and two levels of water availability (well-watered, with a soil water potential > -0.5 MPa, and water stressed, with a soil water potential of -2.0 MPa). Enhanced UV-B radiation reduced leaf area, total plant dry weight, and net photosynthesis under well-watered conditions, but no UV-B effects were detected in water-stressed plants (Murali and Teramura, 1986b). This elimination of UV-B effectiveness by water stress was associated with anatomical and biochemical changes in leaves exposed to water stress. Under water stress, plants are unaffected by UV-B radiation due to the masking influence of growth reduction, increases in UV-B-absorbing compounds in leaves, and anatomical changes resulting in leaf thickening. Furthermore, stomatal limitations to photosynthesis are only significantly affected by the combination of UV-B radiation and drought, as is reduction in apparent quantum efficiency (Sullivan and Teramura, 1990).

Thus, UV-B radiation may significantly affect plant productivity when water availability is high, while UV-B effects will be obscured or masked by drought, when productivity is already reduced. The UV-B effectiveness is also strongly influenced by temperature, precipitation patterns, and visible radiation. For example, in the hot, dry years of a 6-year field study, UV-B radiation had no effect on soybean yield, but in years of high water availability UV-B radiation significantly reduced soybean yield (Teramura *et al.,* 1990a).

11.3. Interactions with Carbon Dioxide

Increases in atmospheric CO_2, which have been monitored for the last three decades, appear to be primarily related to the burning of fossil fuels and to a lesser extent forest clearing. Due to these anthropogenic activities, it is anticipated that CO_2 concentrations will increase from 360 ppm at present and reach 600 ppm sometime within the next 30–75 years. Provided that nutrients are not limiting, it is likely that increases in CO_2 will result in substantial increases in the photosynthesis and growth of C_3 plants, which account for 95% of all known plant species. The combined effects of CO_2 and UV-B radiation on changes in plant productivity and photosynthesis have been separately determined for a large number of species. However, it is also worthwhile to determine the response of plants to the combination of these factors.

Wheat (*Triticum aestivum* L. cv. Bannock), rice (*Oryza sativa* L. cv. IR-36), and soybean (*Glycine max* L. cv. Essex) were grown in a greenhouse (PAR 80–85% of ambient levels) with two levels of UV-B radiation (ambient levels received at midtemperate latitudes and that equivalent to a 10% ozone depletion over the equator, 8.8 and 15.7 kJ m^{-2} day^{-1} UV-B$_{BE}$, respectively) and two levels of CO_2 (350 and 650 ppm; Teramura *et al.,* 1990b). Overall, the relative effects of UV-B radiation were greater under elevated levels of CO_2. Thus while there was no significant UV-B radiation effect on seed yield in wheat, rice, and soybean under ambient (350 ppm) CO_2 conditions, UV-B radiation resulted in a significant reduction in yield in wheat and rice under elevated (650 ppm) CO_2. Although not as pronounced, a similar pattern was found for net photosynthesis.

Seed production and total plant biomass significantly increased in the above-mentioned species as CO_2 was increased from 350 to 650 ppm. The greatest CO_2 response was found in wheat, and the least in rice, with soybean intermediate. However, when elevated levels of UV-B radiation were simultaneously given with increased CO_2, there no longer were significant increases in total biomass or yield for rice, nor was there an increase in yield for wheat. In contrast, CO_2-induced increases in yield and total plant biomass remained in soybean grown under the combination of UV-B radiation and CO_2.

In a similar study with two rice cultures, IR-36 and Fujiyama-5, CO_2 enhancement resulted in a significant increase in net photosynthesis, total biomass, and yield (Ziska and Teramura, 1992). This CO_2 effect was entirely eliminated in IR-36, but only partially reduced in Fujiyama-5, when elevated levels of UV-B radiation were applied simultaneously. It seems that under higher ambient CO_2 levels in the future, UV-B radiation may result in a relatively greater reduction in rice yields than under present levels of CO_2. Also the response of different cultivars will be important. For example, in the rice cultivar IR-36, UV-B radiation appeared to reduce the capacity for ribulose bisphosphate (RuBP) regeneration since the quantum yield F_v/F_m was significantly affected, while in Fujiyama-5, UV-B radiation did not affect F_v/F_m, but this cultivar did show a significant decline in carboxylation efficiency, thus reflecting a change in Rubisco capacity.

11.4. Temperature

Global climate is likely to be influenced in different ways by an increase in temperature, with the main contributing factors being the increased CO_2 concentration together with anthropogenic gases such as CFCs, N_2O, and CH_4, which prevent emission of thermal radiation to space (Dickinson and Cicerone, 1986). With an increase in UV-B radiation, one might expect further changes in the physiological capacity of plants. There are indications of both positive and negative interactive effects between UV-B radiation and high temperature. In a study conducted on four different plant species subjected to either 28° or 32°C together with enhanced UV-B radiation, it appeared that the higher temperature had an ameliorating effect on some of the species with respect to plant height, leaf area, and dry weight, although there was also differential species response (Teramura *et al.*, 1991).

A negative effect of higher temperatures may also occur. It has been shown that photolyase, the photoreactivating enzyme which uses light energy to repair cyclobutane pyrimidine dimers, is very temperature sensitive (Pang and Hays, 1991). This sensitivity occurs within a relatively narrow temperature range, from 22° to 30°C, which may have important consequences given the predicted rise in global temperatures (Schneider, 1989).

11.5. Interactions with Mineral Deficiency

Plants require some 16 essential elements for normal growth and development, each of which has various physiological roles. Inadequate supply of these essential elements impairs these physiological functions, resulting in poor growth and development. With regard to the effects of a combination of elevated levels of UV-B radiation and mineral deficiency, one might intuitively hypothesize that plants growing in nutrient-limiting situations would be even more susceptible to the damaging effects of UV-B radiation.

In short-term, greenhouse studies, UV-B radiation was more effective in inhibiting net photosynthesis in lettuce (*Lactuca sativa*) at high mineral concentrations, while the radiation effectiveness in native alpine sorrel (*Rumex alpinus*) was unaffected by mineral supply. In longer-term growth studies, the relative effectiveness of UV-B radiation in terms of limiting biomass production diminished as nutrient concentrations increased in both species (Bogenrieder and Doute, 1982). The combination of UV-B radiation and phosphorus deficiency has also been investigated. Response of greenhouse-grown soybean (*Glycine max* L. cv. Essex, a UV-sensitive cultivar) to different levels of phosphorus (P) supply (6.5, 13, 26, and 52 μM) and supplemental UV-B radiation (11.5 kJ m^{-2} day^{-1} UV-B$_{BE}$, equivalent to a 16% ozone depletion over midlatitudes during the summer maximum) showed that plants experiencing P deficiency were less sensitive to UV-B radiation than plants at optimum P levels (Murali and Teramura, 1985a,b, 1987). Although both P supply and UV-B radiation significantly reduced plant biomass, these effects were non-additive (Murali and Teramura, 1985b). UV-B radiation had little effect on total biomass, leaf area (Murali and Teramura, 1985a,b) or photosynthesis (Murali and Teramura, 1987) at the lowest P level, while a large effect was observed with UV-B radiation at the highest P level. The combination of enhanced UV-B radiation and low P was associated with maximum flavonoid accumulation and increases in leaf thickness, which may have ameliorated the effectiveness of the UV-B radiation. Additionally, the general growth reduction by low P may have helped mask the deleterious effects of UV-B radiation, since both low P and UV-B radiation individually resulted in a similar magnitude of growth and photosynthetic reduction.

Therefore, in short-term studies plants grown at high mineral availability may be more sensitive to UV-B radiation, a trend seen also for water stress (see Section 11.2). However, there are indications that in longer-term studies the opposite may occur; i.e., with increased mineral supply UV radiation becomes less effective. Further investigations are required to establish this more definitively.

11.6. Interactions with Diseases or Insects

Ultraviolet-B radiation may alter secondary plant chemicals, and this in turn may affect plant sensitivity to disease or susceptibility to insects. Esser (1980) reported that the number of aphids per bean plant was significantly decreased after 11 days of UV-B radiation, while in an experiment where spider mites were applied to the plants, there was no significant difference in their population after 7 days of irradiation. Although the results suggest that UV-B radiation can have potentially beneficial effects on pest control, these conclusions must be viewed cautiously, since the observation period was quite short (less than 2 weeks). Whether there would be any longer-term differences in pest attack is still unknown. Field studies detailing crop growth under enhanced UV-B radiation do not report any difference in natural pest attack among UV-B radiation treatments (e.g. Esser, 1980; Teramura et al., 1990a), although further work is needed in this area. One of the plant defense mechanisms that inhibits fungal development is the production of phytoalexins, although excess production can be toxic due to free-radical formation (Beggs et al., 1986).

The severity of some crop diseases has been studied under laboratory, greenhouse, and field conditions. In vitro spore germination of six fungal pathogens showed that hyaline spores are more sensitive to UV-B radiation than are pigmented spores (Carns et al., 1978). Tests conducted on plants grown in growth chambers also showed a similar relationship. However, this does not hold for *Cladosporium cucumerinum,* a pigmented spore, whose survival was decreased with UV-B radiation (Owens and Krizek, 1980). This was due more to a delay in germ tube emergence than complete inhibition of growth. Other studies also indicate that there is no clear relationship between spore coloration and UV-B radiation effectiveness (Esser, 1980; Biggs *et al.,* 1984).

The effects of UV-B radiation on plant diseases vary with pathogen, plant species, cultivar, and age of the plants. Results from greenhouse studies can also differ from those in the field (Biggs *et al.,* 1984; Biggs, 1985). When leaf rust-resistant and rust-sensitive cultivars of wheat were tested, the sensitive cultivar showed no differences up to 60 days after planting, while at 119 days after planting, disease severity increased with UV-B radiation. In the resistant cultivar, there were no differences in disease severity.

Using various filters in the field (leaves, cheesecloth, and UV filter), Rotem *et al.* (1985) found that natural solar UV-B radiation increased spore mortality by 6- to 30-fold in *Peronospora tabacina, Uromyces phaseoli* and *Alternaria solani.*

Semeniuk and Goth (1980) found that UV-B radiation significantly reduced potato virus infection on *Chenopodium quinoa,* and at high irradiances of UV-B no infection occurred. In this study, virus extract was exposed to UV-B radiation immediately upon its application over the leaf surface. Intuitively, viruses should be highly susceptible to UV-B radiation, since they only contain nucleic acids covered with proteins, both of which have high UV-absorption properties. Furthermore, viroids, which are devoid of a protein coat, may be more susceptible despite their small size. These effects would probably be greatest when the viruses or viroids are directly exposed to UV-B radiation, as may happen during mechanical transmission. Viruses or viroids transmitted through seed, pollen, insects, mites, nematodes, and fungi may not be as susceptible to UV-B radiation, due to the additional cellular screening offered by the host tissues.

In plant and fungal interaction, disease severity may increase when the plants are irradiated with enhanced levels of UV-B radiation. For example, several cucumber (*Cucumis sativus*) varieties were irradiated with UV-B radiation in a greenhouse study (11.6 kJ m^{-2} day^{-1} UV-B$_{BE}$) prior to and after infection with two fungal diseases (*Colletotrichum laganarium* and *Cladosporium cucumerinum*) (Orth *et al.,* 1990). Two of the cultivars were disease resistant and the third was susceptible. In the susceptible cultivar and one of the disease-resistant ones, exposure to UV-B radiation prior to infection led to increased disease severity. Exposure after infection with disease had no effect on disease development, suggesting that UV-B radiation may modify the leaf cuticle or epidermis in a manner which facilitates fungal penetration into the plant.

There seems to be more than one type of effect of UV-B on the interaction between plant and pathogen, since the response varies with species and time of application of a virus or fungus. While certain diseases may be alleviated with enhanced levels of UV-B radiation, others may sustain a deleterious additive effect; for example, in a study on sugar beet (*Beta vulgaris*) irradiated with UV-B radiation (6.9 kJ m^{-2} day^{-1} UV-B$_{BE}$) after infection with *Cercospora beticola* (Panagapoulos *et al.*, 1992), plants which were infected and exposed to UV-B radiation showed much larger reductions in biomass accumulation in leaves, petioles, and storage roots than those receiving UV-B alone or infection alone. Lipid peroxidation was also much higher in leaves exposed to combined infection and UV-B radiation than either separately, suggesting an increase in free radicals. This was further supported by high levels of ultraweak luminescence (see Section 6.4).

11.7. Influence of Metals

High concentrations of metals such as lead, nickel, and cadmium arise mostly from human activities, and constitute yet another stress to which plants in certain habitats can be subjected. It appears that enhanced cadmium levels may substantially affect plants exposed even to small amounts of UV-B radiation (Dubé and Bornman, 1992). In *Picea abies* this combination resulted in reduced rates of CO_2 assimilation, and indications of altered energy distribution between PS I and II, as well as decreases in total chlorophyll because of a decreased chlorophyll *b* content. By contrast, reductions in dry weight were caused by cadmium alone. An increase in UV-B-screening pigments, including flavonoids, with the combined exposure to UV and cadmium also suggests that the plants were under stress, despite a low UV-B supplement (6.2 kJ m^{-2} day^{-1}).

UV-B radiation also seems to exert an influence on the role of the necessary trace metal zinc. The response of UV-irradiated cotton plants is characterized by reduced leaf expansion, red pigmentation in the petioles, and decreased transport of ^{65}Zn from the cotyledons to the shoot. Thus, by reducing zinc translocation, UV-B radiation appeared to have affected mobilization of essential nutrients to developing leaves (Ambler *et al.*, 1975).

A more positive effect of solar UV radiation is the reduction of Fe^{3+} to Fe^{2+}, the form in which it is assimilated by plants. It has been suggested that under conditions of iron deficiency, solar UV may induce the reduction of iron (Pushnik *et al.*, 1987).

12. Prospects for the Future: Research Priorities

After reports in 1985 of the unexpected seasonal occurrence of the Antarctic ozone hole, followed by the detection of ozone depletion in the Arctic, models of future scenarios should be viewed more critically. It seems that we will experience a decreasing stratospheric ozone layer and increasing ambient UV-B radiation reaching the earth for at least several decades to come, given the altered chemistry of the upper atmosphere due to human activities. Since decreases in stratospheric ozone are not uniform in time and distribution over the earth, localized damage to biological systems is likely to have a ripple effect, upsetting the ecological balance of natural ecosystems. Global, long-term effects on living organisms may be more difficult to characterize in the immediate future.

In order to understand the significance of changes due to increased UV-B radiation, future research on plant effects should focus on three broad areas:

1. Basic research including physical, physiological, biochemical, and molecular mechanisms conducted in the laboratory, greenhouse, and growth chamber.
2. Field validation of some or all of the aspects covered in (1) in order to test the significance of these responses under a full solar spectrum and under natural environmental conditions.
3. Ecophysiological studies using natural ecosystems.

In the first category, more action spectra are needed for characterizing different plant responses. At the level of the gene, it is important to understand the consequences and adaptive mechanisms which are set in motion by different types of elicitors. The elicitor is not only UV radiation, but also other environmental factors which may interact in a complex way with UV-B radiation. In the second category, while not all laboratory studies lend themselves to application in the field, it is still desirable to carry out field tests where possible. One of the strengths of more basic research is that it points to the direction in which investigations may be carried out even on field plants. In addition, while conditions may be very artificial in the laboratory, these results may still serve to help elucidate many mechanisms of action as a result of UV radiation. The ability of certain crop cultivars to show heritable traits after UV selection pressure may help to develop tolerance to UV radiation. However, any subsequent change in the chemistry of leaf, stem, or root may have different consequences for herbivores and decomposers, and thus affect natural nutrient cycles. The third category is of obvious importance, although conducting ecophysiological studies in natural environments may pose many difficulties. More information on elevational and latitudinal differences of natural or nonagricultural species is needed. In this regard, low-latitude tropical areas have to date received little attention. Long-lived tree species, with the ability to show accumulative effects of environmental change should be examined. Mechanisms of both repair and protection may be different along elevational and latitudinal gradients. Also, the role of plant growth regulators in plant development as a result of environmental stress may be complicated by both direct and indirect effects of UV-B radiation on these regulators themselves. Finally, the importance of present, ambient levels of UV-B radiation with regard to plant productivity warrants more study both in field validation studies and in natural ecosystems.

References

Abeles, F. B., 1986, Plant chemiluminescence, *Annu. Rev. Plant Physiol.* 37:49–72.

Ambler, J. E., Krizek, D. T., and Semeniuk, P., 1975, Influence of UV-B radiation on early seedling growth and translocation of ^{65}Zn from cotyledons in cotton, *Physiol. Plant.* 34:177–181.

Barber, J., 1987, Rethinking the structure of the photosystem two reaction centre, *Trends Biochem. Sci.* 12:123–124.

Barnes, P. W., Jordan, P. W., Gold, W. G., Flint, S. D., and Caldwell, M. M., 1988, Competition, morphology and canopy structure in wheat (*Triticum aestivum* L.) and wild oat (*Avena fatua* L.) exposed to enhanced ultraviolet-B radiation, *Funct. Ecol.* 2:319–220.

Barnes, P. W., Flint, S. D., and Caldwell, M. M., 1990, Morphological responses to crop and weed species of different growth forms to ultraviolet-B radiation, *Am. J. Bot.* 77:1354–1360.

Bartholic, J. F., Halsey, L. H., and Garrard, L. A., 1975, Field trials with filters to test for effects of UV radiation on agricultural productivity, in: *Climatic Impact Assessment Program* (CIAP), Monograph 5 (D. S. Nachtwey, M. M. Caldwell, and R. H. Biggs, eds.), U.S. Dept of Transportation, Report no. DOT-TST-76-55. Natl. Tech. Info. Serv., Springfield, VA, pp. 61–71.

Basiouny, F. M., and Biggs, R. H., 1975, Photosynthetic and carbonic anhydrase activities in Zn-deficient peach seedlings exposed to UV-B radiation, in: *Impacts of Climate Change on the Biosphere* (CIAP) Monograph 5, Part 1 (Appendix B), Dept of Transportation, Washington, DC, DOT-TST-75-55.

Basiouny, F. M., Van, T. K., and Biggs, R. H., 1978, Some morphological and biochemical characteristics of C_3 and C_4 plants irradiated with UV-B, *Physiol. Plant.* 42:29–32.

Becwar, M. R., Moore, F. D., and Burke, M. J., 1982, Effects of depletion and enhancement of ultraviolet-B (280–315 nm) radiation on plants grown at 3,000-m elevation, *J. Am. Soc. Hortic. Sci.* 107:771–774.

Beggs, C. J., Schneider-Ziebert, U., and Wellmann, E., 1986, UV-B radiation and adaptive mechanisms in plants, in: *Stratospheric Ozone Reductions, Solar Ultraviolet Radiation and Plant Life* (R. C. Worrest and M. M. Caldwell, eds.), NATO ASI Series G. 8:235–250. Springer-Verlag, Berlin.

Bell, A. R., and Nalewaja, J. D., 1968, Competition of wild oat in wheat and barley, *Weed Sci.* 16:505–508.

Biggs, R. H., 1985, Effects of enhanced ultraviolet-B radiation (280–320 nm) on soybean, wheat, corn, rice, citrus, and duckweed, *EPA Progress Report, CR-811216,* Washington, DC.

Biggs, R. H., and Kossuth, S. V., 1978, Effects of ultraviolet-B radiation enhancements under field conditions on potatoes, tomatoes, corn, rice, southern peas, peanuts, squash, mustard and radish, in: *UV-B Biological and Climatic Effects Research* (BACER), Final Report, EPA, Washington, DC.

Biggs, R. H., Webb, P. G., Garrard, L. A., Sinclair, T. R., and West, S. H., 1984, The effects of enhanced ultraviolet-B radiation on rice, wheat, corn, soybean, citrus and duckweed, Year 3 interim Report, *Environ. Protection Agency Report. 808075-03,* EPA, Washington, DC.

Bogenrieder, A., and Doute, Y., 1982, The effect of UV on photosynthesis and growth in dependence of mineral nutrition (*Lactuca sativa* L. and *Rumex alpinus* L.), I. Biological Effects of UV-B Radiation, in: *Biological Effects of UV-B Radiation* (H. Bauer, M. M. Caldwell, M. Tevini, and R. C. Worrest, eds.), Gesellschaft für Strahlen- und Umweltforschung mbH, München, Germany, pp. 164–168.

Bogenrieder, A., and Klein, R., 1982, Does solar UV influence the competitive relationship in higher plants? in: *The Role of Solar Ultraviolet Radiation in Marine Ecosystems* (J. Calkins, ed.), Plenum, New York, pp. 641–649.

Bornman, J. F., 1989, Target sites of UV-B radiation in photosynthesis of higher plants, *J. Photochem. Photobiol.* 4:145–158.

Bornman, J. F., and Vogelmann, T. C., 1988, Penetration of blue and UV radiation measured by fiber optics in spruce and fir needles, *Physiol. Plant.* 72:699–705.

Bornman, J. F., and Vogelmann, T. C., 1991, The effect of UV-B radiation on leaf optical properties measured with fibre optics, *J. Exp. Bot.* 42:547–554.

Bornman, J. F., Evert, R. F., and Mierzwa, R. J., 1983, The effect of UV-B and UV-C radiation on sugar beet leaves, *Protoplasma* 117:7–16.

Brandle, J. R., Campbell, W. F., Sisson, W. B., and Caldwell, M. M., 1977, Net photosynthesis, electron transport capacity, and ultrastructure of *Pisum sativum* L. exposed to ultraviolet-B radiation, *Plant Physiol.* 60:165–169.

Bridge, M. A., and Klarman, W. L., 1973, Soybean phytoalexin, hydroxyphaseollin, induced by ultraviolet irradiation, *Phytopathology* 63:606–609.

Bruns, B., Hahlbrock, K., and Schäfer, E., 1986, Fluence dependence of the ultraviolet-light-induced accumulation of chalcone synthase mRNA and effects of blue and far-red light in cultured parsley cells, *Planta* 169:393–398.

Caldwell, M. M., 1968, Solar ultraviolet radiation as an ecological factor for alpine plants, *Ecol. Monogr.* 38:243–268.

Caldwell, M. M., 1971, Solar irradiation and the growth and development of higher plants: In *Photophysiology* (A. C. Giese, ed.), Academic Press, New York, Vol. 6, pp. 131–171.

Caldwell, M. M., 1977, The effects of solar UV-B (280–315 nm) on higher plants: Implications of stratospheric ozone reduction, in: *Research in Photobiology* (A. Castellani, ed.), Plenum, New York, pp. 597–607.

Caldwell, M. M., and Robberecht, R., 1978, Leaf epidermal transmittance of ultraviolet radiation and its implications for plant sensitivity to ultraviolet-radiation induced injury, *Oecologia* 32:277–287.

Caldwell, M. M., Robberecht, R., Holman, S., Nowak, R., Camp, L. B., Flint, S. D., Harris, G., and Teramura, A. H., 1979, Higher plant responses to elevated ultraviolet irradiance, *Annual Report 1978 NAS-9-14871,* NASA.

Caldwell, M. M., Teramura, A. H., and Tevini, M., 1989, The changing solar ultraviolet climate and the ecological consequences for higher plants, *Trends Ecol. Evol.* 4:363–367.

Campbell, W. F., Caldwell, M. M., and Sisson, W. B., 1975, Effect of UV-B radiation on pollen germination, in: *Impacts of Climatic Change on the Biosphere,* CIAP Monograph 5, U.S. Dept of Transportation, Washington, DC, 4-227 to 4-276.

Carlson, G. J., 1976a, Mitotic effects of monochromatic ultraviolet radiation at 225, 265, and 280 nm on eleven stages of the cell cycle of the grasshopper neuroblast in culture. 1. Overall retardation from the stage irradiated to nuclear membrane breakdown, *Rad. Res.* 68:57–74.

Carlson, G. J., 1976b, Mitotic effects of monochromatic ultraviolet radiation at 225, 265 and 280 nm on eleven stages of the grasshopper neuroblast in culture. 2. Changes in progression rate and cell sequence between the stage irradiated and nuclear membrane breakdown, *Rad. Res.* 68:75–83.

Carns, H. R., Grahm, J. H., and Ravitz, S. J., 1978, Effects of UV-B radiation on selected leaf pathogenic fungi and on disease severity, *EPA-IAG-D6-0168 (BACER Program) Environ. Protection Agency,* Washington, DC.

Cen, Y.-P., and Bornman, J. F., 1990, The response of bean plants to UV-B radiation under different irradiances of background visible light, *J. Exp. Bot.* 41:1489–1495.

Cen, Y.-P., and Bornman, J. F., 1993, The effect of exposure to enhanced UV-B radiation on the penetration of monochromatic and polychromatic UV-B radiation in leaves of *Brassica rapus, Physiol. Plant.* 87:249–255.

Chamberlain, J., and Moss, S. H., 1987, Lipid peroxidation and other membrane damage produced in *Escherichia coli* K1060 by near-UV radiation and deuterium oxide, *Photochem. Photobiol.* 45:625–630.

Chappell, J., and Hahlbrock, K., 1984, Transcription of plant defence genes in response to UV light or fungal elicitor, *Nature* 311:76–78.

Clark, J. B., and Lister, G. R., 1975, Photosynthetic action spectra of trees. II. The relationship of cuticle structure to the visible and ultraviolet spectral properties of needles from four coniferous species. *Plant Physiol.* 55:407–413.

Dai, Y.-R., and Galston, A. W., 1981, Simultaneous phytochrome controlled promotion and inhibition of arginine decarboxylase EC 4.1.1.19 activity in buds and epicotyls of etiolated peas (*Pisum sativum*) cultivar Alaska, *Plant Physiol.* 67:266–269.

Darwin, C., 1880, *The Power of Movement in Plants,* William Clowes and Sons, London.

Day, T. A., Vogelmann, T. C., and DeLucia, E. H., 1992, Are some plant life forms more effective than others in screening out ultraviolet-B radiation? *Oecologia* 92:513–519.

Demeter, S., Neale, P. J., and Melis, A., 1987, Photoinhibition impairment of the primary charge separation between P-680 and pheophytin in photosystem II of chloroplasts, *FEBS Lett.* 214:370–374.

Demmig-Adams, B., 1990, Carotenoids and photoprotection in plants: a role for the xanthophyll zeaxanthin, *Biochim. Biophys. Acta* 1020:1–24.

Dhindsa, R. S., Plumb-Dhindsa, P., and Thorpe, T. A., 1981, Leaf senescence: Correlated with increased levels of membrane permeability and lipid peroxidation, and decreased levels of superoxide dismutase and catalase, *J. Exp. Bot.* 32:93–101.

Dickinson, R. E., and Cicerone, R. J., 1986, Future global warming from atmospheric trace gases, *Nature* 319:109–115.

Dixon, R. A., Dey, P. M., and Lamb, C. J., 1983, Phytoalexins: Enzymology and molecular biology: In *Advances in Enzymology* (A. Meister, ed.), John Wiley, New York, pp. 1–136.

Douglas, C., Hoffmann, H., Schulz, W., and Hahlbrock, K., 1987, Structure and elicitor or UV-light-stimulated expression of two 4 coumarate coenzyme ligase genes in parsley, *EMBO* 6: 1189–1196.

Drumm-Herrel, H., and Mohr, H., 1981, A novel effect of UV-B in a higher plant (*Sorghum vulgare*), *Photochem. Photobiol.* 33:391–398.

Dubé, S. L., and Bornman, J. F., 1992, Response of spruce seedlings to simultaneous exposure to ultraviolet-B radiation and cadmium, *Plant Physiol. Biochem.* 30:761–767.

Duell-Pfaff, N., and Wellmann, E., 1982, Involvement of phytochrome and a blue light photo-receptor in UV-B induced flavonoid synthesis in parsley (*Petroselinum hortense* Hoffm.) cell suspension cultures, *Planta* 156:213–217.

Ehleringer, J. R., and Forseth, I. N., 1980, Solar tracking by plants, *Science* 210:1094–1098.

Ensminger, P. A., and Schäfer, E., 1992, Blue and ultraviolet-B light photoreceptors in parsley cells, *Photochem. Photobiol.* 55:437–447.

Esser, G., 1980, Einfluss einer nach Schadstoffimmission vermehrten Einstrahlung von UV-B-Licht auf Kulturpflanzen, 2 Versuchjahr, *Bericht Battelle Institut E.V. Frankfurt, BF-R-63*, 984-1.

Flint, S. D., and Caldwell, M. M., 1983, Influence of floral optical properties on the ultraviolet radiation environment of pollen, *Am. J. Bot.* 70:1416–1419.

Flint, S. D., and Caldwell, M. M., 1984, Partial inhibition of in vitro pollen germination by simulated solar ultraviolet-B radiation, *Ecology* 65:792–795.

Flint, S. D., and Caldwell, M. M., 1986, Comparative sensitivity of binucleate and trinucleate pollen to ultraviolet radiation. A theoretical perspective, in *Stratospheric Ozone Reductions, Solar Ultraviolet Radiation and Plant Life* (R. C. Worrest, and M. M. Caldwell, eds.), NATO ASI Series G, Springer-Verlag, Berlin, Vol. 8, pp. 211–222.

Flint, S. D., Jordan, P. W., and Caldwell, M. M., 1985, Plant protective response to enhanced UV-B radiation under field conditions: Leaf optical properties and photosynthesis, *Photochem. Photobiol.* 41:95–99.

Flores, H. E., 1990, Polyamines and plant stress, in: *Stress Responses in Plants: Adaptation and Acclimation Mechanisms* (R. G. Alscher and J. R. Cumming, eds.), Vol. 12, Wiley-Liss, New York, pp. 217–239.

Forseth, I. N., 1990, Function of leaf movements, in: *The Pulvinus: Motor Organ for Leaf Movement* (R. L. Satter, H. L. Gorton, and T. C. Vogelmann, eds.), American Society of Plant Physiologists, Rockville, MD, pp. 238–261.

Fox, F. M., and Caldwell, M., 1978, Competitive interaction in plant populations exposed to supplementary ultraviolet-B radiation, *Oecologia* 36:173–190.

Gausman, H. W., Rodriguez, R. R., and Escobar, D. R., 1975, Ultraviolet radiation reflectance, transmittance, and absorptance by plant leaf epidermises, *Agron. J.* 67:720–724.

Gjerstad, D. H., and Barber, B. L., 1987, Forest vegetation problems in the south, in: *Forest Vegetation Management for Conifer Production* (J. D. Walstad, and P. J. Kuch, eds.), Wiley, New York, pp. 55–75.

Gold, W. G., and Caldwell, M. M., 1983, The effects of ultraviolet-B radiation on plant competition in terrestrial ecosystems, *Physiol. Plant.* 58:435–444.

Greenberg, B. M., Gaba, V., Canaani, O., Malkin, S., Mattoo, A. K., and Edelman, M., 1989a, Separate photosensitizers mediate degradation of the 32-kDa photosystem II reaction center protein in the visible and UV spectral regions, *Proc. Nat. Acad. Sci. USA* 86:6617–6620.

Greenberg, B. M., Gaba, V., Mattoo, A. K., and Edelman, M., 1989b, Degradation of the 32 kDa photosystem II reaction center protein in UV, visible and far red light occurs through a common 23.5 kDa intermediate, *Z. Naturforsch.* 44:450–452.

Hahlbrock, K., and Grisebach, H., 1979, Enzymic controls in the biosynthesis of lignin and flavonoids, *Annu. Rev. Plant Physiol.* 30:105–130.

Hahlbrock, K., and Scheel, D., 1989, Physiology and molecular biology of phenylpropanoid metabolism, *Annu. Rev. Plant Physiol. Plant Molec. Biol.* 40:347–369.

Halliwell, B., 1984a, Oxygen-derived species and herbicide action, *Physiol. Plant.* 15:21–24.

Hruszkewycz, A. M., 1988, Evidence for mitochondrial DNA damage by lipid peroxidation, *Biochem. Biophys. Res. Commun.* 153:191–197.

Hugly, S., Kunst, L., Browse, J., and Somerville, C., 1989, Enhanced thermal tolerance of photosynthesis and altered chloroplast ultrastructure in a mutant of *Arabidopsis* deficient in lipid desaturation, *Plant Physiol.* 90:1134–1142.

Hull, H. M., Morton, H. L., and Wharrie, J. R., 1975, Environmental influences on cuticle development and resultant foliar penetration, *Bot. Rev.* 41:421–452.

Imlay, J., and Linn, S., 1988, DNA damage and oxygen radical toxicity, *Science* 240:1302–1309.

Iwanzik, W., Tevini, M., Dohnt, G., Voss, M., Weiss, W., Gräber, O., and Renger, G., 1983, Action of UV-B radiation on photosynthetic primary reactions in spinach chloroplasts, *Physiol. Plant.* 58:401–407.

Jahnen, W., and Hahlbrock, K., 1988, Differential regulation and tissue-specific distribution of enzymes of phenylpropanoid pathways in developing parsley seedlings, *Planta* 173:453–458.

Jordan, B. R., Chow, W. S., Strid, Å., and Anderson, J. M., 1991, Reduction in *cab* and *psb* A RNA transcripts in response to supplementary ultraviolet-B radiation, *FEBS Lett.* 284:5–8.

Jordan, B. R., He, J., Chow, W. S., and Anderson, J. M., 1992, Changes in mRNA levels and polypeptide subunits of ribulose 1,5-bisphosphate carboxylase in response to supplementary ultraviolet-B radiation, *Plant Cell Environ.* 15:91–98.

Kasperbauer, L. W., and Loomis, W. E., 1965, Inhibition of flowering by natural daylight on an inbred strain of *Melilotus, Crop Sci.* 5:193–194.

Kaufman, M. R., 1978, The effect of ultraviolet (UV-B) radiation on Engelmann spruce and lodgepole pine seedlings, in: *E.PA-IAG-D6-0168,* BACER Program, Environ. Protection Agency, Washington, DC.

Kitada, M., Igarashi, K., Hirose, S., and Kitagawa, H., 1979, Inhibition by polyamines of lipid peroxide formation in rat liver microsomes, *Biochem. Biophys. Res. Commun.* 87:388–394.

Kossuth, S. V., and Biggs, R. H., 1981, Ultraviolet-B radiation effects on early seedling growth of Pinaceae species, *Can. J. For. Res.* 11:243–248.

Kramer, G. F., Norman, H. A., Krizek, D. T., and Mirecki, R. M., 1991, Influence of UV-B radiation on polyamines, lipid peroxidation and membrane lipids in cucumber, *Phytochemistry* 30:2101–2108.

Kramer, G. F., Krizek, D. T., and Mirecki, R. M., 1992, Influence of photosynthetically active radiation and spectral quality on UV-B-induced polyamine accumulation in soybean, *Phytochemistry* 31:1119–1125.

Kuć, J., 1984, Phytoalexins and disease resistance mechanisms from a perspective of evolution and adaptation, in: *Origins and Development of Adaptation* Pitman Books, London, pp. 100–118.

Kulandaivelu, G., and Noorudeen, A. M., 1983, Comparative study of the action of ultraviolet-C and ultraviolet-B radiation on photosynthetic electron transport, *Physiol. Plant.* 58:389–394.

Larson, R. A., 1988, The antioxidants of higher plants, *Phytochemistry* 27:969–978.

Larson, R. A., Garrison, W. J., and Carlson, R. W., 1990, Differential responses of alpine and non-alpine *Aquilegia* species to increased ultraviolet-B radiation, *Plant Cell Environ.* 13:983–987.

Levall, M. W., and Bornman, J. F., 1993, Selection *in vitro* for UV-tolerant sugar beet (*Beta vulgaris*) somaclones, *Physiol. Plant.* 83:37–43.

Lichtscheidl-Schultz, I., 1985, Effects of UV-C and UV-B on cytomorphology and water permeability of inner epidermal cells of *Allium cepa, Physiol. Plant.* 63:269–276.

Lydon, J., Teramura, A. H., and Coffman, C. B., 1987, UV-B radiation effects on photosynthesis, growth and cannabinoid production of two *Cannibis sativa* chemotypes, *Photochem. Photobiol.* 46:201–206.

Mantai, K. E., and Bishop, N. I., 1967, Studies on the effects of ultraviolet irradiation on photosynthesis and on the 520 nm light-dark difference spectra in green algae and isolated chloroplasts, *Biochim. Biophys. Acta* 131:350–356.

McClure, J. W., 1976, Secondary metabolism and coevolution, *Nova Acta Leopoldina* 7:463–496.

McWhorter, C. G., and Patterson, D. T., 1980, Ecological factors affecting weed competition in soybeans, in *World Soybean Research Conference II: Proceedings* (F. T. Corbin, ed.), Westview Press, Boulder, CO, pp. 371–392.

Melis, A., Nemson, J. A., and Harrison, M. A., 1992, Damage to functional components and partial degradation of photosystem II reaction center proteins upon chloroplast exposure to ultraviolet-B radiation, *Biochim. Biophys. Acta* 1100:312–320.

Mirecki, R. M., and Teramura, A. H., 1984, Effects of ultraviolet-B irradiance on soybean. V. The dependence of plant sensitivity on the photosynthetic photon flux density during and after leaf expansion, *Plant Physiol.* 74:475–480.

Murali, N. S., and Teramura, A. H., 1985a, Effects of ultraviolet-B irradiance on soybean VI. Influence of phosphorus nutrition on growth and flavonoid content, *Physiol. Plant.* 63:413–416.

Murali, N. S., and Teramura, A. H., 1985b, Effects of ultraviolet-B irradiance on soybean. VII. Biomass and concentration and uptake of nutrients at varying P supply, *J. Plant Nutr.* 8:177–192.

Murali, N. S., and Teramura, A. H., 1986a, Effects of supplemental ultraviolet-B radiation on the growth and physiology of field-grown soybean, *Environ. Exp. Bot.* 26:233–242.

Murali, N. S., and Teramura, A. H., 1986b, Effectiveness of UV-B radiation on the growth and physiology of field-grown soybean modified by water stress, *Photochem. Photobiol.* 44:215–219.

Murali, N. S., and Teramura, A. H., 1987, Insensitivity of soybean photosynthesis to ultraviolet-B radiation under phorphorus deficiency, *J. Plant Nutr.* 10:501–515.

Murali, N. S., Teramura, A. H., and Randall, S. K., 1988, Response differences between two soybean cultivars with contrasting UV-B radiation sensitivities, *Photochem. Photobiol.* 48:653–657.

Murphy, T. M., 1983, Membranes as targets of ultraviolet radiation, *Physiol. Plant.* 58:381–388.

Negash, L., 1987, Wavelength-dependence of stomatal closure by ultraviolet radiation in attached leaves of *Eragrostis tef:* Action spectra under backgrounds of red and blue lights, *Plant Physiol. Biochem.* 25:753–760.

Noorudeen, A. M., and Kulandaivelu, G., 1982, On the possible site of inhibition of photosynthetic electron transport by ultraviolet-B (UV-B) radiation, *Physiol. Plant.* 55:161–166.

Orth, A. B., Teramura, A. H., and Sisler, H. D., 1990, Effects of ultraviolet-B radiation on fungal disease development in *Cucumis sativus, Am. J. Bot.* 77:1188–1192.

Owens, O. V. H., and Krizek, D. T., 1980, Multiple effects of UV radiation (265–330 nm) on fungal spore emergence, *Photochem. Photobiol.* 32:41–49.

Panagopoulos, I., Bornman, J. F., and Björn, L. O., 1989, The effect of UV-B and UV-C radiation on *Hibiscus* leaves determined by ultraweak luminescence and fluorescence induction, *Physiol. Plant.* 76:461–465.

Panagopoulos, I., Bornman, J. F., and Björn, L. O., 1990, Effects of ultraviolet radiation and visible light on growth, fluorescence induction, ultraweak luminescence and peroxidase activity in sugar beet plants, *J. Photobiochem. Photobiol.* 8:73–87.

Panagopoulos, I., Bornman, J. F., and Björn, L. O., 1992, Response of sugar beet plants to ultraviolet-B (280–320 nm) radiation and *Cercospora* leaf spot disease, *Physiol. Plant.* 84:140–145.

Pang, Q., and Hays, J. B., 1991, UV-B-inducible and temperature-sensitive photoreactivation of cyclobutane pyrimidine dimers in *Arabidopsis thaliana, Plant Physiol.* 95:536–543.

Peak, M. J., and Peak, J. G., 1990, Hydroxyl radical quenching agents protect against DNA breakage caused by both 365-nm UVA and gamma radiation, *Photochem. Photobiol.* 51: 649–652.

Pfündel, E., Pan, R.-S., and Dilley, R. A., 1992, Inhibition of violaxanthin deepoxidation by ultraviolet-B radiation in isolated chloroplasts and intact leaves, *Plant Physiol.* 98:1372–1380.

Pushnik, J. C., Miller, G. W., von Jolley, D., Brown, J. C., Davis, T. D., and Barnes, A. M., 1987, Influences of ultra-violet (UV)–blue light radiation on the growth of cotton. II. Photosynthesis, leaf anatomy, and iron reduction, *J. Plant Nutr.* 19:2283–2297.

Quaite, F. E., Sutherland, B. M., and Sutherland, J. C., 1992a, Quantitation of pyrimidine dimers in DNA from UVB-irradiated alfalfa (*Medicago sativa* L.) seedlings, *Appl. Theor. Electrophor.* 2:171–176.

Quaite, F. E., Sutherland, B. M., and Sutherland, J. C., 1992b, Action spectrum for DNA damage in alfalfa lowers predicted impact of ozone depletion, *Nature* 358:576–578.

Renger, G., Rettig, W., and Graber, P., 1991, The effect of UVB irradiation on the lifetimes of singlet excitons in isolated photosystem II membrane fragments from spinach, *J. Photobiochem. Photobiol.* 9:201–210.

Renger, G., Voss, M., Gräber, P., and Schulze, A., 1986, Effect of UV irradiation on different partial reactions of the primary processes of photosynthesis, in: *Stratospheric Ozone Reductions, Solar Ultraviolet Radiation and Plant Life* (R. C. Worrest and M. M. Caldwell, eds.), NATO ASI Series G. 8:171–184. Springer-Verlag, Berlin.

Renger, G., Völker, M., Eckert, H. J., Fromme, R., Hohm-Veit, S., and Gräber, P., 1989, On the mechanism of photosystem II deterioration by UV-B irradiation, *Photochem. Photobiol.* 49: 97–105.

Rich, P. R., and Bonner, W. D., 1978, The sites of superoxide anion generation in higher plant mitochondria, *Arch. Biochem. Biophys.* 188:206–213.

Robberecht, R., and Caldwell, M. M., 1978, Leaf epidermal transmittance of ultraviolet radiation and its implications for plant sensitivity to ultraviolet-radiation induced injury, *Oecologia* 32:277–287.

Robberecht, R., Caldwell, M. M., and Billings, W. D., 1980, Leaf ultraviolet optical properties along a latitudinal gradient in the arctic-alpine life zone, *Ecology* 61:612–619.

Rotem, J., Wooding, B., and Aylor, D. E., 1985, The role of solar radiation, especially ultraviolet, in the mortality of fungal spores, *Phytopathology* 75:510–514.

Schneider, S. H., 1989, The greenhouse effect: Science and policy, *Science* 243:771–781.

Schulze, E.-D., 1986, Carbon dioxide and water vapor exchange in response to drought in the atmosphere and in the soil, *Annu. Rev. Plant Physiol.* 37:247–274.

Schulze-Lefert, P., Dangl, J. L., Becker-André, M., Hahlbrock, K., and Schulz, W., 1989, Inducible *in vivo* DNA footprints define sequences necessary for UV light activation of the parsley chalcone synthase gene, *EMBO J.* 8:651–656.

Semeniuk, P., 1978, Biological effects of ultraviolet radiation on plant growth and development in florist and nursery crops, in: *UV-B Biological and Climatic Effects Research (BACER), FY 77–78 Research Report on the Impacts of Ultraviolet-B Radiation on Biological Systems: A Study Related to Stratospheric Ozone Depletion.* Final Report, Vol. III, SIRA File No. 142. 210. EPA-IAG-D6-0168, USDA-EPA, Stratospheric Impact Research and Assessment Program (SIRA), US Environmental Protection Agency, Washington, DC, 18 pp.

Semeniuk, P., and Goth, R. W., 1980, Effect of ultraviolet irradiation on local lesion development of potato virus S on *Chenopodium quinoa* cv. Valdivia leaves, *Environ. Exp. Bot.* 20:95–98.

Seyfried, M., and Fukshansky, L., 1983, Light gradients in plant tissue, *Appl. Opt.* 22:1402–1408.

Siefermann, D., and Yamamoto, H. Y., 1975, Light-induced deepoxidation of violaxanthin in lettuce chloroplasts. IV. The effects of electron-transport conditions on violaxanthin availability, *Biochim. Biophys. Acta* 387:149–158.

Siggel, U., Renger, G., Stiehl, H., and Rumberg, B., 1972, Evidence for electronic and ionic interaction between electron transport chains in chloroplasts, *Biochim. Biophys. Acta* 256: 328–335.

Sinclair, T. R., N'Diaye, O., and Biggs, R. H., 1990, Growth and yield of field-grown soybean in response to enhanced exposure to ultraviolet-B radiation, *J. Environ. Qual.* 19:478–481.

Staxén, I., Bergounioux, C., and Bornman, J. F., 1993, Effect of ultraviolet radiation on cell division and microtubule organisation in *Petunia hybrida* protoplasts, *Protoplasma* 173:70–76.

Steinmüller, D., and Tevini, M., 1985, Action of ultraviolet radiation (UV-B) upon cuticular waxes in some crop plants, *Planta* 164:557–564.

Strid, Å., Chow, W. S., and Anderson, J. M., 1990, Effects of supplementary ultraviolet-B radiation on photosynthesis in *Pisum sativum, Biophys. Biochem. Acta* 1020:260–268.

Sullivan, J. H., and Teramura, A. H., 1988, Effects of ultraviolet-B irradiation of seedling growth in the Pinaceae, *Am. J. Bot.* 75:225–230.

Sullivan, J. H., and Teramura, A. H., 1989, The effects of ultraviolet-B radiation on loblolly pines I. Growth, photosynthesis and pigment production in greenhouse grown saplings, *Physiol. Plant.* 77:202–207.

Sullivan, J. H., and Teramura, A. H., 1990, Field study of the interaction between solar ultraviolet-B radiation and drought on photosynthesis and growth in soybean, *Plant Physiol.* 92:141–146.

Sullivan, J. H., and Teramura, A. H., 1992, The effects of ultraviolet-B radiation on loblolly pines 2. Growth of field-grown seedlings, *Tree Structure and Function* 6:115–120.

Sullivan, J. H., Teramura, A. H., and Ziska, L. H., 1992, Variation in UV-B sensitivity in plants from a 3,000-m elevation gradient in Hawaii, *Am. J. Bot.* 79:737–743.

Tadolini, B., 1988, Polyamine inhibition of lipoperoxidation. The influence of polyamines on iron oxidation in the presence of compounds mimicking phospholipid polar heads, *Biochem. J.* 249:33–36.

Takahama, U., 1982, Suppression of carotenoid photobleaching by kaempferol in isolated chloroplasts, *Plant Cell Physiol.* 23:859–864.

Takahama, U., 1983, Redox reactions between kaempferol and illuminated chloroplasts, *Plant Physiol.* 71:598–601.

Takahama, U., Egashira, T., and Wakumatsu, K., 1989, Hydrogen peroxide-dependent synthesis of flavonols in meosphyll cells of *Vicia faba* L., *Plant Cell Physiol.* 30:951–955.

Takahashi, A., Takeda, K., and Ohnishi, T., 1991, Light-induced anthocyanin reduces the extent of damage to DNA in UV-irradiated *Centaurea cyanus* cells in culture, *Plant Cell Physiol.* 32:541–547.

Takeuchi, Y., Akizuki, M., Shimizu, H., Kondo, N., and Sugahara, K., 1989, Effect of UV-B (290–320 nm) irradiation on growth and metabolism of cucumber cotyledons, *Physiol. Plant.* 76:425–430.

Teramura, A. H., 1980, Effects of ultraviolet-B irradiances on soybean. I. Importance of photosynthetically active radiation in evaluating ultraviolet-B irradiance effects on soybean and wheat growth, *Physiol. Plant.* 48:333–339.

Teramura, A. H., 1983, Effects of ultraviolet-B radiation on the growth and yield of crop plants, *Physiol. Plant.* 58:415–427.

Teramura, A. H., and Caldwell, M. M., 1981, Effects of ultraviolet-B irradiances on soybean. IV. Leaf ontogeny as a factor in evaluating ultraviolet-B irradiance effects on net photosynthesis, *Am. J. Bot.* 68:934–941.

Teramura, A. H., and Murali, N. S., 1987, Intraspecific differences in growth and yield of soybean exposed to ultraviolet-B radiation under greenhouse and field conditions, *Environ. Exp. Bot.* 26:89–95.

Teramura, A. H., and Sullivan, J. H., 1987, Soybean growth responses to enhanced levels of ultraviolet-B radiation under greenhouse conditions, *Am. J. Bot.* 74:975–979.

Teramura, A. H., Biggs, R. H., and Kossuth, S., 1980, Effects of ultraviolet-B irradiances on soybean. II. Interaction between ultraviolet-B and photosynthetically active radiation on net photosynthesis, dark respiration, and transpiration, *Plant Physiol.* 65:483–488.

Teramura, A. H., Tevini, M., and Iwanzik, W., 1983, Effects of ultraviolet-B irradiance on plants during mild water stress. I. Effects on diurnal stomatal resistance, *Physiol. Plant.* 57:175–180.

Teramura, A. H., Perry, M. C., Lydon, J., McIntosh, M. S., and Summers, E. G., 1984a, Effects of ultraviolet-B radiation on plants during mild water stress. III. Effects on photosynthetic recovery and growth in soybean, *Physiol. Plant.* 60:484–492.

Teramura, A. H., Forseth, I. N., and Lydon, J., 1984b, Effects of ultraviolet-B radiation on plants during mild water stress. IV. The insensitivity of soybean internal water relations to UV-B radiation, *Physiol. Plant.* 62:384–389.

Teramura, A. H., Sullivan, J. H., and Lydon, J., 1990a, Effects of UV-B radiation on soybean yield and seed quality: A 6-year field study, *Physiol. Plant.* 80:5–11.

Teramura, A. H., Sullivan, J. H., and Ziska, L. H., 1990b, Interaction of elevated ultraviolet-B radiation and CO_2 on productivity and photosynthetic characteristics in wheat, rice, and soybean, *Plant Physiol.* 94:470–475.

Teramura, A. H., Tevini, M., Bornman, J. F., Caldwell, M. M., Kulandaivelu, G., and Björn, L. O., 1991, Terrestrial plants, in: *Environmental Effects of Ozone Depletion: 1991 Update,* UNEP Environmental Effects Panel Report, Nairobi, Kenya, pp. 25–32.

Tevini, M., and Steinmüller, D., 1987. Influence of light, UV-B radiation, and herbicides on wax biosynthesis of cucumber seedlings, *J. Plant Physiol.* 131:111–121.

Tevini, M., and Teramura, A. H., 1989, UV-B effects on terrestrial plants, *Photochem. Photobiol.* 50:479–487.

Tevini, M., Iwanzik, W., and Thoma, U., 1981, Some effects of enhanced UV-B irradiation on the growth and composition of plants, *Planta* 153:388–394.

Tevini, M., Iwanzik, W., and Teramura, A. H., 1983, Effects of UV-B radiation on plants during mild water stress. II. Effects on growth, protein and flavonoid content, *Z. Pflanzenphysiol.* 110:459–467.

Tevini, M., Steinmüller, D., and Iwanzik, W., 1986, Über die Wirkung erhöhter UV-B-Strahlung in kombination mit anderen Stressfaktoren auf Wachstum und Funktion von Nutzpflanzen, *BPT-Bericht, Gesellschaft für Strahlen- und Umweltforschung, München* 6/86:1–172.

Tevini, M., Braun, J., and Fieser, G., 1991, The protective function of the epidermal layer of rye seedlings against ultraviolet-B radiation, *Photochem. Photobiol.* 53:329–333.

Van, T. K., Garrard, L. A., and West, S. H., 1977, Effects of 298-nm radiation on photosynthetic reactions of leaf discs and chloroplast preparations of some crop species, *Environ. Exp. Bot.* 17:107–112.

Van de Staaij, J., Rozema, J., and Stroetenga, M., 1990, Expected changes in Dutch coastal vegetation resulting from enhanced levels of solar UV-B, in: *Expected Effects of Climatic Change on Marine Coastal Ecosystems* (J. J. Beukema, W. J. Wolff, and J. J. W. M. Brouns, eds.), Kluwer Academic, the Netherlands, pp. 211–217.

Vierstra, R. D., John, T. R., and Poff, K. L., 1982, Kaempferol 3-*o*-galactoside, 7-*o*-rhamnoside is the major green fluorescing compound in the epidermis of *Vicia faba, Plant Physiol.* 69:522–525.

Vu, C. V., Allen, L. H., and Garrard, L. A., 1984, Effects of UV-B radiation (280–320 nm) on ribulose-1,5-bisphosphate carboxylase in pea and soybean, *Environ. Exp. Bot.* 24:131–143.

Warner, C. W., and Caldwell, M. M., 1983, Influence of photon flux density in the 400–700 nm waveband on inhibition of photosynthesis by UV-B (280–320 nm) irradiation in soybean leaves: Separation of indirect and immediate effects, *Photochem. Photobiol.* 38:341–346.

Weissenböck, G., Plesser, A., and Trinks, K., 1976, Flavonoidgehalt und Enzymaktivitäten isolierter Haferchloroplasten (*Avena sativa* L.), *Ber. Dtsch. Bot. Ges.* 89:457–472.

Wellmann, E., 1974, Regulation der Flavonoidbiosynthese durch ultraviolettes Licht und Phytochrom in Zellkulturen und Keimlingen von Petersilie (*Petroselinum hortense* Hoffm.), *Ber. Deutsch. Bot. Ges.* 87:267–273.

Wellmann, E., 1975, Der Einfluss physiologischer UV-Dosen auf Wachstum und Pigmentierung von Umbelliferenkeimlingen, in: *Industrieller Pflanzenbau* (E. Bancher, ed.), Technische Universität Wien, pp. 229–239.

Whittaker, R. H., 1975, *Communities and Ecosystems,* 2nd ed., MacMillan, Inc. New York.

Wingate, V. P. M., Lawton, M. A., and Lamb, C. J., 1988, Glutathione causes a massive and selective induction of plant defense genes, *Plant Physiol.* 87:206–210.

Yatsuhashi, H., and Hashimoto, T., 1985, Multiplicative action of a UV-B photoreceptor and phytochrome in anthocyanin synthesis, *Photochem. Photobiol.* 41:673–680.

Yatsuhashi, H., Hashimoto, T., and Shimizu, S., 1982, Ultraviolet action spectrum for anthocyanin formation in broom sorghum first internodes, *Plant Physiol.* 70:735–741.

Zamansky, G. B., and Chou, I.-N., 1987, Environmental wavelengths of ultraviolet light induce cytoskeletal damage, *J. Invest. Dermatol.* 89:603–606.

Zamansky, G. B., Perrino, B. A., and Chou, I.-N., 1991, Disruption of cytoplasmic microtubules by ultraviolet radiation, *Exp. Cell. Res.* 195:269–273.

Zaremba, T. G., LeBon, T. R., Millar, D. B., Smejkal, R. M., and Hawley, R. J., 1984, Effects of ultraviolet light on the in vitro assembly of microtubules, *Biochemistry* 23:1073–1080.

Ziska, L. H., and Teramura, A. H., 1992, CO_2 enhancement of growth and photosynthesis in rice (*Oryza sativa*): Modification by increased ultraviolet-B radiation, *Plant Physiol.* 84:269–276.

Ziska, L. H., Teramura, A. H., and Sullivan, J. H., 1992, Physiological sensitivity of plants along an elevational gradient to UV-B radiation, *Am. J. Bot.* 79:863–871.

Index

RETURN DATE